SHELF SEDIMENT TRANSPORT:
Process and Pattern

SHELF SEDIMENT TRANSPORT:
Process and Pattern

Edited by **Donald J. P. Swift**

Atlantic Oceanographic and Meteorological Laboratories

David B. Duane

Coastal Engineering Research Center

Orrin H. Pilkey

Duke University

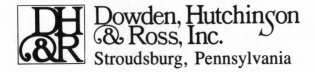

Dowden, Hutchinson & Ross, Inc.
Stroudsburg, Pennsylvania

Manufactured in the United States of America.

74　75　76　　5　4　3　2

Exclusive distributor outside the United States and Canada:
John Wiley & Sons, Inc.

Contributors

RICHARD W. BOEHMER, *Marine Affairs Program, School of Oceanography, University of Rhode Island, Kingston*

BARBARA A. BUSS, *Department of Geology, University of Illinois, Chicago*

DAVID A. CACCHIONE, *Office of Naval Research, Boston, Massachusetts*

DONALD J. COLQUHOUN, *Department of Geology, University of South Carolina, Columbia*

JOE S. CREAGER, *Department of Oceanography, University of Washington, Seattle*

DAVID E. DRAKE, *Department of Geological Sciences, University of Southern California, Los Angeles*

DAVID B. DUANE, *Coastal Engineering Research Center, Washington, D.C.*

PETER FENNER, *Governor's State University, Park Forest South, Illinois*

MICHAEL E. FIELD, *Coastal Engineering Research Center, Washington, D.C.*

PETER J. FISCHER, *Department of Geology, San Fernando Valley State College, Northridge, California*

DONN S. GORSLINE, *Department of Geological Sciences, University of Southern California, Los Angeles*

D. J. GRANT, *Department of Geological Sciences, University of Southern California, Los Angeles*

D. L. HARRIS, *Coastal Engineering Research Center, Washington, D.C.*

J. D. HOWARD, *Skidaway Institute of Oceanography, Savannah, Georgia*

T. S. HOPKINS, *Department of Oceanography, University of Washington, Seattle*

GILBERT KELLING, *Department of Geology, University of Wales, Swansea*

JOHN W. KOFOED, *Atlantic Oceanographic and Meteorologic Laboratories, Virginia Key, Miami, Florida*

RONALD L. KOLPACK, *Department of Geological Sciences, University of Southern California, Los Angeles*

PAUL D. KOMAR, *Department of Oceanography, University of Oregon, Corvallis*

L. P. KULM, *Department of Oceanography, University of Oregon, Corvallis*

J. C. LUDWICK, *Institute of Oceanography, Old Dominion University, Norfolk, Virginia*

I. N. McCAVE, *School of Environmental Sciences, University of East Anglia, Norwich, Great Britain*

EDWARD P. MEISBURGER, *Coastal Engineering Research Center, Washington, D.C.*

DEAN A. McMANUS, *Department of Oceanography, University of Washington, Seattle*

R. H. MEADE, *U.S. Geological Survey, Woods Hole, Massachusetts*

STEPHEN P. MURRAY, *Coastal Studies Institute, Louisiana State University, Baton Rouge*

D. D. NELSON, *Division of Sedimentology, Smithsonian Institution, Washington, D.C.*

R. H. NEUDECK, *3000 La Luz, Atascadero, California*

G. F. OERTEL II, *Skidaway Institute of Oceanography, Savannah, Georgia*

AKIRA, OKUBO, *Chesapeake Bay Institute, Johns Hopkins University, Baltimore, Maryland*

J. W. PIERCE, *Division of Sedimentology, Smithsonian Institution, Washington, D.C.*

ORRIN H. PILKEY, *Department of Geology, Duke University, Durham, North Carolina*

KELVIN S. RODOLFO, *Department of Geology, University of Illinois at Chicago Circle, Chicago*

FRANCIS P. SAULSBURY, *Atlantic Oceanographic and Meteorologic Laboratories, Virginia Key, Miami, Florida*

J. R. SCHUBEL, *Chesapeake Bay Institute, The Johns Hopkins University, Baltimore, Maryland*

PHILLIP SEARS, *Institute of Oceanography, Old Dominion University, Norfolk. Virginia*

J. D. SMITH, *Department of Oceanography, University of Washington, Seattle*

JOHN B. SOUTHARD, *Department of Earth and Planetary Sciences, Massachusetts Institute of Technology, Cambridge*

DANIEL J. STANLEY, *Division of Sedimentology, Smithsonian Institution, Washington, D.C.*

RICHARD W. STERNBERG, *Department of Oceanography, University of Washington, Seattle*

A. H. STRIDE, *National Institute of Oceanography, Wormley, Godalming, Surrey, Great Britain*

DONALD J. P. SWIFT, *Atlantic Oceanographic and Meteorologic Laboratories, Virginia Key, Miami, Florida*

PAUL G. TELEKI, *Coastal Engineering Research Center, Washington, D.C.*

J. RICHARD WEGGEL, *Coastal Engineering Research Center, Washington, D.C.*

S. JEFFRESS WILLIAMS, *Coastal Engineering Research Center, Washington, D.C.*

Preface

This volume contains papers presented in a symposium on Shelf Sediment Transport conducted at the annual meeting of the Geological Society of America held in Washington, D.C., November 1971. Persons who did not take part in the actual meeting but who are actively conducting research in this multidisciplinary field were asked to prepare introductory papers for each section. In addition, several papers not delivered at the symposium have been added to broaden the scope of this volume. This volume has been organized into three sections concerned with shelf hydraulic regime and mechanics of sediment transport, suspended sediment dispersal, and bottom sediment dispersal. Because of the relationships among water motion, processes of sediment entrainment, and sediment distributive pattern, some overlap of content inevitably occurs between sections within this volume. Our goal in assembling these papers is to document some fundamental aspects of the shelf milieux and its sediment dispersal systems, to indicate the nature and scope of present knowledge and the direction of ongoing research, and to provide some perspective on the problems associated with understanding, determining, and predicting sediment transport on continental shelves.

Many sedimentological studies in the geological literature are in reality studies of fossil sediment transport systems. These have been resolved in a general sense by a combined analysis of physical stratigraphy and petrography. In the past two decades, this approach has resulted in an extension of knowledge of the evolution and history of continental-shelflike platforms. However, this method generally fails to resolve short-term characteristics of sediment transport preserved in the stratigraphic record. Therefore, increased interest and effort is being directed toward monitoring of time-variant characteristics of present-day shelf sedimentation and systematic gathering of time series of data on transport processes. Geological oceanographers and marine geologists will hopefully never lose their unique sense of the vastness of geologic time, which gives them a special insight into their studies, but they stand to gain much from the increased sensitivity to short-term processes (illustrated by most of the papers in this volume) which when integrated through geologic time and preserved, yield the stratigraphic record.

Emerging concepts of coastal zone management are usually reflected in studies which deal with specific technology and specific, often local, problems of environmental impact. However, there is an increasing need for more general knowledge concerning the natural processes which affect the seaward margin of the coastal zone, i.e., the continental shelf. In response to human needs, increased use of the continental shelf is foreseen in the immediate future. Effective planning and management of the coastal zone—so as to permit multiple use for recreation, extraction of living and nonliving resources, and waste disposal—must directly or indirectly involve an understanding of sediment transport (entrainment and distribution) on the shelf. Types and results of studies collected in this volume are germane to this understanding.

Though concerned with the mobile sediment surface in the geologic present, we and the authors whose papers are collected here would certainly be pleased to see results of our research used to interpret the geologic past. We are equally prepared to use the stratigraphic record to help illuminate modern transport systems, for, as indicated by the last paper of this volume, a fundamental goal of sedimentology is to construct from both ancient and modern data a general model for sedimentary processes applicable through geologic time. A tested and acceptable model of this type should be expected to have predictive aspects applicable to the solution of problems today associated with management of the coastal zone for the maximum benefit of all users.

We truly hope that this volume will contribute toward a better understanding of the geologic past, and toward an equitable and effective use of the coastal zone. To those who participated in the symposium, contributed papers to this volume, and especially you, too numerous to name, who critically read manuscripts for us, we say, thanks.

November, 1972

DONALD J. P. SWIFT
DAVID B. DUANE
ORRIN H. PILKEY

Contents

II. Patterns of Fine Sediment Dispersal

III. Patterns of Coarse Sediment Dispersal

I. Water Motion and Process of Sediment Entrainment

CHAPTER 1

An Introduction to Oceanic Water Motions and Their Relation to Sediment Transport

J. Richard Weggel

Coastal Engineering Research Center
5201 Little Falls Rd., N.W.
Washington, D.C. 20016

ABSTRACT

Knowledge of fluid motions that agitate and transport marine sediments is a prerequisite to understanding erosion, deposition, and transport of those sediments. Likewise, understanding the physical principles that govern fluid flow must precede study of the flow itself. A brief discussion of those aspects of flow important in sedimentation studies is presented as an introduction to discussion of the physical principles governing fluid flows. Examples of how these principles manifest themselves in the oceans, the assumptions made in simplifying the governing equations, and in some cases how the flow is related to sediment movement, are presented. Wave motions, particularly wave-induced near-bottom velocities, are discussed with regard to their increasing ability to agitate bottom materials as waves move shoreward across the continental shelf. Ocean currents also provide a mechanism for transporting marine sediments. Examples of observed current phenomena and the assumptions made to simplify the governing equations are presented. The deflection of wind driven turbulent currents caused by Coriolis effects results in the well-known Ekman spiral. Ekman's result for deep water is given. The important implication for shelf sediment transport studies is that care must be exercised in extrapolating surface wind and current observations to the near-bottom currents that are important in moving sediments.

INTRODUCTION

Perhaps one of the most complex problems to which an oceanographer can address himself concerns the study of sediment motion on the continental shelf. Sediment motion is difficult to describe because each of the related problems of describing the water-motion and the water-sediment interaction are themselves complex. First, an understanding of the fluid motions which both entrain and transport sediment is required; however, the task of obtaining an adequate description of these important fluid motions is often the most difficult aspect of the problem. Natural flows are controlled by complex and sometimes poorly defined physical conditions. Reasonable simplifications are those that result in a description of a flow that is at least qualitatively observed. Second, sediment motion studies require a knowledge of how flow and sediment interact. How will sediments of given characteristics react to given flow conditions? What sediment characteristics are significant in determining whether particles will move? What relationships exist between flow conditions and bottom conditions (i.e., bed forms)? For some sediments and some flow conditions, the effect of sediment concentrations altering the local flow conditions is significant. By virtue of the complexity of each component part of the sediment transport problem, the erosion, transport, and deposition of sediment in the continental shelf environment is an extremely complex phenomenon. Adequate, quantitative description of sediment movement on the shelf requires a description of the important fluid motions. Conversely, to determine which aspects of the fluid motion are important, their effectiveness in moving sediment must be assessed. The feedback of one problem to the other often requires that studies of fluid motion, sediment characteristics, and the interaction at the fluid-sediment interface be advanced simultaneously. With the exception of a few necessary introductory and illustrative remarks on the water-sediment interaction, only a general description of the fluid motions will be considered here. Discussion of the fluid-sediment interaction is the subject of many of the papers that follow.

GENERAL

In a way similar to alluvial sediments, ocean sediments can move either as bedload in the near-bottom region or as suspended load which moves essentially with the mean flow. On the average, suspended sediments are smaller than are bedload materials. They range in size from fine sands down to colloidal materials suspended in the flow and may be removed from suspension by flocculation induced by changes in the physical and chemical characteristics of the transporting fluid. The largest particles in motion usually move as bedload, settling from the fluid between intermittent periods of motion.

Regardless of the mode of transport involved, several prerequisites for *net sediment movement* can be stated. First, a source of movable sediment must be readily available. The degree of mobility depends on the hydraulic regime and on the physical characteristics of the sediments. Second, a mechanism for initiating sediment movement is required. Motion is initiated when flow induced drag and inertia forces, arising from fluid velocities and accelerations, act on sediment particles and exceed the forces that tend to restrain their motion. For suspended sediments, a mechanism for entraining sediments into the mean flow is required. Flow separation at the crests of bed forms and turbulent velocity fluctuations serve to move sediments from the bottom into suspension. Particle motion by itself is not sufficient for net sediment movement. Therefore, third, the presence of an asymmetry in the sediment motion is required to move more sediment in one direction than another. A purely oscillatory particle motion for which the net transport is zero is theoretically possible; however, such conditions do not commonly occur in nature. Consequently, the near bottom residual currents of the continental shelf and mass transport velocities induced by surface water waves, both of which act to transport sediments, are the rule rather than the exception. Wave motion is more significant in the shallow nearshore region where the wave-induced fluid motion extends to the bottom. On the Pacific shelf where long period waves are common, the depths to which wave motion extends are greater than for the Atlantic shelf. Komar et al. (1972) observed bottom ripples, presumably caused by surface waves, in depths of up to 200 m. The effectiveness of wind driven currents in moving bottom materials also increases with decreasing depth because bottom flows respond more quickly to surface wind stresses in shallow water. The amount of time sediment is in motion will, therefore, depend on depth as well as on sediment size. At the seaward edge of the shelf, periods of sediment motion are relatively infrequent. Sternberg and McManus (1972) indicate that for a region off the coast of Washington, in a depth of 80 m, current velocities of sufficient strength to transport sediment occur only about 2% of the time. At the landward end of the shelf, the beach, periods of sediment motion are nearly constant although direction and rate are variable. Real time synoptic measurement and patterns and conditions of sediment movement in the littoral zone along portions of the California coast have been discussed by Duane (1970). He attributes the direction and speed of longshore sediment transport to the direct effect of waves shoaling and breaking in combination with longshore currents, either wave or wind generated. The task of quantifying the volume of sediment transport in the littoral zone attributable to a given set of environmental conditions has not been unequivocally determined although numerous papers on the topic have been published.

Because the motion of sediment on the continental shelf is inextricably tied to the hydraulic phenomena occurring there, an understanding of the flow governing principles is of paramount importance in understanding the movement of sediments. A relatively cursory introduction to these basic principles along with some examples of observed flows are presented here. Flows which are deemed significant in moving sediment are considered; consequently, to establish which flows are significant, the fluid-sediment interaction and the point of view of the investigator will be touched upon.

THE PROBLEM

The usual sediment transport problem can often be stated quite simply: Determine the direction and rate of significant sediment motion and the rate of erosion and deposition at a particular location.

Information available to obtain solutions to the above problem will vary considerably with geographical location. Information available at the initiation of a study, if not completely nonexistent, is usually limited to qualitative information on surface water circulation and possibly some indication of meteorological and wave conditions at the site. If supplementary data are required, it must be obtained from a program of field measurements.

To determine which aspects of the hydraulic regime are of importance to the problem, it is first necessary to define what is implied by important or significant sediment motion. The definition depends critically on the time scale of interest and, hence, on the point of view of the investigator. From a geological point of view, changes occurring over long periods of time are important. From an engineering point of view, changes occurring over considerably shorter time periods are important. The engineer is rarely concerned with sedimentation or erosion rates of a few inches in a thousand years while such changes may be of great importance to a geologist. Consequently, the temporal scale of motion must be defined before a particular aspect of the hydraulics can be deemed as either important or insignificant.

In addition to differences in time scale, different length scales are encountered. Geologically, large scale or regional variations are frequently under consideration while some specific studies may be concerned with phenomena of smaller areal extent. Reasons for viewing the problem on one scale or another may, in part, be explained by the purpose of a specific investigation. A qualitative regional approach may be satisfactory when only a general description and understanding of the transport phenomenon and related problems such as continental shelf sedimentation and marine sediment origins are desired. On the other hand, when detailed information on sediment behavior and the mechanics of the transport process is desired, a quantitative study of a limited area or a laboratory study is more practical. Such small scale quantitative studies may serve to explain, often in a qualitative way, sedimentation phenomena observed on a larger regional scale. The two scales of approach therefore complement each other. Examples of regional studies, field studies over limited areas and laboratory studies are included in this volume.

Considering the broad range of spatial and temporal scales used to describe sedimentation processes on the continental shelf, most water motions observed to occur on the shelf are important in moving materials. Since long-term changes involving small particles at small rates of transport may be of geologic significance, fluid flows of relatively low intensity cannot be ignored. Wave-induced fluid motion and currents, meteorologically generated currents, and currents generated by fluid density differences, are all, under certain conditions, potential sediment movers.

INFORMATION REQUIRED

Data required to determine sediment transport quantities can be established as follows: (a) the physical characteristics of the sediment, e.g., grain size and shape, (b) physical characteristics of the fluid, e.g., viscosity and density, (c) kinematics of the fluid motion, and (d) the dynamic interaction processes between the moving fluid and the sediments. Fluid motions of primary interest here are those that occur over the continental shelf. When considering suspended sediments, the entire flow field is important since the sediments may be moved at any level in the flow. However, since suspended sediment concentrations increase near their source (the bottom), fluid flows at lower levels are of increased importance. When considering bedload transport, flow velocities very close to the bottom are important since it is the vertical velocity gradient which establishes the shear stresses that move sediment in the near-bottom region.

Near-bottom velocities and flow conditions in the boundary layer are, because of their location, the least accessible; consequently, available data usually pertains to the surface motion of the sea from which subsurface and near-bottom conditions must be obtained (that is, surface wave profiles and surface current circulation patterns).

The various levels of flow, their relative significance in moving sediment and the region wherein the bottom significantly modifies the flow (boundary layer) are shown schematically on Fig. 1. Flow acting directly on bottom sediments is boundary layer flow; the flow in this relatively thin fluid layer is in turn determined by a downward transfer of momentum supplied by the near-bottom fluid velocities. Near-bottom velocities outside of the boundary layer establish mathematical boundary conditions that, along with the bottom boundary itself, determine flow profiles in the boundary layer. When near-bottom velocities are unidirectional and relatively steady (that is, the acceleration or time rate of change of velocity is small), vertical velocity distributions within the boundary layer and bed forms that develop can be estimated with some confidence. When near-bottom velocities are oscillatory or irregular, as induced by wave motion, determination of flow conditions within the boundary layer from data on near-bottom velocities and the prediction of bed forms becomes difficult. The difficulties encountered are discussed by Teleki (1972) for the case of oscillatory flow in the absence of bed forms. As bed forms increase in amplitude, their

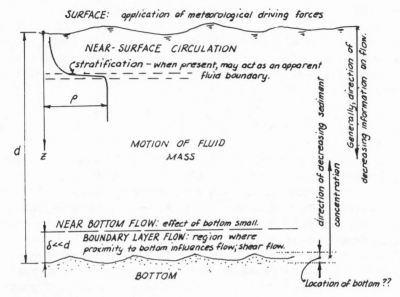

Figure 1.
Regions of flow—vertical section schematic (not to scale).

effect dominates the boundary layer flow. Under such conditions, flow in the boundary layer is complex and describing the mechanics of sediment movement on a microscale, (i.e., describing what happens to a particular grain) becomes meaningless. Similarly, the size of individual grains becomes less important in modifying the near-bottom flow as bed form influence on the flow increases. In this case, bottom effects are averaged over an area to obtain an average bottom roughness which depends primarily on bed form geometry rather than on grain size.

Flow perturbations introduced by bed forms and turbulence are important in entraining sediments into the mean flow some distance above the bottom. This is particularly evident under oscillating flows of the type induced by surface waves feeling bottom when water depth is shallower than one-half of a wavelength. The entrainment process is important since it leads to a vertical sediment concentration gradient that makes mean fluid flows some distance above the bottom important for transporting sediment. Sediment concentration gradients in conjunction with fluid velocity gradients are significant in establishing quantities of sediment moved at a particular level in the flow.

For any particular investigation, data available or readily obtainable determine the level in the flow at which a study can begin and how far the flow data must be extrapolated to estimate erosion, deposition, and transport rates. All too frequently, the only data readily obtainable without an extensive field measurement program are the surface circulations; less often, the structure of the mass flow is known (either from direct measurement or by inference from surface measurements) and can be used to estimate bottom flows. Rarely are near-bottom flows known. The farther from the region of interest (either the bottom, in the case of bedload transport or the mean and near-bottom flow in the case of suspended transport) flow conditions are known, the less reliable are flow condition extrapolations to elevations or regions where they are desired and the less reliable is sediment information derived from the flow data.

PRINCIPLES INVOLVED IN DESCRIBING FLUID FLOWS

The principles governing hydraulic phenomena on the continental shelf are quite elementary. They merely manifest themselves in complex ways. Unfortunately, the complexity of the equations often makes one lose sight of the simplicity of the principles they describe. Conservation of mass, momentum, and energy are the principles that must be satisfied. Detailed derivations and discussion of the equations expressing these principles are given by Rouse (1938; not considering Coriolis effects), Ippen (1966), and Neumann and Pierson (1966) while an elegant mathematical discussion of the equations and their application to the oceans is available in Phillips (1966).

Conservation of mass is expressed by the continuity equation. For an elemental volume of fluid, $\delta \nabla = \delta x \cdot \delta y \cdot \delta z$, the continuity equation can be written,

$$-\underbrace{\frac{\partial \rho}{\partial t}}_{\text{I}} = \underbrace{\frac{\partial(\rho u)}{\partial x}}_{\text{II}} + \underbrace{\frac{\partial(\rho v)}{\partial y}}_{\text{III}} + \underbrace{\frac{\partial(\rho w)}{\partial z}}_{\text{IV}} \tag{1}$$

where u, v, and w are the fluid velocities in the x, y, and z coordinate directions, ρ is the fluid density, and t is time. The equation states that the time rate of change of fluid mass enclosed in the elemental volume (term I) equals the difference between the rate at which mass is entering and the rate at which mass is leaving the volume in each of the x, y, and z directions (terms II, III, and IV, respectively).

Conservation of momentum may be expressed by three equations, one for each coordinate direction, representing a balance of the forces acting on an elemental volume against the change in momentum in that direction. If a Cartesian coordinate system is selected so that the x-y plane is tangent to level surfaces (approximately tangent to the earth's surface) with the y axis oriented in the north-south direction and the z direction positive upward, the three equations corresponding to the force-momentum balance in the x, y, and z directions are,

$$\underbrace{\frac{\partial u}{\partial t}}_{\text{I}} + \underbrace{u\frac{\partial u}{\partial x} + v\frac{\partial u}{\partial y} + w\frac{\partial u}{\partial z}}_{\text{II}} = \underbrace{-\frac{1}{\rho}\frac{\partial p}{\partial x}}_{\text{III}} + \underbrace{2\Omega(v\sin\theta - w\cos\theta)}_{\text{IV}}$$

$$\underbrace{+\frac{1}{\rho}\left\{\frac{\partial\tau_{xx}}{\partial x} + \frac{\partial\tau_{xy}}{\partial y} + \frac{\partial\tau_{xz}}{\partial z}\right\}}_{\text{V}} \quad (x\text{-direction}) \tag{2}$$

$$\underbrace{\frac{\partial v}{\partial t}}_{\text{I}} + \underbrace{u\frac{\partial v}{\partial x} + v\frac{\partial v}{\partial y} + w\frac{\partial v}{\partial z}}_{\text{II}} = \underbrace{-\frac{1}{\rho}\frac{\partial p}{\partial y}}_{\text{III}} - \underbrace{2\Omega u\sin\theta}_{\text{IV}}$$

$$\underbrace{+\frac{1}{\rho}\left\{\frac{\partial\tau_{yx}}{\partial x} + \frac{\partial\tau_{yy}}{\partial y} + \frac{\partial\tau_{yz}}{\partial z}\right\}}_{\text{V}} \quad (y\text{ direction}) \tag{3}$$

and

$$\underbrace{\frac{\partial w}{\partial t}}_{\text{I}} + \underbrace{u\frac{\partial w}{\partial x} + v\frac{\partial w}{\partial y} + w\frac{\partial w}{\partial z}}_{\text{II}} = \underbrace{-\frac{1}{\rho}\frac{\partial p}{\partial z}}_{\text{III}} - \underbrace{2\Omega u\cos\theta}_{\text{IV}}$$

$$\underbrace{+\frac{1}{\rho}\left\{\frac{\partial\tau_{zx}}{\partial x} + \frac{\partial\tau_{zy}}{\partial y} + \frac{\partial\tau_{zz}}{\partial z}\right\}}_{\text{V}} \underbrace{-g}_{\text{VI}} \quad (z\text{ direction}) \tag{4}$$

These force-momentum balance equations or equations of motion are termed the Navier-Stokes equations and mathematically represent Newton's second law. Terms on the left side of Eqs. (2), (3), and (4) represent accelerations experienced by the fluid volume. The local accelerations [term I in Eqs. (2), (3), and (4)] plus the convective accelerations (terms II in the equations) describe the total acceleration when combined. Term III in each equation is the net force per unit mass of fluid resulting from the difference in pressure in going from one side of the elemental volume to the other. Term IV is an apparent force per unit mass of fluid which arises because the coordinate system chosen is fixed with respect to the earth's surface and is therefore rotating about the earth's axis at Ω rad/sec ($\Omega = 7.29 \times 10^{-5}$ rad/sec). The magnitude of the Coriolis force depends on the inclination of the x-y plane with respect to the earth's axis of rotation and, therefore, on the angle of latitude θ. The stress terms (excluding the normal pressure stresses which have been considered separately) are given by term V of the equations. It is through these terms that kinetic energy of the flow is converted to heat, sometimes referred to as "energy dissipation." The first subscript of each stress term indicates the coordinate direction in which the stress is acting. The second subscript refers to the plane in which the stress acts by indicating the coordinate direction normal to the plane. For example, τ_{xy} is the stress acting in the x direction on the x-z plane (the plane normal to the y direction). The stress terms can be related to velocity gradients by empirically determined coefficients of viscosity. Term VI, which appears only in Eq. 4, because of the choice of coordinate system, is the force (weight) per unit mass exerted on the fluid by the earth's gravity.

The Navier-Stokes equations of motion apply to both laminar and turbulent flows; however, to be useful for describing turbulent flows, the equations must be modified by introducing velocities and pressures which have a mean component and a turbulent, fluctuating component. Time averaging each term in the resulting Navier-Stokes and continuity equations gives equations which are similar to Eqs. (1), (2), (3), and (4) with time-average (mean) velocities in place of instantaneous velocities. Several additional terms involving turbulent velocity fluctuations remain in Eqs. (2), (3), and (4) when averaged in this way. These terms, called Reynolds stresses, are usually assumed proportional to the mean velocity gradients in much the same way that the shear stresses of Eqs. (2), (3), and (4) are related to the instantaneous velocity gradients. The proportionality coefficient, analogous to the coefficient of molecular viscosity, is termed the eddy viscosity.

In addition to the continuity equation and equations of motion, an expression for fluid density in terms of pressure, temperature, and salinity is required before the set of equations is complete. This relationship for sea water is complex and most readily available in tabulated form (U.S. Navy, 1966). Quite often an assumption of constant density can be made without introducing significant errors into a particular flow solution. In general, however,

$$\rho = \rho(p, S, T_0) \tag{5}$$

where S is salinity and T_0 is temperature.

The preceding equations, when coupled with the appropriate boundary conditions, can in theory be used to describe most flows encountered in the oceans. In fact, the equations must be simplified before solutions can be obtained. If the fluid is assumed inviscid (zero viscosity) and the flow, therefore, irrotational, the simplified equations of motion can be integrated to give an expression for the conservation of energy (note that the "energy dissipation" is precluded by not considering the viscous terms by which energy is converted to a mechanically unrecoverable form). Conservation of energy is expressed by the Bernoulli equation, which for steady, irrotational flow is

$$\frac{u^2 + v^2 + w^2}{2} + \int \frac{dp}{\rho} + gz = \text{constant} \tag{6}$$

Equation (6) often provides a convenient method for describing flows. Its use, however, requires that the rate of energy dissipation be either known or capable of description in terms of flow velocities and empirical coefficients (friction factors).

The preceding equations are in general form and must be simplified to obtain solutions. The amount of simplification permissible depends on the problem, that is, which terms in the equations must be retained to describe the desired phenomenon. Some common assumptions that are often valid for oceanic flows are incompressibility and irrotationality when the motion is some distance from boundaries (outside of the boundary layer). This implies that the viscous stress terms of Eqs. (2), (3), and (4) are negligible. However, the irrotationality assumption is not possible for boundary layer flows or for flows where momentum transfer by turbulence (mixing) is important. Many flow phenomena are viewed two dimensionally, that is, for long crested wave theories the fluid motion takes place in a vertical plane and for describing some ocean currents, motion is assumed to take place on a horizontal plane. For some laterally constrained flows, Coriolis terms can be neglected. Specific example solutions to simplified forms of the above equations are presented below.

SOME EXAMPLE SOLUTIONS
FOR OCEANIC FLUID MOTIONS

The preceding section outlined principles governing the motion of all fluids; some examples of how the governing equations describe fluid motions observed in the ocean follow.

One way in which oceanic flows can be classified is according to the nature of their generating force; see Fig. 2. Important generating forces are as follows: (a) meteorological, e.g., atmospheric pressure gradients, and wind stresses on the water surface that generate waves and storm tides; (b) astronomical, such as lunar and solar attraction that produce tides (long waves) which in turn give rise to tidal currents through inlets and estuary entrances; (c) impulsive forces which generate flows such as tsunamis caused by earthquakes and other cataclysmic events and submarine landslides (turbidity currents); and (d) forces arising because of fluid density gradients generated by temperature and salinity differences. Such temperature gradients owe their origin to uneven heating of the earth's surface by the sun. Currents caused by temperature and salinity induced density gradients are termed thermohaline circulation. Thermoclines (steep, vertical density gradients) essentially separate two distinct layers of fluid of different densities and provide a fluid interface on which internal waves may propagate. Breaking of internal waves and their effect on sediment motion has been investigated experimentally by Southard and Cacchione (1972). Thermoclines may also provide an apparent fluid boundary separating two regions having distinctly different flow conditions.

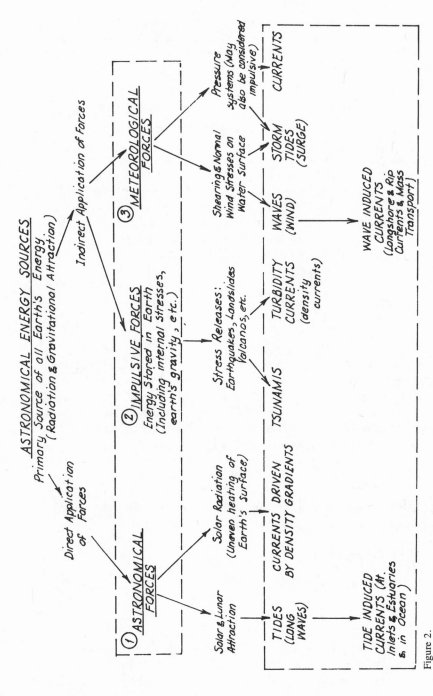

Figure 2.

Classification of flow according to driving forces.

Regardless of the mechanism initiating motion, flow is governed by the preceding equations. The terms retained (or neglected) in the equations provide an additional means for classifying flows. These are illustrated by the examples that follow.

Wave Motion

Assumptions commonly made to obtain solutions to the equations of motion descriptive of wave motion include incompressibility (ρ = constant), irrotationality (viscous terms are neglected), and that Coriolis accelerations are negligible. With these terms omitted the Navier-Stokes equations are termed the Euler equations. It is often further assumed (although not always) that the flow is two dimensional in a vertical plane normal to the wave crests. These assumptions allow the fluid motion to be described by a potential function ϕ. The directional partial derivatives of the potential function give the fluid velocities. In terms of the potential function, the equations of motion and the continuity equation combine to give the Laplace equation,

$$\frac{\partial^2 \phi}{\partial x^2} + \frac{\partial^2 \phi}{\partial z^2} = 0 \tag{7}$$

where $\dfrac{\partial \phi}{\partial x} = u$ and $\dfrac{\partial \phi}{\partial z} = w$

Boundary conditions for the wave problem are that water particles on the free surface remain there and that no fluid is transported across the boundaries (velocities normal to the bottom and free surface are zero). For a sinusoidal, progressive wave moving in the $+x$ direction in water of uniform depth, a solution (which only approximately satisfies the boundary conditions at the free surface and is limited to waves of small height) is given by,

$$\phi = \frac{HgT}{4\pi} \frac{\cosh\left[2\pi(d+z)/L\right]}{\cosh\left[2\pi d/L\right]} \cos\left(\frac{2\pi x}{L} - \frac{2\pi t}{T}\right) \tag{8}$$

where H is the wave height, L is the wave length, T is the wave period, and d is the depth. L, T, and d are related by

$$L = \frac{gT^2}{2\pi} \tanh\left(2\pi d/L\right) \tag{9}$$

and the wave profile, η, is simply,

$$\eta = \frac{H}{2} \sin\left(\frac{2\pi x}{L} - \frac{2\pi t}{T}\right) \tag{10}$$

The *net horizontal fluid particle displacement* over a wave period is zero; consequently, the net movement of sediments suspended in the flow is zero for this approximate solution. Solutions which more accurately satisfy the free-surface boundary conditions (higher order solutions), predict a net particle displacement or mass transport in the direction of wave advance. Suspended sediments can be transported by this component of the flow if a mechanism for moving sediment from the bottom into the flow exists. The oscillatory bottom velocities often accomplish this.

Both the simple theory presented above and higher order wave theories allow finite velocities tangent to the bottom. This is in contrast to the actual physical phenomenon which requires that the velocity tangent to the bottom, as well as the normal velocity, be zero (a no-slip condition). For the sinusoidal wave described by Eq. (8), the horizontal velocity at the bottom is predicted as

$$u_b = \frac{\pi H}{T \sinh(2\pi d/L)} \sin\left(\frac{2\pi x}{L} - \frac{2\pi t}{T}\right) \tag{11}$$

In relatively deep water ($d/L > 1/2$), practically no wave induced fluid motion extends to the bottom; however, as waves move into shallow water maximum bottom velocities increase until eventually they exceed the velocity required to initiate sediment motion. The increase in maximum bottom velocity with decreasing depth is shown on Fig. 3 in dimensionless form. Combined with data on sediment eroding velocities, the depth in which bottom sediments are first disturbed by wave motion can be estimated. Figure 4 illustrates this relationship for waves having a period of 10 sec.

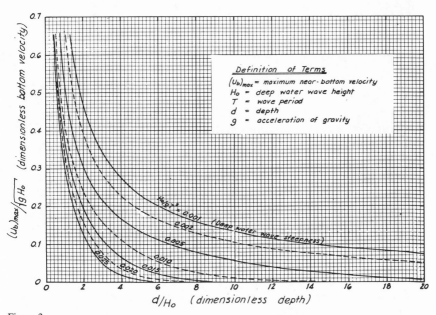

Figure 3.

Dimensionless, maximum near-bottom velocity.

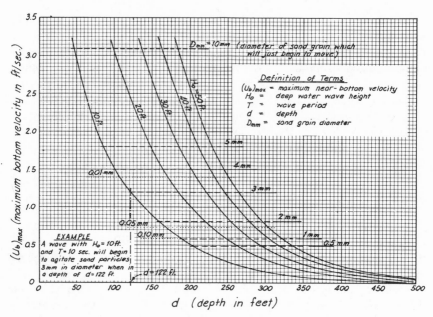

Figure 4.

Initiation of sediment motion under a wave with period, T = 10 sec.

Since the velocity must be zero at the bottom, Eq. (11) actually (in an approximate way) predicts the velocity some small distance away from the boundary. The layer below which potential flow solutions to Eq. (7) are no longer valid (where the velocity goes from u_b to zero) is the boundary layer mentioned earlier. Flow equations describing flow within the boundary layer must include some viscous effects, either through molecular viscosity in the Navier-Stokes equations for laminar flows or through the eddy viscosity in the time-averaged Navier-Stokes equations for turbulent flows.

It is apparent to anyone who has ever observed waves in nature that the preceding equations do not describe what is normally observed. The actual sea surface is quite irregular and made up of a spectrum of waves having different heights, periods, directions of travel and phases. The sea surface and fluid motions, however, can often be described by a summation of ϕ terms, one for each component wave. Harris (1972) describes the irregularity of waves observed in coastal regions and their spectral representation.

Ocean Currents

The simplest solutions to the equations of motion for ocean currents are for frictionless currents where, as in the wave motion described above, shear stress terms are neglected. The flows are two dimensional in the horizontal plane and are classified as inertia currents, geostrophic currents, gradient currents, or cyclostrophic currents, depending on which of the remaining terms of the equations are retained.

Inertia currents are free currents that are no longer under the influence of their generating force. Only acceleration terms (terms I and II) and Coriolis terms (term IV) of Eqs. (2) and (3) are retained so they reduce to

$$\frac{\partial u}{\partial t} + u\frac{\partial u}{\partial x} + v\frac{\partial u}{\partial y} = \frac{du}{dt} = 2\Omega v \sin\theta \tag{12}$$

and

$$\frac{\partial v}{\partial t} + u\frac{\partial v}{\partial x} + v\frac{\partial v}{\partial y} = \frac{dv}{dt} = -2\Omega u \sin\theta \tag{13}$$

The flow is taken uniform in the vertical direction hence changes occurring in the z direction are zero. For flows described by Eqs. (12) and (13), fluid particles move in circular paths having a radius that depends on the angle of latitude, θ, given by,

$$r = \frac{\sqrt{u^2 + v^2}}{2\Omega \sin\theta} \tag{14}$$

Where the current velocity $C = \sqrt{u^2 + v^2}$. The time required for a fluid particle to make one revolution around the inertia circle is the circumference divided by the current velocity, or

$$T_p = \frac{2\pi r}{C} \tag{15}$$

Such circular motions are characteristic of oceanic fluid motions influenced by Coriolis effects. The predominance of clockwise rotating currents in the Northern Hemisphere and counterclockwise rotating currents in the Southern Hemisphere can be attributed to Coriolis effects. The radius of the inertia circle and the time required to make one revolution are shown on Fig. 5 as a function of latitude for a current velocity of 1 knot.

When pressure gradient forces are balanced against Coriolis forces, the equations for two dimensional flow in the x-y plane reduce to,

$$0 = \frac{1}{\rho}\frac{\partial p}{\partial y} - 2\Omega v \sin\theta \tag{16}$$

and

$$0 = \frac{1}{\rho}\frac{\partial p}{\partial y} + 2\Omega u \sin\theta \tag{17}$$

They express a force balance termed a geostrophic equilibrium. Equations (16) and (17) describe a geostrophic current which is in a direction normal to the

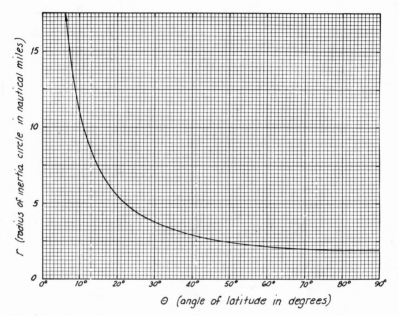

Figure 5.

Radius of inertia circle for a 1 knot current velocity $[T_p(hr) = 2\pi r(nm)]$.

direction in which the pressure gradient acts; to the right for an observer looking down the pressure gradient from the high pressure to low-pressure region in the Northern Hemisphere and in the opposite sense in the Southern Hemisphere. The current velocity in terms of the pressure gradient is

$$C = \sqrt{u^2 + v^2} = \frac{1}{2\rho\Omega \sin \theta} \sqrt{\left(\frac{\partial p}{\partial x}\right)^2 + \left(\frac{\partial p}{\partial y}\right)^2} \qquad (18)$$

where $\sqrt{(\partial p/\partial x)^2 + (\partial p/\partial y)^2}$ is the magnitude of the pressure gradient.

The geostrophic current solution assumes a uniform velocity in the z direction. To consider changes in density in the vertical direction, the third equation of motion must be included. A simple form of Eq. 4 is the hydrostatic equation obtained when only terms III and VI of the equation are retained ($\partial p/\partial z = -\rho g$), that is, the increase in pressure at a point with increasing distance of the point below the free surface is equal to the increase in fluid weight above the point.

Other frictionless ocean currents can be described by balancing convective accelerations, pressure gradient forces, Coriolis forces and centrifugal forces on a fluid particle. Known as gradient currents these currents describe the counter-clockwise rotation about low-pressure regions and the clockwise rotation about high-pressure regions in the Northern Hemisphere and vice versa in the Southern Hemisphere. (In meteorology this motion is quite discernable in wind fields.)

Small scale, circular flows are relatively unaffected by Coriolis forces if the radius of the circular motion is much smaller than the radius of the inertia circle given by Eq. (14). One such two-dimensional flow unaffected by Coriolis forces balances centrifugal forces against pressure gradient forces and is called a cyclostrophic current.

Stratification in the ocean modifies the flow patterns of currents described above. In some cases, the behavior of surface currents is not indicative of flow in lower fluid layers where sediment transport is important.

Internal shear stresses, boundary shear stresses, and/or turbulent transfers of momentum from one layer of flow to another all act to complicate actual flows. These frictional effects cause velocity gradients in the vertical direction which, when coupled with Coriolis effects, may cause flow in different fluid layers to be in different directions. One classical solution for turbulent flow (Ekman, 1905) results in the well-known Ekman spiral. Since the flow is turbulent, the time-averaged Navier-Stokes equations in terms of average velocities are used and the stress terms (Reynolds stresses) are expressed in terms of an empirical eddy viscosity. Ekman considered shear stresses on a horizontal plane (vertical transfer of momentum) so the equations of motion become,

$$2\Omega\bar{v}\sin\theta = \frac{1}{\rho}\left\{\frac{\partial}{\partial z}\left(\epsilon\frac{\partial\bar{u}}{\partial z}\right)\right\} \tag{19}$$

$$2\Omega\bar{u}\sin\theta = \frac{1}{\rho}\left\{\frac{\partial}{\partial z}\left(\epsilon\frac{\partial\bar{v}}{\partial z}\right)\right\} \tag{20}$$

where $\bar{u}$ and $\bar{v}$ are time average velocities and ϵ is the eddy viscosity (the terms $\epsilon\,\partial\bar{u}/\partial z$ and $\epsilon\,\partial\bar{v}/\partial z$ are the Reynolds stresses). Ekman's solution for a current in deep water driven by a surface wind stress, τ, with constant eddy viscosity is given by

$$\bar{u} = V_0 e^{\frac{-\pi z}{D}} \cos\left(\frac{\pi}{4} - \frac{\pi z}{D}\right) \tag{21}$$

$$\bar{v} = V_0 e^{\frac{-\pi z}{D}} \sin\left(\frac{\pi}{4} - \frac{\pi z}{D}\right) \tag{22}$$

where $D = \pi \sqrt{\epsilon/(\rho\Omega \sin\theta)}$ and V_0 is the surface current velocity equal to $\tau/\sqrt{2\rho\epsilon\Omega \sin\theta}$. The current direction at the surface is deflected $\pi/4$ radians to the right of the direction of the driving wind stress in the Northern Hemisphere. Moving downward through the fluid, the magnitude of the velocity decreases exponentially. For each succeeding fluid layer, the velocity vector is deflected more to the right. In shallow water, the bottom modifies the flow predicted by Eqs. 21 and 22. The important implication is that the direction and magnitude of wind driven surface currents may not be indicative of flow conditions near the bottom.

Other currents are also important transporting mechanisms. In nearshore regions, longshore currents induced by breaking waves and tidal currents at inlet and estuary entrances are extremely efficient sediment movers compared to shelf currents farther from shore. These currents are also solutions of the equations of motion presented above. However, because of their location near the shoreline in shallow water, the bottom and shoreline geometry (shoreline configuration and bottom topography) strongly influences these flows. Conversely, these flows also have a strong influence on boundaries and cause relatively rapid changes in bathymetry and shoreline location.

CONCLUSION

The foregoing discussion of oceanic fluid motions has been rather incomplete and quite superficial; however, the flows described are characteristic of what is actually observed. Their governing principles apply to even the most complex fluid motions. From a practical point of view, the preceding equations are useful primarily for explaining observations rather than for predicting fluid motion (and subsequently sediment motion) at actual locations on the continental shelf. In reality, the sea's motion is made up of combinations of the motions described above, each interacting with the other in numerous complex ways. Unless coupled with a substantial measurement program the equations can provide only a qualitative understanding of the sea's motion at a specific site. A wave or current climatology for a particular location can only be established by field measurement of the waves and currents over long time periods. Then, combined with measurements of bottom conditions and driving forces, flow phenomena might possibly be explainable in terms of solutions to the governing equations. If long records of measurement (either wave or current) are available, statistical analyses of the flow data may provide an alternative to solving the equations of motion. When and if statistics of the flow climate at a site can be established, the problem reduces to one of translating flow information into sediment transport rates and rates of erosion and deposition: the next component problem in the complex phenomenon of sediment motion on the continental shelf.

ACKNOWLEDGMENT

This paper was prepared as a part of the civil works program, U.S. Army Corps of Engineers at the Coastal Engineering Research Center.

LIST OF SYMBOLS

Symbol	Definitions	Dimensions
C	Current velocity	L/T
d	Water depth	L
D	$\pi\sqrt{\epsilon/\rho\Omega\sin\theta}$	L
D_{mm}	Sand grain diameter in millimeters	L
g	Acceleration due to gravity	L/T^2
H	Wave height	L
H_0	Wave height in deep water (when $d/L > \frac{1}{2}$)	L
L	Wave length	L
p	Pressure	F/L^2
r	Radius of inertia circle	L
S	Salinity	
t	Time	T
T	Wave period	T
T_0	Temperature	deg
T_p	Time required for a water particle to make one revolution about the inertia circle	T
u	Fluid velocity in the x direction	L/T
$\bar{u}$	Time averaged fluid velocity in the x direction	L/T
u_b	Wave induced horizontal velocity at the bottom predicted by linear wave theory	L/T
$(u_b)_{max}$	Maximum value of u_b	L/T
v	Fluid velocity in the y direction	L/T
$\bar{v}$	Time averaged fluid velocity in the y direction	L/T
V_0	Surface current velocity	L/T
w	Fluid velocity in the z direction	L/T
x	Cartesian coordinate direction	L
y	Cartesian coordinate direction	L
z	Cartesian coordinate direction	L
$\delta\nabla$	Elemental volume equal to $\delta x \cdot \delta y \cdot \delta z$	L^3
δx	Incremental length in the direction	L
δy	Incremental length in the direction	L
δz	Incremental length in the direction	L
ϵ	Eddy viscosity	FT/L^2
θ	Angle of latitude	deg
η	Ordinate of free water surface	L
ρ	Mass density	FT^2/L^4
ϕ	Potential function	L^2/T
τ_{xy}	Shear stress (subscripts denote direction in which	F/L^2

stress acts and direction normal to plane on which
stress acts respectively)

Ω Angular rotation of earth ($\Omega = 7.29 \times 10^{-5}$ rad/sec) $1/T$

REFERENCES

Duane, D. B. (1970). "Synoptic observations of sand movement," *Proc. 12th Coastal Engineering Conf., ASCE,* 799-813.

Ekman, V. W. (1905). "On the Influence of the Earth's Rotation on Ocean Currents," Vol. 2, No. 11, Arkiv. f. Matem., Astr. O. Fysik, Stockholm, Sweden.

Harris, D. L. (1972). Wave estimates for coastal regions. *In* "Shelf Sediment Transport: Process and Pattern" (D. J. P. Swift, D. B. Duane, and O. H. Pilkey, eds.), Dowden, Hutchinson & Ross, Stroudsberg, Pennsylvania.

Ippen, A. T., ed. (1966). "Estuary and Coastline Hydrodynamics," Engineering Societies Monographs, McGraw-Hill, New York.

Komar, P. D., Neudeck, R. H., and Kulm, L. D. (1972). Observations and significance of deep water oscillatory ripple marks on the Oregon continental shelf. *In* "Shelf Sediment Transport: Process and Pattern" (D. J. P. Swift, D. B. Duane, and O. H. Pilkey, eds.), Dowden, Hutchinson & Ross, Stroudsberg, Pennsylvania.

Neumann, G. and Pierson, W. J., Jr. (1966). "Principles of Physical Oceanography," Prentice-Hall, Englewood Cliffs, New Jersey.

Phillips, O. M. (1966). "The Dynamics of the Upper Ocean," Cambridge Univ. Press, London and New York.

Rouse, H. (1938). "Fluid Mechanics for Hydraulic Engineers," Engineering Societies Monographs, McGraw-Hill, New York, reprinted by Dover Publications, Inc., 1961.

Smith, J. D., and Hopkins, T. S. (1972). Sediment transport on the continental shelf off of Washington and Oregon in light of recent current measurements. *In* "Shelf Sediment Transport: Process and Pattern" (D. J. P. Swift, D. B. Duane, and O. H. Pilkey, eds.), Dowden, Hutchinson & Ross, Stroudsberg, Pennsylvania.

Sternberg, R. W. (1972). Predicting initial motion and bedload transport of sediment particles in the shallow marine environment. *In* "Shelf Sediment Transport: Process and Pattern" (D. J. P. Swift, D. B. Duane, and O. H. Pilkey, eds.), Dowden, Hutchinson & Ross, Stroudsberg, Pennsylvania.

Teleki, P. G. (1972). Wave boundary layers and their relation to sediment transport. *In* "Shelf Sediment Transport: Process and Pattern" (D. J. P. Swift, D. B. Duane, and O. H. Pilkey, eds.), Dowden, Hutchinson & Ross, Stroudsberg, Pennsylvania.

U.S. Navy (1966). "Handbook of Oceanographic Tables," Special Publication # 68, U.S. Naval Oceanographic Office, Washington, D.C.

CHAPTER 2

Wave Boundary Layers and Their Relation to Sediment Transport

P. G. Teleki

Coastal Engineering Research Center
5201 Little Falls Rd., N.W.
Washington, D.C. 20016

ABSTRACT

Sediment transport in the ocean is examined from the viewpoint of oscillating flows. Principles of both steady and unsteady boundary layers are reviewed and updated from recent experimental results. In the potential flow region the forcing function is represented by the combined effect of waves and currents. This paper is concerned mainly with wave-induced effects. A unifying theory for energy dissipation and sediment transport under shoaling, breaking, and transformed waves does not exist because the nonlinearity present in wave motion and in turbulence has not been appraised. Study of the response function (sediment motion) can be conducted more efficiently through research on boundary layers in oscillating flows, enabling the confirmation of universal relationships for the distribution of velocity and shear stress in periodic flows. These relationships are dependent on the characteristic frequency and amplitude of oscillation shown experimentally in shear stress measurements of increasing magnitude shoreward outside the zone of breaking waves. As bottom friction is a measure of viscous dissipation in the boundary layer, the principle of energy conservation prior to wave breaking based on this evidence becomes unacceptable, especially for natural conditions, where other dissipation mechanisms may be equally significant.

INTRODUCTION

When a progressive gravity wave propagates from deep water onshore, the decreasing depth of water produces a continuous deformation of the wave profile, measurable in change of wave steepness and attenuation of wave height. During shoaling some of the energy is transmitted, and a portion is used for the transport of fluid and sediment. There is increasing nonlinearity associated with the motion of fluid as the wave approaches breaking conditions. This nonlinearity results not only from instability of the wave form, but from increasing levels and duration of turbulence. Initially turbulence will be an intermittent phenomenon even in steady flows; in the unsteady reversing flows under waves comparatively higher input of energy is needed to reach an equivalent turbulent state. Maintenance of the turbulent state will depend on the wave frequency, the associated acceleration effects and the rate of vortex generation resulting from instability.

Nonlinear wave theories of Stokes (1880), Korteweg-deVries (1895), and Dean (1965) attempt solving the wave equation in terms of the free-surface configuration, for a given water depth d, whereby the irrotational solution to the internal velocity-pressure field

$$[u,v,w,p] \ (x,y,z,t) = F(\eta,\omega,k,d)$$

becomes a function of the parameters $a = \eta_{max}$, which is the variation in the free-surface elevation with wave amplitude $a = H/2$ for a harmonic wave of height H, and $\omega = 2\pi/T$, $k = 2\pi/L$, which are the characteristic wave numbers of wave period T and length L; t is time; u, v, w are the velocity components corresponding to the coordinates x,y,z; and p is pressure. These solutions are restricted by certain imposed boundary conditions; a mathematical model universally applicable to the entire region of wave transformation prior to breaking does not exist. At best, one can match, piecewise, existing analytical solutions of increasing nonlinearity, where the continuous if not linear energy dissipation due to various causes is taken into account [Johnson (1970)].

In the region of breaking and broken waves no analytical solution exists because the contributions of nonlinear effects of fluid motion due to convective inertia and turbulence, changes in fluid properties due to aeration and the influence of longshore currents are difficult to separate and consequently to evaluate. This is the environment, however, where the rate and volume of sediment transport is most intense and where the variations in transport mode and quantity are quite substantial. It is, therefore, quite disheartening to find that analytical or even empirical approximation of transport of sediment in the littoral zone is predicated upon linear approximations of wave-induced flow outside this area, resulting in unsupported extrapolation to the phenomena in the surf zone.

An example of a linear estimating technique is computing the onshore and alongshore components of wave energy flux, based on the principle of energy conservation up to wave breaking, assuming no energy loss due to interaction with the bottom. From the works of Putnam and Johnson (1949), Putnam (1949), Reid and Kajiura (1957), Savage (1953), and Iwagaki and Tsuchiya (1967), it is evident that bottom friction and percolation are plausible and in many cases significant mechanisms of energy dissipation. Relating linear energy flux to the rate of sediment transport in the nearshore environment consequently produces discrepancies in this rate often by three orders of magnitude. Such variation is not the singular result of neglecting certain mechanisms, it is also caused by the inaccuracy inherent in the collection of field data on wave parameters, which commonly are the height, period, and length of the progressive wave, and the local water depth. Modification of the flow field in presence of mass transport [Longuet-Higgins (1953, 1956); Noda (1969); Johns (1969)], reflection [Caldwell (1949); Noda (1969)], and wave or wind setup along the shore should be manifested in the equation of motion; the difficulty lies in the evaluation of their mutual interaction in spatial and temporal frames of reference.

An alternate approach to the evaluation of the near-bottom flow field and the response of sediment is represented by boundary layer research. Although the phenomenon of boundary layers may be considered to be on the microscopic scale, and only an intermediate step in a process-response model when considering the wave-sediment system as a whole, their study holds very specific promises of advancement in sediment transport research. First, it allows one to forego the use of questionable wave theories and assumptions on the flow distribution near the bottom. In the zone of high sediment transport rates, most nonlinear wave theories break down. In presence of a current superimposed on wave motion (which typically occurs on the inner continental shelf), the combined momentum of currents and waves in an unsteady flow field is difficult to estimate. Second, material transported in traction moves within the boundary layer, or when suspended by periodic uplift, particles are transmitted across the layer. The rate of exchange between bedload and suspended load, and the local distribution and concentration of suspended sediment under a wave depends on the structural composition of the wave boundary layer. Third, whether the flow in the layer is laminar or turbulent influences the rate and pattern of sediment dispersal.

In the domain of unsteady oscillating boundary layers, concepts, methods and results owe their origin to developments in aerodynamics. There is paucity of literature on wave-induced flows in boundary layers, especially where this pertains to progressive waves in deep water, progressive gravity waves transforming on a slope, or waves coupled with currents. Nonetheless, it seems reasonable to advance the thought of investigating the process operative at the water-sediment interface in both the laboratory and in field experiments, in the hope that, as information accumulates, accurate modeling of the process-response exchange characterizing sediment in motion will later enable extension of its results to the observable wave parameters.

In the analytical description of the kinetics of sedimentary particles certain requirements must first be met. For the fluid these are:

1. accurate description of the velocity distribution in the free stream, including the time-dependent fluctuations, especially near the bottom boundary,

2. evaluation of resistance to the flow including those due to grain friction and bed form-induced drag, and

3. description of changes in the physical properties of the fluid.

For the sediment, accurate characterization of its physical properties and definition of criteria for equilibrium are necessary. Near the water-sediment interface this characterization must account for large variations in both the dynamics of the system and in properties of fluid and sediment (or its mixture) over small vertical distances resulting from variations in sediment concentration, bottom topography and due to the presence of the boundary layer.

Studies of unsteady boundary layers draw liberally upon both theories and experimental procedures developed for steady state conditions. For this reason, the major aspects of steady boundary layer theory are reviewed in the following section, in which boundary layer structure, the concept of equilibrium, and universal relationships for the velocity distribution, are discussed. Following the review of the steady boundary layer theory, unsteady boundary layer theories are discussed which are shown to fit into three categories: shear wave solution, perturbation method, and turbulent flow analogy. Because energy dissipation in oscillating flows is the principal mechanism of sediment entrainment, it follows that exact definitions and precise measurements for the boundary shear stress and the boundary layer thickness are requisites for the understanding of sediment motion. But, using published data and data from the author's experiments, it is demonstrated in the Section titled ''Boundary Shear and Energy Dissipation'' that both quantities are difficult to predict. The consequence of this is that sediment transport dynamics on the inner continental shelf can now be realized only in its larger aspects.

BOUNDARY LAYERS—STEADY CASE

In review of historical developments, reference must be made to Prandtl's (1904) contribution to the theory of classical hydrodynamics. On observing the discrepancy between his experimental results and Euler's equation of motion, he suggested that the theory had neglected fluid friction. He also suggested that any flow about a solid object, irrespective of the shape of the object should be divided into two regions, a very thin layer in the immediate vicinity of the body where friction due to the viscosity of the fluid plays an essential part, and the remainder of the fluid outside this region where viscous effects are negligible. The thin layer is the boundary layer. A no-slip condition imposed on this flow requires the velocity at the solid wall to be zero. Successive layers, outward from the boundary, moving with increasing velocities, create a transverse velocity gradient until the velocity of the flow at some level becomes nearly constant, i.e., reaches the value present in the free stream. The upper limit of the boundary layer, therefore, can be defined by that elevation where $du/dy = 0$ for the logarithmic velocity profile, and in general where $u = 0.99\ U_\infty$, given u as the velocity in the boundary layer and U_∞ the velocity at the edge of the layer.

There is considerable internal shearing between adjacent layers of fluid which move with different velocities. This shearing is tangential, i.e., parallel to the solid wall. At the boundary the resistance is generated by viscous retardation, and the resulting shear stress distribution becomes a function of the velocity distribution. Inside the boundary layer the friction and inertia forces are of the same order of magnitude, while in the free stream forces of inertia predominate.

Boundary Layer Structure

In classical experiments, boundary layers are often generated on a flat plate, and their development observed from the so-called leading edge. The thickness of the boundary layer increases with distance from the edge, shown in Fig. 6, which is attributable to more and more fluid slowing down due to the retarding effects of the plate, until a state of instability is reached and the flow becomes turbulent. The thickness of a turbulent layer is generally greater than in laminar flow and so is its rate of growth. In the fully turbulent case a laminar sublayer develops, which also thickens as the velocity and turbulent intensity of the

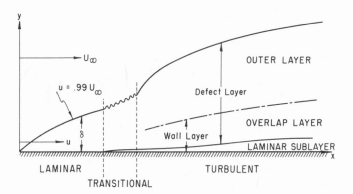

Figure 6.
Schematic of a boundary layer developing on a flat plate.

flow increases. The character of the outer flow does not effect the velocity distribution in the laminar region. It is the outer flow, however, in which small eddies are generated, those carried toward the boundary become damped by the laminar sublayer, and those carried outward gain momentum in transport while diffusing throughout the fluid.

This vortex generation is the underlying principle of turbulent flows, and the process by which the boundary layer acquires additional turbulent fluid is known as entrainment. As Reynolds (1969) states:

"Whereas a laminar boundary layer thickens by outward viscous diffusion, the growth of the turbulent boundary layer seems to be determined by the rate of entrainment of non-turbulent fluid by the large eddy structure at the outer edge of the layer. The physical thickness of a laminar boundary layer is a poorly defined quantity; in contrast, the super-layer, which separates turbulent and non-turbulent fluid, is quite sharply defined, and its average height can be taken as a very real measure of the physical thickness of the turbulent boundary layer."

The threefold division of the turbulent boundary layer (Fig. 6) is a direct consequence of the behavior of the fluid in these regions. In the inner layer the character of the flow is governed by the molecular viscosity and the shear stress at the wall. Although turbulent eddies do not penetrate its lower portion, the laminar sublayer, there seems to be some relationship between the intermittency of vortex shedding in the outer layer and the intermittent existence of the sublayer, a phenomenon discovered by Einstein and Li (1956).

The distance between bed and the vortex street was experimentally determined to be about 5 mm for wave period of 3.5 sec and oscillation amplitude of 11.0 cm by Carstens and others (1969). Generally, the inner layer responds quickly to changes in wall conditions, such as boundary roughness. The outer layer is fully dependent upon conditions imposed by the mean flow, such as the streamwise pressure gradient $\partial p/\partial x$, and by the shear stress at the wall.

Velocity Distribution in Steady Boundary Layers

Flow in the inner and the outer layers can be described by the so-called wall law and the defect law respectively, and these relationships are both applicable in the overlap layer in Fig. 7. The law of the wall is of the form

$$\frac{u}{u_*} = f\left(\frac{yu_*}{\nu}\right) \tag{1}$$

with f as an unspecified function and ν the kinematic viscosity. The shear velocity $u_* = (\tau/\rho)^{1/2}$ is the square root of the ratio of the shear stress τ, and the fluid density ρ.

For the laminar sublayer Eq. 1 simply reads

$$\frac{u}{u_*} = \frac{yu_*}{\nu} \tag{2}$$

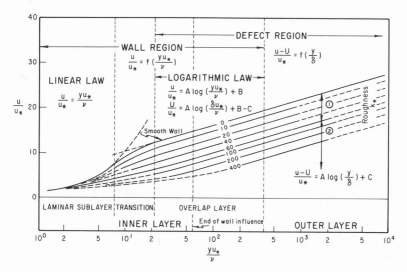

Figure 7.

Universal velocity profile in turbulent boundary layers of incompressible steady flow for $0 < k_* = u_* k/\nu < 400$. Velocity distribution in the tripartite layer is described by the law of the wall and the velocity defect law, and matched in the overlap region by the logarithmic law. The shear velocity is expressed as $u_* = (\nu \partial u/\partial y)^{1/2}$ in the laminar sublayer, $u_* = (\nu \partial u/\partial y - \overline{u'v'})^{1/2}$ in the transitional sublayer, where $\overline{u'v'}$ is the Reynolds stress, and $u_* = [K_z \partial u/\partial y]^{1/2}$ elsewhere, given K_z as the eddy viscosity. Values of the constants are approximately: $A = 5.85$, $B = 4.1$, $C = 2.0$. Region (1) represents turbulent, partly rough flow; region (2) turbulent, fully rough flow conditions [after Hama (1954), Clauser (1956), van Driest (1958), Granville (1958)].

and the quantity on the right side is known as the shear Reynolds number. The inner region profile gives way to what is often referred to as the logarithmic velocity distribution or the Kármán-Prandtl equation:

$$\frac{u}{u_*} = \frac{1}{\kappa} \ln \frac{yu_*}{\nu} + B$$

$$= A \log \frac{yu_*}{\nu} + B$$

where A and B are constants and $\kappa = 0.41$ is Kármán's universal constant, all of which are independent of the pressure gradient.

For the outer flow the law of wake [Coles (1956)] is written as

$$\frac{u-U}{u_*} = \frac{1}{\kappa} \ln\left(\frac{y}{\delta}\right) - \frac{\Pi}{\kappa}\left[2 - \omega\left(\frac{y}{\delta}\right)\right]$$

$$= A \log\left(\frac{y}{\delta}\right) + C \tag{3}$$

where δ is the boundary layer thickness, and the left-hand side of Eq. 3 is the defect velocity relationship.

Equation 3 states that the tangential stress at the wall causes a reduction in the differential velocity $U - u$ as a function of the distance from the wall, and this is independent of the nature or origin of the stress. The response in the outer layer is slow relative to the inner portion, on account of the larger scale eddy structure [Harris (1970)].

The universal wake function

$$\omega\left(\frac{y}{\delta}\right) = 1 - \cos\left(\Pi\,\frac{y}{\delta}\right)$$
$$= 2\sin^2\left(\frac{\Pi y}{2\delta}\right) \tag{4}$$

contains the "wake parameter" Π, which is constant for any profile in equilibrium [Clauser (1956)] but varies in the direction of the flow for nonequilibrium flows. Some oscillating flows may be quasi-steady, but will not be in equilibrium, therefore, β must be evaluated for oscillatory boundary layers, preferably as function of the characteristic frequency ω.

The condition of equilibrium can be tested with

$$\beta = \frac{\delta^*}{\tau_0}\,\frac{\partial p}{\partial x} \tag{5}$$

where the bottom shear stress

$$\tau_0 = \nu\rho\left|\frac{\partial u}{\partial y}\right|_{y=0} \tag{6}$$

for the laminar case, and the displacement thickness

$$\delta^* = \int_0^\infty \left(1 - \frac{u}{U}\right)dy \tag{7}$$

is a function of the velocity defect with U the velocity just outside the layer.

For small values of β, the velocity profile of the defect layer is logarithmic and the inner and outer profiles are assumed to overlap asymptotically (Fig. 7); an estimate of the friction at the wall can then be made from

$$\frac{U}{u_*} = \frac{1}{\kappa}\ln\left(\frac{\delta u_*}{\nu}\right) + B + \left[\frac{2\Pi}{\kappa} - \frac{\omega\Pi}{\kappa}\left(\frac{y}{\delta}\right)\right]$$
$$= A\log\left(\frac{\delta u_*}{\nu}\right) + B - C \tag{8}$$

The shear stress is now expressed in terms of u_*, δ, and Π. In the laminar sublayer Eq. 6 applies; in the transition region

$$u_*^2 = \frac{\tau}{\rho} = \nu\frac{\partial u}{\partial y} - \overline{u'v'} \tag{9}$$

where $-\rho\overline{u'v'}$ is the Reynold stress originating from the turbulent fluctuations. In fully turbulent flows

$$u_*{}^2 = \frac{\tau}{\rho} = K_z \frac{\partial u}{\partial y} \tag{10}$$

where K_z, the eddy viscosity, depends on the magnitude and distribution of Reynolds stresses in the boundary layer, and consequently is a variable quantity not only in the vertical profile at a given station, but in time as well, influenced by the rate of vortex generation upstream of the point of measurement. Although the distribution of K_z is poorly known for all types of boundaries and varying intensities of turbulence, in practice it is approximated with either a constant value, or through Prandtl's "mixing length" method [for a summary see Shepard (1963), pp. 119-123] if the shape of the flow field is taken into consideration. Departure from the theoretical velocity distribution (smooth wall profile, Fig. 7) occurs with increasing bottom roughness. The boundary becomes rough (fully turbulent) when the size of roughness on the bottom (grains of sediment) exceed the thickness of the laminar sublayer. With k_s denoting Nikuradse's representative roughness length, the velocity profile will be shifted downward by increments of

$$k_* = \frac{u_* k_s}{\nu} \tag{11}$$

as shown on Fig. 7, suppressing the range for which the laminar sublayer could exist. The vertical shift on the profile is equivalent to $-\Delta u/u_*$, which can be determined only experimentally. In the overlap layer, for example, Eq. 8 modifies to

$$\frac{U}{u^*} = A \log\left(\frac{\delta u_*}{\nu}\right) + B - C - \frac{\Delta u}{u_*} \tag{12}$$

and $\dfrac{\Delta u}{u_*} = f(k_*)$.

Returning to the problem of equilibrium flows, the presence or absence of the pressure gradient is the criterion of nonequilibrium or equilibrium conditions, respectively. For $\partial p/\partial x \geq 0$, Mellor (1966) established that the true shear stress τ is influenced by the local pressure gradient, and

$$(u_*)^2 = (u_o^*)^2 + \frac{1}{\rho}\frac{\partial p}{\partial x} \tag{13}$$

which gives

$$\tau = \tau_o + \frac{\partial p}{\partial x} \tag{14}$$

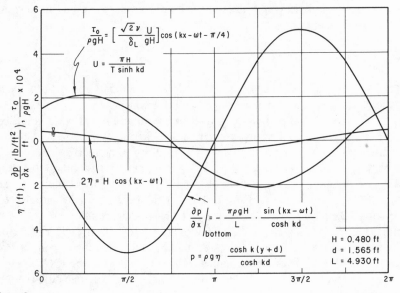

Figure 8.
Distribution of the wave profile, pressure gradient, and bottom shear stress over one wave period calculated using linear theory.

in the viscous or laminar sublayer. Mellor's results for adverse gradients were extended by Herring and Norbury (1967) to negative (favorable) pressure gradients.

This development is similar to that of Cebeci (1970), who assumed that the Reynolds stress can be neglected in the wall region, and the shear stress distribution is directly related to the pressure term. Positive and negative pressure gradients alternate in progressive gravity waves, shown in Fig. 8 for the distribution of $\partial p/\partial x$ over a wave period at the bottom boundary. The presence of nonnegligible pressure makes it difficult to evaluate the bottom shear stress under high frequency waves in terms of a time-averaged value. The only condition satisfying $\partial p/\partial x = 0$ is at the instant a wave crest or trough passes. The shear stress related to these two events in the wave cycle can be evaluated in its instantaneous expression, but must be corrected for the frequency-dependent phase differences which exist between the free surface of the fluid and the wave-induced velocity and shear stress at the bottom.

THE OSCILLATING BOUNDARY LAYER

Equations of Motion

The principal characteristics of oscillating boundary layers are that the motion is unsteady, periodic in time or space or both, and that the kinematic viscosity gives a measure of the rate of diffusion of vorticity and transport of momentum. The layer dominated by vorticity is the oscillating boundary layer. In the presence of waves with a free surface the characteristics of an oscillating boundary layer are governed by the outer, potential flow. Certain basic differences exist between steady and oscillating flow. As illustrated in Fig. 9 the parabolic and logarithmic velocity distributions of steady flow are replaced, except for low-frequency quasi-steady tidal flow [Yalin and Russell (1966)], by the velocity profiles of oscillating laminar and turbulent flow.

For the motion of an incompressible viscous fluid the pressure and the terms corresponding to the orthogonal velocity components can be determined from only two equations, one for the momentum, the other for continuity, because the energy equation has been replaced by $d\rho/dt = 0$, on the assumption of homogeneity in fluid density. Flow in the boundary layer in the direction of wave propagation is described by the momentum equation

$$\frac{\partial u}{\partial t} + u\frac{\partial u}{\partial x} + v\frac{\partial u}{\partial y} + \frac{1}{\rho}\frac{\partial p}{\partial x} - \nu\nabla^2 u = 0 \tag{15}$$

$$\frac{\partial v}{\partial t} + v\frac{\partial v}{\partial y} + u\frac{\partial v}{\partial x} + \frac{1}{\rho}\frac{\partial p}{\partial y} - \nu\nabla^2 v = 0 \tag{16}$$

in the two-dimensional case. When continuity is satisfied in the potential flow region outside the boundary layer, we write

$$\nabla^2\phi = \frac{\partial^2\phi}{\partial x^2} + \frac{\partial^2\phi}{\partial y^2} = 0 \tag{17}$$

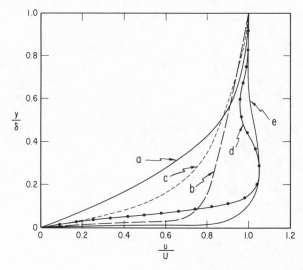

Figure 9.

Typical velocity profiles in steady (a) laminar flow, (b) turbulent flow, sediment-free fluid, and (c) turbulent flow, sediment-laden fluid; and unsteady (oscillating) (d) laminar flow, and (e) turbulent flow.

given $\phi = Ae^{i(kx-\omega t)}$ as the potential function (A is a function of y only), as the mass of a given element in incompressible flow must be conserved. In Eqs. 15 and 16 velocity gradients in respect to v and the pressure variation across the boundary layer are assumed to be negligible. Because of the lack of viscous effects in the free stream, the flow in the potential region can be described by

$$\frac{\partial U_\infty}{\partial t} + U_\infty \frac{\partial U_\infty}{\partial x} + \frac{1}{\rho}\frac{\partial p}{\partial x} = 0 \tag{18}$$

for the unsteady case, where U_∞ is the velocity in the free stream.

By assuming the pressure at the edge of the layer to equal the pressure within, we combine Eqs. (15) and (18) and get

$$\frac{\partial u}{\partial t} + u\frac{\partial u}{\partial x} + v\frac{\partial u}{\partial y} = \frac{\partial U_\infty}{\partial t} + U_\infty \frac{\partial U_\infty}{\partial x} + v\frac{\partial^2 U}{\partial y^2} \tag{19}$$

which is the equation of motion for the total flow, and is in itself a boundary condition. Other appropriate boundary conditions are

$u = v = 0$ when $y = 0$
$u = U$ for $y = \delta$
$U \to U_\infty$ as $y \to \infty$

where y is the coordinate measured from the bottom boundary upwards, u and v are the boundary layer velocity components in the x and y planes respectively. U is the velocity at the edge of the boundary layer and δ is the thickness of the layer.

The Navier-Stokes Eqs. (15, 16) contain a fundamental difficulty represented by the nonlinear convection terms $u(\partial u/\partial x)$, $v(\partial u/\partial y)$, and conventionally some method of linearization must be applied to obtain a solution for the expression.

Several solutions are based on the numerical negligibility of the convective terms of Eq. (15) and (16). Basic assumptions are that the vertical component of velocity is numerically very small and the motion time-dependent by require-ment of continuity. These solutions are applicable to flows in equilibrium or quasi-equilibrium, that is without appreciable pressure gradients, consequently the numerical calculations become simplified. Application is limited to long waves of small amplitude, such as shallow water waves or tidal motion. This model is based on the following:

Consider an infinite flat plate above which the velocity in the free stream has the form

$$U = U(x) \cos(\omega t) \tag{20}$$

which is the linear expression of the velocity in an oscillating fluid. The corresponding boundary layer velocity

$$u = U(x) \left[\cos \omega t - e^{-\eta} \cos (\omega t - \eta) \right] \qquad (21)$$
$$\eta = y/\delta$$

derived by Stokes (1851) and known as the "shear wave" solution is an adequate description of the flow if the ratio of wave amplitude to water depth and wave length is small, i.e. $(a/d, ka) \ll 1$ [Grosch (1962)].

The vertical component of velocity is given by [Rosenhead (1963), p. 383]

$$v = - \delta \frac{\partial U}{\partial x} \left[\eta \cos \omega t + \frac{1}{\sqrt{2}} \left(\omega t + \frac{3\pi}{4} \right) \right.$$
$$\left. + \frac{e^{-\eta}}{\sqrt{2}} \cos \left(\omega t - \eta - \frac{\pi}{4} \right) \right] \qquad (22)$$

and for the shear stress

$$\tau = \nu \rho \delta U(x) \cos \left(\omega t + \frac{\pi}{4} \right) \qquad (23)$$

where only the real part of the complex variable is taken into account. Eqs. (22) and (23) show the presence of phase advance of the vertical velocity and the skin friction in relation to the horizontal velocity component. Fig. 9.d illustrates the corresponding velocity profile. This model has been used by Kalkanis (1964), Abou-Seida (1965), Manohar (1955), Iwagaki and others (1965) among others for the study of wave-induced sediment transport.

For progressive waves, Longuet-Higgins (1956) has introduced

$$u = U(x) \left[\cos (kx - \omega t) - e^{-\eta} \cos (kx - \omega t - \eta) \right] \qquad (24)$$

The model has been extended to include turbulent flows (Abou-Seida, 1965) by modifying Eq. (21) to

$$u = U(x) \left[\cos \omega t - f_1(y) \cos (\omega t - f_2(y)) \right] \qquad (25)$$

in which $f_1(y)$, the velocity amplitude function and $f_2(y)$ the phase shift function, are related to the turbulent velocity component and have been experimentally evaluated.

Empirical expressions for $f_1(y)$, $f_2(y)$ for smooth bottom, two and three dimensional roughness can be found in the review article of Einstein (1971).

This approach, which employs a plate oscillating in its own plane in still fluid, becomes physically reasonable because the solution, based on symmetry, does not depend on distance from the wall. Linearization is carried out assuming the quadratic terms to be smaller than the rest in Eq. (15) and (16) by a factor of *ka* (Kalkanis, 1964). Noda (1971) criticized this approximation technique on the grounds that experimental results of Kalkanis (1957) did not support his simplified model.

A different approach is presented by solutions which evaluate the convective inertia terms in the equations of motion. Inclusion of these terms is unavoidable where the frequency of oscillation is high, or the wave profile is sufficiently steep (e.g., in finite amplitude waves) to originate nonnegligible acceleration effects [Abou-Seida (1965)], or a combination of both.

Solution is found in the perturbation of the boundary layer equations, obtaining first a so-called zero order solution by setting $U(U\partial/\partial x) = u(u\partial/\partial x) = v(\partial u/\partial y) = 0$. These results are subsequently used to compute the nonlinear terms in the first order approximation corresponding to the secondary flow field.

This iterative scheme was first used by Schlichting (1932) for the case of the oscillating cylinder in a fluid at rest. Assume that the free stream velocity $U(x,t)$ is given by irrotational flow theory and that u and U experience small perturbations about a steady mean, so that

$$u = u_0(x,y) + \epsilon u_1(x,y) e^{i\omega t} + \epsilon^2 u_2(x,y) e^{i2\varphi t}. \tag{26}$$

and

$$U = U_0(x) + \zeta U_1(x) e^{i\omega t} + \zeta^2 U_2(x) e^{i2\omega t} \cdots. \tag{27}$$

If only terms of $0(\epsilon,\xi)$ are considered, for the oscillating part we get

$$i\omega u_1 + u_0 \frac{\partial u_1}{\partial x} + u_1 \frac{\partial u_0}{\partial x} - i\omega U_1 - \frac{d}{dx}(U_0 U_1) = \nu \frac{\partial^2 u_1}{\partial y^2} \tag{28}$$

The second approximation yields a steady secondary flow, which does not vanish far from the boundary. This effect has been confirmed for high frequency waves in shallow seas by Hasselman (1970). It results from Reynolds stresses created in the flow field by the oscillatory motion, therefore is independent of viscosity. A particular case of solving Eq. (28) is known as the Kármán-Pohlhausen Method, used by Lighthill (1953) and by Eagleson (1959) in the investigation of skin friction. The resulting boundary layer profiles are shown in Fig. 10. This model is extendable to include higher order terms, i.e., $0(\epsilon^2)$, $0(\epsilon^4)$, etc. to compensate for the increasing nonlinearity of fluid motion. Numerical calculations quickly become tenuous, as shown by Kestin and others (1967), at the second approximation.

This technique was applied by Iwagaki, Tsuchiya and Chen (1967) to progressive waves using Airy theory for the potential flow, by Noda (1969) for standing waves, and by Noda (1971) for the oscillating turbulent boundary layer.

A particular linearized solution was performed by Lin (1956) using *turbulent*

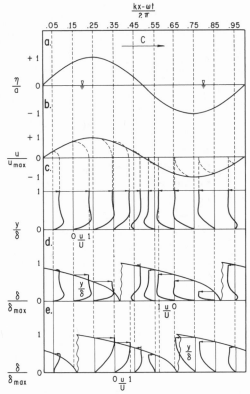

Figure 10.
Characteristic velocity profiles in the oscillating boundary layer [Eagleson (1959)] for a wave cycle; (a) is the elevation of free surface propagating with the phase velocity C, (b) is the amplitude of the velocity in the boundary layer, (c) is the velocity profile for the shear wave solution, unseparated flow, (d) is the velocity profile for the shear wave solution with separation, and (e) is the velocity profile for the Kármán-Polhausen solution with separation.

flow analogy, in which the time-averaged and oscillating components of flow were separated

$$
\begin{aligned}
u(x,y,t) &= \bar{u}(x,y) + \tilde{u}(x,y,t) \\
U(x,t) &= \bar{U}(x) + \tilde{U}(x,t) \\
p(x,t) &= \bar{p}(x) + \tilde{p}(x,t)
\end{aligned}
\tag{29}
$$

where the second terms on the right side of Eq. (29) denote the oscillating components. This approach was used by Karlsson (1958) in connection with unsteady turbulent flows.

An important aspect of boundary layer research pertains to establishing the value of the phase lag among velocity components and between these and the shear stress. Calculated phase differences can vary considerably, depending on which theory is employed. Experimentally, phase is a difficult parameter to measure accurately. What is certain, however, is that phase differences are controlled by frequency of the oscillation (Yalin and Russell, 1966). Experiments of Horikawa and Watanabe (1968) and Teleki (1970) attempted to establish the distribution of phase advance in the boundary layer for both smooth and rough boundaries, the results of which confirmed that oscillations in the boundary layer precede free stream oscillations. As pointed out by Karlsson (1958) this is a consequence of the fluctuating pressure gradient acting more rapidly on the slower moving fluid in the boundary layer.

Phase relationships between the elevation of the free surface η, the velocities U and u, and δ_{max} were examined by Eagleson (1959) for separating and non-separating shear flow and the Kármán-Polhausen solution. These are shown in Fig. 10, but experimental verification is lacking.

The Thickness of the Boundary Layer

Conventionally, the thickness of the boundary layer is defined to be the vertical distance between $y = 0$ and that level ($y = \delta$) where $u = 0.99\, U\infty$ for both steady and unsteady flows (Fig. 6).

For oscillatory flows, the boundary layer thickness is defined by Schlichting (1968) as

$$\delta_L = \left(\frac{2\nu}{\omega}\right)^{\frac{1}{2}} \tag{30}$$

According to Li (1954) the region in which shear flow is effective is

$$\delta_l = 6.5\left(\frac{\nu}{\omega}\right)^{\frac{1}{2}} \tag{31}$$

or $\delta_l = 4.6\, \delta_L$.

In other reports, the layer is considered to extend only to where u/U'_∞ reaches its maximum [Eagleson and Dean (1961)] or where $u = U_{max}$ [Jonsson, (1965b)].

Proper evaluation of the boundary layer thickness is important to sediment transport because the nature of the flow within and without the layer differ. Consequently, the local value of δ is a valuable indicator, along with the prevailing flow regime, of the localized mode of sediment transport. Boundary layer flow regime and the thickness of the layer influence entrainment of sedimentary particles. The relationship between δ and the incipient particle size was shown by Cacchione (1970). Subsequent transport is similarly actuated by these properties, as is discussed in the section titled "Sediment Transport."

A mathematical definition does not exist for the thickness of the turbulent boundary layer in oscillating flow. Contrary to the laminar case, where δ_L is governed by the local viscosity and the frequency of the harmonic motion, the thickness of the turbulent boundary layer is a function of the rate of vortex generation and the value of the intermittency, namely whether the vortex shedding is burstlike or steady. In the presence of roughness the boundary layer has been shown by Liu and others (1966) and Teleki and Anderson (1970) to thicken much faster than on a smooth wall, other conditions being identical.

Jonsson (1963) expressed the turbulent boundary layer thickness as

$$30\frac{\delta}{k_s} \log\left(\frac{30\delta}{k_s}\right) = 1.2\frac{\xi\,\text{max}}{k_s} \tag{32}$$

where k_s is Nikuradse's roughness and $\xi_{\text{max}} = U_{\text{max}}/\omega$ is the maximum horizontal amplitude of particle motion. Increasing values of ξ_{max}/k_s were found to be correlatable to increased thickness of the overlap layer; that is, the effect of coarser sediment on the boundary expedites the onset of turbulence for lower Reynolds criteria [Jonsson (1966)].

Comparison of theory and data obtained from experimental velocity profiles under progressive waves with a free surface transforming on a slope [Teleki and Anderson (1970)] and beneath internal waves in a stratified fluid transforming on a slope (Cacchione, 1970) indicate that Eq. (30) underestimates the measured values of δ in each case (Fig. 11). For the data of Teleki and Anderson (1970), agreement is much closer if Eq. (31) proposed by Li (1954) is used. The various definitions for δ attest to the fact that accurate measurement of the dimensions of the layer is difficult. Nevertheless, the maximum thickness can be shown

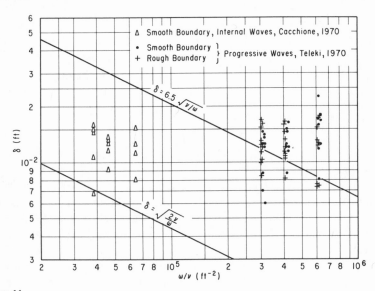

Figure 11.

Comparison of measured boundary layer thicknesses in progressive internal and gravity waves to Eqs. 13 and 14.

to occur immediately following the onset of the maximum horizontal velocity in the boundary layer. The two maxima for δ_L, corresponding to the forward and reverse portions of the cycle, will only be equal for harmonic wave motion on a flat bottom.

Because wave transformation on a steep slope was the principal phenomenon studied by both Cacchione (1970) and Teleki (1970) certain aspects of these experiments are comparable. The upslope increases in boundary layer thicknesses can be ascribed to the increasing contribution of the nonlinear convection. But, sorting out the viscous and nonlinear effects is difficult in the region of shoaling waves. Both studies also indicate that close to the region of breaking waves the experimental boundary layer thickness decreases. Although it is not clear what the nature of the responsible mechanism is—it could be ascribed to the onset of fully developed turbulence—the corresponding shear stress at the bottom in progressive waves is less than expected. This aspect merits further research.

BOUNDARY SHEAR AND ENERGY DISSIPATION

The total energy contained in a wave is equally partitioned between kinetic and potential energy

$$E = E_p + E_k = \int_0^L \int_0^{-d} \frac{\rho}{2} (u^2 + v^2)\, dxdy + \int_0^L \frac{\rho H^2}{4} \cos^2 (kx - \omega t)\, dx$$

$$= \frac{\rho g H^2 L}{8} \tag{33}$$

and E is the surface energy density per unit width of wave crest.

The rate at which this energy is propagated in the direction of wave advance is

$$P = ECn = \frac{\rho g H^2 L}{8 T} \left\{ \frac{1}{2} \left(1 + \frac{2kd}{\sinh kd} \right) \right\} \tag{34}$$

or the energy flux per unit wave; Cn is the group velocity and $n = \{\ \}$ is the shoaling coefficient.

Up to the breaking wave, the principle of conservation of energy flux is an approximate method for calculation of wave transformation, assuming there is no friction at the boundary nor reflection of wave energy due to the sloping bottom (Koh and LeMéhauté, 1966). Under transforming waves energy loss is a real and significant phenomenon, however, and its origin can be found in the following mechanisms: (a) bottom friction; (b) bottom percolation; (c) wave reflection; (d) conversion to heat in turbulent flow; and (e) increase in potential energy due to wave runup. Of these, the first item contributes the most to the total effect, and has been, therefore, most often investigated. Bottom friction results from resistance offered by the granular material on the sea bottom to the flow above it; the resulting force, which acts tangentially on the boundary is called the shear stress. In the presence of bedforms, the total force includes the drag due to the undular surface; the resistance to flow gains momentum from the wavy bed, and the corresponding energy dissipation is higher than for flat surfaces.

If we assume the energy dissipation to be the result only of viscous effects in the boundary layer on a flat, impermeable surface, we can write

$$E_f = \nu \rho \left(\frac{\partial u}{\partial y}\right)^2 = \tau_0 \frac{\partial u}{\partial y} \tag{35}$$

where E_f is the dissipation function, and τ is the shear stress in the boundary layer.

The average rate of energy dissipation per unit length of wave crest per unit area can be expressed as

$$-\frac{dP}{dxdy} = -\overline{E}_{fo} \approx \frac{1}{L}\int_0^L \int_0^\delta E_f dy dx$$

$$\approx \frac{\nu\rho}{L}\int_0^L \int_0^\delta \left(\frac{\partial u}{\partial y}\right)^2 dy dx \tag{36}$$

and $\dfrac{dE}{dt} \approx \dfrac{1}{T}\displaystyle\int_0^T \tau_0 u\, dt$ per unit wave period T.

For the shear wave case, Iwagaki and others (1965) have shown

$$-\overline{E}_{fo} \approx \frac{\nu\rho}{2\delta}\left(\frac{\pi H}{T}\right)^2 \operatorname{cosech}^2 kd \tag{37}$$

to the first approximation. Subsequently, Iwagaki and others (1967) reported extending Eq. (36) to the second approximation which modified the value of $\overline{E}_{fo}$ by only 2%.

Putnam and Johnson (1949) calculated the average bottom friction representative of the continental shelf on the assumption that the effect of bottom roughness is numerically more significant than viscosity, and obtained

$$-E_{fo} = \frac{4\pi^2\, T^3}{3 \sinh^3 kd} \tag{38}$$

In this expression the dissipation function is independent of water velocity. The effect of bottom friction on wave height and, therefore, on the potential energy was found to be about 21% for a typical continental shelf slope of 1/300.

Putnam (1949) examined the effect of percolation on energy dissipation and found it to be small in comparison to bottom friction. In fact, its numerical significance was further reduced by Reid and Kajiura (1957) when it was found Putnam overestimated the dissipation function by a factor of four. Experimentally, this was indicated much earlier in a laboratory study by Savage (1953), who introduced identical wave conditions to smooth permeable, smooth impermeable, and rippled surfaces, confirming that the effect of percolation is negligible for sand sizes less than 2 mm. On the other hand Murray (1965) contends that for porous sands of 1.0 mm size dissipation through percolation is more important than viscous dissipation at the boundary for uprush type flows. However, Murray did not solve for the case of local intermittent fluidization, which characteristically occurs in the swash-backwash zone.

Experiments by Bagnold (1946) indicated that the mean bottom drag on rippled surfaces is a function of ω^2 for all oscillation amplitudes. Duplicating test conditions, Savage (1953) has shown that friction due to the rippled surface is greater than on a smooth sandy bottom and that energy loss is greatest during ripple formation. This is to say that excess energy loss occurs when the flow and the boundary are not in equilibrium. Energy dissipation under these conditions can exceed 50%.

Bottom Friction

A measure of energy loss is the bottom shear stress τ_0, whose definition for the laminar case under waves is identical to Eq. (6) except that u is calculated from Eq. (20). In view of the unknown phase differences and partial understanding of the mechanism of flow separation, it is a difficult task to determine the shear stress of the mechanics of flow separation, determining the shear stress distribution for all phases in periodic flow is a difficult task. The problem can be simplified and shear stress, i.e., when the influence of flow "history" is least felt. When this postulate is satisfied, the use of steady-state analogy may, in many cases, be justified. For example, the drag coefficient under oscillatory flow is time-variant, but when the shear stress is near its maximum value, the drag coefficient is found to be very nearly the same as in steady open channel flow given the same boundary conditions.

In consequence of the foregoing, we write the equation for the maximum shear stress under waves according to linear theory

$$\frac{\tau_{0\ max}}{\rho g H} = \frac{\sqrt{2\nu}}{g \sinh kd} \left(\frac{\pi}{T} \right)^{3/2} \tag{39}$$

$$\tau_{0\ max} = \rho U_{max} (2\pi\nu/T)^{1/2} \tag{40}$$

and the left side of Eq. (39) is a measure of wave height attenuation due to friction. Distribution of Eq. (39) for a wave cycle is depicted in Fig. 8.

The corresponding Reynolds number

$$\mathrm{Re} = \frac{\delta_L}{\sqrt{2}} \frac{U}{\nu} = \frac{U \delta_L'}{\nu} \tag{41}$$

$$\delta'_L = \left(\frac{\nu}{\omega}\right)^{\frac{1}{2}}$$

(42)

where

$$U = \frac{\pi H}{T} \frac{\cosh \ k(y+d)}{\sinh \ kd} \cdot \cos \ (kx - \omega t)$$

(43)

is the velocity due to potential motion.

In steady flow the mean velocity is related to the bottom shear stress through the friction factor, f:

$$\tau_o = f \rho \overline{U}^{\,2}$$

(44)

For a given roughness and for small Reynolds numbers these quantities are also related through the coefficient of friction

$$C_f = 2\left(\frac{u*}{U}\right)^2$$

(45)

with C_f as the steady value corresponding to τ_{0max}. Although $\frac{1}{2}C_f$ and f should be numerically the same, Iwagaki and others (1967) report the relationship

$$f = \frac{3\pi^2}{64} \sqrt{2}C_f = 0.654 \ C_f$$

(46)

For wave motion C_f is more difficult to obtain than Eq. (45) indicates. It will be dependent on bottom roughness, the Reynolds and Froude numbers, and the level of turbulence present in the flow [Zhukovets (1963)].

Because of the uncertainties attached to the form of the velocity distribution in oscillatory boundary layers, the correct magnitude of τ_0 and its phase relationship with u_0 are open to experimental evaluation, especially in anharmonic, nonmonochromatic waves. Several experiments in the past were designed to forego measuring the velocity profile, or compared results to several theoretical expressions. Among these, Eagleson (1959) evaluated boundary shear according to three different analytical solutions of the equations of motion. Iwagaki and others (1965), using the linear (shear wave) solution found Eagleson's measured friction coefficients to be too high. In 1967, Iwagaki and others used a perturbation method to extend the 1965 computations to the second order, and found the contribution of $0(\epsilon, \epsilon^2)$ to be negligible.

When the flow becomes turbulent in the wave boundary layer and the effect is due to roughness rather than instability near breaking, the Reynolds number

$$RE = \frac{UD}{\nu}$$

(47)

applies, where D is the particle size, assumed to equal $k_s/30$. Kajiura (1968) has shown that the appropriate friction coefficient must be defined in respect to the wave frequency; thus

$$C_f = 1.7\left(\frac{U}{\omega k_s}\right)^{-2/3} \tag{48}$$

In the turbulent oscillating boundary layer the boundary shear stress is assumed to be of the same form as in unidirectional flows

$$\frac{\tau_o}{\rho} = K_z\left(\frac{\partial u}{\partial y}\right)_{y=0} \tag{49}$$

where K_z is the coefficient of the vertical eddy diffusivity. Kajiura (1968) assumed the distribution of K_z across the layer to be an analogue of the structure of the boundary layer, therefore,

	Boundary			
	Smooth	Rough	Layer	
	ν	$0.369\,\kappa u_o^*\,(15\,k_s)$	Inner	
$K_z =$	$\kappa u_o^* y$	$\kappa u_o^* y$	Overlap	(50)
	$0.2U\delta'_L$	$0.2u_o^*\delta'_L \dfrac{U}{u_o}$	Outer	

for the average state of turbulence over one wave period, where $\kappa = 0.41$ as before, and u_o* is the shear velocity corresponding to τ_o.

Experiments of Horikawa and Watanabe (1968) indicated the distribution of K_z to be time-dependent, and experimentally rather scattered about the predicted value. The vertical distribution of K_z must also be expected to be influenced by the vertical sediment concentration gradient in suspensions [Horikawa and Watanabe (1970)].

The friction coefficeint used by Jonsson (1963) for turbulent rough flow

$$C_f = 2\left(\frac{u*}{U}\right)^2 = \frac{0.0604}{\left(\log\frac{30\,\delta}{k_s}\right)^2} \tag{51}$$

leads to the expression

$$\frac{1}{4\sqrt{C_f}} + \log\frac{1}{4\sqrt{C_f}} = 0.21 + \log\frac{\xi_{max}}{k_s} \tag{52}$$

derived from the Prandtl-Kármán velocity distribution for unidirectional flows. The corresponding maximum shear stress in turbulent flow

$$\tau_{o\,max} = \frac{0.0604}{\left(\log\frac{30\,\delta}{k_s}\right)^2} \cdot \frac{\rho U^2_{max}}{2} \tag{53}$$

with the boundary layer thickness $\delta = 0.04 \, \pi \xi_{max}/\ln(30\delta/k_s)$.

The corresponding energy dissipation

$$-\bar{E}_{fo} = 0.21 \, \rho C_f U_{max}^2 \tag{54}$$

In their study of low frequency waves, Yalin and Russell (1966) equated the pressure gradient to the slope of the free surface and the shear stress distribution, such that

$$\frac{\partial p}{\partial x} = \rho g \frac{\partial \eta}{\partial x} = -\frac{\partial \tau}{\partial y} \tag{55}$$

for the laminar case.

Their expression for the bottom shear stress,

$$\tau_o = \alpha \rho U^2 + \beta \rho g \frac{\partial \eta}{\partial x} \, \delta \tag{56}$$

contains the empirical constants α and β, which they evaluated for laminar and rough turbulent flow in the boundary layer. With U given by Eq. (43), the constant α is implicitly a function of time and explicitly that of RE, whereas β is a constant for laminar flow. For the rough turbulent case

$$\alpha = f\left(\frac{u}{u_*}, \frac{yu^*}{v}\right) = \left[2.5 \ln\left(\frac{11d}{k_s}\right)\right]^{-2} \tag{57}$$

where the water depth is measured from still water level. Similarly $\beta = f(k_s/d)$ near the boundary.

When a wave crest or trough passes, the instantaneous value of $\partial \eta/\partial x = 0$. If the pressure phase difference between surface and bottom is very small $\partial p/\partial x$ also equals zero at that instant, and Eq. (56) reads

$$\tau_o = \frac{\rho U^2}{\left[2.5 \ln\left(\frac{11d}{k_s}\right)\right]^2} \tag{58}$$

When $u = 0$, such as near separation in the boundary layer,

$$\tau_o = \beta \rho g \delta \frac{\partial \eta}{\partial x} \tag{59}$$

When $\partial p/\partial x > 0$, the true magnitude of the bottom shear is described by Eq. (14). Writing the equation for the subsurface pressure

$$p = \rho g \eta \frac{\cosh k(y+d)}{\cosh kd} \cos(kx + \omega t) - \rho g y \tag{60}$$

where $\eta = \frac{H}{2} \cos(kx - \omega t)$ \hfill (61)

the pressure variation in the direction of wave advance at the bottom is

$$\frac{\partial p}{\partial x} = -\frac{\pi \rho g H}{L \cosh kd} \sin (kx - \omega t) \tag{62}$$

at the bottom of the ocean. The shear stress for the laminar case now becomes $\tau = \tau_o + \partial p/\partial x$ as in Eq. (14), and we write

$$\tau = \rho (\nu \omega)^{1/2} \frac{\pi H}{T \sinh kd} \sin (kx - \omega t - \pi/4) -$$

$$\rho g \frac{\pi H}{L \cosh kd} \sin (kx - \omega t - \pi/2) \tag{63}$$

$$= \rho \frac{\pi H}{T} \left[\frac{(\nu \omega)^{1/2}}{\sin kd} \sin (kx - \omega t - \pi/4) - \frac{g}{C \cosh kd} \sin (kx - \omega t - \pi/2) \right]$$

Figure 8 illustrates the phase relationships between the free surface, the pressure gradient and the bottom shear stress in a cosine wave.

The shear stress may be evaluated indirectly from a measured velocity profile. Studies with turbulent oscillatory boundary layers by Jonsson (1963) compared the measured velocity profile obtained with a small current meter with the logarithmic portion of Fig. 7; this gave reasonable agreement for the distribution of boundary shear over the wave cycle.

Horikawa and Watanabe (1968) used the hydrogen bubble method to obtain the instantaneous velocity profile for both hydrodynamically smooth and rough boundaries in transitional to turbulent regimes, where the criteria of $25 < \mathrm{Re} < 658$ for the smooth-bottom case and $100 < \mathrm{RE} < 1000$ for the rough bottom case were used. The shear stress obtained using Eq. (49) was consistently higher than the theoretical value derived from the defect velocity relationship in and near the boundary layer.

McDowell (1965) reported that the magnitude of the shear stress, distributed over a wave cycle, was slightly higher than for an equivalent steady flow, although the low frequencies of oscillation used in his experiments created a quasi-steady condition.

The author, using a Preston probe, obtained the distribution of instantaneous velocity and maximum bottom shear stress corresponding to the crest and trough of progressive waves transforming on a slope in a wave tank. In laminar to transitional flows, this indirect method was found to underestimate the value of τ_{0max} obtainable from linear theory [Eq. (39)]. Justification for the use of the Preston probe in not fully turbulent flows was put forth by Teleki and Anderson (1970), following the developments of Hsu (1955). Comparison of four experiments indicates, when based on a laminar analogue, that use of

Figure 12.

Comparison of boundary shear measured in a wavetank with a Preston probe to its theoretical value, $\nu \rho (\partial u/\partial y)$. Results are shown for smooth boundary high amplitude waves (a), and low amplitude waves (b). Congruence is best between theory and case (b). Deviation from the theoretical values in case (a) is due to the neglect of nonlinear effects in waves of high steepness.

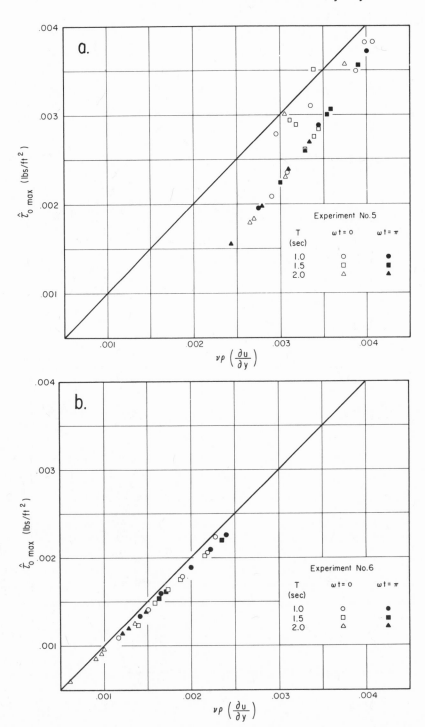

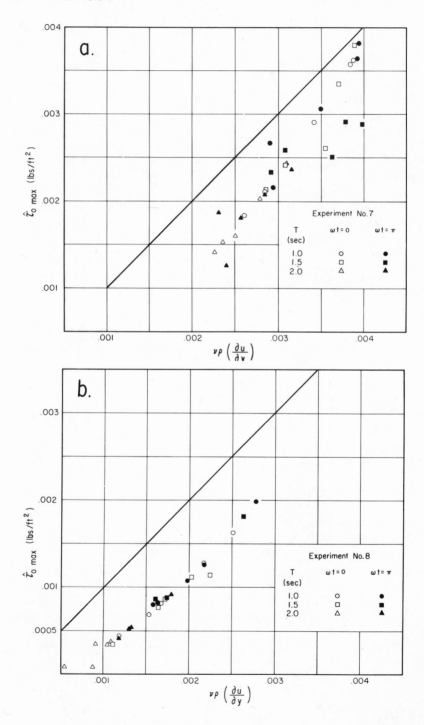

this technique is justified only for small amplitude waves (Fig. 12.b) and smooth boundaries. The effect of roughness (Fig. 13.a) or the additional nonlinearity gained by increasing wave steepness (Fig. 12.a) or both (Fig. 13.b) result in a shift of the experimental values. The nearly linear variation of τ_{0max} along the slope, with τ_{0max} increasing toward the breaker zone, was found to be sensitive to wave amplitude and bottom roughness. Figure 14 illustrates that $\partial\tau/\partial x$, the shear stress gradient, is larger for low-amplitude waves than for high-amplitude waves. The effect of (fixed) roughness was found to reduce the local value of τ_{0max} below its equivalent smooth-boundary value.

The large diversity in numerical results for the shear stress can be attributed to the techniques used to collect the data. Comparing direct methods of Eagleson (1959), Iwagaki and others (1965), Matsunashi and Kawadani (1965), and Yalin and Russell (1966), and the indirect method used by Teleki and Anderson (1970) shows considerable overestimation of Eq. (39) by the shear plate measurement and underestimation of similar magnitude by the Preston probe (Fig. 15). Admittedly, sinusoidal, monochromatic waves were present in none of the experiments.

Field experiments on the measurement of bottom friction have been very few. Bowden and others (1959) and Charnock (1959) measured tidal velocities in estuaries and obtained a measure of the boundary shear by extrapolation of the profile to the boundary.

For the shallow waters of Atchafalaya Bay, Bretschneider (1954) calculated $\bar{f} = 0.08$, which is comparable to $0.033 < f < 0.09$ calculated by Iwagaki and Kakinuma (1963), who applied Bretschneider and Reid's (1954) methods for estimating the combined dissipation due to percolation and bottom friction of a refracted wave train for the Akita Coast, Japan. These values of f are considerably higher than the so-called steady-state value of $f = 0.01$, used first by Putnam and subsequently by many others. Sternberg (1968 and this volume) assumed the logarithmic law to apply for the velocity distribution in Puget Sound and the Strait of Juan de Fuca; the relationship between friction factors thus obtained and the Reynolds criteria show considerable scatter. Nece and Smith (1970) used a field version of the Preston probe also in quasi-steady tidal flows of the Columbia estuary, where reasonable agreement was found between shear stresses measured with the probe and those obtained independently from current meters.

What we may conclude is that bottom friction cannot be estimated from linear theory because the conventionally applied small amplitude theory breaks down in the region of the bottom boundary for both high and low frequency oscillations. Consequently one must independently obtain the velocity profile for given wave conditions, appraise the nonlinear contributions, and then use

Figure 13.

Rough boundary equivalents of experimental data shown in previous illustration, high amplitude waves are used in (a), low amplitude waves in (b). The latter illustrates the shift of the reference datum resulting from the rough boundary.

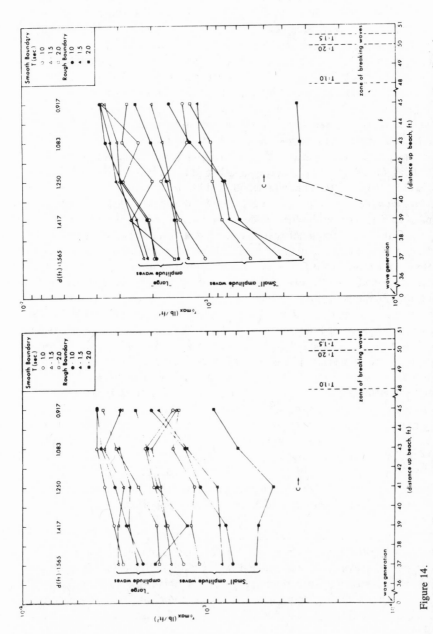

Figure 14.
Experimental shear stress data obtained on a 1:12.5 slope for small and large amplitude progressive waves on smooth and rough boundaries. On the left the instantaneous shear stress values correspond to a wave crest passing, on the right to the wave trough. Measurements are outside the region of wave breaking, indicated in the lower right of the diagrams.

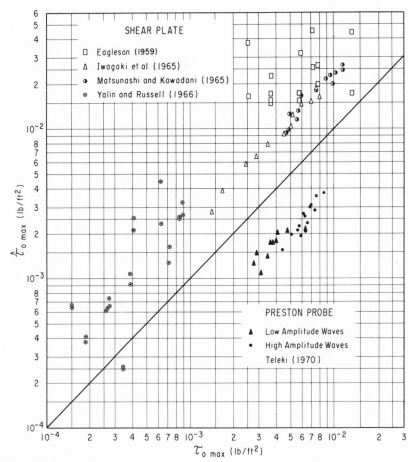

Figure 15.

Comparison of measured shear stress data, obtained with direct and indirect methods, to Eq. (40). Shear plate data are represented for $0.77 < T < 240$ sec, $0.018 < H < 0.3$ ft, and $0.35 < d < 1.31$ ft.; Preston probe data encompass $1.0 < T < 2.0$ sec., $.071 < H < .432$ ft., $1.56 < d < 0.92$ ft. Smooth boundary conditions.

the results to compute a corrected energy loss on the continental shelf bottom. Subsequently, results should be correlated with finite amplitude wave theories.

Other Mechanisms of Energy Dissipation

The phenomenon of a standing wave was discussed by Lettau (1932) and Noda (1969) as a means of bar generation in the offshore zone. The requirement is that there be reflection of the incident wave energy whose magnitude will depend on the beach slope and permeability of the bottom, assuming energy loss due to bottom friction to be minimal. Caldwell (1949) has shown reflection to be negligible for solitary waves on a beach slope of less than 4.5 deg. According to Machemehl (1970) the reflecting capability of a roughened slope decreases as the wave energy density increases, which for increasing wave heights implies more energy to remain contained in the nearshore zone. The probable cause of proportionally greater energy containment is in the gradually greater loss due to turbulence generated by the higher amplitude waves. Therefore, reduction in potential energy at the beach by its transfer to the offshore in reflected waves is probably not significant for most coastal segments of the United States, at least not for waves of high amplitude.

Another means of offshore energy propagation is a consequence of momentum transfer across the breaker zone, resulting in wave set-up near the shore which is relieved periodically by longshore and rip currents. Seaward of the breakers, water will return to the offshore some distance above the boundary layer, which was shown by Longuet-Higgins (1953) for progressive waves and by Noda (1969) for standing waves. For large d/L this mass transport of fluid is onshore at the bottom and at the surface and offshore in between; for small d/L it is onshore at the bottom and offshore at the surface. It is also sensitive to the bottom roughness: for Re < 160 the effect of roughness is to increase the mass transport in laminar boundary layers, whereas exceeding this critical value of Collins (1963) produces a turbulent boundary layer in which the mass transport will have increased over its laminar value [Brebner and others (1967)].

Previous discussion showed percolation to be a negligible cause of energy loss for fine (or poorly sorted) sands. However, Sleath (1968) has shown in laboratory tests that bottom percolation could be the largest cause of wave attenuation (a measure of energy loss) for beds of high permeability. Stratification in the substratum also influences the amount of energy lost, because in graded beds permeability is nonisotropic.

Other mechanisms are poorly understood, and therefore remain unappraised. One is energy conversion to heat in a breaking wave and the other is loss associated with wave runup. Because of increased percolation through fluidization in the swash zone, the amount of energy returned by the backwash should be measurably less than that represented in the uprush.

SEDIMENT TRANSPORT

Flow conditions at and near the sea bottom control the frequency of entrainment of sedimentary particles, provide conditions for incipient development and maintenance of bed forms, and govern the quantity and rate of sediment transport and the mode of transportation. Among these, sediment entrainment is particularly important because of its response to changes in boundary layer structure, associated absolute and nominal boundary layer thickness, and prevailing flow regime.

What is necessary for correct appraisal of sediment transport? One is the measurement of the instantaneous critical shear force near the bottom. This critical shear may be regarded as the forcing function, the resulting sediment motion as the response function. Under many natural conditions, the average boundary shear may be near zero, although the instantaneous maximum shear force in a given direction may be high.

Entrainment of sediment will be a reaction to the instantaneous force. On the other hand, the measurement of local instantaneous velocities is equally useful, given the dependency between velocity and shear stress. However, the measurement of the Eulerian velocity alone is inadequate for ascertaining the rate and direction of sediment transport, for this purpose the Lagrangian frame of reference, namely following the path of a fluid particle, must also be used. When both are measured simultaneously, their difference indicates the magnitude of the Stokian mass transport velocity on the continental shelf [Longuet-Higgins (1969)] and a physically significant model for sediment motion is established.

The second task concerns establishing the form of the vertical velocity distribution u/u_* for oscillating flows as a function of a characteristic wave Reynolds number, say $a^2\omega/\nu$, analogously to Fig. 7 which is not applicable to bidirectional flows. Determination of the similarity profiles not only would enable the summary description of the oscillatory boundary layer, but also would establish the Reynolds criteria for laminar, transitional, and turbulent regimes in the wave-boundary layer. In addition to wave frequency this universal relationship should be expressed in terms of bottom roughness and wave steepness (a measure of the horizontal pressure gradient). Alternately, if the entire velocity profile is known, measurement of $\partial p/\partial x$ and its use in correcting the shear stress distribution may be superfluous.

One must keep in mind that the source of turbulence and the origin of suspension is the boundary layer [Einstein (1971)] and by and large the velocity field in boundary layers under ocean waves will be turbulent, and flow near the bottom will be modified as the concentration of entrained sediment increases. Although uplift and dispersal due to wave motion provides for a discontinuous diffusion through the periodic boundary layer, one can make use of the work of Einstein and Chien (1955), who showed that the velocity distribution in a sediment-laden flow in the boundary layer

$$\frac{u}{u_*} = 17.66 + \frac{2.3}{\kappa} \log \frac{y}{35.45 \, k_s} \tag{64}$$

is a variation on the logarithmic Prandtl-Kármán formula for rough boundaries (Fig. 9.c). The sediment-laden boundary layer will have greater thickness and lower velocities in the inner layer because of damping of turbulent eddies by the entrained sediment and higher potential velocities in the outer layer in comparison to its clear-water analogue.

Under natural conditions sediment motion can be viewed to be in equilibrium if a sufficiently long time span is chosen for its study. Entrainment of additional sediment or different grainsizes requires additional force to upset the equilibrium. Several previously published papers have dealt with the relationship between the intensity of turbulence, initiation of grain transportation in a unidirectional turbulent flow field, and this force: the critical shear stress. Typical of these is the study of Shields (1936) which is based on the concept that shear stress is a measure of the momentum exchange between flow and bed. For wave-boundary layers Einstein (1971) disclaims the appropriateness of momentum exchange and suggests vorticity exchange be used to evaluate conditions at the sediment-water interface.

When a particle is entrained by the shear force the vertical distance to which the sediment is elevated will depend first on the lift force, originating from the rotational behavior of grains on the bed, then on the vortex size and rate of dissipation in the turbulent portion of the boundary layer. Subsequent dispersal is related to the impinging fluid force. Müller and others (1971) have recently shown that the critical lift force for the dislodgement of a particle from its equilibrium position is 0.7 times the submerged particle weight. This is probably equally applicable to oscillating flows, considering Mashima and others (1966) confirmation of the rotational motion of sand grains under wave-induced flow.

As discussed earlier, not only the tangential shear force, but also the pressure force at the bed must be taken into consideration in finite amplitude waves. If the bed is permeable, and the added moments due to the mean shear and mean pressure exceed the moment of resistance, the grains begin to move. The critical force for entrainment of sediment under shoaling waves, should, therefore, be based on the model represented in Eqs. (12) and (17).

On a larger scale, the concept of equilibrium in sand transport should also be reexamined. Briefly the hypothesis states that material transport in the nearshore zone is a response to changes which take place in the environment. The idea that disequilibrium causes entrainment and displacement is based more on observation than measurement, although it is known from experiments with radioactive tracers that a sudden increase in the rate of movement is associated with a change in the wave field. If a wave train of given obliquity with the shore and given wave height, period, and length is replaced by another of different characteristics, the resulting disequilibrium at the ocean bottom will act to readjust the profile and the bedforms until a new equilibrium is attained. Experiments by Bagnold (1946) bear out this hypothesis; Bagnold found sediment entrainment to increase when the new condition introduced changed the equilibrium represented by the ratio of the water particle excursion distance to the ripple pitch.

CONCLUSIONS

There is no adequate theory for the mechanics of turbulent flow at this time. For oscillating flows, where turbulence generated at a given location may return as a modified disturbance, a universal velocity distribution has not been established, partly because of disagreement on Reynolds criteria characterizing laminar, transitional and turbulent regimes, partly because intermittent turbulence prevails over a wide Reynolds number range, and partly due to the dependence of the similarity profiles on the characteristic frequency of oscillation. Steady flow principles and assumptions, especially for the velocity distribution and the critical shear stress, are not applicable to wave-induced flows, except where quasi-steady conditions prevail, i.e., where the frequency of oscillation is low and the wave amplitude is much smaller than the wavelength of oscillation.

There is no adequate theory for the mechanics of energy dissipation and sediment motion in the region of shoaling, breaking and broken waves. Consequently, the tacit assumption of wave energy conservation up to wave breaking is open to reappraisal because it ignores the physical evidence presented by tracer experiments of sediment moving offshore of the breaker zone.

In view of the prevalence of nonequilibrium flows in nearly all of the inner continental shelf, the definition of critical shear stress for the entrainment of sediment must be based on the criterion of the added moments due to shear and pressure overcoming the local moment of resistance.

Considering the difficulties associated with recognition and description of the principal elements of fluid motion, and the deficiency of universal relationships, experimental study of boundary layers is the proper tool in the analysis of the kinetics of sediment motion, especially in respect to sediment entrainment. A comprehensive effort is needed, directed toward the description of the distribution of velocity and shear stress in the boundary layer, and assessment of the influence of boundary friction and form roughness, bottom slope, and local sediment concentration. Which of these divergent theoretical solutions will be found pertinent to a particular environment and its process will depend on the boundary conditions chosen.

The alternate means of establishing relationships between commonly employed wave parameters, such as height, period and length, and sediment motion in the nearshore zone of the ocean is fraught only with additional assumptions and generalizations, most of which attempt to circumvent the inclusion and evaluation of nonlinearity present in the physical process. Nearshore wave motion, currents and turbulence of the flow field are all nonlinear phenomena, and therefore for the time being this process-response model is best studied at the exclusion of theories for the potential flow field.

A final observation is that the effect of waves in deep water, specifically on continental shelves, is negligible in the transportation of sediments; however, fluid displacement due to wave motion can be important to the extent that it can be coupled to a weak current whose origins may be found in mass transport under waves, geostrophic or tidal forces, local baroclinic pressure, or internal temperature-density gradients. Boundary layer research of such conditions holds particular promise for the understanding of shelf processes.

ACKNOWLEDGMENT

Support for the experimental part of this research was provided by the Departments of Geology and Civil Engineering, Louisiana State University. Thanks are due Dr. M. W. Anderson for his guidance in the formulation of the problem. Additional support was provided by the U.S. Army, Coastal Engineering Research Center. Permission to publish was granted by the Chief of Engineers.

LIST OF SYMBOLS

Symbol	Definitions	Dimensions
A	Potential function	
a	Wave amplitude	L
C	Wave celerity	LT^{-1}
C	Coefficient of friction	
D	Grain diameter	L
d	Local water depth	L
E_f	Wave energy, potential	MLT^{-2}
E_{f_0}	Wave energy, kinetic	MLT^{-2}
$\underline{E}$	Dissipation function	$ML^{-1}T^{-3}$
$\overline{E}$	Average rate of energy dissipation	MT^{-3}
e	Exponential	
f	Friction factor	
F	Function, unspecified	
f_1	Velocity amplitude function	
f_2	Phase shift function	
g	Gravitational acceleration	LT^{-2}
H	Wave height	L
i	$\sqrt{-1}$	
k	Wave number	L^{-1}
k_s	Bottom roughness	L
K_z	Vertical eddy diffusion coefficient	L^2T^{-1}
L	Wavelength	L
n	Coefficient of shoaling	
o	Subscript denoting boundary conditions	
P	Wave energy flux, power	ML^2T^{-3}
p	Pressure	$ML^{-1}T^{-2}$
Re, RE	Reynolds numbers	
T	Wave period	T
t	Time	T
U	Free stream velocity	LT^{-1}

U	Velocity at the outer edge of the boundary layer	LT^{-1}
u	Wave orbital velocity, horizontal component	LT^{-1}
u	Velocity in the boundary layer	LT^{-1}
u_0	Velocity at the bottom boundary	LT^{-1}
u	Shear velocity	LT^{-1}
$\tilde{u}$	Oscillating component of mean velocity	LT^{-1}
u'	Turbulent velocity component	LT^{-1}
$\overline{u'v'}$	Reynolds stress	L^2T^{-2}
v	Wave orbital velocity, vertical component	LT^{-1}
w	Wave orbital velocity, normal component	LT^{-1}
x	Horizontal distance or coordinate	L
y	Vertical distance or coordinate	L
z	Normal distance or coordinate	L
α	Dimensionless parameter	
β	Clauser's equilibrium parameter	
δ	Boundary layer thickness	L
δ_L, δ_l	Boundary layer thickness, laminar flow	L
δ_L'	Wave displacement thickness	L
δ^*	Displacement thickness, steady flow	L
ϵ	Perturbation parameter	
ζ	Perturbation parameter	
η	Vertical displacement of water surface from mean surface elevation	L
κ	Kármán's universal constant	
ν	Kinematic viscosity	L^2T^{-1}
ξ	Amplitude of horizontal particle excursion	L
Π	Wake parameter	
ρ	Density of fluid	ML^{-3}
τ	Horizontal shear stress	$ML^{-1}T^{-2}$
τ_0	Boundary shear stress	$ML^{-1}T^{-2}$
ω	Wave number	
ω	Wake function	
ϕ	Velocity potential	
max	Maximum value	
—	Average value	
∇^2	Laplacian operator	

REFERENCES

Abou-Seida, M. M. (1965). Bed load function due to wave action. *Univ. Calif., Berkeley, Hydr. Eng. Lab. Rept.* **HEL-2-11**, 78.

Bagnold, R. A. (1946). Motion of waves in shallow water, interaction between waves and sand bottoms, *Proc. Royal Soc. London* [A], **187**, 1-18.

Bowden, K. F., Fairbairn, L. A., and Hughes, P. (1959). The distribution of shearing stresses in a tidal current. *Geophys. J.* **2**, 288-305.

Brebner, A., Askew, J. A., and Law, S. W. (1967). The effect of roughness on the mass transport of progressive gravity waves. *10th Coastal Eng. Conf. (Tokyo)* **1**, 175-184.

Cacchione, D. A. (1970). Experimental study of internal gravity waves over a slope, *Ph.D. Thesis, Mass. Inst. Tech. and Woods Hole Oceanog. Inst., Cambridge, Mass.*, 226 pp. (unpublished).

Caldwell, J. M. (1949). Reflection of solitary waves. *Beach Erosion Board, Tech. Memo.* **11**, 35.

Carstens, M. R., Neilson, F. M., and Altinbilek, H. D. (1969). Bed forms generated in the laboratory under an oscillatory flow, analytical and experimental study. *Coastal Engineering Res. Center, Tech. Memo.* **28**, 93.

Cebeci, T. (1970). Behavior of turbulent flow near a porous wall with pressure gradient, *AIAA J.* **8**[12], 2152-2156.

Charnock, H. (1959). Tidal friction from currents near the sea bed. *Geophys. J.* **2**[3], 215-221.

Clauser, F. (1956). The turbulent boundary layer. *In* "Advances in Applied Mechanics" (H. L. Dryden and Th. v. Karman, eds.), Vol. 4, pp. 1-51. Academic Press, New York.

Coles, D. (1956). The law of the wake in the turbulent boundary layer. *J. Fluid Mechanics* **1**[2], 191-226.

Collins, J. I. (1963). Inception of turbulence at the bed under periodic gravity waves. *J. Geophys. Res.* **68**[21], 6007-6014.

Dean, R. G. (1965). Stream function representation of nonlinear ocean waves. *J. Geophys. Res.* **70**[18], 4561-4572.

Eagleson, P. S. (1959). The damping of oscillatory waves by laminar boundary layers. *Beach Erosion Board, Tech. Memo.* **117**, 38.

Eagleson, P. S., and Dean, R. G. (1959). Wave induced motion of bottom sediment particles. *Proc. ASCE, Hydraulics Div.* **85**[HY10], 53-79.

Einstein, H. A. (1971). A basic description of sediment transport on beaches. *Univ. Calif., Berkeley, Hydr. Eng. Lab. Rept.* **HEL 2-34**, 37.

Einstein, H. A., and Chien, N. (1955). Effects of heavy sediment concentration near the bed on velocity and sediment distribution. *Univ. Calif., Berkeley, and U.S. Army Eng. Div., Omaha, Neb., M.R.D. Sediment Series* **8**, 76.

Einstein, H. A., and Li, H. (1958). The viscous sublayer along a smooth boundary. *Trans. ASCE* **123**, 293-313.

Granville, P. S. (1958). The frictional resistance and turbulent boundary layer of rough surfaces. *U.S. Navy, David W. Taylor Model Basin, Rept.* **1024**, 45.

Grosch, C. E. (1962). Laminar boundary layer under a wave. *Phys. of Fluids* **5**, 1163-1167.

Hama, F. R. (1954). Boundary layer characteristics for smooth and rough surfaces. *Trans. Soc. Naval Arch. and Mar. Eng.* **62**, 333-358.

Harris, J. E. (1970). Numerical solution of the compressible laminar, transitional and turbulent boundary layer equations with comparisons to experimental data. *Ph.D. Thesis in Aerospace Eng., Va. Polytechnic Inst., Blacksburg,* 172 pp. (unpublished).

Hasselman, K. (1970). Wave-driven inertial oscillations. *Geophys. Fluid Dynamics* **1**[4], 463-502.

Herring, H. J., and Norbury, J. F. (1967). Some experiments on equilibrium turbulent boundary layers in favorable pressure gradients. *J. Fluid Mechanics* **27**[3], 541-549.

Horikawa, K., and Watanabe, A. (1968). Laboratory study on oscillatory boundary layer flow. *Coastal Eng. in Japan* **11**, 13-28.

Horikawa, K., and Watanabe, A. (1970). Turbulence and sediment concentration due to waves. *Proc. 12th Coastal Eng. Conf., Washington, D.C.*, 751-766.

Hsu, E. Y. (1955). The measurement of local turbulent skin friction by means of surface Pitot tubes. *U.S. Navy, David W. Taylor Model Basin, Rept.* **957**, 15.

Iwagaki, Y., and Kakinuma, T. (1963). On the bottom friction factor of the Akita coast. *Coastal Eng. in Japan* **6**, 83-91.

Iwagaki, Y., Tsuchiya, Y., and Chen, H. (1967). On the mechanism of laminar damping of oscillatory waves due to bottom friction. *Bull. Disaster Prev. Res. Inst. (Kyoto Univ., Japan)* **16**[116], 49-75.

Iwagaki, Y., Tsuchiya, Y., and Sakai, M. (1965). Basic studies on the wave damping due to bottom friction. *Coastal Eng. in Japan* **8**, 37-49.

Johns, B. (1969). On the mass transport induced by oscillatory flow in a turbulent boundary layer. *J., Fluid Mechanics* **43**[1], 177-185.

Johnson, R. J. (1970). Characterization of the shallow water wave environment, prediction techniques and modeling facilities. *Naval Ship Res. and Dev. Center, Rept.* **3401**, 40.

Jonsson, I. G. (1963). Measurements in the turbulent wave boundary layer. *Proc. 10th IAHR Conf., London*, 85-92.

Jonsson, I. G. (1965a). Determination of the maximum bed shear stress in oscillatory turbulent flow. *Coastal Eng. Lab., Tech. Univ. of Denmark, Copenhagen Basic Res. Prog. Rept.* **9**, 14-20.

Jonsson I. G. (1965b). Friction factor diagrams for oscillatory boundary layers. *Coastal Eng. Lab., Tech. Univ. of Denmark, Copenhagen Basic Res. Prog. Rept.* **10**, 10-21.

Jonsson, I. G. (1966). On the existence of universal velocity distribution in an oscillatory turbulent boundary layer. *Coastal Eng. Lab., Tech. Univ. of Denmark, Copenhagen Basic Res. Prog. Rept.* **12**, 2-10.

Kajiura, K. (1968). A model of the bottom boundary layer in waves. *Bulletin Earthquake Res. Inst., Tokyo Univ.* **46**, 75-123.

Kalkanis, G. (1957). Turbulent flow near an oscillating wall. *Beach Erosion Board, Tech. Memo.* **97**, 36.

Kalkanis, G. (1964). Transportation of bed material due to wave action. *Coastal Engineering Res. Center, Tech. Memo.* **2**, 38.

Karlsson, S. K. F. (1958). An unsteady turbulent boundary layer. *Ph.D. Thesis, Johns Hopkins Univ., Baltimore*, 59 pp. (unpublished).

Kestin, J., Persen, L. N., and Shah, V. L. (1967). The transfer of heat across a two-dimensional oscillating boundary layer. *Zeitschrift für Flugwissenschaften* **15**[8/9], 277-285.

Koh, R. C. Y., and LeMéhauté, B. (1966). Wave shoaling. *J. Geophys. Res.* **71**[8], 2005-2012.

Korteweg, D. J., and Vries, G. de (1895). On the change of form of long waves advancing in a rectangular canal, and on a new type of long stationary waves. *Phil. Mag.* [5], **39**, 422-443.

Lettau, H. (1932). Stehende Wellen als Ursache und Gestaltender Vorgange in Seen. *Ann. d. Hydrogr. u. Mar. Met.* **60**, 385-388.

Li, H. (1954). Stability of oscillatory laminar flow along a wall. *Beach Erosion Board, Tech. Memo.* **47**, 48.

Lighthill, M. J. (1954). The response of laminar skin friction and heat transfer to fluctuations in the stream velocity. *Proc. Royal Soc. London, Series A* **224**[1156], 1-23.

Lin, C. C. (1957). Motion in the boundary layer with a rapidly oscillating external flow. *Proc. 9th Internat. Cong. Appl. Mech., Bruxelles Univ.* **4**, 155-167.

Liu, C. K., Kline, S. J., and Johnson, J. P. (1966). An experimental study of turbulent boundary layer on rough wall. *Thermo-sciences Div., Mech. Eng. Dept., Stanford Univ., Stanford, Calif. Rept.* **MD-15**.

Longuet-Higgins, M. S. (1953). Mass transport in water waves. *Phil. Trans., Royal Soc. London* [A]**245**[903], 535-581.

Longuet-Higgins, M. S. (1958). The mechanics of the boundary layer near the bottom in a progressive wave. (App. to: R. C. H. Russell and J. D. C. Osorio, an experimental investigation of drift profiles in a closed channel.) *Proc. 6th Conf. Coastal Eng., Gainesville, Fla.*, 184-193.

Longuet-Higgins, M. S. (1969). On the transport of mass by time varying ocean currents. *Deep-Sea Res.* **16**, 431-447.

Machemehl, J. L. (1970). Effects of slope roughness on regular and irregular wave run up on composite slopes. *Ph.D. Thesis, Texas A&M Univ., College Station*, 454 pp. (unpublished).

Manohar, M. (1955). Mechanics of bottom sediment movement due to wave action. *Beach Erosion Board, Tech. Memo.* **75**, 100.

Mashima, Y., Ikeuti, M., and Shigemura, T. (1966). The effect of wave action on sand grain. *Memoirs, Defense Academy, Japan* **6**[1], 77-83.

Matsunashi, J., and Kawadani, T. (1965). Basic research on transformation of waves due to changes of the ocean bottom, measurement of friction stress on bottom surfaces. *Proc. 12th Coastal Eng. Conf. in Japan*, 29-34 (in Japanese).

McDowell, D. M. (1965). Some effects of friction on oscillating in laboratory channels. *Proc. IAHR, 11th Cong., Leningrad* **3**[3.4], 14.

Mellor, G. L. (1966). The effects of pressure gradients on turbulent flow near a smooth wall. *J., Fluid Mechanics* **24**[2], 255-274.

Müller, A., Gyr, A., and Dracos, T. (1971). Interaction of rotating elements of the boundary layer with grains of a bed, a contribution to the problem of the threshold of sediment transportation. *J. Hydr. Res.* **9**[3], 373-411.

Murray, J. D. (1965). Viscous damping of gravity waves over a permeable bed. *J. Geophys. Res.* **70**[10], 2325-2331.

Nece, R. E., and Smith, J. D. (1970). Boundary shear stress in rivers and estuaries. *Proc. ASCE, Journal Waterways and Harbors Div.* **96**[WW2], 335-358.

Noda, H. (1969). A study on mass transport in boundary layers in standing waves. *Coastal Eng. in Japan* **12**, 57-68.

Prandtl, L. (1904). Über Flüssigkeitsbewegung bei sehr kleiner Reibung. *Proc. 3rd Internat. Math. Cong., Heidelberg*, 484-491.

Putnam, J. A. (1949). Loss of wave energy due to percolation in a permeable sea bottom. *Trans., Am. Geophys. Union* **30**[3], 349-356.

Putnam, J. A., and Johnson, J. W. (1949). The dissipation of wave energy by bottom friction. *Trans., Am. Geophys. Union* **30**[1], 67-74.

Reid, R. O., and Kajiura, K. (1957). On the damping of gravity waves over a permeable sea bed. *Trans., Am. Geophys. Union* **38**[5], 662-666.

Reynolds, W. C. (1969). A morphology of the prediction methods. *In* "Proceedings, Computation of Turbulent Boundary Layers, AFOSR-IFP—Stanford Conference, Stanford, Calif." (S. J. Kline, M. V. Morkovin, G. Sovran, and D. J. Cockrell, eds.), Vol. 1, pp. 1-15, Stanford Univ. Press, Stanford, California.

Rosenhead, L. (ed.) (1963). "Laminar Boundary Layers." Clarendon Press, Oxford.

Savage, R. P. (1953). Laboratory study of wave energy losses by bottom friction and percolation. *Beach Erosion Board, Tech. Memo.* **31**, 25.

Schlichting, H. (1932). Berechnung ebener periodischer Grenzschichtströmungen. *Phys. Zeitung* **33**, 327.

Schlichting, H. (1968). "Boundary Layer Theory." 6th ed., McGraw-Hill, New York.

Shepard, F. P. (1963). "Submarine Geology." Harper and Row, New York.

Shields, A. (1936). Anwendung der Ahnlichkeitsmechanik und der Turbulenzforschung auf die Geschiebewegung. *Mitt. Preus. Versuchanst. f. Wasserbau u. Schiffbau, Berlin* **26**.

Sleath, J. F. A. (1968). The effect of waves on the pressure in the bed of sand in a water channel and on the velocity distribution above it. *Ph.D. Thesis, St. John's College, Cambridge,* 141 pp. (unpublished).

Sternberg, R. W. (1968). Friction factors in tidal channels with differing bed roughness. *Marine Geol.* **6**, 243-260.

Stokes, G. G. (1851). On the effect of the internal friction of fluids on the motion of pendulums. *Trans. Cambr. Phil. Soc.* **9**[2], 8-106.

Stokes, G. G. (1880). On the theory of oscillatory waves. *In* "Math. and Phys. Papers" Vol. I, Cambridge Univ. Press, London and New York.

Teleki, P. G. (1970). Measurement of boundary shear in oscillating flow in presence of roughness. *Ph.D. Thesis, Louisiana State University, Baton Rouge,* 192 pp. (unpublished).

Teleki, P. G., and Anderson, M. W. (1970). Bottom boundary shear stresses on a model beach. *Proc. 12th Conf. on Coastal Eng., Washington, D.C.,* 269-288.

van Driest, E. R. (1958). Boundary layer transition. *In* "Grenzschichtforschung; Symp., Internat. Union for Theor. & Appl. Mech., Freiburg, 1957" (H. Görtler, ed.), pp. 180-184, Springer Verlag, Berlin.

Yalin, M. S., and Russell, R. C. H. (1966). Shear stresses due to long waves. *J. Hydr. Res.* **4**[2], 55-98.

Zhukovets, A. M. (1963). The influence of bottom roughness on wave motion in a shallow body of water. *Izv., Geophys. ser.* **10**, 1561-1570.

CHAPTER 3

Predicting Initial Motion and Bedload Transport of Sediment Particles in the Shallow Marine Environment

Richard W. Sternberg

Department of Oceanography
University of Washington
Seattle, Washington 98195

ABSTRACT

A compilation of existing boundary-layer and sediment-transport measurements from the marine environment is used to develop a procedure for predicting grain movement for sand-sized particles over the sea floor. Given a knowledge of sediment texture and mean velocity one meter from the seabed it is possible to: (a) estimate the boundary-shear stress, (b) predict the initiation of grain movement, and (c) estimate the mass transport of sediment as bedload. The Bagnold bedload equations are used for the bedload transport predictions, however, they have been modified to fit the results of previous oceanographic field experiments. The advantages of this procedure are its simplicity with respect to the independent variables (mean velocity and sediment size) and the fact that every step in the technique has been developed from some *in-situ* field measurements; hence, it is specially suited for application in the marine environment.

INTRODUCTION

A consistent goal of geological oceanographers has been to predict the occurrence of sediment movement and the mass transport of sediment over the sea floor. A continued emphasis has been placed on the need to be able to estimate the occurrence of motion, its mode, and rate of mass transport of sedimentary particles given a single measurement of bottom velocity and some knowledge of sediment texture. It is questionable whether the status of fluid and sediment dynamics in the near or distant future will provide precise techniques for predicting sediment motion, but nevertheless, a relatively crude, semiempirical method remains a useful tool for geologists and oceanographers interpreting ancient and modern environments, respectively.

In the past several years oceanographic field techniques and observations have proceeded in such a way that a body of data and knowledge exists from *in-situ* measurements made in the marine environment that will allow limited predictions of sediment motion. Prediction of sediment movement of sandy material would require several steps depending on the type of data available. In its simplest form the procedure can be arranged in a hierarchy as shown in Fig. 16. Beginning with a mean velocity measurement $(\overline{U}_{100})$ the first step (1) is to estimate the boundary shear stress (τ_0) exerted by the fluid moving over a sandy bottom. Once this is accomplished, the magnitude of τ_0 and the sediment textural characteristics can be used to make further estimates regarding the occurrence of grain movement under the prevailing conditions (Step 2) and the bedload mass transport of sedimentary grains (Step 3). Each of the levels in Fig. 16 and the associations between Steps 1 and 3 are briefly presented in this paper. The data used are primarily from *in-situ* measurements and large-scale flume studies that compare favorably with equivalent field measurements so the results obtained are specifically oriented toward the marine environment, however, they may have wider application. It should be emphasized that the relationships between each step exhibits significant variance, hence, the sum of steps involved in the total prediction represents, at best, a rough approximation.

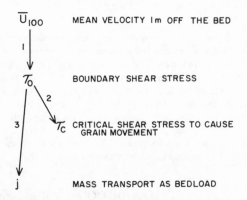

Figure 16.

Hierarchical arrangement of steps leading to the prediction of initial grain movement and mass transport of sediment as bedload.

By comparison, however, the techniques outlined in this paper appear to predict the movement of sand-size particles in the marine environment more accurately than other methods (Kachel and Sternberg, 1971). Also, the method has the advantages of requiring the measurement of few variables and ease of application.

SAMPLING PROCEDURE

The field data used in this analysis is part of a continuing study of marine boundary-layer flows and sediment transport carried out in recent years at the University of Washington. Data were collected by means of a large instrumented tripod that measures the flow conditions within the boundary layer (Sternberg and Creager, 1965), and thus makes it possible to estimate bedload transport.

Data collection consists of two phases: continuous measurement of the velocity distribution within 1.5 m of the seabed, and visual observation of the sediment-water interface. The velocity distribution within the boundary layer is measured with six miniature-Savonius-rotor current meters (10 cm high) rigidly mounted on the underwater tripod. All velocity measurements are telemetered as analog signals to shipboard, converted to digital format, and recorded on paper tape.

Visual observations are made in two ways, direct viewing through an underwater television system and oriented stereophotography. The television system observes continuously the sediment-water interface (e.g., bed configuration and the initiation and mode of sediment movement), the position of the tripod, and the condition of the sensing elements. The stereocameras provide high-resolution photographs for measuring the bed geometry by stereophotogrammetric techniques; ripple migration rates are measured by time-lapse analysis of the bottom photographs. By knowing the geometry of the bed and the ripple migration rate an estimate of mass transport of sediment as bedload can be obtained. A complete description of the data analysis procedures for the velocity measurements is given by Sternberg (1968) and for the time-lapse analysis by Kachel and Sternberg (1971).

ESTIMATING BOUNDARY SHEAR STRESS

Assuming steady, two-dimensional, turbulent flow in an open channel several techniques exist for estimating the fluid shear stress on the boundary. These techniques are the (a) Reynolds stress or eddy correlation, (b) velocity profile, and (c) quadratic stress law. The velocity profile and quadratic stress law methods have received the most widespread application in oceanographic research and results from these techniques are presented in this paper. Measurements of shear stress by the eddy-correlation method requires more sophisticated measurement and analytical procedures and, therefore, is not commonly used. Comparisons of the eddy-correlation and velocity-profile methods for estimating shear stress in the boundary layer of the lower atmosphere have shown that the results are very similar (Bowden and Fairbairn, 1956; Frizzola and Singer, 1967; Goddard, personal communication, 1971), hence, only methods (2) and (3) will be reviewed.

Velocity Profile Method

The Karman-Prandtl velocity profile equations relate the mean velocity $(\overline{U_z})$ at a given distance (z) from the boundary to the boundary shear stress:

$$\bar{U}_z = U_* \frac{1}{k} \ln (z + z_0)/z_0 \tag{1}$$

where U_* is the friction velocity $(\tau_0/\rho)^{12}$; k is von Karman's constant; and z_0 is called the roughness length. Assuming that $z_0 << z$, k = 0.4, and converting from 1n to $\log_{10}$, Eq. (1) may be written:

$$\bar{U}_z = 5.75 \, U_* \log (z/z_0) \tag{2}$$

Equation 2 was derived on the assumption that the shear stress within the boundary layer is constant; hence, the boundary shear stress (τ_0) may be obtained from the slope of the velocity profile as shown in Eq. (3).

$$\tau_0 = \rho \, \frac{\bar{U}_{z_2} - \bar{U}_{z_1}}{5.75 \, (\log z_2 - \log z_1)} \tag{3}$$

Mean velocity profiles (consisting of 5-10 min time averages) have been measured over the sea floor by numerous researchers and results have generally verified the applicability of the Karman-Prandtl equations in marine boundary layers (summarized in Bowden, 1962; Sternberg, 1968). This technique of estimating τ_0 requires a minimum of three current meters placed within 1.5 m of the seabed, hence, a considerable degree of instrumentation is required.

Quadratic Stress Law

It has been experimentally shown that in a turbulent flow the boundary-shear stress is proportional to the fluid density and the square of the mean velocity:

$$\tau_0 \propto \rho \bar{U}_z^2 \tag{4}$$

or, introducing a proportionality coefficient:

$$\tau_0 = C_D \, \rho \bar{U}_z^2 \tag{5}$$

where C_D is called the drag coefficient which relates the mean velocity near the seabed to the force exerted by the fluid per unit area of the bed. If the velocity is measured at a standard distance from the bed (generally 100 cm in oceanographic research) then Eq. (5) becomes:

$$\tau_{o} = C_{100}\rho\,\bar{U}_{100}{}^{2} \tag{6}$$

or, in terms of the friction velocity:

$$U_{*} = C_{100}{}^{1/2}\,\bar{U}_{100} \tag{7}$$

For fully rough turbulent flows the Quadratic Stress Law offers the advantages of simplicity of measurement and application. Investigations in the laboratory (Nikuradse, 1933) and in the sea (Sternberg, 1968) have shown that for hydrodynamically rough flows the drag coefficient assumes a constant value related to the bed configuration; hence, given a representative value of C_{100} the boundary-shear stress can be estimated from a single measurement of mean velocity within the boundary layer.

The relationship between C_{100} and Reynolds number for a variety of natural bed configurations is shown in Fig. 17. The bed configurations are classified as rocky (A); gravelly (B); rippled sand (D,F); and indistinctly roughened sand (E,C). The current velocity one meter from the bed in these areas attained values of 40 cm sec^{-1}. These results show that at values of Reynolds number greater than about 1.5×10^{5} (equivalent to $\bar{U}_{100} = 15$ cm sec^{-1}) the magnitude of C_{100} for each particular bed configuration is essentially constant suggesting the condition of hydrodynamically rough flow. The value of C_{100} for the various bed types ranges from about 2×10^{-3} to 4×10^{-3}. The condition of hydrodynamically rough flow occurs at relatively low velocities ($\bar{U}_{100}$ 15 cm sec^{-1}) because of the existence of compound roughness elements on the bed (e.g., ripples and gravel) that are as much as two orders of magnitude larger than the sand grains comprising the bed. Smooth sandy beds are not often encountered in the natural marine environment and, hence, at the flow speeds required to initiate sediment motion the marine boundary layer is usually hydrodynamically rough which justifies the use of C_{100} as a constant.

The relationship between the fully developed value of C_{100} (for large Reynolds number), roughness length (z_{0}), and bed configuration is schematically depicted in Fig. 18. Although there appear to be a difference of as much as a factor of two between the magnitude of C_{100} associated with various bed configurations (2×10^{-3} to 4×10^{-3}), the difference is of the same order of magnitude as the variance of C_{100} for some of the individual experiments. This fact, coupled with the notion that in most cases of field measurements the bed configuration is not accurately known suggests that all data should be grouped in order to present an average value of C_{100} regardless of bed configuration. In Fig. 19, all data are superimposed on the same coordinate system. This illustration shows the mean and dispersion of C_{100} related to a wide variety of bed types found in the shallow marine environment. These values can be considered as representa-

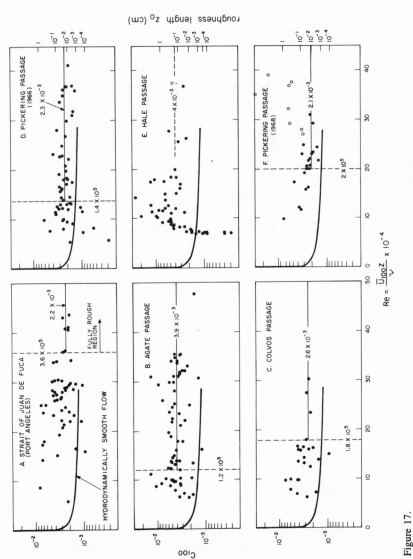

Figure 17.

Drag coefficient (C_{100}) as related to the Reynolds number. Here $\overline{U}_{100}$ is the mean velocity one meter from the bed, z is 100 cm, and ν is the kinematic viscosity. Mean values of C_{100} and the boundary between regions of relatively large and small dispersion are also included. The open circles in (F) refer to measurements made during periods of significant suspended sediment motion. Equivalent values of roughness length (z_0) can be determined by using the right hand vertical axis. (After Sternberg, 1968.)

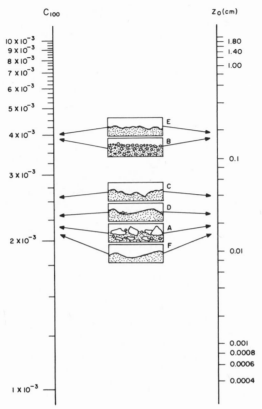

Figure 18.
Bed configuration with respect to the fully developed values of the roughness length (z_0) and drag coefficient (C_{100}). Letters A-F refer to channels as shown in Fig. 17. The relationship between C_{100} and z_0 is obtained by connecting the two scales with a horizontal line. (After Sternberg, 1968.)

tive except perhaps for large-scale ripples, dunes, or sand waves (>5 cm in height) which are thought to present greater resistance to flow than other bed types (Simons and Richardson, 1962).

The mean value for C_{100} for all data and the dispersion expressed as 95% confidence limits is given in Table I. Variance analysis indicates that as the value of Reynolds number increases, the dispersion of the grouped data decreases in such a way that a boundary placed at Reynolds number equal to 1.5×10^5 approximates the division between hydrodynamically transitional and fully rough-flow conditions (Sternberg, 1968). The mean and 95% confidence limits of C_{100} for each hydrodynamic region are also included in Table I.

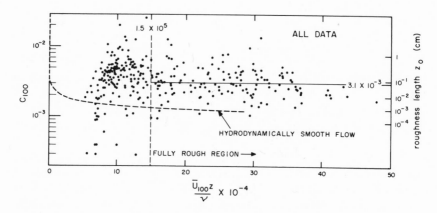

Figure 19.

Drag coefficient (C_{100}) as related to the Reynolds number for all data. Here $\bar{U}_{100}$ is the mean velocity one meter from the bed, z equals 100 cm, ν is the kinematic viscosity. Equivalent values of roughness length (z_0) can be determined by using the right-hand vertical axis. (After Sternberg, 1968.)

Table I

Mean C_{100} and 95% Confidence Limits for All Boundary-Layer Data Regardless of Bed Configuration[a]

	All data	Re<1.5×10^5	Re>1.5×10^5
C_{100}	3.1×10^{-3}	3.2×10^{-3}	3.0×10^{-3}
Upper limit	1.1×10^{-2}	1.6×10^{-2}	7.8×10^{-3}
Lower limit	8.7×10^{-4}	6.6×10^{-4}	1.1×10^{-3}

[a]After Sternberg, 1968.

These results indicate that for naturally sorted sand and gravel sediment textures an estimate of boundary shear stress (τ_0) from measurements of $\bar{U}_{100}$ is best obtained from the following equation:

$$\tau_0 = 3 \times 10^{-3} \rho \, \bar{U}_{100}{}^2 \tag{8}$$

or, in terms of the friction velocity:

$$U_* = 5.47 \times 10^{-2} \, \bar{U}_{100} \tag{9}$$

It is noteworthy that the use of the quadratic stress law depends on a logarithmic velocity distribution similar to the Karman-Prandtl method. Sternberg (1968) found that in various tidal channels in Puget Sound, Washington logarithmic velocity profiles occurred from 62% to 100% of the time, averaging about 85%. Thus, by using the quadratic stress law for estimating τ_0 a small to moderate degree of uncertainty occurs because it is not known from a single velocity measurement if a logarithmic velocity profile exists. For the reasons stated above the velocity profile method should be used if practicable, because the multiple velocity measurements will indicate the existence of a logarithmic velocity profile.

THRESHOLD OF GRAIN MOTION

Research regarding the critical velocity, or force per area of bed, required to erode unconsolidated sedimentary particles has been carried on for many years and has yielded a variety of "competency curves" that form a basis for predicting sediment motion, given some knowledge of sediment size, specific gravity, and water velocity or boundary-shear stress. Noteworthy results are those of Gilbert (1914), Shields (1936), Hjulstrom (1935, 1955), Bagnold (1963), and Sundborg (1956).

Competency curves are used by oceanographers and have been published in the oceanographic literature (e.g., Sverdrup et al., 1942; Bagnold, 1963; Inman, 1963). These curves represent results obtained primarily from laboratory investigations and they have not been well verified for the flow conditions and sediment mixtures of various marine environments. Sternberg (1967, 1971) has measured the conditions for initial movement from various tidal channels in Puget Sound, Washington, and Novak (1971) has measured the competent velocity for initial motion of cobble sized particles in the swash zone. The results of Sternberg are summarized in Fig. 20 which also includes the commonly used and published competency curves relating: (A) threshold mean velocity one meter from the bed (from Sundborg, 1956), (B) threshold shear velocity, and (C) Shields' entrainment function (from Inman, 1963) versus mean grain diameter of the sediment.

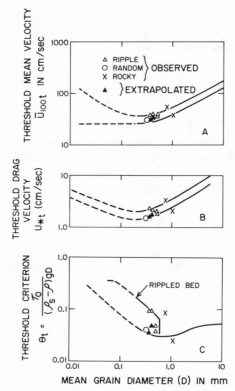

Figure 20.

Threshold criteria for initiation of grain movement as a function of grain diameter. The definition of symbols is given in the text. A. Threshold mean velocity at one meter B,C. Threshold drag velocity and Shields entrainment function for rippled smooth beds. (From Inman, 1963; after Sternberg, 1971.)

The different bed configurations included in Sternberg's data are classified as rippled, random, and rocky. Rippled beds are composed of sandy material deformed into regular wavelike features. Random beds consisted of sandy material deformed into larger roughness elements that did not exhibit a systematic pattern. Rocky beds refer to those areas in which pebble-sized material (1-3 cm) covered the bed. In these areas, the observation of sediment motion and the textural characteristics of the bed are related to the sandy matrix rather than to the pebbles.

In general, the field measurements agree closely with the competency curves. In Fig. 20A, B, and C, all data points fall within the error bands or very close to the existing curves except the data from channels with rocky beds. Besides the pebble-sized material on the bed, these areas also exhibited the poorest degree of sediment sorting, and thus it is encouraging that the observed data deviate so slightly from these curves which were initially constructed from data obtained for uniform sediments.

BEDLOAD TRANSPORT

The most frequently quoted theory of bedload transport presented in recent oceanographic literature is that of Bagnold (1963), which was field tested by Inman et al. (1966) in the coastal sand dunes of Guerreo Negro, Baja California and in a tidal channel in Puget Sound, Washington (Kachel and Sternberg, 1971). Bagnold's theory relates the rate of mass transport of sediment as bedload to the power expended by the fluid moving over the boundary. This relationship is expressed as:

$$\frac{\rho_s - \rho}{\rho_s} \, gj = K\omega \tag{10}$$

where j = mass discharge of sediment (gm cm^{-1}sec^{-1}); ρ_s = density of sediment; ρ = fluid density; g = acceleration due to gravity; K is a proportionality coefficient that expresses the ability of a flow to transport sediment; and ω is a measure of the power expended on the bed by the fluid.

The value of K for bedload transport may be expressed as [Bagnold, 1963, Eq. (4)]:

$$K = \frac{\epsilon_b}{\tan \phi - \tan \beta} \tag{11}$$

where ϵ_b is an efficiency factor which must be less than unity; tan ϕ is a "friction angle" which in the case of a granular mass, is a measure of the mean angle of contact between the grains; and β is the inclination of the bottom. Various attempts have been made to evaluate K. Inman (1966) quotes Bagnold (1941) who experimentally showed that K was a constant whose value depended on sand size and sorting. Bagnold (1963) also suggests from available evidence in flumes and rivers that K reaches a constant value, which is a function of the relative roughness h/D, where h is the depth of the flow and D is the mean diameter of the transported grains.

Fluid power ω can be expressed in several ways. Bagnold (1963) used the product of the boundary shear stress (τ_0) and the mean fluid velocity ($\overline{U}$) near the boundary:

$$\omega = \tau_0 \, \overline{U} \tag{12}$$

Inman and others (1966) used the friction velocity, U_*, instead of some mean velocity. Thus, Eq. (11) becomes:

$$\omega = \rho U_*^3 \tag{13}$$

and the Bagnold bedload equation becomes:

$$\frac{\rho_s - \rho}{\rho_s} \, gj = K\rho U_*^3 \tag{14}$$

Equation (12) presented by Bagnold (1963) and Eq. (13) as given in Inman et al. (1966) differ by the value of C_D, the drag coefficient. Since the drag

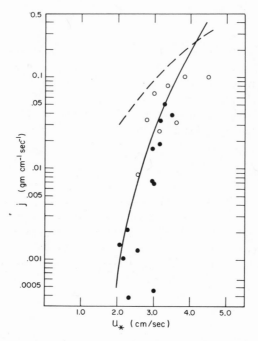

Figure 21.

Reproduction of Fig. 8 from Kachel and Sternberg (1971) comparing field results with flume experiments. Broken line is Bagnold's equation using constant K. Solid line is the modified version of Bagnold's equation. Open circles are field measurements made during accelerating or steady periods of the flow. Solid circles are the results of flume experiments of Guy et al. (1966).

coefficient for hydrodynamically rough flows is constant then these equations are directly proportional to one another and in Eq. (14) the drag coefficient is encompassed by the proportionality coefficient K.

Recent field measurements by Kachel and Sternberg (1971) have indicated that Eq. (14) is applicable in the marine environment provided that there are certain modifications. From field data it appears that the coefficient K does not approach a constant value for a given flow condition and has been shown empirically to vary as a function of the excess boundary stress $(\tau_0 - \tau_c)/\tau_c$. Since the measurements cited previously were conducted in only one field location it was not possible to relate K to sediment characteristics.

The final result of the paper by Kachel and Sternberg is shown in Fig. 21. The open circles represent the field data during steady or accelerating portions of the tidal flow and the solid circles are flume data for nearly equivalent sized sand as summarized by Guy and others (1966)* The dashed line is the mass

*Kachel and Sternberg measured absolute mass transport which represents the total transport of sand associated with ripple migration. In the process of ripple migration, sediment grains are alternately deposited and reeroded so that net transport rate along the floor is less than the absolute rate. The measurements reported by Guy and others are in terms of net mass transport and the equation they used is equal to one-half the absolute rate as computed by Kachel and Sternberg

transport prediction according to Eq. (14) using the suggested value for K of 2.2 (Inman et al., 1966), and the solid line is the prediction based on the modified version of Eq. (14) where K is equal to $f(\tau_0 - \tau_c)/\tau_c$ for that particular sand size. This result is noteworthy in that the modified Bagnold relationship is in excellent agreement with the flume experiments suggesting that these modifications may provide a better potential for predicting net bedload transport than existing techniques.

Since bedload measurements from other marine environments are not presently available, the flume data from Guy and others has been analyzed further to investigate the dependence of K on $(\tau_0 - \tau_c)/\tau_c$ and mean sediment size (D). These particular flume experiments are unique in that the channel was very large (2.4 m wide, 0.61 m deep, 45.7 m long), the bed was composed of natural sands, and sufficient data were collected (velocity profiles and bedload transport rates) so that the analytical procedures used by Kachel and Sternberg could be applied to the flume data. Independent experimental runs were reported with sand having a median fall diameter of 0.19 mm, 0.27 mm, 0.45 mm, and 0.93 mm, respectively.

The flume data were analyzed in a fashion similar to the oceanographic field experiments. Velocity profiles were plotted to determine values of U_* and τ_0. The lower limit of Fig. 20B was used to estimate τ_c for each particular sand size. Values of mass transport associated with each flume run were also reported. These experiments were carried out under steady, uniform flow conditions, naturally sorted sand beds, and essentially no sediment suspension. Although these conditions are not completely representative of the natural marine environment, the close agreement between the field and flume data shown in Fig. 21 suggests that results may still have wide application.

The basic data used for this analysis is plotted in Fig. 2) which shows the relationship between mass discharge (j) of bedload and friction velocity (U_*) for 0.19 mm, 0.27 mm, 0.45 mm and 0.93 mm sands, respectively. Values of the proportionality coefficient (K) were calculated by substituting the measured values of U_* and j into Eq. (14) and the results are plotted in Fig. 23. Although the data are somewhat disperse it is evident by the varying slopes of the regression lines in Fig. 23 that the value of K is not constant but shows a strong dependence on both sediment size (D) and excess shear stress $(\tau_0 - \tau_c)/\tau_c$.

The procedure for predicting the magnitude of mass transport is to first determine a value for K using Fig. 24. If the mean sediment diameter is similar to those shown in Fig. 24A, then K can be related directly to the magnitude of the excess shear stress $(\tau_0 - \tau_c)/\tau_c$ by choosing the appropriate curve. If the mean sediment size falls somewhere between the curves in Fig. 24A, then Fig. 24B can be used. It relates K to D for various values of $(\tau_0 - \tau_c)/\tau_c$. Once the magnitude of K has been obtained, the bedload transport rate can be calculated from Eq. (14) or graphically determined with the aid of the nomogram shown in Fig. 25.

___
(1971). In Fig. 21 the results of Kachel and Sternberg have been converted to net mass transport to allow comparison with the flume data..

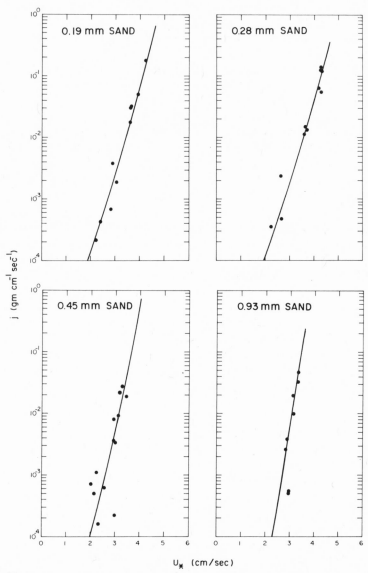

Figure 22.

Plot of the flume data of Guy et al. (1966) used for the bedload analysis. Data points refer to runs in which suspended sediment motion did not occur.

DISCUSSION

The total procedure for graphically predicting sediment movement has been presented in steps. Each step has been kept more or less independent so that, depending on the type of data available, any aspect of the technique may be

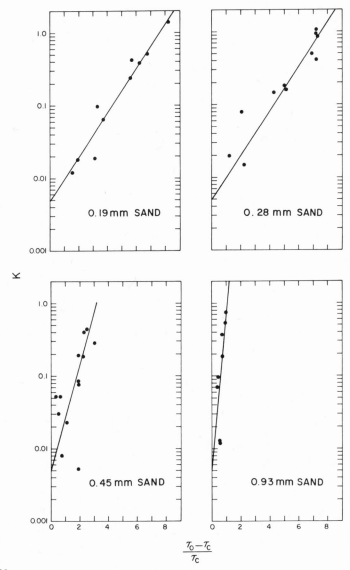

Figure 23.

Comparison of the excess shear stress $(\tau_0 - \tau_c)/\tau_c$ with the coefficient K for various sand sizes. (Data from Guy et al., 1966.)

used. For example, if only $\overline{U}_{100}$ and D were known, it would be necessary to (a) apply the representative value of C_{100} to calculate τ_0 $\left[\text{Eq. (8)}\right]$; (b) graphically estimate τ_c from Fig. 20; (c) determine a value for K from Fig. 24 and then use Eq. (14) or Fig. 25 to estimate the mass transport (j). On the other hand, if velocity profiles were available, the Kármán-Prandtl relationship

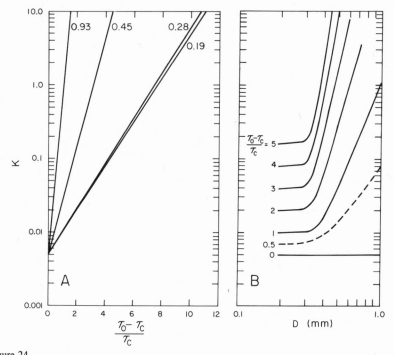

Figure 24.

(A) Relationship between excess shear stress $(\tau_0 - \tau_c)/\tau_c$ and the coefficient K for various sediment sizes. (B) Relationship between sediment diameter (D) and the coefficient K for various values of excess shear stress $(\tau_0 - \tau_c)/\tau_c$.

[Eq. (3)] may be used to estimate τ_0 rather than the quadratic stress law but the subsequent steps for estimating initial-grain motion and mass transport would be the same. In other words, it is not necessary to begin the procedure at the first step. Each step stands on its own and can be used as an independent entity.

It is also important to realize the practical limitations of the procedures. Limitations exist regarding sediment size, flow conditions, and mode of sediment transport.

Sediment Size

The analysis presented above was carried out for sediment sizes ranging from 0.19 mm (2.38 ϕ) to 0.93 mm (0.08 ϕ). Sizes are reported as median fall diameter. The sediment samples used are naturally-sorted river deposits so in this respect they more closely resemble marine deposits than artificial combinations of sizes.

Extrapolation of the prediction technique to sediment sizes smaller than those used in the experiments would probably yield questionable results. For diameters

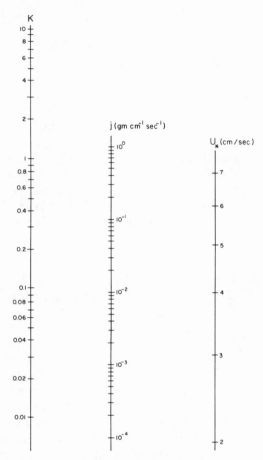

Figure 25.

Nomogram relating U_*, j, and K according to Eq. (14). By connecting any two known variables with a straight edge, the magnitude of the third variable can be read directly from the point of intersection on the scale.

finer than 0.19 mm, sedimentary particles often begin to exhibit special properties of shape, packing, and cohesiveness which may significantly offset or bias the results obtained from the use of larger particles. An example of the additional complications arising from the unique features of fine particles is seen in Fig. 20A where the incipient motion curve begins to spread and becomes dashed showing the inconclusive nature of the curve in this size range.

For diameters greater than 0.93 mm some extrapolation may be possible. The curves in Fig. 20 tend to show a consistent or almost linear increase in $\overline{U}_{100}$ and U_* with increasing sediment size. This suggests that extrapolating of the prediction technique to the very coarse sand size may be permissible.

Flow Conditions

Most of the field data used in this report were collected in tidal channels within Puget Sound, Washington. Due to the inherent difficulties of anchoring on station, data collection, water clarity, etc., a large portion of the data were collected during the early part of a tidal cycle, i.e., the accelerating or steady part of the tidal excursion of velocity.

In the comparison of various bedload formulas presented by Kachel and Sternberg (1971), it was observed that a rather large "hysteresis" effect occurred with respect to ripple height and migration rates. Ripples tend to maintain their height and migration rate even though the mean flow begins to decelerate. The causes of this phenomenon are discussed in that paper but the conclusions from that research and the close agreement observed in Fig. 21 dealt with data collected during accelerating or steady portions of the flow.

Mode of Transport

The influence that suspended sediment has on the flow conditions and bedload transport rate has not been investigated in the marine environment. It is known from flume studies that relatively high concentrations of suspended sediment affect the characteristics of the flow such that an apparent decrease in von Kármán's constant of as much as 50% occurs (Einstein and Chien, 1952; Vanoni, 1952). With high concentrations of suspended sediment the direct use of the Karman-Prandtl equations or the quadratic stress law for evaluating τ_0 is subject to question, hence the procedures outlined in this paper may be subject to further modification. The data cited from both field and flume measurements were collected in the absence of significant suspended load; hence, the results of this paper are limited to (a) sandy beds consisting of relatively small quantities of fine material subject to suspended motion and (b) flow conditions such that for the sediment size being transported, the primary mode of transport is as bedload. Conditions (a) and (b) above are graphically depicted in Fig. 26 which is taken from Sundborg (1967). This figure is the competency curve of Fig. 20A on which the relative percentage of suspended material has been superimposed. The figures depicting relative concentration refer to a ratio of suspended sediment concentrations between mid-depth of the flow and a reference height near the bed. This figure is included merely to delineate the range of flow conditions to which the prediction technique is thought to apply. The use of Fig. 26 to estimate suspended-load transport in the marine environment should not be attempted for it relates to river flows in which the boundary layer and suspended load extend throughout the whole depth and may not be applicable. Zone 3 in Fig. 26 represents that part of the curve within which sediment size and velocity would dictate a predominance of bedload motion; thus, it delimits the suggested area of applicability of the prediction procedures outlined above.

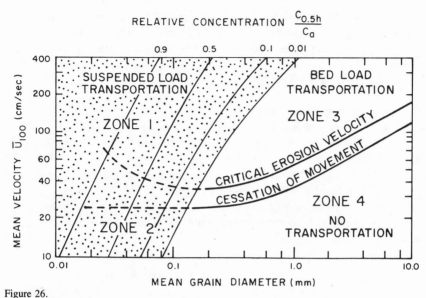

Figure 26.

The relation between flow velocity, grain size and state of sediment movement. The sediment is presumed to have a density of 2.65 g cm^{-3} (quartz and feldspar). (After Sundborg, 1967.)

CONCLUSIONS

A procedure for predicting initial motion and bedload transport of sediment in the marine environment has been presented. The required measurements are (a) the mean velocity 1 m off the seabed ($\bar{U}_{100}$) or the mean velocity profile within 1.5 m of the boundary and (b) the mean sediment diameter (D). In general, the velocity measurements are used to estimate the magnitude of boundary shear stress (τ_0) associated with the flow. The sediment size is used to determine the critical boundary shear stress (τ_c) that must be exceeded to cause initial motion. The parameters (τ_0, τ_c, and D) are then used to estimate the bedload transport of sand. The specific steps of this procedure are:

1. Estimate the boundary shear stress τ_0 from the velocity measurements. If the velocity profile is known, τ_0 can be estimated from Eq. (3):

$$\tau_0 = \rho \left[\frac{U_{z_1} - U_{z_1}}{5.75 \left(\log z_2 - \log z_1 \right)} \right]^2$$

and if the mean velocity at one meter is known then use Eq. (8):

$$\tau_0 = 3 \times 10^{-3} \, \rho \, \bar{U}_{100}{}^2$$

2. Knowing a value of mean sediment diameter (D) estimate from Fig. 20B the critical shear stress (τ_c) required to initiate sediment movement.

3. Compute the excess shear stress $(\tau_0 - \tau_c)/\tau_c$ and together with the value of (D), use Fig. 24 to estimate the magnitude of the coefficient K.

4. The mass transport as bedload (j) can be estimated from Eq. (14):

$$\left(\frac{\rho_s - \rho}{\rho_s}\right) gj = K\rho U_*^3$$

or from the nomogram presented in Fig. 25.

The application of this technique has the following limitations:

1. In a tidal current the flow conditions should be in an accelerating or a relatively steady part of the tidal cycle.

2. The technique is considered to be applicable for mean sediment sizes between approximately 0.20 mm and 1 to 2 mm.

3. There should be little or no transport as suspended load. The combination of flow conditions and sediment sizes that dictate bedload motion as the primary mode is delimited in Fig. 26.

ACKNOWLEDGMENTS

The author wishes to thank the National Science Foundation (Grant GA 28964) and the Atomic Energy Commission (Contract No. AT-45-1-2225 TA 21-1) for their continued support of this research. Gratitude is also expressed to Dr. Joe S. Creager who initiated this program in 1962 and has always been an important part of its operation and success. This paper is University of Washington, Department of Oceanography, Contribution No. 666.

LIST OF SYMBOLS

Symbol	Definitions	Dimensions
C_a	Mass concentration of suspended sediment at a reference height, "a"	ML^{-3}
$C_{0.5h}$	Mass concentration of suspended sediment at mid-depth	ML^{-3}
$\overline{U}_{100t}$	Critical or threshold value of the mean velocity 100 cm off the bed	LT^{-1}
$\overline{U}_{100}$	Mean velocity 100 cm off the bed	LT^{-1}
τ_0	Boundary shear stress	$ML^{-1}T^{-2}$
$\overline{U}_z, \overline{U}_{z_1,}$ $U_{z_2,}$	Mean velocity at a height z off the bed	LT^{-1}
U_*	Friction velocity	LT^{-1}
k	von Karman's constant	
z_1, z_2, z	Distance above the sea bed	L

z_0	Roughness length	L
C_D	Drag coefficient	
C_{100}	Drag coefficient related to $z = 100$ cm	
ρ	Fluid density	ML^{-3}
ρ_s	Sediment density	ML^{-3}
j	Bedload discharge of sediment	$ML^{-1}T^{-1}$
ω	Fluid power per unit area	MT^{-3}
g	Acceleration of gravity	LT^{-2}
K	Proportionality coefficient	
e_b	Efficiency factor	
ϕ	Friction angle	
β	Bed inclination	
D	Mean grain diameter	L
h	Water depth	L
τ_c	Threshold or critical value of the boundary shear stress	$ML^{-1}T^{-2}$
θ_t	Shields criterion for grain movement	
Re	Reynolds number	$\overline{U}_{100} z/\nu$
ν	Kinematic viscosity	L^2T^{-1}
U_{*_t}	Critical or threshold value of the friction velocity	LT^{-1}

REFERENCES

Bagnold, R. A. (1941). "Physics of Blown Sands and Desert Dunes." Methuen, London.

Bagnold, R. A. (1963). Mechanics of marine sedimentation. *In* "The Sea: Ideas and Observations" (M. N. Hill, ed.), Vol. III, Interscience, New York.

Bowden, K. F. (1962). Turbulence. *In* "The Sea: Ideas and Observations" (M. N. Hill, ed.), Vol. III, Interscience, New York.

Bowden, K. F., and Fairbairn, L. A. (1956). Measurements of turbulent fluctuations and Reynolds stresses in a tidal current. *Proc. Royal Soc., London* [A], **237**, (1210), 422-438.

Einstein, H. A., and Chien, N. (1952). Second approximation to the solution of the suspended load theory. *Inst. of Engineering Research, Univ. of Calif.*, **2**[47].

Frizzola, J. A., and Singer, I. A. (1967). Representation of stress from various wind profile equations. *Trans. Am. Geophysical Union* **48**, (1), 121.

Gilbert, K. G. (1914). "The Transportation of Debris by Running Water." U.S. Geol. Survey Prof., Paper 86, Washington, D.C.

Guy, H. R., Simons, D. B., and Richardson, E. V. (1966). "Summary of Alluvial Channel Data from Flume Experiments, 1956-1961." U.S. Geol. Survey Prof., Paper 462-I, Washington, D.C.

Hjulstrom, F. (1935). Studies of the morphological activity of rivers as illustrated by the river Fyris. *Bull. Geol. Inst. Uppsala,* **25**, 221-527.

Hjulstrom, F. (1955). Transportation of detritus by moving water. *In* "Recent Marine Sediments, A symposium, Society of Economic Paleontologists and Mineralogists" (Parker D. Trask, ed.), Special Publication No. 4, Tulsa, Oklahoma.

Inman, D. L. (1963). *In* "Submarine Geology" (F. P. Shepard, ed.), Chapter V, 2nd Ed., Harper and Row, New York.

Inman, D. L., Ewing, G. C., and Corliss, J. B. (1966). Coastal sand dunes of Guerrero Negro, Baja California, Mexico. *Geol. Soc. Am. Bull.* **77**, 787-802.

Kachel, N. B., and Sternberg, R. W. (1971). Transport of bedload as ripples during an ebb current. *Marine Geol.* **19**, 229-244.

Nikuradse, J. (1933). "Laws of Flow in Rough Pipes." Natl. Advisory Comm. Aeronautics Tech. Mem. 1292 (translation from German, 1950).

Novak, Irwin D. (1971). "Origin, Distribution, and Transportation of Gravel on Broad Cove Beach, Appledore Island, Maine." Ph.D. Dissertation, Cornell Univ. Press, Ithaca, New York.

Shields, A. (1936). Anwendung der Ahnlichkeits Mechanik und der Turbulenzforschung auf die Geschiebe Bewegung. Preussische Versuchanstalt fur Wasserbau und Schiffbau, Berlin.

Simons, D. B., and Richardson, E. V. (1962). The Effect on Bed Roughness on Depth-Discharge Relations in Alluvial Channels. U.S. Geological Survey Water Supply Paper 1498-E.

Sternberg, R. W. (1967). Measurements of sediment movement and ripple migration in a shallow marine environment. *Marine Geol.* **5**, 195-205.

Sternberg, R. W. (1968). Friction factors in tidal channels with differing bed roughness. *Marine Geol.* **6**, 243-260.

Sternberg, R. W. (1971). Measurements of incipient motion of sediment particles in the marine environment. *Marine Geol.* **10**, 113-119.

Sternberg, R. W., and Creager, J. S. (1965). An instrument system to measure boundary-layer conditions at the sea floor. *Marine Geol.* **3**, 475-482.

Sundborg, A. (1965). The River Klaralven: A study of fluvial processes. *Geograf. Ann.* **38**, 127-316.

Sundborg, A. (1967). Some aspects on fluvial sediments and fluvial morphology, I. General views and graphic methods. *Geograf. Ann.* **49A**, 333-343.

Sverdrup, H. U., Johnson, M. W., and Fleming, R. H. (1942). "The Oceans, Their Physics, Chemistry, and General Biology." Prentice-Hall, Englewood Cliffs, New Jersey.

Vanoni, V. A. (1952). Some effects of suspended sediment on flow characteristics. Proc., Fifth Hydraulic Conference, Bull. 34, State University of Iowa, Iowa City.

CHAPTER 4

Experiments on Bottom Sediment Movement by Breaking Internal Waves

John B. Southard* and David A. Cacchione†

ABSTRACT

To gain a qualitative idea of bottom-sediment movement produced by oceanic internal gravity waves as they shoal on the continental shelf or upper slope, exploratory experiments on breaking of interfacial waves over a planar sloping bottom mantled with lightweight acrylic sediment were made in a two-layer medium in a large wave tank.

The waves break abruptly as they shoal, producing a breaker in the form of a turbulent and rapidly dissipating vortex. There is a compensating return flow of mixed fluid along the bed between breakers. Sediment is moved upslope by the breakers, partly in suspension, and downslope by the return flow, as bedload. Predominant flow at the bed, and, therefore, also net sediment transport, is downslope. Sediment movement is strongest just upslope of the point of breaking and decreases to zero gradually upslope and sharply downslope. Sediment ripples form and migrate downslope toward the point of breaking despite disruption by each breaker. There is a broad band of slow net erosion upslope and a narrower band of more rapid net deposition near the point of breaking. Runs could not be continued long enough to determine the equilibrium slope profile after prolonged wave breaking.

The experiments do not represent a true scale model of the oceanic case, mainly because (a) sediment size is far too large relative to wave size, and (b) the Reynolds number and Froude number based on wave amplitude and wave period are too small. Nonetheless, the experiments should be a guide to possible major effects of sediment transport on the natural scale.

*Dept. of Earth and Planetary Sciences, Massachusetts Institute of Technology, Cambridge, Massachusetts 02139.

†U.S. Navy, Office of Naval Research, Boston, Massachusetts.

INTRODUCTION

Internal gravity waves are common in both the main thermocline and the seasonal thermocline in the oceans (see summaries by LaFond, 1962, and Wunsch, 1971). It has been suggested that internal waves, either standing or progressive, might transport and suspend bottom sediment (Emery, 1956; Cartwright, 1959; LaFond, 1961), but except for the work of LaFond (1965) there seem to have been no direct observations of interaction between internal waves and bottom sediment in the oceans to establish whether appreciable volumes of sediment can be moved in this way.

Under certain conditions, internal waves break as they propagate over a sloping bottom, as do surface waves. It seems likely that internal waves would affect the sediment bottom most strongly during breaking, because turbulence and boundary shear stress are much greater then than before the waves break. Thorpe (1966) and Cacchione (1970) have described breaking of internal waves in both a two-layer medium and a linearly stratified medium due to propagation over a smooth planar slope, but no experiments have been made with sediment-mantled slopes.

We here report the results of several exploratory runs involving breaking of interfacial waves on a sloping sediment bottom in a large wave tank. Though too weak to move quartz sand, these waves caused substantial movement of sand-size acrylic beads (density 1.18 g/cm^3). The experiments thus provide a qualitative idea of the patterns of sediment transport produced by breaking internal waves. In the following two sections we describe the fluid and sediment motions in the experiments, and in the final section we discuss the relevance of the experiments to the oceans.

PROCEDURE

The wave tank is 11 m long, 110 cm wide, and 70 cm deep, with Plexiglas walls and bottom. The wave generator (Fig. 27), modeled after a device used by Keulegan and Carpenter (1961), consists of two horizontal compartments, open only in the downchannel direction, separated by a partition at the interface level. Connected to the partition by a rubber gasket is a square plate which is driven up and down, approximately in simple harmonic motion, by a rotating eccentric device. The movable plate forces the upper and lower fluid layers alternately back and forth past a rounded pier projecting from the partition, thereby generating a train of nearly sinusoidal interfacial waves (Fig. 28). Wave amplitudes ranged from one half cm to 2 cm, and wave periods ranged from 5 to 15 sec. At the other end of the tank, a 1:10 planar slope extended upward from the bottom to a horizontal platform 30 cm above the bottom (Fig. 29).

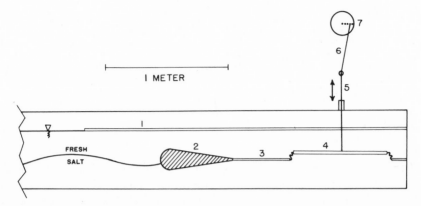

Figure 27.
Schematic side view of the wave generator. (1) free-surface plate; (2) pier; (3) stationary partition at interface level; (4) movable plate in partition; (5) drive shaft; (6) connecting rod; and (7) rotating disk.

Figure 28.
Side view of interfacial waves propagating over the horizontal portion of the tank bottom. The undisturbed air-water interface shows faintly just above the meter stick.

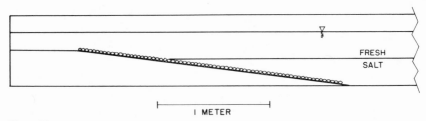

Figure 29.
Slope geometry.

To prepare a run, first a deep layer of salt water, dyed blue for contrast, was mixed in the tank; density was not closely controlled from run to run but ranged from 1.03 to 1.05 g /cm^{-3}. A planar sediment bed was then prepared on the slope by adding acrylic beads and screeding the sediment surface underwater. The bed was made deep enough, about a centimeter, so that the substratum would nowhere become exposed as bed features developed. Two sizes of beads were used, 0.1 mm and 0.5 mm. (A wetting agent had to be added to the salt layer to prevent the beads from clumping together.) The upper layer was emplaced by running fresh water slowly onto a spreading board floating at the free surface. Some mixing at the interface is inevitable; after filling, the mixed fluid was siphoned off all along the interface, thus reducing the thickness of the mixed layer to less than 2 cm. Finally, excess water was siphoned from both layers to bring the interface and free surface to the desired levels, about 20 cm and 40 cm from the bottom, respectively. Figure 30 shows a typical density profile prior to a run.

As described below, mixing of the fresh and salt layers by breaking of waves on the slope generates fluid with intermediate density, which penetrates back along the interface as a distinct and internally stratified layer. As this mixed layer grows, waves become weaker and more complex, and sediment movement on the slope degenerates within 10 to 15 min after the start of the run, long before bed features become fully developed. To establish steady conditions necessary for longer runs, growth of the mixed layer was arrested by siphoning off mixed fluid continuously through a thin horizontal pipe with numerous small ports, fixed at the mean level of the interface about midway between wave generator and slope. This limited the thickness of the mixed layer to approximately two wave amplitudes without noticeably disrupting the incoming waves. In order to maintain a steady mean position of the interface, new fluid was added to the fresh and salt layers at requisite rates. In this way, runs could be continued for 3 to 4 hr. Extraneous motions of the added fluid near the point of wave breaking were held to a minimum by placing the point of addition far from the slope and by making the temperature of the fluid the same.

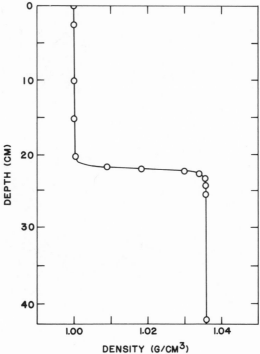

Figure 30.
Typical vertical profile of fluid density in the tank before the start of a run.

FLUID MOTION

Thicknesses of the upper and lower layers are great enough relative to wave size that fluid velocities near the free surface and bottom are much smaller than fluid velocities near the interface. (Near-bottom velocities are nonetheless great enough to produce "oscillation ripples" in the form of straight bands of nearly neutrally buoyant flocs of unidentifiable fine debris on the tank bottom, evenly spaced 1 to 2 cm apart.) Over most of the slope, motions of both fluid and sediment are nearly two-dimensional. Only in a narrow zone along each sidewall does secondary circulation affect the patterns of motion.

The waves become steeper as they propagate over the slope, by decreasing in wavelength and increasing in amplitude. As could be observed by watching the orbits of bits of extraneous material suspended near the bottom, near-bottom velocities produced by passage of the waves are amplified progressively upslope. The waves break abruptly at a position on the slope that is invariably well

Figure 31.

Oblique side view of advancing breakers (upslope is to the right). The breaker to the left has just developed; the preceding breaker, to the right, has dissipated almost entirely.

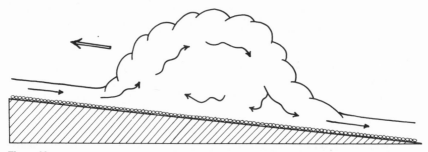

Figure 32.

Sketch of a breaker vortex advancing upslope, showing thin return current upslope and downslope of the vortex and sense of circulation in the vortex itself.

downslope of the original position of the interface. The resulting breaker consists of a large and highly turbulent vortex, initially 5 to 10 cm high (Fig. 31). The sense of the overall circulation in an advancing breaker vortex is such that the fluid motion is upslope at the bottom of the vortex and downslope at the top (Fig. 32). The breaker vortex entrains overlying fluid as it moves upslope; the mixed fluid penetrates back along the interface toward the wave source at surprisingly high speeds, up to 5 cm sec^{-1}.

Compensating for the upslope mass flux due to movement of the vortex there is a thin downslope current of fast-moving fluid 1 to 3 cm thick that flows at speeds of several centimeters per second along the bed between breakers. This return current, because it has lower density due to the progressive mixing upslope, seems to be diverted away from the bed as it flows into oncoming breakers (Fig. 32). Thus, flow at the bed is briefly upslope during passage of a breaker and downslope the rest of the time; over one wave period the predominant flow at the bed is downslope. Judging by the nature of the sediment movement (described below), turbulence near the bed is much stronger during the brief interval of upslope flow than during the more protracted downslope flow.

The breakers dissipate so rapidly as they advance upslope that only the strongest waves produce breakers that reach the horizontal segment at the upper end of the slope. Even in the absence of wave-absorbing material, the weak reflections produced at the end of the tank in these cases have no noticeable effect on the behavior of the waves breaking on the slope.

SEDIMENT TRANSPORT

Even the largest waves were barely strong enough to produce slight movement of fine quartz sand, but all waves except those with very low amplitudes and very short periods caused general movement of both sizes of acrylic sediment. A striking feature of the pattern of sediment movement produced by the breaking waves is that the zone of sediment movement lies entirely upslope of the point of breaking: despite the progressive amplification of near-bottom velocities upslope from the base of the slope, even the largest waves moved no sediment before they broke. As can be seen in Fig. 33, there is an abrupt onset of sediment movement at the point of breaking; intensity of sediment movement reaches a maximum just upslope of the point of breaking, and then decreases gradually upslope from there. With the larger waves, the band of sediment movement is a few tens of centimeters wide, more than an order of magnitude greater than the wave amplitude; this width should also vary with slope angle, which was not changed in the experiments.

For some distance above the point of breaking, sediment movement is bidirectional: sediment is thrown into suspension violently and carried upslope a short distance, partly in suspension and partly along the bed, by the highly turbulent breaker vortex, and then is transported downslope, chiefly as bedload, by the downslope return current between passages of a breaker. For most of the time at a given point, sediment is transported downslope, even at points not far upslope of the point of breaking. Farther upslope, the breaker becomes too weak to move any sediment, but the return current still transports sediment downslope.

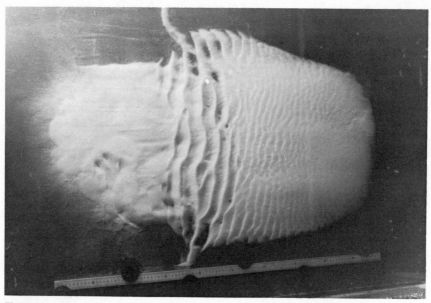

Figure 33.

Top view of rippled bed in 0.1 mm acrylic sediment. Upslope is to the right; the meter stick below gives the scale. Only part of the width of the sloping bottom was initially covered with sediment. The waves first broke at a position coincident with the transition from undisturbed bed to rippled bed. Bed features downslope of the point of breaking are not related to the wave action.

At this stage in the experiments we have not studied in any greater detail the modes of grain movement or the nature of the fluid forces acting on the grains.

Sediment ripples form only a few minutes after waves begin to break on an undisturbed bed and rapidly grow to a steady size (Figs. 33 and 34). Ripple spacing, from 2 to 6 cm, is greater for the coarse beads than for the fine beads, and is mostly independent of input wave conditions. Spacing is greater near the point of breaking than high up on the slope, where sediment movement is weakest. Near the point of breaking the ripples are almost symmetrical, like oscillation ripples produced by surface waves, but farther upslope the ripples are distinctly asymmetrical, with steep downslope faces and gentle upslope faces, reflecting the predominance of downslope sediment transport. These asymmetrical ripples migrate very slowly downslope.

No very fine sediment was present in the bed in interstices between plastic beads, but it is evident from the strong suspensive action of the advancing breaker vortices that fine sediment would be entrained by the vortices, mixed into overlying fluid, and then transported horizontally away from the slope by the backflow of the mixed layer.

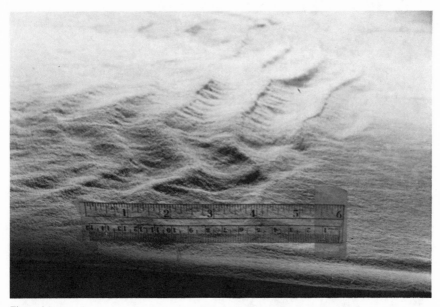

Figure 34.
Oblique side view of ripples in 0.5 mm acrylic sediment. Upslope is to the left; the scale on the tank wall is in inches and centimeters. Note that the bed is wholly undisturbed downslope of the point of breaking.

The form and surface sculpture of the ripples reflect the differing nature of upslope and downslope flow. Strong turbulent eddies in the passing breaker vortex lift swirls of sediment from the surface of the ripples, in the troughs as well as near the crests, and leave the ripples rounded in profile and covered with shallow irregular streamwise ridges and furrows produced by the erosive effects of individual eddies as they move upslope [Fig. 35 (A)]. The prolonged return flow reconstructs the asymmetrical ripple profile by erosion and bedload transport on the backs of the ripples and deposition at the crests to rebuild the slipfaces. The streamwise ridges and furrows produced by the breaker are immediately smoothed out on the backs of the ripples but are left intact in the troughs, so that the slipfaces are built out over the furrowed trough surfaces [Fig. 35 (B)].

Our visual impression is that net downslope sediment transport rate increases downslope from zero on the undisturbed bed above the zone of breaker action to a maximum at a point more than halfway to the point of breaking and then decreases more sharply to zero at the point of breaking [Fig. 36 (A)]. (This variation in net downslope sediment transport rate is much more nearly symmetrical than that of overall intensity of sediment movement, because movement near the point of breaking is largely oscillatory.) By the conservation-of-mass relation for sediment transport,

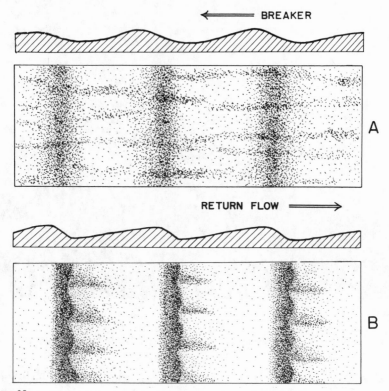

Figure 35.

Sketches of rippled bed configuration (illumination from lower left) and streamwise vertical profile (A) immediately after passage of a breaker vortex and (B) just before passage of a breaker vortex, after return flow has reconstructed the ripples. Upslope direction is to the left. Ripple spacing is about 4 cm.

$$\frac{\partial h}{\partial t} = -(1-\lambda)\frac{\partial q_s}{\partial x} \qquad (1)$$

where h is bed height measured normal to the slope, q_s is sediment transport rate per unit width, and λ is bed porosity (h and q_s are considered to be smoothed to eliminate the effects of passage of individual waves and movement of sediment in the form of ripples), this pattern of sediment transport implies the existence of a narrow band of relatively rapid net deposition just upslope of the point of breaking and a wider band of slower net erosion higher up on the slope (Fig. 36). This is borne out by measured longitudinal bed profiles after a few hours of wave breaking (Fig. 37), which show a broad shallow depression upslope and a narrow zone of greater average bed height, complicated by the presence of large ripples, downslope. Unfortunately, runs could not be continued long enough for development of bed relief more substantial than that shown

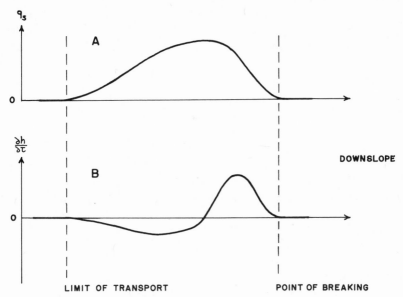

Figure 36.

(A) Nonquantitative visual estimate of q_s, net downslope sediment transport rate per unit width, versus position on slope, showing an increase in q_s downslope to a maximum at a point more than halfway from the upslope limit of sediment transport to the point of breaking and a gentler decrease down to the point of breaking. (B) Consequent rate of change of bed height $\partial h/\partial t$ versus position on slope, by Eq. (1), the conservation-of-mass relation for sediment transport, showing a zone of relatively rapid net deposition downslope ($\partial h/\partial t$ positive) and a broader zone of relatively slow net erosion upslope ($\partial h/\partial t$ negative), as observed in the experiments. Both q_s and h are here considered to be smoothed (see text).

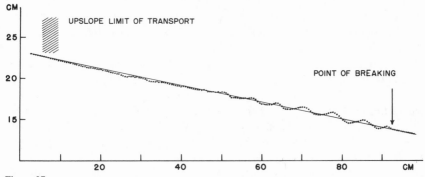

Figure 37.

Longitudinal profile of the sediment bed after 4 hr of wave breaking (dotted curve; dots represent spacing of measurements). The straight line shows the originally planar bed surface. The vertical scale is exaggerated 2 : 1.

in Fig. 37. Presumably the profile would continue to evolve toward some equilibrium state in which net sediment transport is zero at all points. We have little basis for speculating what the equilibrium profile is like, but in any case it seems unlikely that internal waves in the ocean would break in one position long enough for an equilibrium profile to develop.

DISCUSSION

What bearing do these experiments have on sediment movement by shoaling internal waves in the ocean? This question leads to three interrelated and more specific questions: (a) Are common oceanic internal waves strong enough to move bottom sediment at all? (b) Do our experiments represent a dynamic model of the interaction between waves and sediments on the oceanic scale? (c) Are experiments involving a two-layer medium representative of waves in a continuously stratified medium?

Cacchione (1970) has used solutions developed by Wunsch (1969) for velocity in the viscous boundary layer beneath shoaling internal waves, as verified by experiments on shoaling waves in a linearly stratified medium in a small wave tank, to develop a criterion for incipient movement of bottom sediment by shoaling internal waves based on the balance of moments due to fluid drag force and gravity force on bottom-sediment particles. Application of this criterion to the oceanic case (Cacchione, 1970) has shown that upslope amplification of near-bottom velocities beneath internal waves with realistic periods and amplitudes shoaling on the continental margin should cause incipient movement of a wide range of sediment sizes found on the outer shelf and upper slope even prior to breaking of the waves. Since it was observed both in Cacchione's experiments, made in a linearly stratified medium, and in the experiments reported here, made in a two-layer medium, that near-bottom velocities in the zone of breaking are much greater than the already amplified velocities just downslope, it seems likely that breaking internal waves move oceanic bottom sediment, although no direct observations are available.

To what extent do the laboratory experiments represent a scale model of sediment movement by breaking waves in the ocean? Neglecting such effects as thickness of the zone of density gradation, presence of higher wave modes, and sorting and grain shape of the sedimentary material, 11 variables can be viewed as governing the behavior of the system: wave amplitude a and wave frequency ω; thickness h_1, density ρ_1, and viscosity μ_1 in the upper layer; thickness h_2, density ρ_2, and viscosity μ_2 in the lower layer; slope S; sediment density ρ_s and mean sediment diameter d; and acceleration due to gravity, g. Dimensional analysis then provides the following 9 dimensionless modeling parameters equivalent to the original 12 variables:

$$\frac{\rho_1}{\rho_2}, \frac{\rho_s}{\rho_1}, \frac{\mu_1}{\mu_2}, S, \frac{h_1}{a}, \frac{h_2}{a}, \frac{d}{a}, \frac{\rho_1 a^2 \omega}{\mu_1}, \frac{a\omega^2}{g}$$

These are, respectively, two density ratios, a viscosity ratio, the slope (already dimensionless), three ratios involving the four length variables, and a Reynolds number and a Froude number based on wave amplitude and frequency. If all the important variables have been included, and if each of the dimensionless variables has the same value in ocean and laboratory, and if no other forces such as surface tension or cohesion are important in the model system, then the experimental system is a dynamic model of the oceanic system, and all effects observed in the model can be scaled up to the ocean (Buckingham, 1914).

The laboratory values of the first four modeling parameters above are not radically different from the oceanic values, except that slope is too great. The laboratory values of $h_1 a^{-1}$ and $h_2 a^{-1}$ correspond to oceanic cases in which the effects of free surface and bottom boundary on the waves before they shoal are small; however, cases involving shallow seasonal thermoclines in which the free-surface effect might be appreciable should also be considered. The ratio d/a is much too large, and except for oceanic waves with a very low amplitude and high frequency, the Reynolds number is at least one or two orders of magnitude too small. Finally, the Froude number is at least a few orders of magnitude too small. In common with most scale models of natural sedimentary processes, none of these last difficulties can be remedied. First, if the sediment size were to be scaled down properly, by something of the order of 100 : 1, it would be in the fine clay size range at the very coarsest, and the various cohesive effects would make sediment transport in the model unrealistic; second, no fluids with the requisite combinations of density and viscosity are available to make the Reynolds number and Froude number realistically high in the model system. Thus, the experiments are not, and cannot be, a scale model of sediment movement by breaking waves in the ocean.

Finally, there is no reason to expect that internal waves in a continuously stratified medium would show the same patterns of breaking as interfacial waves in a two-layer medium. However, the patterns of breaking observed in our experiments in a two-layer medium are similar to those observed by Cacchione (1970) in experiments on internal waves in a linearly stratified medium. This lends support to our extrapolation of the results of the two-layer experiments to the wider natural range of continuous stratification conditions. (It would seem desirable to make similar runs in a linearly stratified medium as well. The difficulty is that mixing in the zone of breaking degenerates the stratification rapidly, as noted above, and in the case of continuous density distribution there is no way to stabilize the process by removal of mixed fluid and replenishment of properly stratified fluid.)

Despite the major uncertainties involving modeling and nature of density stratification, we believe that the basic aspects of wave motion and sediment transport in the experiments are likely to be qualitatively the same as in the ocean, and that the experiments are thus a guide to the effects of breaking internal waves on the continental shelf or upper continental slope, in the same way that small-scale experiments on turbidity currents and beach profiles have been valuable guides to important effects on the natural scale without actually being dynamic models. Accordingly, we suggest that several dominant effects in the experiments might be observed when internal waves impinge on the continental margin:

1. Breaking of the waves, with advance of large decaying vortices upslope and return flow of mixed fluid downslope, creating a band of sediment movement with width at least an order of magnitude greater than the wave amplitude;

2. Abrupt onset of sediment movement at the point of breaking and gradual decrease in intensity downslope;

3. Predominant downslope movement of sediment except near the breaking point, causing deposition in a narrow band just upslope of the point of breaking and erosion in a wide band farther upslope;

4. Generation of downslope-migrating bed forms in the wide band of net downslope sediment transport; and

5. Suspension of fine sediment by the advancing breakers and horizontal transport away from the slope in the mixed layer.

In regard to generation of bed forms, the downslope flow would produce trains of asymmetrical sediment ripples migrating downslope provided only that the current is strong enough to transport at least some sediment as bedload (and that the sediment is not coarser than about one-half millimeter). If the maximum downslope velocity, attained not far upslope of the point of breaking, is great enough, dunes or sand waves migrating downslope might also be produced (Simons and Richardson, 1963, and Southard, 1971). In regard to suspension of fine sediment, horizontal spreading of mixed fluid away from the slope at the appropriate density level might contribute to the development of turbidity maxima observed in oceanic thermoclines (Costin, 1970).

ACKNOWLEDGMENTS

Roger D. Flood aided significantly in the experiments during the 1971 undergraduate Independent Activities Period at MIT. This work was supported by the Office of Naval Research under contract no. N00014-67-A0204-0048, NR 083-157.

LIST OF SYMBOLS

Symbol	Definitions
a	Input wave amplitude
d	Diameter of sediment particles
g	Acceleration due to gravity

h	Bed height normal to slope
h_1	Thickness of upper fluid layer
h_2	Thickness of lower fluid layer
q_s	Sediment volume transport rate per unit width
S	Slope of sediment bed
t	Time
λ	Bed porosity (volume of voids/bulk volume of sediment)
μ_1	Dynamic viscosity of upper fluid layer
μ_2	Dynamic viscosity of lower fluid layer
ρ_1	Density of upper fluid layer
ρ_2	Density of lower fluid layer
ρ_s	Density of sediment particles
ω	Input wave frequency

REFERENCES

Buckingham, E. (1914). On physically similar systems; illustrations of the use of dimensional equations. *Phys. Rev.* **4**[4], 345-376.

Cacchione, D. A. (1970). Experimental study of internal gravity waves over a slope. *Mass. Inst. Technol., Dept. Earth and Planetary Sciences, Rept.* **70-6**, 226 pp.

Cartwright, D. E. (1959). On submarine sand waves and tidal lee-waves. *Proc. Roy. Soc. London,* [*A*], **253**(1273), 218-241.

Costin, J. M. (1970). Visual observations of suspended-particle distribution at three sites in the Caribbean Sea. *J. Geophys. Res.* **75**[21], 4144-4150.

Emery, K. O. (1956). Deep standing internal waves in California basins. *Limnol. Oceanogr.* **1**[1], 35-41.

Keulegan, G. H., and Carpenter, L. H. (1961). An experimental study of internal progressive oscillatory waves. *Natl. Bur. Stand. Rept.* **7319**, 34 pp.

LaFond, E. C. (1961). Internal wave motion and its geological significance. *In* "Mahadevan Volume," pp. 61-77, Osmania Univ. Press.

LaFond, E. C. (1962). Internal waves. *In* "The Sea," Part I (M. N. Hill, ed.), Vol. 1. Interscience, New York.

LaFond, E. C. (1965). The U.S. Navy Electronics Laboratory's oceanographic research tower. *U.S. Navy Electronics Lab., Rept.* **1342**.

Simons, D. B., and Richardson, E. V. (1963). Forms of bed roughness in alluvial channels. *Trans. Am. Soc. Civil Eng.* **128**, (I), 284-302.

Southard, J. B. (1971). Representation of bed configurations in depth-velocity-size diagrams. *J. Sed. Petrol.* **41**(4), 903-915.

Thorpe, S. A. (1966). "Internal Gravity Waves." 164 pp. Ph.D. Dissertation, Trinity College, Cambridge.

Wunsch, C. I. (1969). Progressive internal waves on slopes. *J. Fluid Mech.* **35**(1), 131-144.

Wunsch, C. I. (1971). Internal waves. *Trans. Am. Geophys. Union* **52**(6), 233-235.

CHAPTER 5

Wave Estimates for Coastal Regions

D. Lee Harris

Research Division
U.S. Army Coastal Engineering Research Center
Washington, D.C. 20016

ABSTRACT

Significant information about wave climate can be obtained from aerial photographs and instrument records, visual observations, and wave hindcasts based on weather charts.

It is found that estimates of wave height data from all of these are reasonably well correlated. Observations of wave period are considerably less satisfactory. Data obtained from aerial photographs and the spectrum analysis of instrument records obtained in the coastal zone indicate that two or more wave trains with distinct periods are generally present. However, visual observations and most manual types of analyses of instrument records give only one wave period for each observation. This appears to be largely the result of a tendency by observers to assign a single value to the wave period even though several distinct periods are important. In addition, reported values of wave period from shipboard generally give little indication of the period experienced near the beach. This results both from masking of long period waves by short period seas, and from peaking up of long period waves in the breaker zone.

Much of the available data concerning wave direction are of doubtful quality individually. It may be possible that overall statistical representations will be useful.

Several sources of wave data are identified.

INTRODUCTION

The purpose of this paper is to describe the types of wave data that are presently available, or could be made available by established techniques. The principal concern is with observation and analysis procedures that are being or have been used extensively, not with ideal techniques, or recent research results. It will be obvious to many readers, that the procedures described do not satisfy all of their requirements for wave information.

A study of wave climate may be based on records made by instruments which respond to some aspect of the wave motion, on visual estimates by observers based on ship or shore, or on wave hindcasts based on meteorological analysis. Aerial photographs, although too limited in number to serve as the basis for a wave climatology, can provide insight not available from any other source. Even though data of the highest quality are too limited in quantity to serve as the basis for a study of wave climate, a review of the best information available about each aspect of the problem will provide information about the characteristics of waves in the coastal region and the extent to which these characteristics are represented by data from each available source.

AERIAL PHOTOGRAPHS

The most unambiguous qualitative information about two-dimensional wave patterns is provided by aerial photographs. Figure 38, from Harris (1972), is reproduced from a mosaic of aerial photographs of Pt. Mugu, California. Notice that in the lower right part of the picture, the dominant waves are relatively

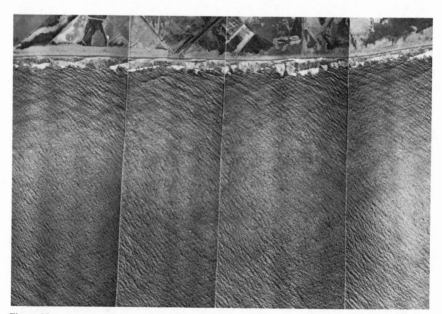

Figure 38.

A mosaic of aerial photographs of Pt. Mugu, California. From Harris (1972).

short, crest to crest, and that they are traveling from about 135° clockwise from the shore. In the upper right portion of the picture, the dominant waves are much longer, crest to crest, and are traveling from about 45° clockwise from the shore. The latter wave train is responsible for most of the breakers, but it could hardly be identified by a shipboard observer more than a few wavelengths from shore. Other wave trains with distinctly different wave length and direction can be clearly identified in the picture.

Wave patterns are of secondary interest in most aerial photographs of the coastal region, and in many cases the only waves that can be clearly distinguished in the photographs are the breakers in very shallow water. Whenever waves can be recognized in a large part of the photograph, however, it is often possible to identify two or more distinct wave trains with distinctly different directions of propagation and distinctly different wavelengths.

The above statements need to be emphasized because it is easy to lose sight of the complexity of the real wave field when one can see only a small area from an observation point on shore or shipboard. Complexities are very easily forgotten when one is working with tabulations of wave height, wave period, and possibly wave direction, and generally only one value for each of these is provided by the available wave data.

ANALYSIS OF INSTRUMENTAL RECORDINGS
OF WAVE CONDITIONS

Figure 39 is a sample of wave records from Atlantic City, New Jersey, obtained simultaneously from two surface gages (one a continuous wire type and the other a step resistance gage) and two pressure gages of the same type but installed at different depths. Notice that the records from the pressure gages are much smoother than those of the surface gages. The amplitudes of the individual waves as shown by the records vary greatly.

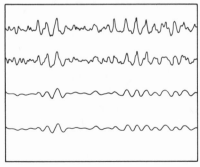

Figure 39.
Simultaneous records from four wave gages at Atlantic City, New Jersey (Harris, 1970).

Ocean-wave theory was well advanced before the first useful records of ocean waves were obtained, and thus it is natural that the available theory has been used in developing procedures for the analysis of wave gage records and for reporting visual wave observations.

In the most elementary wave theory it is assumed that the profile of the water surface is given by

$$h(x,t) = (H/2) \cos (2\pi x/L - 2\pi t/T) \tag{1}$$

where h is the displacement of the water surface from its mean elevation, H is the wave height, L the wave length, and t the wave period. The relation between L and T is given by

$$L = (g/2\pi) T^2 \tanh 2\pi d/L \tag{2}$$

where g is the acceleration of gravity and d is the water depth. Since there is a direct relationship between L and T, this elementary wave is fully specified by its height and either the wavelength or the period.

Several terms similar to Eq. (1) involving many frequencies would be necessary to provide a reasonably accurate reproduction of Fig. 39. Many terms which include the effect of wave direction would be required to reproduce Fig. 38.

Two general procedures for the analysis of wave records have been widely adopted. In the first, an attempt is made to "fit" Eq. (1) to the complex sea by determining a "significant wave height and period" that would capture the most significant aspects of the complete records while retaining the simplicity of the elementary theory.

In the second procedure, the wave energy is partitioned into a large number of frequency bands. The resulting function is called the wave energy spectrum. The second procedure permits a much better description of the wave field, but it loses the inherent simplicity of Eq. (1). Furthermore, the computation effort required for an evaluation of the spectrum made it prohibitively expensive when wave records first began to be generally available in the early 1940's. However, by 1970, computational technology had advanced to the point that the spectrum approach is actually the most economical, provided one is set up to use it. Therefore, the spectrum approach will be discussed first.

SPECTRUM ANALYSIS

The simplest approach to spectrum analysis is to note that any bounded function h(t) can be expressed in the form

$$h(t) = \sum_{n=1}^{N} A_n \cos (2n\pi t/P - \phi_n) \tag{3}$$

where P is the time interval over which the function is defined. All wave records must satisfy this requirement. It can be shown (Kinsman, 1965, Chapter 3) that for quite general conditions the average potential energy of the waves, E_p, is proportional to the average value of $[h(t)]^2$, that is, to the variance of the wave record, V_p. It can also be shown that the variance of the wave record is given by

$$V_p = (1/2) \sum A_n^2$$
$$= (1/2) \sum A^2 (f) \qquad (4)$$

where $A^2 (f) = A^2$, and $f = n/p$.

Equation (4) provides a means of partitioning the variance of the wave record into components due to various frequency bands. The kinetic energy of progressive gravity waves is equal to the potential energy under many conditions and approximately equal under most conditions. Hence, the function $A^2(f)$ is often called the energy spectrum, but it is usually given with dimensions of length squared, that is without the proportionality constant required to give the proper dimensions for energy.

A direct evaluation of Eqs. (3) and (4) may provide more detail about the spectrum than one can conveniently use. Therefore, the spectrum is usually smoothed by averaging over a number of the harmonics obtained in (3), and A^2 is often treated as a continuous function of f.

More rigorous derivations of Eq. (4) along the lines given here may be found in Taylor (1938, who credits it to Lord Rayleigh), and Harris (1972). Other derivations may be found in Blackman and Tukey (1958), Kinsman (1965), and Bingham, Godfrey, and Tukey (1967).

When spectrum techniques are used for wave analysis, the square root of V_p, that is the standard deviation of the wave record, is the natural measure for wave height, and the period corresponding to the maximum energy value is the natural choice for the characteristic wave period.

ANALYSIS OF SIGNIFICANT WAVE HEIGHT AND PERIOD

While developing an early system for the prediction of wave conditions from weather data, Munk (1944) and Sverdrup and Munk (1947) recommended that wave records be characterized by a "significant wave height" defined as the average height of the one-third highest waves and a "significant wave period" defined as the average period of the significant waves.

Estimates of the significant height and period of waves have been the most common product of wave record analysis since 1947. Rigorous application of the definition is rare because of the difficulty of deciding just which waves should be counted. Several quasi-objective procedures for estimating the significant height and period without examining all of the individual waves have been introduced. Most of these are discussed by Draper (1967). The system used by the Coastal Engineering Research Center (CERC) from 1960-1970 had not been published when Draper's paper was prepared. As this system was used in obtaining some of the data discussed below it will be described here. The average period of a few of the best formed waves is selected as the significant wave period. It is assumed that the Rayleigh distribution function describes the actual distribution of individual wave heights. According to the Rayleigh distribution function, the probability that an individual wave height will exceed the average of the one-third highest waves is 0.135. A few of the highest waves are ranked in order of wave height, with the highest having rank 1. The rank of the wave whose height will be a good estimate of the average of the one-third highest wave is obtained as 0.135 times the estimated number of waves in the record. The number of waves to be considered is obtained by dividing the duration of the observation by the significant period. The significant wave height obtained in this manner is about four times the standard deviation of the wave record.

Harris (1970) compared the height and period estimated obtained by the CERC procedure and the procedures described by Draper with the results of spectrum analysis for the November 1966 wave records obtained from the step-resistance wave gage at Atlantic City, New Jersey. The correlation matrix obtained for wave-height estimates is shown in Table II. The correlation matrix for period estimates is given in Table III. Note that all of the correlations for estimates of wave height are 0.868 or higher. It seems likely that any of these procedures will be satisfactory for estimating wave height. Note too that both the estimates used by CERC and the one proposed by Tucker are more highly correlated with the root mean square height, that is with the square root of the total

Table II
Correslation Matrix for Wave Height Estimates[a]

	H_{RMS}[b]	H_{CERC}[c]	H_{Tucker}[d]	$H_{1/3}$[e]
H_{RMS}	1.000	0.949	0.982	0.938
H_{CERC}	0.949	1.000	0.945	0.868
H_{Tucker}	0.982	0.945	1.000	0.918
$H_{1/3}$	0.938	0.868	0.918	1.000

[a] Atlantic City, N. J., November 1966 (129 records).

[b] H_{RMS} is the standard deviation of the record.

[c] H_{CERC} is the value obtained from the analysis system used at CERC since 1965, based on a 7 min. record. Other estimates are based on records of 1024 sec. The correlation between records of 420 sec and 1024 sec duration is approximately 0.98.

[d] H_{Tucker} is the estimate of H_{RMS} based on the highest crest and lowest trough as recommended by Tucker (1961).

[e] $H_{1/3}$ is the average value of the one-third highest waves as obtained by a digital computer.

Table III
Correlation Matrix for Wave Period Estimates[a]

	T_{FM}[b]	T_{CERC}[c]	T_{PM}[d]	T_{ZUC}[e]	T_{all}[f]
T_{FM}	1.000	0.725	0.581	0.493	0.131
T_{CERC}	0.725	1.000	0.615	0.552	0.236
T_{PM}	0.581	0.615	1.000	0.648	0.260
T_{ZUC}	0.493	0.552	0.648	1.000	0.710
T_{all}	0.131	0.236	0.260	0.710	1.000

[a] Atlantic City, N. J., November 1966 (129 records).
[b] T_{FM} is the period corresponding to the frequency of maximum energy density per unit frequency.
[c] T_{CERC} is the most prominent period as determined by the CERC method (see text); based on a seven-minute record. All other measurements are based on a record on 1024 sec.
[d] T_{PM} is the period of maximum energy density per unit period.
[e] T_{ZUC} is the average period of zero up-crossings.
[f] T_{all} is the average period between maxima in the record.
$T_{1/3}$ is the average period of the one-third highest waves.

wave energy, than with a careful computation of the average height of the one-third highest waves. No two of the estimates for wave period were highly correlated.

INTERPRETATION OF WAVE RECORDS
OBTAINED WITH PRESSURE GAGES

At some locations for which wave records are needed, it is not possible to install an instrument which will directly sense the position of the free surface of the water. In such cases, it is often possible to obtain a record of pressure variations beneath the water surface which may be used to obtain estimates of water surface elevations.

According to elementary (Airy) wave theory, the pressure due to a pure wave at a single frequency is given by

$$P(x,z,t) = \rho g \frac{\cosh 2\pi(z + D)/L}{\cosh 2\pi D/L} h(x,t) \tag{5}$$

where z is the depth at which the pressure is to be evaluated. The origin of the z axis is taken in the mean water surface. Below this surface z is negative. Equation (5) can be rearranged slightly to provide

$$h(x,t) = \frac{\cosh 2\pi D/L}{\cosh 2\pi(z + D)/L} \frac{P(x,z,t)}{\rho g} \tag{6}$$

When pressure sensors are used as wave gages, they are usually calibrated directly in height units. The procedure most commonly used for the interpretation of wave data obtained from pressure records has been to select a wave period, characteristic of the record, which would be combined with (2) to obtain the wavelength and to use this wavelength with (6) to compensate for the depth of the gage. This procedure generally leads to an underestimation of the surface wave height by about 25 to 30% because it does not give adequate consideration to the shorter waves that are almost always present (Hom-ma, Horikowa, and Komori, 1966; Grace, 1970).

Hom-ma et al. (1966), Esteva and Harris (1971), and others have shown that a more accurate procedure is to compute the spectrum of the pressure record and to compensate each frequency band of the spectrum for the effect of gage submersion by a factor based on (6). That is,

$$E_{sfc}(f) = \left(\frac{\cosh 2\pi D/L}{\cosh 2\pi(z + D/L)} \right)^2 \frac{E_{pres}(f,z)}{(\rho g)^2} \tag{7}$$

where the subscripts "sfc" and "pres" are used to indicate the surface spectrum and the pressure spectrum, respectively. The correction must be made for small frequency intervals. Summing overall frequency intervals and extracting the square root of the total variance, often called "energy" in this connection, gives a computed value for the root mean square wave height. A comparison of many computations made in this manner with direct measurements of the surface waves is shown in Fig. 40 and described in a later section.

Table IV (from Harris, 1972) provides some perspective for the accuracy obtainable when (6) is applied at the significant wave period only. This table shows a distribution function for the ratio of the wave height obtained from (7) to the wave height computed directly from the pressure record without compensation for 232 observations at Pt. Mugu, California. The value obtained from (6) is shown in the column headed "Theor Ratio." These records were obtained with a pressure gage mounted about 2 ft above the bottom and about 26 ft below mean lower low water with a mean tide range of about 3.7 ft. The wide scatter in these ratios is readily apparent. The largest ratios are associated with sea states in which the period of maximum energy density in the surface spectrum was much shorter than the corresponding period in the pressure record.

This ratio and the average difference between the calculations based on (6) and those based on (7) increase with the submersion of the gage. The average differences shown in Table IV are larger than the averages cited by Hom-ma et al. (1966) and Grace (1970) because the depth of submersion of the pressure gages was greater than that used in most comparisons of wave parameters evaluated directly from pressure-gage records with the parameters evaluated from records of the free-surface displacement.

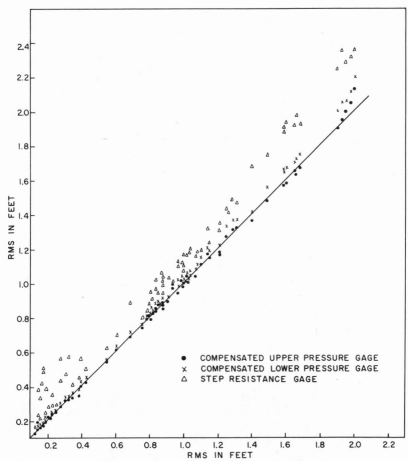

Figure 40.
A comparison of the standard deviations of the wave records based on simultaneous records by four gages at Atlantic City, New Jersey; horizontal axis corresponds to continuous-wire gage. (From Esteva and Harris, 1970.)

VARIABILITY IN WAVE MEASUREMENTS
DUE TO GAGE DESIGN OR INSTALLATION

Records obtained from the four-gage installation which produced Fig. 39 have been analyzed in detail to provide a field test for the validity of Eq. (7) and a comparison of the characteristics of the gage and its support.

Table IV

Ratio of Surface Wave Height to the Wave Height Obtained from the Pressure Record Without Compensation[a]

Freq (Hz) upper class limit	Period	1.0	1.1	1.2	1.3	1.4	1.5	1.6	1.7	1.8	1.9	2.0	2.1	2.2	Row totals	Row average	Theor[b] ratio
0.0484	20.7		2	7	2	4	1	3							19	1.33	1.03
0.0591	16.9		5	26	27	19	15	6	2	6	2		2	1	111	1.40	1.05
0.0699	14.3		1	26	45	39	35	21	14	7	6	2	2	4	202	1.46	1.06
0.0806	12.4			6	20	22	16	12	5	4		2	2	5	94	1.50	1.09
0.0949	10.5			1	1	3	5	2	1	1	1				15	1.52	1.12
0.1021	9.8					1	4	3	1	1	1				11	1.53	1.15
0.1128	8.9					6	6	1	1		1				15	1.51	1.19
0.1236	8.1						9	1	2	3	2				17	1.63	1.23
0.1343	7.4						3	2	1	2	4	5	1		18	1.90	1.32
0.1562	6.4									1	1	1		1	4	2.12	1.43

[a]Upper limits of class intervals shown for frequencies; lower limits of class intervals shown for periods.

[b]The theoretical ratio is obtained from E. (6) with a single value for the frequency.

Estimates of the standard deviation of the water surface elevation obtained by the spectrum procedure from the records of two pressure gages are compared with the simultaneous measurements obtained from the records of a continuous wire staff gage in Fig. 40, based on data published by Esteva and Harris (1970). The consistently good agreement among these three sets of data, based on two different principles, indicates that both are rather good.

Estimates of wave height obtained from the simultaneous records from a step-resistance wave gage are also shown in Fig. 40. The estimates from the step-resistance gage records are consistently higher than the other three and show more scatter than the estimates from the pressure gages, when the continuous-wire gage data are taken as a standard. All four gages were installed at the end of the Steel Pier, Atlantic City, New Jersey, about 1000 ft from shore. The step-resistance gage and the continuous-wire gage were both designed to record the elevation of the free water surface. Figure 40 shows that one should not assume that all wave-gage records have equal validity. The type of gage and the precise method of installation may have a significant effect on the records obtained.

VARIABILITY IN WAVE SPECTRA

Figure 41 from Harris (1972), shows that spectra computed from the records of two pressure gages of the type used in the above study located at a depth of about 26 ft and about 200 ft apart horizontally at Pt. Mugu, California.

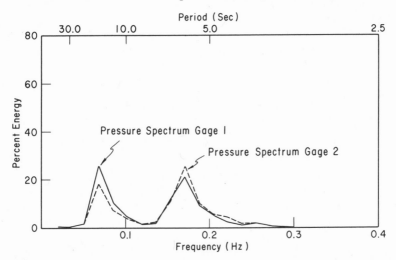

Figure 41.

The spectra from two pressure gages at Pt. Mugu, California for the 1024-sec period beginning at 0307 PST, June 23, 1970 (Harris, 1972).

The highest concentration of energy corresponds to a period of 14.7 sec at gage 1 and to a period of 5.9 sec at gage 2. Thus, in this case, the characteristic wave period, when only one period is reported, could change by nearly 9 sec with a change of about 200 ft in the position at which the period is evaluated.

Comparison of 43 records from five gages within a radius of 100 ft showed that the period of maximum energy density was the same at all five pressure gages 93% of the time.

An examination of several hundred spectra from the Atlantic, Pacific, and Gulf of Mexico coastal zones shows that multinodal spectra occur more than half the time. When two or more nodes are of nearly the same magnitude (and this is not uncommon) their relative magnitudes may be expected to vary enough within short distances to cause a shift in the frequency of the maximum energy level, as illustrated in Fig. 41.

Equations for transforming the spectrum of wave height or pressure at a specified depth into spectra for the velocity components at any designated depth or the pressure at some other depth can be developed from linear theory. Comparisons of the surface spectra derived from pressure spectra generally give encouraging results. Pressure spectra computed from surface spectra are even more satisfactory because the transformation tends to suppress the high-frequency components in the surface spectrum.

Measurement of particle velocities due to waves is very difficult, and few comparisons between observed and computed wave-velocity spectra have been published. Nevertheless, it is interesting to examine such computed spectra because this may provide the best means of estimating the perturbation velocities near the sea bed.

The spectrum of the pressure record from gage 2 of the Pt. Mugu array for the 1024-sec period beginning at 1808 PST, June 25, 1970, is shown in Fig. 42. The surface spectrum, computed from the pressure spectrum by Eq. (7), and the spectrum for horizontal velocity at the pressure gage, computed by the relation

$$E_{vel}\,(f,z) = (T/L)\,E_{pres}\,(f,z)/\rho^2 \tag{8}$$

where the subscript vel is used to indicate the horizontal velocity, are also shown. Notice that both the peak in the surface spectrum, and the peak in the horizontal velocity spectrum at the same depth as the gage, occur at a distinctly shorter period than the peak in the pressure spectrum. The peak in the spectrum for horizontal velocity at the surface may occur at an even shorter period. It appears from an analysis of about 280 readings involving both winter and summer conditions at Pt. Mugu that about 30% of the time the peak of the pressure spectrum occurs at a distinctly different period than the peak of the velocity spectrum or the peak of the surface spectrum.

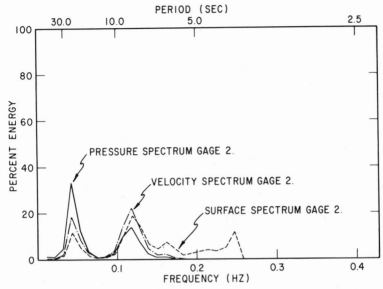

Figure 42.

An example of surface and horizontal velocity spectra computed from pressure spectra at Pt. Mugu, California, for the 1024-sec period beginning at 1808 PST, June 25, 1970 (Harris, 1972).

SHIPBOARD WAVE OBSERVATIONS

From 1949 through 1967, merchant ship officers were asked to report the direction of both sea and swell to the nearest 10 degrees, the average height of sea and swell in one-half-meter units, and wave period in a code which lumped all periods less than five seconds into one group and assigned longer periods to code groups with two-second intervals. In 1968 these instructions were changed to report the period directly and to omit the direction of the sea. It is now assumed that wind and sea direction are always identical.

Digital magnetic tapes containing all ship reports, filed by Marsden squares, 10 degrees of latitude or longitude on a side, can be obtained from the National Weather Records Center, Asheville, North Carolina. The records from each Marsden square can be readily subdivided into subsquares one degree of latitude or longitude on a side. The locations of several subsquares being discussed in this report are shown in Fig. 43.

A typical distribution function for reported wave directions in a small region, as calculated from the observations from subsquares 36, 37, 46, and 47 is shown in Fig. 44. It is clear that the cardinal and semi-cardinal directions are

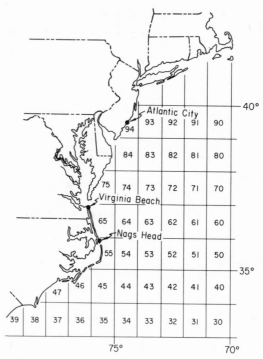

Figure 43.

The location of several subsquares of Marsden square 116 being referenced in this report (Thompson and Harris, 1972).

favored over the others, and that the full 10-degree resolution is not achieved in practice.

Joint distribution functions for observations of "significant wave height" as evaluated from a CERC gage and a reported wave height from a single subsquare, when both reports coincide within two hours, have been computed for three CERC wave gages and 24 subsquares. A typical result is shown in Fig. 45. The wide scatter is readily apparent. The data do, however, cluster around a line, thus revealing a significant if low degree of correlation between wave height measured near the shore and shipboard reports of wave heights. Only a trivial improvement is realized when the waves are stratified by reported direction of propagation.

A careful examination of Figure 45 shows that a height of 5.5 m is reported more often than a height of 5.0 m. This is common to nearly all summaries of shipboard wave observations and appears to be an artifact of the code for reporting wave height. Only one digit is allowed in the code. Height values between 5.0 and 9.5 m are to be reported by adding 50 to the direction code and subtracting ten from the height code. Thus, a 4.5 m wave from the west

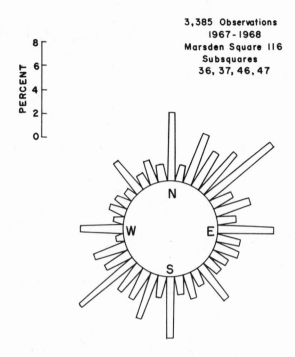

FREQUENCY OF REPORTED WAVE DIRECTION

Figure 44.

The frequency of reported wave direction (Marsden square 11, in subsquares 36, 37, 46 and 47) by 10° arcs.

would be reported as direction code 27, height code 9, but a 5.0 m wave from the west would be reported as direction code 77, height code 0. This must seem inconsistent to shipboard observers because they show a distinct bias against reporting the wave height as 5 m. Wave heights in excess of 9.5 m are reported in plain language.

A similar scatter diagram for wave periods is shown in Fig. 46. Note that periods of 20-21 sec are reported 4 times and periods greater than 21 sec are reported 13 times, but there are no reports of 18-19 sec periods. This is believed to result from the structure of the reporting code. Code Fig. 0 is assigned to periods of 20-21 sec, Code Fig. 1 to periods greater than 21 sec; Code Fig. 2 to all periods less than 6 sec; Code Figs. of 3-9 are assigned to 2-sec period intervals in an orderly fashion. The reports are easier to believe if it is assumed that some observers define Code Fig. 0 and 1 by backward extrapolation as applying to periods of less than 4 sec. It appears that only those shipboard observations which report periods between 6 and 17 sec can be used for establish-

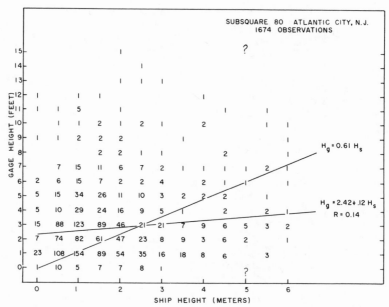

Figure 45.

Comparison of 1674 estimates of significant wave height from the Atlantic City gage and shipboard observations in subsquare 80. Each shipboard observation was taken within 2 hr of the corresponding gage record. Shipboard observations above 6 m are omitted. A "?" is shown at 5 m to indicate that this column is believed to be incorrect. See text.

ing a regression line between the periods based on wave gage records and shipboard reports. The least-squares regression line shows an insignificant but negative correlation. A regression line is also drawn through the origin.

CERC has collected visual wave observations from Coast Guard stations for many years. Only a few of these observation sites are near enough to recording wave gages to permit a direct comparison, and only a small fraction of these have been studied. The results obtained to date, however, are similar to those just shown for the ship data: a wide scatter but with significant correlation for wave height and no clear correlation for period.

Aerial photographs, such as Fig. 38, frequently show breakers longer than the more prominent wind waves a few hundred meters from shore. This phenomenon may be due to amplification by shoaling of long waves whose crests are nearly parallel to the shore, combined with a spreading out by refraction of short waves, whose crests are nearly normal to the shore. Thus modification of the wave field in shallow water, as well as the tendency for observers to assign a single period to waves which actually display several distinct characteristic periods may contribute to the low correlation between independent estimates of the wave period between surf and waves offshore.

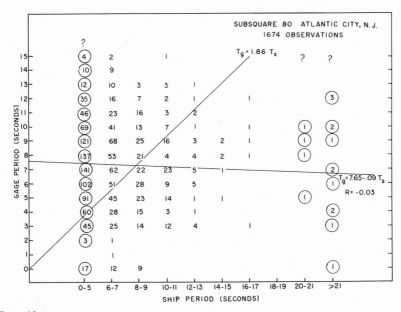

Figure 46.

Significant wave period estimates for 1674 observations corresponding to the height estimates in Fig. 12. Due to nonuniformities in the code for reporting wave periods, observations in the lowest period interval and all entries greater than 19 sec are not suitable for use in linear regression analysis. See text.

It seems likely that the low correlation for independent estimates of wave period results from the attempts to assign a single characteristic period to phenomena which actually display several distinct characteristic periods.

DISTRIBUTION FUNCTIONS
FOR WAVE HEIGHT AND PERIOD

The real concern in many problems is the distribution functions governing wave height and period rather than the values for individual waves, or even values that are characteristic of individual observations. The correlation between different estimates of wave conditions, from the records of nearby gages or from visual observations, shows a more satisfactory structure when examined in this way.

The joint distribution function for wave height and period at Atlantic City, New Jersey for the year 1967 shown in Table V is typical of those based on the CERC system for analyzing wave records from the Atlantic, Gulf of Mexico, and Pacific coastal regions of the contiguous United States, with the following exceptions.

Table V
Wave Climatology for Atlantic City, New Jersey[a]

Period (sec)	Height (ft)										Total	Acc total
	0-1	1-2	2-3	3-4	4-5	5-6	6-7	7-8	8-9	9-10		
0.0–1.9	2.2										2.2	2.2
2.0–2.4	—											2.2
2.5–2.9	—											2.2
3.0–3.4	—	1.1									1.1	3.3
3.5–3.9	—	1.1	0.7								1.8	5.1
4.0–4.9	—	1.8	2.9	0.7							5.5	10.6
5.0–5.9	—	1.8	2.6	1.5	0.7						6.9	17.5
6.0–6.9	0.4	3.3	4.0	1.8	1.5	0.4					12.4	29.9
7.0–7.9	0.4	5.5	6.6	2.9	1.5	0.7					18.2	48.2
8.0–8.9	0.4	8.0	6.2	2.9	1.1	0.7	0.4	0.4	0.4		20.1	68.2
9.0–9.9	0.4	6.9	4.0	1.5	0.7	0.4	0.4	0.4	0.4	0.4	15.7	83.9
10.0–10.9	0.4	4.4	1.8	0.7	0.4	0.4	0.4	0.4	0.4		8.8	92.7
11.0–11.9	0.4	1.8	1.5	0.4	0.4	0.4	0.7	0.4			4.8	97.4
12.0–12.9		1.1	0.7			0.4		0.4			1.8	99.3
13.0–13.9		0.7									0.7	100.0
Total	4.4	37.6	31.0	12.4	6.2	3.3	1.8	1.8	1.1	0.4	100.0	
Acc. total	4.4	42.0	73.0	85.4	91.6	94.9	96.7	98.5	99.6	100.0	100.0	

[a]Total = 12 months, January 1967 to December 1967. Distribution of height (in percent) as a function of period for 2083 observations. Each entry in the table is rounded individually; therefore, the sum and the accumulated total for each row or column may not agree with the figures as shown in the table.

The classifications of wave heights less than one foot and of wave period less than two seconds may contain as many as one-third of all observations at some locations. With the exception of these near-calm conditions, the distribution of wave periods is approximately symmetric about a mean between 4 and 8 sec at all Atlantic and Gulf coast gages and at some Pacific gages. A second node between 12 and 15 sec appears in the summaries for some Pacific coast gages.

In all cases the distribution of wave heights is skewed. From 50 to 75% of all reported wave heights are less than 2 ft, and at most locations few exceed 5-10 ft. The highest waves are rarely associated with the longest wave periods.

An analytical description of these distribution functions would be useful for many purposes. The functions must be bounded at the origin. The node in the height distribution must be near the origin. Three well-known expressions, the exponential, the Rayleigh, and the log-normal or Galton distribution functions, satisfy these requirements.

It is worthwhile to consider the representativeness and the accuracy of the data and the projected use of the statistics in selecting one of these functions for use with empirical data. The growth of waves under wind action is approximately exponential. Thus, low waves can change their characteristics within very short distances, but high waves are representative of much larger areas. Most of the wave data collected by CERC have been obtained with a step-resistance wave gage (Williams, 1969). This gage was designed to achieve dependability in recording the higher storm waves. It does not provide adequate resolution for wave heights less than one foot. Both of these factors seem to indicate that the higher waves should be weighted more heavily than the lower waves in determining the parameters of an analytic distribution function. This consideration favors the exponential and Rayleigh function over the Galton function, for the latter gives more emphasis to the lower, less well-known waves than to the high waves. The Galton function has been used by some workers in this field (Jasper, 1957; and Ploeg, 1971).

Studies at CERC indicate that the Rayleigh distribution provides the better fit to the heights of individual waves within an observation of a few minutes' duration, but that a single modification of the exponential distribution provides a better fit to the distribution of "significant wave heights" determined over a month, a year, or a longer period.

The cumulative form of the modified exponential distribution may be stated in the form

$$P(h > h') = exp\ [-(h' - h_0)/\sigma_{h'}]$$ (9)

where σ_h is the standard deviation of h, h_0 may be interpreted as the minimum wave height likely to occur at the location and h' is an arbitrary value of h. By taking logarithms of both sides and rearranging slightly we obtain

$h = h_0 - \sigma_h \mathrm{Ln}[P(h - h')]$

where Ln(x) indicates the natural logarithm of x,n

$h' = -A\alpha\log [P(h > h')]$ (10)

where A = 2.30258, and log indicates logarithm to the base 10.

It is seen that the distribution should plot as a straight line on semilog paper. Figure 47 shows the distribution functions derived from most of the CERC observations. It is seen that a straight line does provide a good approximation to the empirical distribution function provided the wave height exceeds one to two feet and the probability exceeds 0.001.

Larras (1965, 1967, and 1969) and Mayencon (1969) have used the exponential distribution for climatic studies of waves in the North Atlantic and Mediterranean Sea. The log-normal or Galton distribution has been used by Jasper (1957) for studies of waves in the North Atlantic and Ploeg (1971) for studies of waves in the Great Lakes.

The U.S. Naval Weather Service Command (1970) has compiled summaries of the meteorological observations from ships (SSMO's) for many regions consisting of 5-10 Marsden subsquares adjacent to the coast. Figure 48 is a chart showing the location of those regions near the continental coast of the United States. Summaries have been compiled for many other regions.

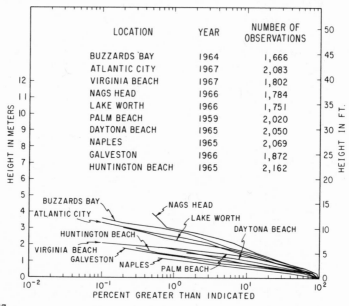

Figure 47.

Plotted distribution functions for significant wave heights from wave gage records as determined by the CERC procedure (Thompson and Harris, 1972).

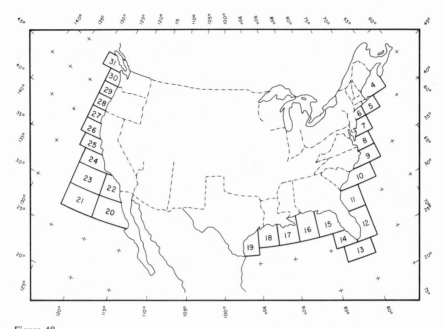

Figure 48.

North American coastal summary of synoptic meteorological observation (SSMO) areas (Thompson and Harris, 1972).

Distribution functions for the wave heights reported during the period 1963-1968 in most of the areas identified in Fig. 48 are shown in Fig. 49. A straight line is found to be a good approximation to the empirical distribution for wave heights greater than 0.5 m up to about 5 m. The kink near 5 m is believed to result from the aversion to the form of the coded report for a wave height of 5 m, as explained heretofore.

WAVE HINDCASTS

Wave data are needed for many locations where no routine observations are available. An attempt is often made to fill this data gap by wave "hindcasting." A wave hindcast is identical with a wave forecast with the exception that it is usually made after the event, and is based on observed wind and pressure fields, whereas predicted wind and pressure fields must be used in wave forecasting.

The earliest wave prediction procedure to be widely used in the United States was developed by Sverdrup and Munk (1947) and revised by Bretschneider (1952, 1958). This is sometimes referred to as the SMB method. The second procedure was given by Pierson, Neumann, and James (1955), and is sometimes called the PNJ method. Saville (1954) and Neumann and James (1955) give

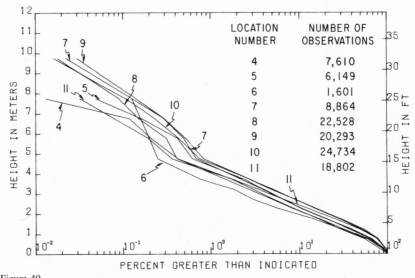

Figure 49.

Wave height distribution functions based on shipboard observations from North American coastal SSMO areas along the Atlantic Coast. Wave heights greater than 12 m and height with less than a 0.01% probability of occurrence are not shown (Thompson and Harris, 1972).

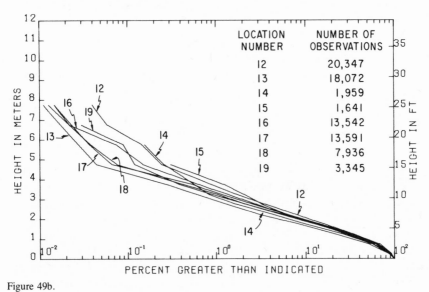

Figure 49b.

Wave height distribution functions for SSMO areas along the Gulf coast (Thompson and Harris, 1972).

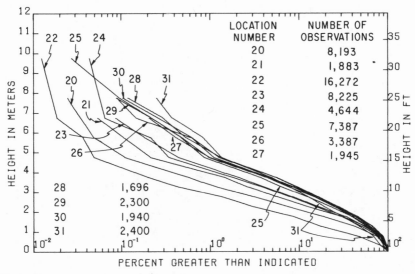

Figure 49c.

Wave height distribution functions for SSMO areas along the Pacific coast (Thompson and Harris, 1972).

wave hindcasts for four locations, identified as stations A-D in Fig. 50. The SSMO areas 4, 5, 6, and 8 are indicated in the same figure. Distribution functions for wave heights as determined from shipboard wave observations and by both hindcasting procedures are given in Fig. 51.

SOURCES OF WAVE DATA

The most extensive and detailed summaries of wave records presently available are provided by the summaries of synoptic meteorological observations (SSMO's) discussed above. These are available through the Federal Clearinghouse for Scientific and Technical Information, Springfield, Virginia 22151.

The published data permit stratification of the data by wind direction and month of the year. Tabulations of the individual shipboard observations are available on magnetic tape or as hard copy from the Environmental Data Service, National Weather Records Center, Asheville, North Carolina 28801.

Visual observations of surf conditions at 27 Coast Guard stations have been collected by CERC since 1954. A listing of most of these observations through 1967 and some later observations is available on loan from CERC in the form of microfilm or microfiche. Much of the material has been summarized, but it has not been subjected to strict quality control. These data have been discussed by Helle (1958) and Darling (1968). Tabulations of significant wave height

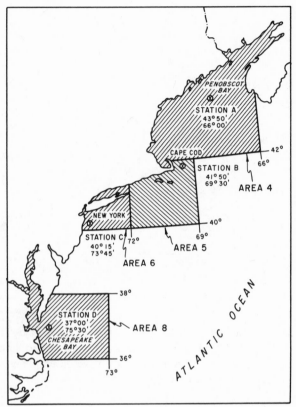

Figure 50.
Location of wave hindcast stations and the SSMO locations for the data displayed in Fig. 14.

and period obtained from wave-gage records can be made available on loan from the CERC library. Tabulations or summaries of wave hindcasts have been given by Saville (1953a,b) for Lakes Michigan, Erie and Ontario, Saville (1954) and Neumann and James (1955) for the North Atlantic coast of the U.S., Bretschneider and Gaul (1956a,e) for the Gulf of Mexico, Wilson (1957) for hurricanes in the Gulf of Mexico, and by many others.

SUMMARY

Aerial photographs of the coast have been used to investigate the two-dimensional structure of the wave field near the coast. Digital wave spectra have been used to investigate the compatibility of wave records obtained with different instrument and analysis procedures. Comparisons of shipboard wave observations with records from nearby shore-based gages have been used to

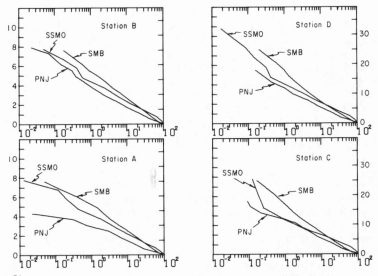

Figure 51.
Comparison of the distribution functions for wave height based on shipboard observations for the years 1963-1968 and on hindcasts by both SMB procedures for the years 1948-1950 and PNJ procedures for the years 1947-1949.

investigate the applicability of shipboard observations to coastal problems. Wave height distributions, graphs based on shipboard observations and two hindcasting procedures have been compared to provide some indication of the value of hindcasts as a source of data about wave climatology.

It is found that most procedures for estimating wave height are reasonably well correlated. Observations of wave period are generally unsatisfactory. This appears to be largely the result of a tendency by observers to assign a single value to the wave period when several distinct periods are important. This difficulty could be removed, as far as wave gage records are concerned, by considering the entire wave spectrum. This would, of course, lead to a significant increase in complexity, but it may be essential to the solution of most problems in which the period of waves is important.

The quality of the data on wave direction presently available appears to be unsatisfactory, but no method has been found for adequate evaluation of reports of wave direction or for obtaining reliable information on wave direction routinely.

Several sources for wave data have been identified.

ACKNOWLEDGMENTS

The observational data and analysis presented herein, unless otherwise noted, were obtained from research conducted under the Coastal Engineering Research Center. The findings in this report are not to be construed as an official Department of the Army position unless so designated by other authorized documents.

REFERENCES

Bingham, C., Godfrey, M. D., and Tukey, J. W. (1967). Modern techniques of power spectrum estimation. *IEEE Transactions on Audio and Electroacoustics* **AA-15**, 56-66.

Blackman, R. B., and Tukey, J. W. (1958). "The Measurements of Power Spectra." Dover, New York.

Bretschneider, C. L. (1952). Revised forecasting relationships. *Proc. of the Second Conference on Coastal Engineering, Houston,* Council on Wave Research, Univ. of California.

Bretschneider, C. L., and Gaul, R. D. (1956a). Wave statistics for the Gulf of Mexico off Brownsville, Texas. *Coastal Engineering Res. Center Techn. Memo.* **85**, Sept. 1956.

Bretschneider, C. L., and Gaul, R. D. (1956b). Wave statistics for the Gulf of Mexico off Caplen, Texas. *Coastal Engineering Res. Center Tech. Memo.* **86** (Sept., 1956).

Bretschneider, C. L., and Gaul, R. D. (1956c). Wave statistics for the Gulf of Mexico off Burrwood, Louisiana. *Coastal Engineering Res. Center Tech. Memo.* **87** (Oct., 1956).

Bretschneider, C. L., and Gaul, R. D. (1956d). Wave statistics for the Gulf of Mexico off Apalachicola, Florida. *Coastal Engineering Res. Center Tech. Memo.* **88** (Oct., 1956).

Bretschneider, C. L., and Gaul, R. D. (1956e). Wave statistics for the Gulf of Mexico off Tampa Bay, Florida. *Coastal Engineering Res. Center Tech. Memo.* **89** (Oct., 1956).

Bretschneider, C. L. (1958). Revisions in wave forecasting; deep and shallow water, *Proc. 6th Conference on Coastal Engineering,* Council on Wave Research, Univ. of California.

Darling, J. M. (1968). Surf observations along the United States coasts. *J. Waterways and Harbors Div., Proc. Am. Soc. Civil Engineers* (Feb., 1968).

Draper, L. (1967). The analysis and presentation of wave data—A plea for uniformity. *Proc. Tenth Conference on Coastal Engineering (ASCE, New York).*

Esteva, D., and Harris, D. L. (1970). Analysis of pressure wave records and surface wave records. *Proc. Twelfth Conference (ASCE, Washington).*

Grace, Robert A. (1970). How to measure waves. *Ocean Industry* **5**, 65-69.

Harris, D. L. (1970). The analysis of wave records. *Proc. Twelfth Conference on Coastal Engineering. (Washington),* 85-100.

Harris, D. Lee (1972). "Characteristics of Wave Records in the Coastal Zone," *Advanced Seminar on Waves and Beaches,* Academic Press, (in press).

Helle, J. R. (1958). Surf statistics for the coasts of the United States. *U.S. Army Corps of Engineers, Beach Erosion Board, Tech. Memo.* **108** (Nov., 1958).

Hom-ma, M., Horikawa, K., and Komori, S. (1966). Response characteristics of underwater wave gage. *Proc. Tenth Conference on Coastal Engineering (Tokyo),* 99-114.

Jasper, N. H. (1957). "Statistical Distribution Patterns of Ocean Waves and of Wave-Induced Ship Stresses and Motions, with Engineering Applications," *Report 921,* Navy Department David Taylor Model Basin, Structural Mechanics Laboratory, Research and Development Report (Oct., 1957).

Kinsman, B. (1965). "Wind Waves." P. 676, Prentice-Hall, Englewood Cliffs, New Jersey.

Larras, J. (1965). Probabilité d'apparition des houles dont 1 amplitude dépasse une valeur donneé. *Comptes Rendus, Acad. Sc. Paris* **260**[10], 3125-3128.*

Larras, J. (1969). Les phénomènes aléatoires et l'ingénieur: propriétés additives et lois de probabilités. *Travaux,* 119-129.*

Mayencon, R. (1969). Etude Statistique des observations de vagues. *Cahiers Océanographiques* **21**, 487-501.*

Munk, W. H. (1944). Proposed uniform procedure for observing waves and interpreting instrument records. S.I.O. Wave Project.

*Translations by D. D. Bidde and R. L. Wiegel can be found in "Translations of Four French Papers on Ocean Wave Climate Distribution Functions," *Technical Report HEL-15-1,* Hydraulic Engineering Laboratory, College of Engineering, University of California, May 1970.

Neumann, G., and James, R. W. (1955). North Atlantic Coast wave statistics hindcast by the wave spectrum method. *Coastal Engineering Res. Center Tech. Memo.* **57**, Feb. 1955.

Pierson, W. J., Jr., Neumann, G., and James, R. W. (1955). Practical methods for observing and forecasting ocean waves by means of wave spectra and statistics. *U.S. Navy Hydrographic Office Pub.* **603**.

Ploeg, J. (1971). Wave climate study Great Lakes and Gulf of St. Lawrence. *Techn. Res. Bull. 2-17*, The Society of Naval Architects and Marine Engineers, May, 1971; also *in* "Mechanical Engineering Report MH-107A," Vol. I, National Research Council of Canada, March, 1971.

Saville, T. (1953a). Wave and lake level statistics for Lake Michigan. *Coastal Engineering Res. Center Tech. Memo.* **36**, March, 1953.

Saville, T. (1953b). Wave and lake level statistics for Lake Erie. *Coastal Engineering Res. Center Tech. Memo.* **37**.

Saville, T. (1954). North Atlantic Coast wave statistics hindcast by Bretschneider-Revised Sverdrup-Munk method. *Coastal Engineering Res. Center Tech. Memo.* **55**.

Taylor, G. I. (1938). The Spectrum of Turbulence. *Proc. Royal Soc.*, **A164**, 476-490; Reprinted *in* S. K. Friedlander and L. Topper, "Turbulence, Classic Papers on Statistical Theory," Interscience, New York.

Thompson, E. F., and Harris, D. L. (1972). A wave climatology for U.S. coastal waters. *Preprints, Fourth Annual Offshore Technology Conference,* May 1-3, 1972, Houston, Texas, Paper No. 1693. (To be published).

U.S. Naval Weather Service Command (1970). "Summary of Synoptic Meteorological Observations - North American Coastal Marine Areas," Vol. 3, Federal Clearinghouse for Scientific and Technical Information, Springfield, Virginia.

Williams, Leo C. (1969). *Coastal Engineering Res. Center Tech. Memo.* **30** (Dec. 1969).

Wilson, B. W. (1957). Hurricane wave statistics for the Gulf of Mexico. *Coastal Engineering Res. Center Tech. Memo.* **98**, June, 1957.

CHAPTER 6

Observations on Wind, Tidal, and Density-Driven Currents in the Vicinity of the Mississippi River Delta

Stephen P. Murray

Coastal Studies Institute
Louisiana State University
Baton Rouge, Louisiana 70803

ABSTRACT

Observations were made on the variation with depth of current and water density at an anchor station in 15 m of water east of the Mississippi River delta for 5 days in March, 1970. After a tidal current with a 15 cm sec^{-1} maximum value was isolated and removed, the residual current data showed close correlation with wind and density gradient effects. For more than a day onshore storm winds produced a vertical circulation pattern, with onshore flow near the surface and offshore flow near the bottom. Removal of the wind stress combined with density-gradient-driven currents completely reversed this flow pattern. Currents associated with the storm persisted for at least 3 days, producing on the average a relatively stable current.

INTRODUCTION

During the period March 20-25, 1970, detailed observations of current velocity,

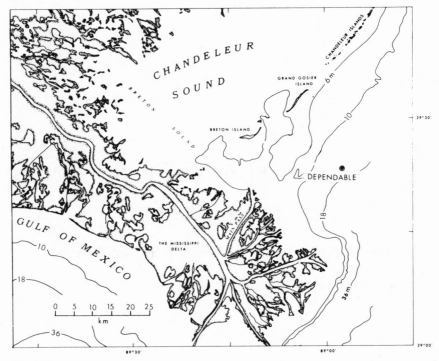

Figure 52.
Location of the observation site, the "Dependable" station. Depth contours are in meters.

wind velocity, and water density structure were conducted continuously from the 63-m-long U.S. Coast Guard cutter "Dependable" at an anchor station 16 km off the eastern flank of the Mississippi delta (Fig. 52). The currents on this section of the continental shelf result from the combined effects of the tide, the winds, and the strong density gradients associated with the discharge of fresh water from the eastern outlets of the Mississippi River. Despite their bearing on the important fisheries and petroleum industries in this area, our knowledge of the currents produced by the combination of the above forces remains poor. The objectives of the present paper are (a) to determine the time-variant velocity profiles of the nontidal component of the observed current and (b) to qualitatively estimate the role played by the density gradient forces in determining the current structure.

DESCRIPTION OF STUDY AREA

The observation site (Fig. 52) is situated about 16 km east of Main Pass of the Mississippi River delta in an area where the local water depth is 15.6 m.

To the north lies the Chandeleur Island chain, which partially impounds the shallow (~4 m deep) Breton-Chandeleur Sound. Bottom topography near the station trends north-northeast to south-southwest.

Tides in the area are mainly diurnal and have an average range of approximately 30 cm. Semidiurnal effects appear in the tide only during times of equatorial tides, when tidal ranges (and tidal currents) are smallest. *In situ* measurements (Murray, Smith, and Sonu, 1970) showed that the local tidal heights agreed quite well with the nearest U.S. Coast and Geodetic Survey reference station, at Pensacola, Florida. Tide range decreased from 40 cm on the first day of observations to 25 cm on the last day of observations.

The winds during the study interval were dominated by the passage of a storm with winds out of the southeast at speeds up to 25 knots which lasted most of March 16 and suddenly dropped off in intensity mid-day of March 17. March 18, 19, and 20 were characterized by a 4-hr lull which was followed by moderate winds (10-20 knots) turning from northeast to southeast.

Density determinations from data taken with a Beckman RS-5 salinometer usually showed a layer of relatively light water with a steep density gradient occupying the upper 4 m of the water column. The density of the heavier water in the lower layer decreased less rapidly with depth. Representative values for the density gradients are $\rho \simeq 1.020$ gr cm^{-3}, $\Delta\rho/\Delta z \simeq 1 \times 10^{-5}$ gr cm^{-4} in the upper layer and $\rho \simeq 1.026$ gr cm^{-3}, $\Delta\rho/\Delta z \simeq 2 \times 10^{-6}$ gr cm^{-4} in the lower layer. These densities correspond roughly to salinities of 27 and 35 ppt, respectively. The lighter surface water is clearly associated with large freshwater sources on the adjacent coasts. The Mississippi River itself is in full flood in the month of March, and so strong density gradients should be expected.

The most significant temporal change in the vertical density structure (Fig. 53) was associated with the storm of March 16 and 17. During the period of strong onshore winds from the southeast (0800 March 16-1000 March 17) the density at the station decreased markedly in the depths below 4 m and also decreased, but to a much lesser extent, above that depth. After the abrupt cessation of these strong winds at noon March 17, the column regained its prestorm density structure within about 15 hr.

As deduced earlier by Longard and Banks (1952), it appears that when vertical density structure is present strong onshore winds drive onshore relatively light surface water which is eventually advected, and probably to a lesser extent diffused, downward, with the effect of reducing the density throughout the column. Accordingly, when the driving force ceases, currents driven by the strong onshore density gradients in the lower layer soon advect back in heavier offshore water, restoring the prestorm balance. Similarly, offshore winds will drive light surface waters away from the coast, resulting in the upwelling of heavier bottom water in replacement. The horizontal density gradient which results in the offshore wind case is mainly confined to the surface layer and

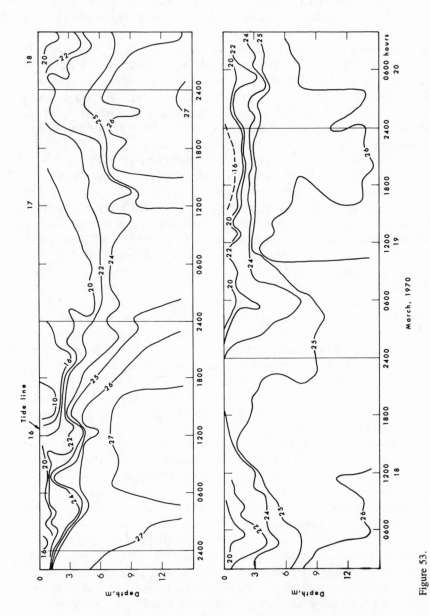

Figure 53.

Serial plot of water density σ_t at the "Dependable" station during the observation interval, March 15-March 20, 1970.

will result in a density current directed offshore. Longard and Banks describe strong onshore winds filling the coastal water prism off Nova Scotia with light surface water down to depths of 90 m.

The remainder of the time history of the density structure in Fig. 53 shows several more such events of considerably less magnitude, but their interpretation is questionable. The complex lateral density gradients sometimes found in this area are described by Chew et al. (1962a).

OBSERVED CURRENTS

On station (water depth of 15.6 m) currents and salinity-temperature were successively observed for a 5-minute period at depths of 1.5, 3, 4.5, 7.5, 10.5, and 13.5 m. Profiles followed each other continuously, taking about 1 hr for completion. The current meter was a Marine Advisers Model Q-15, which is especially useful for current studies in coastal waters because it electronically smooths out "noisy" wave motions while remaining aligned in the mean current direction (see Murray, 1971).

Figure 54 (a) and (b) show the speed and direction of the currents observed at the near-surface and near-bottom levels, 1.5 m and 13.5 m, respectively.

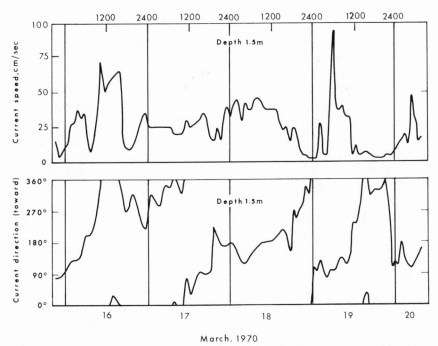

March, 1970

Figure 54a.
Speed and direction of near-surface waters during the observation interval.

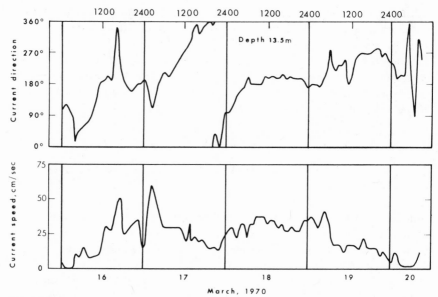

Figure 54b.

Speed and direction of near-bottom waters during the observation interval.

Speeds are generally in excess of 25 cm sec⁻¹, and there are no obvious tidal periodicities present in the data. Also note the persistent southwesterly flow in the bottom waters in the final 48 hr.

For analysis the current measurements at each depth were decomposed into an N-S coordinate system; positive values of u were taken as the current flowing east, and positive values of v, as the current flowing north. These components were then plotted as a function of time and smoothed lines connected the data points. Values of u and v at each depth for each of 96 consecutive lunar hours were then read by a Calma digitizer, which interfaces with the digital computer to furnish the data points for further analysis. In this context hour 1 is 2200 hr March 15.

Figure 55 shows typical u and v components of the velocity profiles for hours 26, 28, and 30. The u component displays a high-speed (>60 cm sec⁻¹) transient near the bottom. These abrupt changes in speed occurred in both components with no apparent preference for depth level and generally lasted 2-3 hr. In the total record five such events are all confined to zones 3 m or less in thickness and exhibit acceleration of ~30 cm sec⁻¹ hr⁻¹.

The v components in Fig. 55 illustrate an example of the reversal from northerly flow in the upper one-quarter of the column to southerly flow in the lower three-quarters of the column which was prevalent during the first 2 days of the total record.

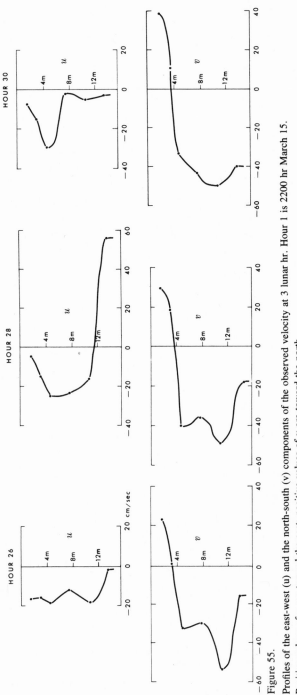

Figure 55.

Profiles of the east-west (u) and the north-south (v) components of the observed velocity at 3 lunar hr. Hour 1 is 2200 hr March 15. Positive values of u are toward the east; positive values of v are toward the north.

During the second half of the record the profiles became considerably smoother but only rarely resembled any simple functional form such as a logarithmic or power law.

TIDAL CURRENTS

To isolate the tidal components (U, V) the 96 lunar hr of u and v speed components at each depth were subjected to harmonic analysis considering only the dominant K1 (diurnal) and M2 (semidiurnal) tides, respectively. Computations showed higher tidal harmonics to be negligible for our purposes.

Figure 56 summarizes the results shown in terms of tidal ellipses at six levels of observation. The average orientation of the ellipses indicates that the tidal wave is propagating toward the northwest, as expected (Defant, 1961, p. 435).

Distortion of the three ellipses nearest the surface is very likely due to the density structure in the upper layer, also discussed by Defant (1961). All depths

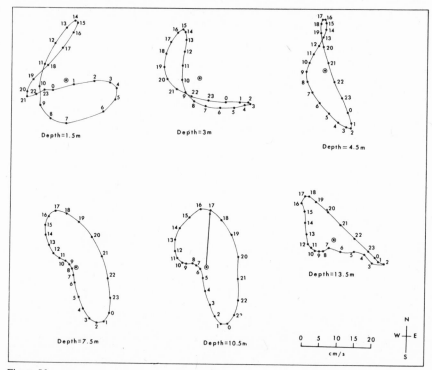

Figure 56.

Tidal current ellipses at the six levels of observation. The current vector at each marked hour is read from the origin to the labeled dot. Mean current has been removed. High tide occurred at hour 17.

except the uppermost show a maximum northward current within 2 hr of predicted high tide, suggesting a progressive wave form for the tide. The interference, however, of the diurnal and semidiurnal tidal waves produces unequal angular velocities of the tidal current. Maximum southward flow occurs only 8 hr after maximum northward flow at high tide, while 16 hr are required for rotation from maximum southward flow to maximum northward flow. It is clear from Fig. 56 that the amplitude of the tidal current is about 15 cm sec^{-1}. For a progressive long wave the maximum tidal current speed is

$$U_{max} = \sqrt{g/h}\ \eta_0$$

With h = 1560 cm, the acceleration of gravity g =980 cm sec^{-2}, and η_0 the average tidal amplitude (one half the range) $\cong$ 20 cm, then $U_{max} \cong$ 16 cm sec^{-1}, in good agreement with the observation.

Except for the density-distorted surface layer, the decelerative phase of the tidal current (Fig. 57a) has north-south component profiles which closely resemble those observed and predicted by earlier workers (Sverdrup et al., 1942) to account for the effect of bottom friction in the tidal current. The accelerative phase (Fig. 57b), however, shows much less resemblance to these simple theoretical models.

NONTIDAL CURRENTS

The nontidal currents with which we are concerned here have two main components, currents driven directly by the wind and currents driven by slopes in the water surface induced by the wind. The question of the presence of a permanent current in the sense of Chew et al. (1962b) remains open.

In deep water far from coasts and with a dynamical balance composed only of the wind stress force and the Coriolis force, Ekman (1905) showed that

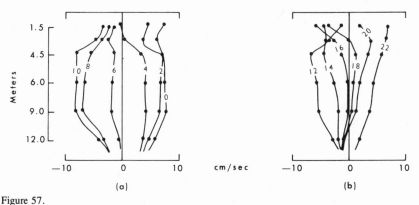

Figure 57.

Profiles of the north-south components of the tidal currents during (a) the decelerative phase of the tidal current and (b) the accelerative phase. Curves are labeled in hours after high tide.

surface currents in the Northern Hemisphere are deflected 45° to the right of the wind direction and that this deflection continues to rotate clockwise as depth increases, forming the well-known logarithmic spiral. In shallow waters far from coasts the same balance of forces again produces a deflection to the right, but the angle between wind and surface current is less than 45°. In water depths of 5-10 m the maximum deflection with depth is only 5-10°.

In the vicinity of coastlines, however, the dynamical balance can be considerably altered, depending mainly on the wind direction with respect to the coastline. For example, an onshore wind will produce a piling up of water against the coast (set-up). The elevation of the water level near the coast induces a seaward-directed pressure gradient acting, in effect, in opposition to the wind force. As the force of the wind is initially a surface effect, it decreases in importance toward the bottom, while the pressure gradient force remains uniform across depth. Thus, according to Ekman, in the case of an onshore wind in shallow water the surface waters will tend to flow with the wind direction while bottom waters tend to flow offshore down the pressure gradient. The flow of bottom water seaward under strong onshore winds was observed by Murray (1970). If the wind is blowing from the right-hand side (while the observer is facing the coastline), the intermediate-depth waters will flow generally toward the left, forming a counterclockwise rotating spiral with increasing depth. If the wind is blowing onshore from the left-hand side (while the observer is facing the coastline), the sense of rotation of the current with depth will be clockwise, the same as in a Coriolis balance, but for quite different reasons.

In nature the presence of other forces in the dynamical balance can alter this simple scheme considerably. For example, if a horizontal density gradient is present in the bottom waters such that lighter water lies near the coastline, the density current would oppose and perhaps reverse the effect of an onshore wind on the current field.

Jeffreys (1923) has further refined Ekman's original ideas in a theoretical analysis which suggests that under gentle winds at an angle to a coastline the water will generally flow parallel to the bottom contours but that under strong winds there will be a decided onshore-offshore component to the flow along the lines suggested by Ekman.

As both Ekman's and Jeffreys' ideas were developed for homogeneous water masses, we might expect to find the density gradient effects observed by Longard and Banks (1952) somehow superimposed on a current system observed in the Mississippi River delta area.

In order to isolate the nontidal current from the present data, the hourly values of the tidal currents were subtracted each hour from the hourly observed currents. Grouping and averaging these data according to well-defined wind episodes clearly brings out the dominant features of the nontidal current profiles which are shown in Fig. 58. Table VI presents the duration, wind speed and direction, and average surface stress for the nine wind episodes.

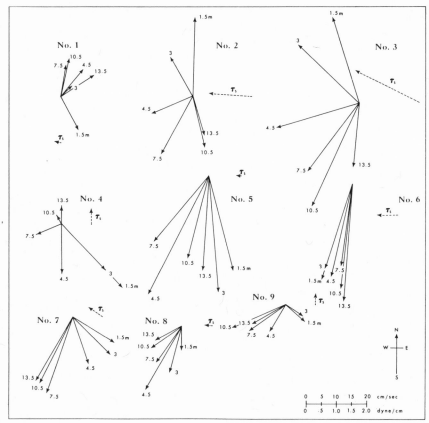

Figure 58.

Profiles of the nontidal currents during the nine wind episodes listed in Table VI. Depths of measurements label each velocity vector. The surface shear stress of the wind is plotted adjacent to each episode.

Episode 1

As the wind stress is quite small during this episode, the current vectors (Fig. 58) are likely the result of prior winds. During the preceding 72 hr the wind first blew from the west at 20 knots for 60 hr and then dropped off sharply to less than 8 knots for the remaining 12 hr. These 20-knot offshore westerly winds very likely significantly increased the density of the surface waters at the station, and the majority of the readjustment of mass had probably already taken place before our observations commenced.

Immediately preceding the episode the wind was moderate from the west, and the bulk of the currents are flowing north-northeasterly along the contours, as predicted by Jeffreys (1923). The surface current appears anomalous, but

Table VI
Average Surface Stress and Corresponding Winds

Episode	Duration (lunar hours)	Wind speed (m sec^{-1})	Wind direction (toward)	Wind stress[a] (dynes cm^{-2})
1	1-9	3.6	274	0.5
2	10-23	9.0	276	1.4
3	24-36	10.9	300	2.3
4	37-52	5.3	001	0.38
5	53-61	4.2	272	0.01
6	62-67	6.3	269	0.59
7	68-83	5.5	300	0.42
8	84-88	3.2	275	0.11
9	89-96	2.7	001	0.29

[a] Calculated from $\tau = 6 \times 10^{-8} W|W|^{3/2}$, Wu (1969).

it is likely associated with density gradient effects because the density is still decreasing during the first 6 hr, as shown in Fig. 53, probably in readjustment to the mass displaced by the preceding strong offshore winds.

Episode 2

During this episode onshore winds from the east to southeast with average surface stresses of about 1.4 dynes cm^{-2} completely alter the local velocity field from that of the previous episode. The velocity vectors (Fig. 58) describe a well-developed counterclockwise-rotating spiral with strong (25 cm sec^{-1}) currents near the surface moving northerly toward the sound and an outward-flowing current ($\sim$15 cm sec^{-1}) near the bottom. The spiral is obviously formed, not by the earth's rotation (as it rotates in the wrong sense), but largely from the interaction between the force vector induced in the water from the wind stress and the pressure gradient force induced in the water column from the water level slope and the redistribution of mass, as discussed earlier.

The directions of the surface and bottom current vectors do not conform exactly to this model, and it is suspected that this is due to strong influence from flow into the Breton-Chandeleur Sound. It is well known in the area that strong easterly and southeasterly winds can raise the water in the sound up to one-half meter or more.

Episode 3

In this episode the surface stress of the wind doubles in magnitude and turns more northwesterly. The spiral structure of the flow remains virtually unchanged (Fig. 58) except for an overall increase in magnitude of the current speeds. The dynamical elements of the velocity structure appear to remain essentially the same: surface waters leaking into the sound, bottom water moving offshore, strongly influenced by the pressure gradient force, intermediate-depth water increasingly affected (as depth lessens) by wind-induced eddy stresses to turn toward the wind direction.

The onshore flow of the surface waters during episodes 2 and 3 produces a decided decrease in density at the station, seen in Fig. 53 as the sharp plunge of the isopycnals from 1200 hr March 16 to 0600 hr March 17.

Episode 4

During this interval the wind stress decreases abruptly by a factor of 4 and turns northerly. Apparently as a response to the effective removal of the surface stress and its resultant set-up against the coast, the bulk of the water column changes direction and flows offshore at considerable speed.

It is interesting to note that the near-bottom waters are actually flowing onshore. The evidence is strong that this is a density-driven current. During a concurrent 18-hr interval the salinity in the lowest 4 m increased by about 3 ppt (Fig. 53). Approximating $\Delta S/\Delta X \simeq \Delta S/u\Delta t$ with $u = 5$ cm sec^{-1} gives $\Delta S/\Delta X \simeq 0.92 \times 10^{-5}$ ppt cm^{-1}. Pritchard (1952) observed an 8 cm sec^{-1} velocity in the bottom layer of a partially mixed estuary driven by a salinity gradient $\Delta S/\Delta X \simeq 0.52 \times 10^{-5}$ ppt cm^{-1}. As our estimate is twice Pritchard's, it is apparent that in this interval the density gradient in the bottom waters is making a significant contribution to the pressure gradient term. The velocities in the upper one-third of the water column appear to be dominated by the change in the water surface slope, while in the near-bottom water the abnormally large density gradient overpowers the surface slope effect and drives the water toward the shore.

Episode 5

During this interval the wind stress is essentially zero and the current (Fig. 58) flows generally to the south-southwest; clockwise rotation of current vectors in the upper half of the column changes to counterclockwise rotation in the lower half. Sufficient information necessary to interpret this current structure dynamically is not available, but it appears, on the basis of experience elsewhere, that we see here a combination of outflow from Breton-Chandeleur Sound and a current produced by a regional slope in the water surface during the storm of March 16-17.

Episode 6

In this interval the wind again has a small but significant stress directed nearly due west. The result is a tight bundle of current vectors (Fig. 58) directed south-southwest and a smooth deflection of the current in the counterclockwise sense as depth increases. This sense of rotation is in accord with the presence of a weak pressure gradient force directed southeasterly (suggesting slight set-up of the water against the coast) superimposed on the larger scale regional slope current discussed in episode 5. Figure 53 also indicates that between 0600 and 1800 hr March 18, which includes the time of this episode, the water density at the station measurably increased (rising isopycnals). This increase in density does not correlate with the onshore-offshore model of Longard and Banks (1952). Rather, it appears that this dense water has advected in from the northeast.

Episode 7

A shift in the wind stress direction to a more northerly position apparently triggers a situation similar to that of episode 4. The decrease of the stress supporting the set-up discussed in episode 6 allows the surface water to run directly offshore from the delta. The sense of rotation of the current with depth is reversed from the previous episode, probably for the same reasons discussed in episode 4, that is, a weak density gradient in the bottom waters directed onshore.

Episode 8

During this episode the wind stress is again directed to the west, as in episode 6, but with only one-fifth the magnitude. Despite the presence, albeit weak, of this westerly directed wind stress, the deflection with depth of the current vectors remains clockwise (Fig. 58), and there is no hint of a seaward-directed pressure gradient.

Episode 9

The wind stress shifts from westerly directed to northerly directed in this interval. Decrease of the stress directed onto the delta shore, as low as it was, apparently again allows the surface waters to flow offshore (Fig. 58) in a readjustment process. The bottom current vectors again rotate in a sense opposite to the surface currents. It is interesting to note that such small changes in the wind stress vector such as from episode 8 to episode 9 could result in current shifts of 45° at 5-8 cm sec^{-1}. The role of density currents in episodes 7, 8, and 9 is likely important but is difficult to evaluate without detailed information on the lateral density gradients.

DISCUSSION AND CONCLUSIONS

In general, the direct observations of currents reported in this study are compatible with the physical oceanographic picture of this area presented by Chew et al. (1962a,b). For 4 out of 5 days, however, our observed currents were associated directly with the southeasterly storm of March 16-17 rather than with a permanent current as envisaged by Chew and his colleagues. Locally, during an average March (Orton, 1964), east and southeasterly winds of 15-20-knot speeds occur nearly 50% of the time. As strong southwesterly setting currents persisted for at least 3 days after the March 16 storm, it seems quite reasonable to assume that a "permanent" current is maintained in the area by the cyclic repetition of storms associated with frontal passages. It would take an unusual (Orton, 1964) east wind like that of March 13 and 14 to drive the inshore currents northeasterly.

The relationship of this inshore current to a permanent eastward flowing Loop current as discussed by Chew et al. (1962b) is not yet clear.

In conclusion, we note the following.

1. Wind effects are of great importance to current structure on the shelf east of the Mississippi River delta. When wind speeds exceed about 20 knots from the southeasterly quadrant, currents driven directly by the wind, in combination with slope currents generated by wind set-up against the coast, produce vertical circulation patterns with onshore flow in the surface waters and offshore flow in the bottom waters.

2. Density structure and the redistribution of mass by winds can alter and even reverse the velocity field produced by the wind effects.

3. Subsequent to a southeasterly storm, strong south to southwesterly setting currents persist for several days, producing in the mean a relatively stable current.

ACKNOWLEDGMENTS

We wish to acknowledge the financial support of this study by the Geography Programs, Office of Naval Research, under Contract No. N00014-69-A-0211-0003, Project NR 388 002, with Coastal Studies Institute, Louisiana State University. The data collection was made possible through the enthusiastic assistance of the United States Coast Guard, Eighth District, and the officers and men of the U.S.C.G. cutter "Dependable." Dr. Choule J. Sonu was especially helpful in helping to identify several of the key points made in this paper.

REFERENCES

Chew, F., Drennan, K. L., and Demoran, W. J. (1962a). On the temperature field east of the Mississippi delta. *J. Geophys. Res.* **67**(1), 271-280.

Chew, F., Drennan, K. L., and Demoran, W. J. (1962b). Some results of drift bottle studies off the Mississippi delta. *Limnology and Oceanography* **7**(2), 252-257.

Defant, A. (1961). "Physical Oceanography." Macmillan (Pergamon), New York.

Ekman, V. W. (1905). On the influence of the earth's rotation on ocean currents. *Ark. Mat. Astro. Fysik* **2**(11), 1-52.

Jeffreys, Harold (1923). The effect of a steady wind on the sea level near a straight shore. *Phil. Mag.* **46**, 114-125.

Longard, J. R., and Banks, R. E. (1952). Wind induced vertical movement of the water on an open coast. *Trans. Am. Geophys. Union* **33**(3), 377-380.

Murray, S. P. (1970). Bottom currents near the coast during Hurricane Camille. *J. Geophys. Res.* **75**(24), 4579-4582.

Murray, S. P. (1971). Turbulence in hurricane-generated coastal currents. *Proc. Twelfth Conf. on Coastal Engineering*, 2051-2068.

Murray, S. P., Smith, W. G., and Sonu, C. J. (1970). Oceanographic observations and theoretical analysis of oil slicks during the Chevron spill, March, 1970. *Coastal Studies Inst., Louisiana State Univ., Tech. Rept.* **87**, 106 pp.

Orton, R. B. (1964). The climate of Texas and the adjacent Gulf waters. *U.S. Dept. of Commerce, Weather Bur.*, 112 pp.

Pritchard, D. W. (1952). Salinity distribution and circulation in the Chesapeake Bay estuarine system. *J. Marine Res.* **11**(2), 133-144.

Sverdrup, H. U., Johnson, M. W., and Fleming, R. H. (1942). "The Oceans, Their Physics, Chemistry, and General Biology." Prentice-Hall, Englewood Cliffs, New Jersey.

Wu, Jin (1969). Wind stress and surface roughness at air-sea interface. *J. Geophys. Res.* **74**(2), 444-455.

CHAPTER 7

Sediment Transport on the Continental Shelf Off of Washington and Oregon in Light of Recent Current Measurements

J. Dungan Smith* and T. S. Hopkins†

ABSTRACT

Prolonged series of direct-current measurements on the central and outer parts of continental shelves are rare, but an understanding of the detailed flow regime in this area is of considerable geological importance. Due to this lack of data, especially in regard to the temporally variable near-bottom velocity field, a direct-current measurement program was initiated at the University of Washington during the summer of 1967. Emphasis was placed upon obtaining a time series of at least a two-year duration at a single location. Data were obtained with current meters located 3 m above the seabed in 50 and 80 m of water. Results indicate that significant sediment transport occurs only during storms, and the near-bottom currents were found to have a substantial offshore component. Calculations based on the current measurements and on analyses of sediment samples taken from the experimental site show that suspended load transport of sediment is extremely important, whereas bedload transport of sediment is not. Although no completely satisfactory theory for suspended sediment transport is available, estimates indicate that a typical winter storm with current speeds up to 60 cm sec^{-1} transports on the order of 6 m^3 hr^{-1} m^{-1} of shelf length; a storm with speeds of up to 70 cm sec^{-1} transports about 15 m^3 hr^{-1} m^{-1} of sediment off of the continental shelf and into deeper water. Such calculations suggest that a severe storm occurring every few years might have more geological significance than a number of less severe storms.

*Department of Oceanography, University of Washington, Seattle, Washington.
†Greek Atomic Energy Commission, Athens, Greece.

INTRODUCTION

Continental shelves are important geological laboratories, for they comprise one of the few physiographic provinces of regional extent in which the imprint of recent processes merges more or less conformably with the stratigraphic record. Furthermore, one of the goals of modern stratigraphy is to relate the geological record to the biological, physical, and chemical processes that were responsible for its salient features, but we can hardly hope to understand these processes as they occurred in ancient seas without being able to explain the predominant and more or less directly observable interactions on recent shelves and without being able to relate these processes to their stratigraphic signature. If the present is the key to the past then it is so by the constancy of physical laws rather than by the similarity of particular environments; nevertheless, the first test of our understanding of geological mechanisms is the successful application of our theories to various modern environments.

An understanding both of sediment-transport mechanics and of the mechanics of the sediment-transporting flows is essential in elucidating the physical characteristics of a particular environment, whether ancient or modern. In this paper an extremely primitive attempt to integrate surface sediment data and current measurements will be made. Neither the sediment data nor the current data are sufficient for a complete examination of the system. Moreover, our understanding of the mechanics of sediment transport and of sedimentary geochemistry is in an extremely elementary state and as such it fails us in critical areas.

In particular the investigation is of the wind-driven currents and sediment transport on the central continental shelf northwest of the Columbia River mouth. The area, shown in Fig. 59, is of particular importance because it is within this region that the silts carried to the sea by the Columbia River are transported offshore by wind-driven coastal currents. The further west of the two black dots on Fig. 59 is the location of a sea bouy that is rented from the Coast Guard by the University of Washington to protect our current-meter arrays. The easterly black dot on Fig. 59 is the location of a mooring from which a current-meter record to be discussed below was recovered. The sediment data to be described are centered around these two moorings.

BASIC PRINCIPLES OF SEDIMENT TRANSPORT

In a turbulent flow it is useful to separate the velocity, pressure, and sediment-concentration fields into mean and fluctuating parts. In this section a bar over a variable will be used to denote its time-mean part and a prime over a variable will denote a purely fluctuating quantity. Moreover, the bar will represent a running average over a suitable interval such as 20 min. In a horizontally uniform flow the mean stress $\bar{\tau}$ is related to the mean velocity gradient $\partial\bar{u}/\partial z$ as follows:

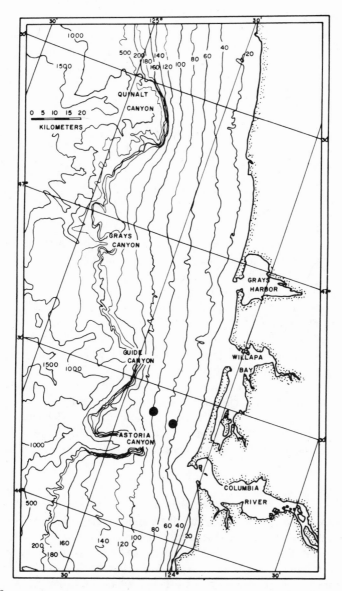

Figure 59.
Bathymetry of the Washington continental shelf. The farther offshore of the two black dots situated northwest of the Columbia River mouth represents the location from which most of the current-meter data, including record O, was obtained.

$$\bar{\tau} = \rho K \left[\frac{\partial \bar{u}}{\partial z} \right] \tag{1}$$

where ρ is the fluid density and $K(z)$ is a kinematic momentum diffusion coefficient or "eddy viscosity." The stress at the boundary $\bar{\tau}_b$ is given by:

$$\bar{\tau}_b = \rho K \left[\frac{\partial \bar{u}}{\partial z} \right]_{z=b} = \rho u_*^2 \tag{2}$$

where u_* is called the shear velocity or friction velocity. If the mean flow is steady or varies with periods of several hours or more then the flow speed in the vicinity of the bed is:

$$\bar{u} = \frac{u_*}{k} \ln \frac{z}{z_0} \tag{3}$$

where $\bar{u}$ is the current speed at level z, $k \cong 0.40$ is von Kármán's constant, and z_0 is a constant of integration with the dimension of length. However, z_0 can be thought of as a measure of the flow speed at some reference level, the flow at that level increasing as z_0 decreases.

If δ represents the thickness of the region, called the viscous sublayer, in which the viscous stress is larger than the turbulent or Reynolds stress and we examine the mean flow at $z > \delta$ then the mean shear $\partial \bar{u}/\partial z$ at level z must be a function of the local stress, the fluid density, and the distance from the boundary; that is, $\partial \bar{u}/\partial z = F_1(\bar{\tau}, \rho, z)$. Moreover, if h is the water depth and we require $z \ll h$ then $\bar{\tau} \cong \bar{\tau}_b$ and $\partial \bar{u}/\partial z = F_1(\bar{\tau}_b, \rho, z)$. Only one unique dimensionless number is possible using these variables and that is $(\partial \bar{u}/\partial z)(z/u_*)$. However, the theory of dimensional analysis then requires $F_2 \left[(\partial \bar{u}/\partial z)(z/u_*) \right] = 0$ or $(\partial \bar{u}/\partial z)(z/u_*) = $ constant. Calling the constant $1/k$ and integrating with respect to z leads to Eq. (3), which has been experimentally verified numerous times. The kinematic eddy viscosity consistent with Eqs. (2) and (3) is:

$$K = k u_* z \tag{4}$$

Nikuradse (1933) first confirmed the validity of Eq. (3) and provided an empirical means of relating z_0/k_s and the roughness Reynolds number $\bar{u}_* k_s / \nu$, where k_s is the median grain diameter of the bed material and ν is the kinematic viscosity of the fluid. If the flow is steady and uniform then the thickness of the viscous sublayer is given by $\delta = 11.6 \nu/u_*$. For the flow to be independent of the nature of the roughness elements, then they should be deeply submerged in the viscous sublayer; that is, $\delta \gg k_s$ or $u_* k_s/\nu \ll 11.6$. Under these conditions, δ is the only length that can be associated with z_0, so dimensional analysis requires that $z_0 \propto \delta$. Nikuradse's experiments showed that for this case, called hydraulically smooth flow, $u_* k_s/\nu < 3$ and $z_0 = 104\delta \cong \nu/(9u_*)$.

On the other hand, if $k_s \gg \delta$ then the viscous sublayer is wrapped around the surfaces of the sand grains and can be neglected. In this case, $u_* k_s/\nu \gg 11.6$ and dimensional analysis requires that $z_0 \propto k_s$. Nikuradse's experiments showed that for this case, called hydraulically rough flow, $u_* k_s/\nu > 90$ and $z_0 = k_s/30$. For $3 < u_* k_s/\nu < 90$ the flow is called hydraulically transitional.

In the case of hydraulically smooth flow the exact nature of the roughness is not particularly important as long as all of its scale lengths are small relative to δ; however, in the case of hydraulically rough flow this is not the case, and Nikuradse's results are valid only for roughness comparable to smooth beds of well sorted sand, that is, for flows characterized by only one length scale, namely the sand grain diameter. These results cannot be used for poorly sorted gravel beds, for rippled beds, or for beds with obstacles on them because in all of these cases k_s does not give a unique description of the boundary geometry.

Nikuradse's data have been criticized by various experimenters, some of whom have obtained different results for hydraulically transitional and hydraulically rough flow; however, in all of the cases examined by the writers, the strict criterion of steady uniform flow characterized by a single length scale is best satisfied by the Nikuradse experiments, and the writers see no reason to doubt his results for hydraulically rough and transitional flow when properly applied. For moderately sorted but smooth beds, Einstein (1950) suggests that k_s can be equated to the grain diameter for which 35% of the sediment sample is coarser.

In a multiphase flow such as in a sediment-bearing river, the mean sediment velocity and sediment concentration can be defined at a point for all particle sizes. Although this approach appears at first glance to be questionable, it can be justified using a fundamental statistical approach; however, the point of this paper is not to examine the philosophical basis of sediment transport theory, but rather to use generally accepted concepts to interpret continental shelf processes. Under the formalism described directly above, the equation for conservation of mass for an incompressible phase denoted by n is:

$$\frac{\partial \epsilon_n}{\partial t} + \nabla \cdot \left[\underset{\sim}{u}_n \epsilon_n \right] = 0 \tag{5}$$

where ϵ_n is the instantaneous volume concentration of phase n and $\underset{\sim}{u}_n$ is the instantaneous velocity of phase n. The volume concentration is defined such that

$$\sum_{n=1}^{N} \epsilon_n = 1 \tag{6}$$

where N is the number of phases and N-1 is the number of particle size classes present. Equations (5) and (6) can be averaged in time to give:

$$\frac{\partial \bar{\epsilon}_n}{\partial t} + \nabla \cdot \underset{\sim}{u}_n \bar{\epsilon}_n = - \nabla \cdot \overline{\underset{\sim}{u}'_n \epsilon'_n} \tag{7}$$

as long as the time constant for the temporal variation of $\bar{\epsilon}_n$ is large relative to the averaging interval. The term $\overline{u_n' \epsilon_n'}$ is the volume flux of phase n due to turbulent diffusion and can be approximated as follows by assuming Fickian type diffusion:

$$\overline{\underset{\sim}{u}'_n \epsilon'_n} = - K_n \nabla \bar{\epsilon}_n \tag{8}$$

Therefore, Eq. 7 can be rewritten as:

$$\frac{\partial \bar{\epsilon}_n}{\partial t} + \nabla \cdot \underset{\sim}{u}_n \bar{\epsilon}_n = \nabla \cdot K_n \nabla \bar{\epsilon}_n \tag{9}$$

Moreover, for horizontally uniform flow

$$\frac{\partial \bar{\epsilon}_n}{\partial t} + \frac{\partial}{\partial z} \left[\bar{w}_n \bar{\epsilon}_n \right] = \frac{\partial}{\partial z} K_n \frac{\partial \bar{\epsilon}_n}{\partial z} \tag{10}$$

If the sediment sample can be treated as a single size class then only two equations are necessary. They are:

$$\frac{\partial \bar{\epsilon}_s}{\partial t} + \frac{\partial}{\partial z} \bar{w}_s \bar{\epsilon}_s = \frac{\partial}{\partial z} K_s \frac{\partial \bar{\epsilon}_n}{\partial z} \tag{11a}$$

$$\frac{\partial \bar{\epsilon}_s}{\partial t} - \frac{\partial}{\partial z} \left[\bar{w}_s (1 - \epsilon_s) \right] = \frac{\partial}{\partial z} K_w \frac{\partial \bar{\epsilon}_s}{\partial z} \tag{11b}$$

where $\bar{\epsilon}_w = 1 - \bar{\epsilon}_s$ has been used in the latter case.

In suspended-sediment problems it is usual to assume that the sediment velocity is equal to the fluid velocity minus the settling velocity of the sediment:

$$\underset{\sim}{u}_s = \underset{\sim}{u} - \underset{\sim}{w}_s \tag{12a}$$

so

$$\underset{\sim}{\bar{u}}_s = \underset{\sim}{\bar{u}} - \underset{\sim}{w}_s \tag{12b}$$

and by subtraction

$$\underset{\sim}{u}'_s = \underset{\sim}{u}' \tag{12c}$$

Combining Eqs. (11a), (11b), and (12b) gives:

$$\frac{\partial \bar{\epsilon}_s}{\partial t} + \frac{\partial}{\partial z} (\bar{w}\epsilon_s - w_s \bar{\epsilon}_s) = \frac{\partial}{\partial z} K \frac{\partial \bar{\epsilon}_s}{\partial z} \tag{13a}$$

$$\frac{\partial \bar{\epsilon}_s}{\partial t} + \frac{\partial}{\partial z} (\bar{w} \, \bar{\epsilon}_s - \bar{w}) = \frac{\partial}{\partial z} K \frac{\partial \bar{\epsilon}_s}{\partial z} \tag{13b}$$

At this point it should be noted that the assumption represented by Eq. (12a) requires the diffusion coefficient for the sediment (K_s) to be the same as the diffusion coefficient for the water (K_w) because $u_s' \equiv u_w'$ so $\overline{u'_s \epsilon_s'} \equiv \overline{u_w' \epsilon_s'}$ and $K_s \nabla \bar{\epsilon}_s \equiv K_w \nabla \bar{\epsilon}_s$ or $K_s = K_w$. Subtracting Eq. (13b) from Eq. (13a) gives $\partial(\bar{w} - w_s\bar{\epsilon}_s)/\partial z = 0$ or $\bar{w} = w_s\bar{\epsilon}_s + $ const. However, for most problems $\bar{w} = 0$ when $\epsilon_s = 0$ requiring that const. $= 0$ and permitting Eq. (13a) to be written as:

$$\frac{\partial \bar{\epsilon}_s}{\partial t} - \frac{\partial}{\partial z} \, w_s \bar{\epsilon}_s (1 - \bar{\epsilon}_s) = \frac{\partial}{\partial z} K \frac{\partial \bar{\epsilon}_s}{\partial z} \tag{14}$$

For the low-frequency motions to be considered in this paper the use of Eq. (3) introduces a negligible error so we can introduce a generalized version of Eq. (4) in the form:

$$K = ku_* h \, f(\xi) \tag{15}$$

where $\xi = z/h$ and rearrange Eq. (14) to get

$$\frac{\partial \bar{\epsilon}_s}{\partial t} = ku_* \frac{\partial}{\partial z} \left[\frac{w_s}{ku_*} \bar{\epsilon}_s (1 - \bar{\epsilon}_s) + hf(\xi) \frac{\partial \bar{\epsilon}_s}{\partial z} \right] \tag{16}$$

or we can rearrange Eq. (16) to get

$$\frac{\partial \bar{\epsilon}_s}{\partial t} = \frac{ku_*}{h} \left\{ \left[\frac{\partial f(\xi)}{\partial \xi} + \left[\frac{w_s}{ku_*} \right] \left[1 - 2\bar{\epsilon}_s \right] \right] \frac{\partial \bar{\epsilon}_s}{\partial \xi} + f(\xi) \frac{\partial^2 \bar{\epsilon}_s}{\partial \xi^2} \right\} \tag{17}$$

Taking $f(\xi) = \xi$ for $\xi < 0.2$ and $f(\xi) = (\beta k)^{-1}$ where $\beta = 15.6$ for $\xi > 0.2$ gives results consistent with the commonly used logarithmic and parabolic channel flow equations. Moreover, if a better representation of $K(\xi)$ were possible for a given flow of interest the above formulation would permit it to be used.

In the latter equation if $f(\xi) = \xi, \upsilon = \int (ku_*/h) \, dt$, and $p_* = p (1 - 2\bar{\epsilon}_s)$ then

$$\frac{\partial \bar{\epsilon}_s}{\partial \upsilon} = (1 + p_*) \frac{\partial \bar{\epsilon}_s}{\partial \xi} + \xi \frac{\partial^2 \bar{\epsilon}_s}{\partial \xi^2} \tag{18}$$

where $p_* \ll 1$ when $p = w_s/(ku_*) \ll 1$ because $|1 - 2\bar{\epsilon}_s| < 1$. For the case of steady flow ($\partial \bar{\epsilon}_s/\partial t = 0$) Eq. (16) can be integrated to give

$$\alpha_0 + p\bar{\epsilon}_s (1 - \bar{\epsilon}_s) + hf(\xi) \frac{\partial \bar{\epsilon}_s}{\partial z} = 0 \tag{19}$$

where α_0 is a constant of integration. However, if there is no flux through the free surface, then the second two terms sum to zero at that surface and $\alpha_0 = 0$. Here the advective sediment flux $w_s(1-\bar{\epsilon}_s)\bar{\epsilon}$, due to the settling sediment must exactly balance the diffusive flux, $ku_*hf(\xi)$ $(\partial\bar{\epsilon}_s/\partial z)$. Equation 19 can be integrated to give:

$$\frac{\bar{\epsilon}_s}{1-\bar{\epsilon}_s} = \left[\frac{\bar{\epsilon}_a}{1-\bar{\epsilon}_a}\right]\left[\exp\int_{\xi_a}^{\xi}\frac{-p}{f(\xi)}d\xi\right] \tag{20}$$

which becomes

$$\frac{\bar{\epsilon}_s}{1-\bar{\epsilon}_s} = \left[\frac{\bar{\epsilon}_a}{1-\bar{\epsilon}_a}\right]\left[\frac{a}{z}\right]^p \tag{21}$$

for $f(\xi) = \xi$. Here $\bar{\epsilon}_a$ is the concentration at some reference level denoted by $z = a$.

When the parameter, $p = w_s/(ku_*)$, is large, then Eq. 21 shows that most of the sediment is found in a layer on the order of a few grain diameters thick near the boundary. Under these conditions the sediment transport is said to be as bedload. Assuming that $a = 2D$ is the appropriate reference level for $\bar{\epsilon}_a$ where D is the grain diameter of the sediment and taking $p = 2$, $z = 10D$ gives $(\bar{\epsilon}_s/\bar{\epsilon}_a) = 0.04$. Therefore, for $p \leq 2$ most of the sediment is transported in a layer near the bed less than 10 grain diameters thick and should be considered bedload. For $p = 1$, $a = 2D$ and $(\bar{\epsilon}_s/\bar{\epsilon}_a) = 0.04$ the layer thickness is 50D and for $p = 0.5$ the layer thickness is 1250D. In practice sediment transport for $p < 0.8$ should be considered as suspended load; whereas, sediment transport for $0.8 < p < 2$ may be considered as bedload, suspended load or transitional depending upon the nature of the problem. For instance, if the suspended sediment transport range is to be associated with high transport rates then it is best to ignore most of the transition region, but if the suspended sediment transport range is to be associated with a particular mechanism of transport then most of the transition range should be included. As $p \rightarrow 0$, Eq. (21) shows that the concentration becomes uniform in z and that the sediment is transported as wash load.

When $|\bar{\epsilon}_s| << 1$ and $p_* = $ const. Eq. (18) can be separated by assuming $\bar{\epsilon}_s = T(t)Z(z)$ where $T(t)$ is a function of t only and $Z(z)$ is a function of z only to give:

$$\frac{dT}{dt} = -ku_*\lambda T \tag{22a}$$

and

$$\xi\frac{d^2Z}{d\xi^2} + (1+p_*)\frac{dZ}{d\xi} + \lambda Z = 0 \tag{22b}$$

A solution to Eq. (22a) is

$$T = T_0 \exp \int - ki\lambda u_* \, dt \cdot \tag{23}$$

where T_0 is a constant of integration, λ is the separation constant, and $i = \sqrt{-1}$. Equation (22b) can be transformed into Bessels' equation and has a solution of the form

$$Z = a_1 (2\sqrt{i\lambda\xi})^p * J_{p_*}(2\sqrt{i\lambda\xi}) + a_2 (2\sqrt{i\lambda\xi})^p * Y_{p_*}(2\sqrt{i\lambda\xi}) \tag{24}$$

where a_1 and a_2 are constants, and J_{p*} and Y_{p*} are Bessel functions of the first and second kinds respectively, and where both are of order p_*. These functions are described and tabulated by Abramowitz and Stegun (1964, p. 358 ff.).

Using Fourier analysis Eqs. (23) and (24) can be combined and manipulated to satisfy the boundary conditions that there is no flux through the free surface, $(\partial \bar{\epsilon}_s/\partial z) = (p/h)(1 - \bar{\epsilon}_s) \, \bar{\epsilon}_s \cong (p/h) \, \bar{\epsilon}_s$ at $t = h$, that there is a constant sediment flux at some level very close to the boundary under equilibrium conditions that the initial concentration field is zero, and that the final concentration field is that given by Eq. (21) when $\bar{\epsilon}_s \ll 1$ is assumed. However, it is just as easy and more useful to solve numerically an equation that replaces (22b) when $f(\xi)$ is an experimentally specified function of ξ. In either case the suspended sediment concentration field is specified only to the extent that the sediment flux at the lower boundary is known, and at present there is no satisfactory way to determine this flux other than to measure the concentration as a function of time at one point in the field.

The proper way to evaluate the suspended sediment transport on the continental shelf is to find the solution to Eq. (18) as outlined above for a measured eddy viscosity profile and a measured near-bed sediment concentration and to evaluate the integral

$$Q_s = \int_{z_0}^{h} \underset{\sim}{u}\bar{\epsilon}_s dz \tag{25}$$

where $\underset{\sim}{u}$ is the measured water velocity profile. However, the data needed to carry out this procedure are not presently available.

TEXTURE AND HYDRAULIC PROPERTIES
OF THE SHELF SEDIMENTS

The general textural characteristics of the surface sediments on the Washington continental shelf have been discussed by Gross, McManus, and Ling (1967),

Kelley and McManus (1969, 1970), and McManus (1972). In addition, Ballard (1964) has described the nature of the nearshore sediments and Harman (1972) has described the distribution of the biogenic sediments and pumice on the shelf. Detailed textural data are available at the University of Washington for over 1000 samples collected at 450 stations on the continental shelf between Tillamook Head, Oregon and Grays Harbor, Washington. Samples were obtained on BROWN BEAR Cruise number 333 as described by Kelley and McManus (1970). The latter authors have treated the size distribution data statistically and have found that the most diagnostic textural parameters are the percent sand, percent silt, and the median grain diameter. Contour maps of the percent clay as well as the latter two parameters are reproduced from McManus (1972) as Figs. 60, 61, and 62. The area of primary concern in the present paper is that between the mouth of the Columbia River and the mouth of Willapa Bay (see Fig. 59), as both current and sedimentological data are available in this region. Figure 60 shows that in this area there is generally less than 10% clay. Moreover, reanalysis of the sediment data throws some doubt on the validity of the high clay values found on the central shelf near 46°22′N, as does the fact that Kelley and McManus (1969) found the variance in percent clay between samples from the same station to be essentially the same as the variance between stations.

According to McManus (personal communication) the sediments with high clay content to the south of the Columbia River are associated with outcrops of Tertiary material and are thought to be caused by erosion of these local clay sources. The cause of the patches of high clay content west of the mouth of Willapa Bay is not known but may be associated with deposition of clay from some northern fluvial source during summer conditions. In this area the clay content is high enough to armor the surface and to prevent erosion of surface sediments even during severe winter storms. However, for sediment distributions of the type found on the Washington shelf, from 20 to 40% clay is necessary before the surface can be successfully protected against erosion, so the armoring process can be ignored over the rest of the region of interest.

Figure 61 shows that there is generally less than 10% silt in the first 15 km from the beach, that is, in water less than 50 m deep. This figure also shows a band of sediment with high silt content situated more or less on the central shelf, but trending from the mouth of the Columbia River to the north-northwest. On the central shelf in this region the maximum silt content ranges from 40% to somewhat in excess of 60%; however, at the shelf edge north and south of the Willapa Canyon the silt content drops to less than 20%. On the other hand, at the head of Willapa Canyon the silt content rises to its central shelf value of 60%. Just south of the Columbia River mouth no high-silt band is found on the central shelf, suggesting that most of the silt emanating from the river mouth is transported to the north. Therefore, careful examination of Fig. 61 suggests that the Columbia River silts are being transported to the north and offshore in a band of increasing width that trends 25° to the west of north. Along the axis of the silt band the

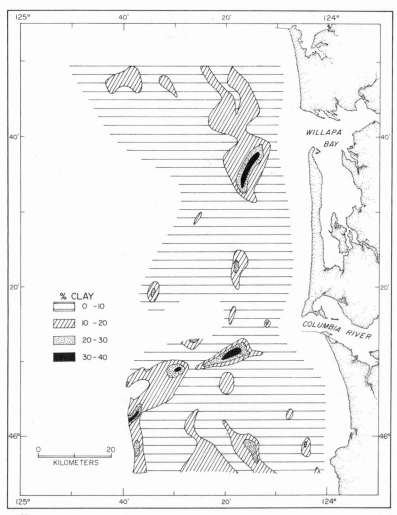

Figure 60.

Map showing the percent clay in the surface sediments on the southern Washington continental shelf. Note that the sediments on most of the shelf north of the Columbia River mouth contain less than 10% clay and that the only area north of the river mouth that has over 40% clay is just south of the entrance to Willapa Bay. (After McManus, 1972.)

relative amounts of silt increase from 40% to 60% or 70% in about 100 km. The Willapa Canyon cuts into the shelf and, therefore, into the silt band around 46°36′N probably diverting some of the silt from the shelf to the slope.

Figure 62 shows the median phi distribution for the shelf sediments. Here the band of high silt content can be seen as a strip of finer material trending 30° to the west of north across the shelf north of the Columbia River mouth. Within 10 km of the coast and over a large area south of the river mouth

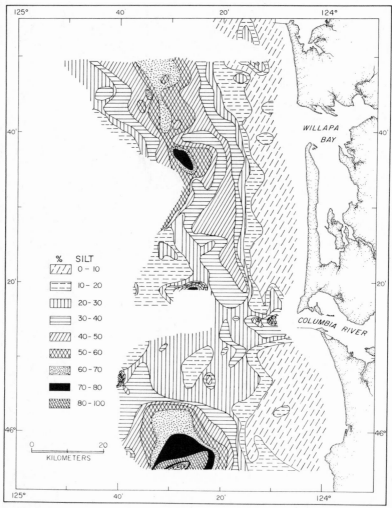

Figure 61.
Map showing the percent silt in the surface sediments on the southern Washington continental shelf. Note the band of high silt content trending to the north northwest from the Columbia River mouth and the band of low silt content near the coast. Also note that Guide Canyon cuts into the high silt band. (After McManus, 1972.)

lies a region covered by sands with median phi of 2.0 to 3.0. At least near the coast this is the area affected by shoaling waves and other nearshore processes.

Somewhat less confidence can be placed in the interpretation of the median phi map than in the percent silt map for reasons to be described below. Nevertheless, the band of high silt content to the north of the river, the zone of 2.0 to 3.0 phi sands associated with nearshore processes, and the low silt area to the south of the Columbia River mouth are all visible.

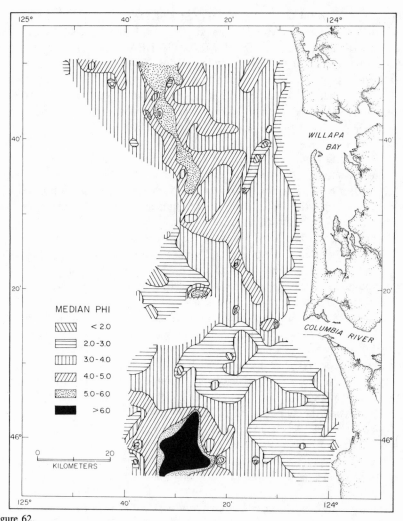

Figure 62.

Map showing the distribution of median phi size of the sand and silt on the southern Washington continental shelf. Note the band of finer material trending north northwest across the central shelf to the north of the Columbia River mouth and the band of 2.0 to 3.0 ϕ sand next to the coast and over a large section of the shelf south of the river mouth. (After McManus, 1972.)

Due to the large number of samples involved, detailed analysis of R. V. BROWN BEAR Cruise 333 sediment data (such as resulted in Figs. 60, 61, and 62) was based on the percent sand, percent silt, percent clay, and the statistical moments of the size distribution data. Moreover, these moments were obtained from the cumulative frequency distributions for each sample using Inman's graphical method (Shepard, 1963, p. 107). Inasmuch as sediment transport is a highly nonlinear phenomenon, such an analysis, even for a unimodal distribution, is useful for sediment transport calculations only if the sediment is so well sorted that it can be treated as a single size class. As this was not expected to be the case for the BROWN BEAR 333 samples, the raw data for the section of the continental shelf just north of the Columbia River mouth were plotted as size distribution curves, and examined station by station. It was found that a distribution curve representative of each station could be constructed. All of these curves, like the primary ones, were polymodal, thus shedding some doubt on the applicability of the Inman method or any other simple statistical method in analyzing them. Moreover, this is probably the reason that Kelley and McManus (1970) found the higher statistical moments not to be particularly good indicators of regional variation. Typical frequency distribution curves for several stations are shown in Fig. 63. Fortunately, a very systematic relationship among the amplitudes of the three primary modes was discovered and is shown in Fig. 64.

At almost every station a well-sorted sand with a median diameter of 0.088 mm (3.5 ϕ) and a second well-sorted sand with a median diameter of 0.125 mm (3.0 ϕ) were found in addition to a broad silt mode with variable median diameter. On the outer shelf the 3.0 ϕ sand dominated over the 3.5 ϕ sand; whereas, on the inner shelf the 3.5 ϕ sand was the dominant sand (Fig. 64). On the central shelf the silt comprised 50% of the sample; whereas, it comprised only 25% of the sample on the outer shelf and less than 10% of the sample on the inner shelf. No systematic pattern was found in the percentage of clay. Moreover, there was a disturbing tendency for the samples with large amounts of clay to come from stations for which other samples had no clay. No explanation for this correlation was found, leaving the dilemma of whether to believe the results or discard them. Fortunately, in the area of interest most of the samples contained less than 10% clay so that it is not a significant factor in the sediment transport and will be ignored in the rest of this paper.

The polymodal nature of the sediment samples is strongly suggestive of several transport mechanisms, each important at some time during a given time interval that includes one or more storms. Inasmuch as the BROWN BEAR samples represent a mixture of the upper centimeter of the shelf sediment it is not possible to determine whether or not the various modes represent separate layers. However, it is very likely that they do.

McManus (personal communication) feels that the sediment at the shelf edge on the north side of Astoria Canyon, like the sediment on the outer shelf to the south of this canyon, is relict and represents the material left on the shelf after the postglacial rise in sea level. This idea is very plausible, but there is no indication that the material comprising the 3.0 ϕ mode in this region is different from the material comprising the 3.0 ϕ anywhere else on the Washington shelf. Therefore, it is possible that the 3.0 ϕ sand, wherever it is found, represents relict material. The increasing importance of this material with distance from the coast as shown in Fig. 64 then would be due to the absence of silt

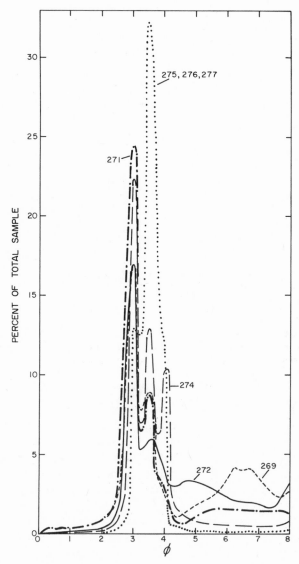

Figure 63.

Typical grain size frequency distributions for a transect across the continental shelf near the Coast Guard Buoy. The relative positions of the stations are shown on Fig. 64. Note that these samples all show 3.0 ϕ and 3.5 ϕ modes. In addition, samples from the central shelf show a broad distribution of silt.

on the shelf edge and the decreasing importance of the 3.5 ϕ sand with distance from its source, which presumably is the nearshore zone.

In order to understand the hydraulic behavior of the 3.0 ϕ and 3.5 ϕ sands

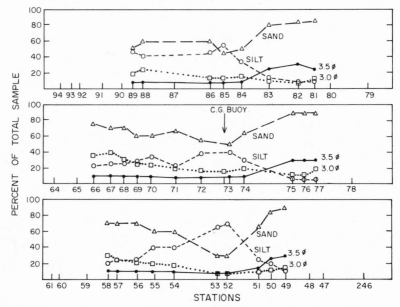

Figure 64.

Variation of pecent sand, percent silt, percent 3.0 ϕ, and percent 3.5 ϕ on three transects on the continental shelf. The right-hand edge of the figure is about half way between stations 246 and the coast, whereas the left-hand edge of the figure is very close to the shelf edge. This figure is based on the average sand-silt size distribution for a given station and cannot be compared directly to Fig. 63 which is based upon typical samples and includes the clay fraction.

and the silt it is necessary to calculate the settling velocity, the critical shear velocity, and the ratio of these two parameters. In the case of the sands, the calculations are straightforward and the results are given in Table VII. In the case of the silts the appropriate median diameter is not known, so that it is difficult to determine the proper settling velocity, and there is no good theory from which the critical shear stress for silts can be obtained. On the other hand, the fact that there is no well-defined, most-probable silt size class suggests that no selective transport process is occurring for these materials and that a single critical shear stress probably represents the whole size range.

For $0.1 < (u_*) D/\nu < 1.0$ Shields' diagram (Graf, 1971, p. 96) can be approximated by:

$$\frac{\tau_c}{(\rho_s - \rho)gD} = 0.1\left(\frac{\nu}{(u_*)_c D}\right) \tag{26}$$

or

$$(u_*)_c = \left[0.1\left(\frac{\rho_s - \rho}{\rho}\right)g\nu\right]^{1/3} \tag{27}$$

Table VII
Settling Velocity, Critical Shear Velocity, and $p_c = w_s[K(u_*)_c]^{-1}$ [a]

ϕ	D	w_s	$(u_*)_c$	p_c
3.0	0.0125	1.00	1.3	1.92
3.5	0.0088	0.49	1.3	0.95
4.0	0.0063	0.25	1.3	0.48
5.0	0.0031	0.061	1.3	0.11
6.0	0.0016	0.016	—	—

[a] In cgs units for various grain diameters assuming a sediment density of 2.65 g/cm^3 and a water temperature of 7°C.

The temperature of the bottom water on the continental shelf averages around 7°C; thus, the kinematic viscosity of this water is approximately 1.4×10^{-2} cm^2 sec^{-1}, and quartz sand with $10^{-2} < D < 10^{-3}$ cm has a critical shear velocity of $(u_*)_c \cong 1.3$ cm sec^{-1}. Therefore, in the case of the 3.0 ϕ and 3.5 ϕ sands in particular and the silty sands found on the continental shelf in general, a critical shear velocity of 1.3 cm sec^{-1} appears to be a suitable approximation.

Taking this value and noting that for silt $w_s < 0.3$ cm sec^{-1} gives $p_c < 0.6$ which means that the silt can never be transported as bedload. Therefore, if the shear velocity exceeds 1.3 cm sec^{-1} the silt will be eroded and transported as suspended load. The sands, on the other hand, can be transported as bed load and the 3.0 ϕ sand will be if u_* is between 1.3 and 2.7 cm sec^{-1}. The 3.5 ϕ sand will be transported as bed load if u'_* is between 1.3 and 1.4 cm sec^{-1}. The important question to be answered now is whether these sands can ever be transported as suspended load, as that is a much more efficient transport process. If they were able to move only as bed load then each particle would have a net movement of about 10^{-2} cm sec^{-1} when transport was taking place. Assuming that some sediments were in motion at least 3% of the year then the surface sands would move on the order of 100 m year^{-1} or 10 km century^{-1}. Under the best of conditions this estimate could be increased only by an order of magnitude. On the other hand, if the sands could be transported as suspended load their net motion might be as high as 10 km year^{-1}.

Due to the polymodal nature of the sediment samples on the continental shelf, choosing a characteristic grain roughness for the bed is not a straightforward procedure. However, the high degree of sorting of the two sand modes suggests that the median diameter of the coarsest mode should prove to be a good choice. Taking $k_s = 0.012$ cm and noting that the critical shear velocity for this material is about 1.3 cm sec^{-1} gives a critical roughness Reynolds number of $(R_*)_c = (u_*)_c k_s/\nu = 1.1$. Moreover, if $u_* < 2.7(u_*)_c$, then $R_* < 3$ so the flow is hydraulically smooth and not dependent on the exact value of k_s anyway.

At this point it should be emphasized that the shear stresses from which the shear velocities used in this paper are calculated, are local values of skin friction and do not include the form drag due to boundary irregularities. This approach is taken because only the skin friction is important in the erosion and transport of sediment. Many other workers such as Sternberg (1972) use the total boundary shear stress (skin friction plus form drag averaged over a suitable area of the boundary) and assume that this parameter is related to the skin friction in a simple, single-valued way. This is done primarily because the total boundary shear stress is what is measured in most field experiments, and because the measurement of skin friction is a difficult task, even in the laboratory, for sediment-bearing flows. Nevertheless, the first author's theoretical and experimental studies of flow over topography (Smith 1969, 1970) have lead him to doubt whether there is either a simple or a single-valued relationship between skin friction and total boundary shear stress.

In the case of the discrimination between hydraulically rough and hydraulically smooth flows, it should be remembered that Nikuradse's results (discussed above) from which the criteria for hydraulically rough and smooth flow were derived are valid only for flow over flat sand beds, or, for the zero-order flow over rippled beds (Smith, 1969). In particular, they are applicable to the present formulation of the problem, but not to the situation where the total shear stress is used instead of the skin friction.

If the amplitudes relative to the depth of water and slopes of the boundary topography are small, then the skin friction on the boundary can be written in the form of a convergent series, $\tau_s = (\tau_s)_0 + (\tau_s)_1 + (\tau_s)_2 + (\tau_s)_3 + \ldots$. The $(\tau_s)_0$ term is the flat-bed or zero-order skin friction; whereas, the other terms are respective functions of the slope, the slope squared, the slope cubed, and so forth. The $(\tau_s)_1$ term is a symmetrical one and averages to zero over a suitably chosen area. The $(\tau_s)_2$ and higher order even terms do not average out but are very small for small slopes. Therefore, the skin friction averaged over a suitable area does not deviate from its flat bed value by very much in most natural flows. Of course this is true only on the average, and local variations of 30% or more due to the symmetrical term $(\tau_s)_1$ are possible (Smith, 1969).

This means that if the velocity at some distance from the boundary were a reasonable estimate of the velocity that would occur at that level if the form drag in the flow were negligible, then the spatially averaged local skin friction could be estimated by applying z_0 calculated from Nikuradse's data through Eq. (3) to the velocity at the reference level. Fortunately, if the current measurement is made a distance from the boundary comparable to the diameter of the region over which the skin friction is being averaged then an error of less than 20% is caused by this procedure and a method permitting at least an estimate of τ_s is available. If the diameter of the area over which the spatial average is taken is not large relative to the largest wave length in the natural topography, then the $(\tau_s)_1$ term is not averaged out and must be accounted for. Following the procedure outlined above, u_c can be found from $(u_*)_c$ using the following expression for hydraulically smooth flow:

$$\frac{\bar{u}_c}{(u_*)_c} = (5.75) \log \left(\frac{(u_*)_c z_1}{\nu} \right) + 5.5 \tag{28}$$

where $(u_*) = 1.3$ cm sec^{-1}, $z_1 = 300$ cm, and $\nu = 1.4 \times 10^{-2}$ cm^2 sec^{-1}. Therefore, at three meters from the bed $u_c = 31$ $(u_*)_c = 40$ cm sec^{-1} and for speeds not too much higher than u_c, we can write $\tau_b = \rho C_{300} u^2$ where $C_{300} \cong 1.04 \times 10^{-3}$. For reference, the critical boundary shear stress is 1.7 dy cm^{-2}, the critical velocity 100 cm from the bed is 36 cm sec^{-1} and at 100 cm, $u_c \cong 28$ $(u_*)_c$ giving a drag coefficient of $C_{100} = 1.28 \times 10^{-3}$.

PHYSICAL OCEANOGRAPHY
OF THE WASHINGTON SHELF

The circulation near the Washington coast is reviewed by Barnes and Paquette (1957) and Hopkins (1971a), and the effects of the Columbia River effluent in this region are reviewed by Budinger, Coachman and Barnes (1964). Away from the shelf the currents are weak and poorly defined due to the divergence of the west wind drift, about 300 mi west of the Washington coast, into the northward-flowing Alaskan Current and the larger southward-flowing California Current. In the summer a large atmospheric high-pressure cell is situated over the East Pacific, whereas in the winter a series of eastward-moving lows causes a mean low-pressure cell to develop over the Gulf of Alaska as shown on Fig. 65. In response to the changing meteorological conditions the California Current originates farther to the north in the summer than in the winter.

Inshore of the California current and generally confined to the continental slope is a poleward flow of a few tens of centimeters per second called the Davidson current. The surface expression of this current develops during autumn and winter and abates in late spring; however, the deeper part of the current is probably a permanent phenomenon. On the continental shelf the bottom water predominantly flows to the north during both the summer and winter, although the surface water responds to the local wind stress and is driven to the north during the winter and to the south during the summer.

Atmospheric motions on the scale of 500 km or larger are very close to being in geostrophic balance, that is the velocity is proportional to the pressure gradient but directed 90° to its left in the northern hemisphere and 90° to its right in the southern hemisphere. Using this observation and referring to Fig. 65 shows that the winds off the Washington-Oregon coast will be southerly in the winter and northerly in the summer. In each case the geostrophic winds approximately parallel the coast, as do the observed surface winds.

In geostrophic flow the Coriolis effect and the pressure gradient force predominate; however, near a boundary the Coriolis effect is balanced by friction.

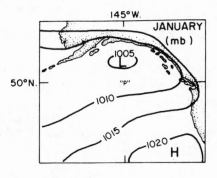

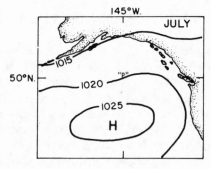

Figure 65.

Atmospheric pressure (in millibars) over the Northeast Pacific showing the winter low pressure cell and the summer high pressure cell that causes the respective southerly and northerly coastal winds. (After Hopkins, 1971a.)

Therefore, when the wind blows over the sea, friction tends to cause a surface flow in the direction of the wind but the Coriolis effect prevents the surface water from moving in that direction and causes the surface current to be deflected to the right of the wind (in the northern hemisphere). Under ideal conditions the surface velocity is deflected 45° to the right of the wind stress and each subsequent layer in the fluid is forced somewhat to the right of the layer above it. The result is a spiraling velocity field, called an Ekman spiral, of decreasing speed and systematic rightward deflection with depth. The net flux of momentum in the Ekman spiral is called the Ekman transport or Ekman drift and is at 90° to the right of the surface stress. Along the Washington-Oregon coast the southerly winter winds cause the surface water over the continental shelf to flow in an onshore direction and pile up next to the coast. However, the piling up of the water next to the coast results in an upward sloping free surface and an onshore pressure gradient in this region. The pressure gradient then causes a northward flowing geostrophic current with an offshore component in the frictional layer near the sea bed. This general type of flow, called downwelling because the surface water is forced down as it approaches the coast, is typical of the winter situation on the Washington-Oregon continental shelf. Under these conditions any stratification that exists in the shelf water will be broken down.

In the summer the surface circulation is reversed. The winds are from the north so the Coriolis effect causes an offshore drift of the surface water and a resulting depression of the free surface near the coast. The depressed free surface causes a southerly geostrophic surface current and, if a sharp pycnocline develops, a northward-flowing baroclinic current near the bottom. Water is brought onshore in the pycnocline to replace the surface water that has been blown offshore, resulting in an upward motion of typically cooler and more dense fluid and a sharpening of the stratification. This situation is called upwelling and is typical along the Washington-Oregon-Northern California coast in the summer.

From a sediment transport point of view the Washington continental shelf can be divided into three regions. The first of these is the inner shelf, on which shoaling waves and tidal currents are at least as important as the wind driven currents. The second is the central shelf, over which wind driven currents are the most important oceanographic motion. The third region is the shelf edge, over which shoaling internal waves may well be as important as the wind driven currents. The inner shelf is covered by medium to fine sands in a band extending from the beach to the 40 or 50 m isobath. Sediment transport by small-amplitude surface gravity waves of moderate steepness begins when $h \, \lambda^{-1}_0 \simeq 1/4$, where λ_0 is the deep water wave length. Using the relationship, $\lambda \simeq 1.56 \, T^2 \simeq 156$ m for the typical swell with 10 sec period gives a deep water length of 156 m and a critical depth of 39 m, making plausible the suggestion that the edge of this zone is determined by the depth of wave interaction with the bottom sediments.

In order to determine the bottom currents on the inner continental shelf, Barnes and coworkers (Morse, Gross and Barnes, 1968; Gross, Morse, and Barnes, 1969; and Barnes, Duxbury and Morse, 1972) have successfully employed sea-bed drifters and have found a general northerly bottom current. In water from 40 to 90 m deep the drifter data suggest a northerly current of 1 to 3 cm sec^{-1} that either parallels the shelf contours or has a very slight offshore component. Many drifters released near the Columbia River mouth in this depth zone have traveled along the shelf and entered the straits of Juan de Fuca and some have even been found on the Vancouver Island and Alaskan coasts. However, drifters released inside of the 40 m contour typically have moved toward the beach under the influence of shoaling waves and wave driven currents. These data are insufficient to prove that the bottom currents are always in the onshore direction, but it is likely that this is the case. Under these conditions the coastal sands would be transported slowly in the onshore direction except during storms sufficiently severe to put the material in suspension. For the

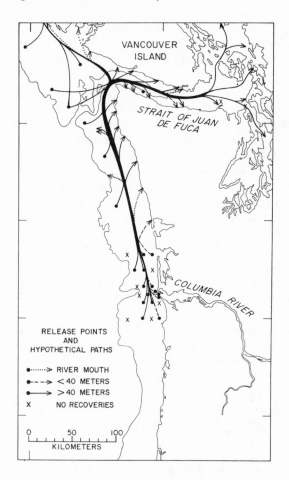

Figure 66.

Seabed drifter tracks showing the general northerly flow of the bottom water on the central continental shelf. Drifters released in less than 40 m of water and that moved directly onto the beach and are not included in this figure. (After Barnes, Duxbury, and Morse, 1972.)

coarser sands this occurrence would be unlikely, but the finer sands may well be moved seaward by this mechanism. In this way the coarser material is trapped next to the coast and continually reworked. Only the sediment fine enough to go into suspension can escape, and presumably this material is represented primarily by the 3.5 ϕ sand on the Washington shelf.

Figure 66 presents a graphical summary of the drifter results and shows the general northerly and onshore drift of water and possibly bottom sediment on the inner shelf.

As the water from the Columbia River enters the sea it is mixed with the surrounding shelf water, thereby diluting it and causing a low-salinity surface layer to be formed. If there were no currents on the continental shelf, this low-density layer would tend to flow outward from the river mouth but would be acted upon by the Coriolis effect and deflected to the right, building up an upward-sloping free surface at the coast and causing a northerly geostrophic flow along the coast. This type of current can be observed in many areas along the east coast of the United States as a weak southward flow.

During the winter, when the surface current on the shelf is northerly, the density-driven flow in the plume and the wind-driven surface currents are in the same direction, and the plume flows northerly in a tight band along the Washington coast as shown in Fig. 67. However, in the summer the wind-driven surface current on the shelf opposes the natural northerly flow of the plume, and the pressure gradients required to decelerate the surface currents force the plume into a southwesterly orientation (Fig. 67). During the upwelling period, the central continental shelf off of the Washington coast is not severely affected by the density anomaly due to the river effluent (see Fig. 67), and in this respect the upwelling conditions may be somewhat different on the Washington and Oregon shelves. The situation is reversed during downwelling when the accumulation of river water on the inner Washington shelf complicates the dynamics of this region relative to the Oregon shelf.

DIRECT CURRENT MEASUREMENTS

In the summer of 1967 a program aimed at making direct measurements of the near-bottom currents on the continental shelf was initiated at the University of Washington by the writers. Previous measurements had been made on the Oregon shelf (see Collins, 1961; Mooers, 1970; Cutchin, 1972; Pillsbury, 1972), but were concentrated in the upper part of the water column and were aimed primarily at the flow in the summer months, when upwelling occurs along this coast. The thrust of the University of Washington program was aimed toward the bottom currents and toward winter conditions. Moreover, in 1967 there was considerable disagreement as to whether recent sediment was being transported on the Washington shelf and as to whether the transport was onshore or offshore if it existed at all.

The current measurements discussed in this paper were made using Braincon 316 and 381 histogram type current meters that were located near the bottom of a taut wire mooring. Figure 68 shows diagrammatically how the moorings were rigged. One or more railroad wheels were used as an anchor, and these were separated from the rest of the mooring on command by an acoustic release. Buoyancy for the mooring was provided by one or more 80 cm diameter subsurface floats. In order to prevent excessive mooring motion, no surface floats were attached.

During Hopkins' early experiments so many moorings were lost to foreign

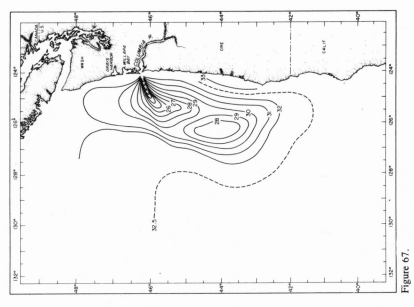

Figure 67.

Surface salinity anomaly due to the Columbia River outflow. The left hand figure shows the summer condition when the plume is driven south and broadens. The right hand figure shows the winter condition when the plume flows in a tight band along the Washington coast. (After Duxbury, Morse, and McGary, 1966.)

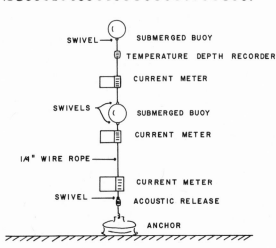

Figure 68.
Schematic of typical current-meter mooring.

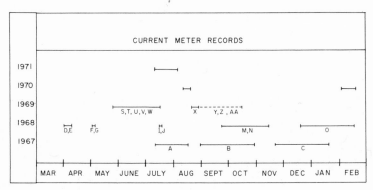

Figure 69.

Graphical representation of the times for which current meter data are available. The letters denote particular records and the dashed line signifies a time when the bottom current meter was inoperative.

and possibly even American fishermen, that it was deemed necessary to choose an experimental site within the 12-mi limit and to rent a sea buoy for this location from the Coast Guard. This was done and all records between July, 1968, and July, 1970, were obtained from this position. Of the records taken prior to July, 1968, two also were taken in this vicinity, one was taken to the east of this location in 50 m of water, and the rest were taken near the Columbia River Light Ship. Figure 69 shows the temporal distribution of all available records. For more details the reader is referred to Hopkins

(1971a-c). Since July, 1970, the experiments have been shorter in duration, but have employed six current meters in a vertical array.

Record C provides good data on the currents 3 m from the sea bed in 50 m of water during winter conditions. The mooring, located on Fig. 59, lasted from November 23, 1967 to January 19, 1968. The highest current speeds that we have measured so far are on this record. Although no simultaneous data from moorings at other locations are available for this time interval, analysis of wind records from the Columbia River Light Ship for this and other winter periods indicates that the high-current velocities were associated with wind velocities similar to those found for other winter storms in later years. For instance, the winds recorded at the Columbia River Light Ship during the storms in early December, 1967, late December, 1967, and mid-January, 1968, and the winds recorded at the same location for storms in late December, 1968, and in early February, 1969, all had peak speeds of approximately 20 m sec^{-1}. Currents driven by the first set of storms are found on record C; whereas, currents driven by the second set of storms are found on record O. However, the current speeds associated with these storms are larger at the record C location (Fig. 59) than at the record O location. Therefore, it appears that the high velocities found in this record are due to the shallower water depth (50 m versus 80 m) rather than being due to more severe storms during the early winter of 1967-1968.

The progressive vector diagram for this record, Fig. 70, shows a typical initial quiet spell of weak southerly flow. The current meters are usually put in place during relatively calm weather as the moorings are difficult to handle in heavy seas; therefore, the first part of any record is likely to represent fair-weather conditions. The quiet spell in this record is followed by a storm that caused high velocities in the north-northwest direction. This storm is followed by a second quiet spell with southerly flow and then another pair of storms separated by a quiet period. Over the duration of the record the net flow is to the northwest, and the net drift, assuming uniform flow field, would be about 360 km. Of course, little if any sediment would have been transported by these currents during the quiet spells, so it is not proper to use the 360 km figure in sediment-transport estimates.

Record C was filtered by applying a 24-hour running mean to each velocity component in order to remove semidiurnal and diurnal tidal currents. The current speed calculated from the filtered components is shown in Fig. 71 together with a vector representation of the filtered velocity field. The critical speed of 40 cm sec^{-1} is exceeded significantly only three times in the record, and in each case the currents can be seen to have been headed from 15 to 30° to the west of north. At this location the bottom contours trend about 7° to the west of north, leaving an 8 to 12° offshore component to the flow. For each kilometer the sediment is transported parallel to the bottom contours it is transported from 120 to 200 m in the offshore direction. Over the course

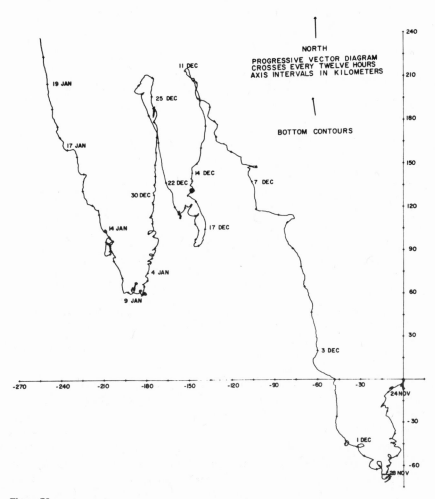

Figure 70.

Progressive vector diagram for record C showing strong northerly drift during storms and southerly drift during meteorologically quiet periods.

of the first storm the longshore distance a coarse silt particle that was eroded when the current speed reached the critical speed and that remained in suspension for the duration of the storm would have traveled would be in excess of 110 km, yielding an offshore transport of approximately 20 km. Of course, this estmated offshore transport is obtained by assuming a nearly uniform offshore component to the velocity field across the shelf seaward of the record C mooring. That this is the case is not at all obvious, but current records from other sites suggest that the assumption of a uniform offshore component is a reasonable one, at

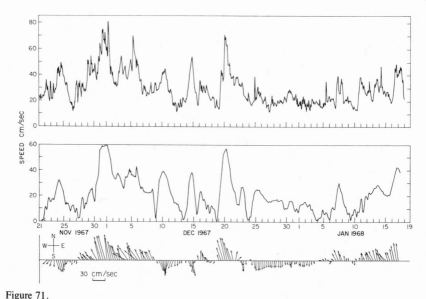

Figure 71.

Current speed and direction for record C. The top graph is the raw time series; whereas, a 24 hr filter has been applied to velocity components in the middle and lower graphs. Note the effects of several storms causing current speeds in excess of 40 cm sec⁻¹.

least for sediment transport estimates. Although the speed during the southerly flow is too small to cause erosion and transport of the shelf sediments, it is noteworthy that the near-bottom flow during these times also has an offshore component of from 8° to 12°.

Figure 71 also presents the unfiltered speed for record C. In this case the tidal, inertial and wind-driven current fluctuations with periods of one to 24 hr are included in the record, and much higher speeds are seen to occur. Even in this case the speed measurements are averaged for 20 min by the current meter and then smoothed using a five-point running mean, so the instantaneous currents may occasionally be as much as 10% higher than the measured mean speed.

Just as record C appears to represent a typical winter velocity field in 50 m of water, record O is believed to represent a typical winter velocity field in 80 m of water. As mentioned above, and as can be seen by comparing Fig. 73 to Fig. 71, the current speed associated with a given storm appears to be greater in shallower water. The progressive vector diagram for record O is shown in Fig. 72. The initial period of southerly flow is followed after a few days by a prolonged period of north-northwesterly flow.

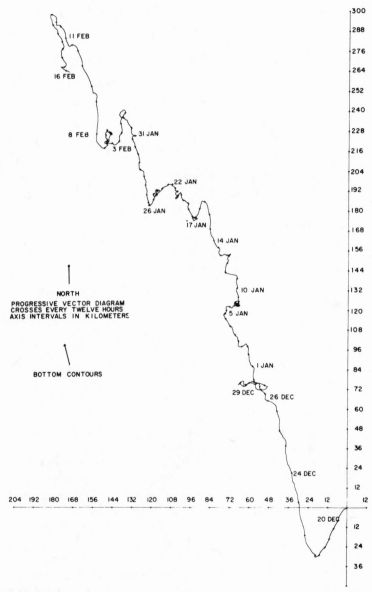

Figure 72.
Progressive vector diagram for record O showing a general north northwesterly drift.

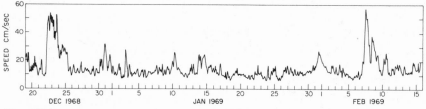

Figure 73.

Time series of current speed for record O showing the effect of two storms during which the current speeds exceed 40 cm sec⁻¹.

The 20-min speed averages for record O smoothed with a five-point running mean are shown in Fig. 73. During most of the record the current speed was well below 30 cm sec; however, twice within this period storms with winds of 20 m sec⁻¹ resulted in currents with peak speeds of from 54 to 58 cm sec⁻². During the first storm the currents 3 m from the sea bed exceeded 40 cm sec⁻¹ for 36 hr and during the second storm they exceeded 40 cm sec⁻¹ for 12 hr. Assuming uniform flow on the shelf, silt eroded by the high currents at the beginning of the first storm would have traveled 54 km by the time the current speed dropped below 40 cm sec⁻¹ and would have traveled another 40 km in the next two days. This material would have been carried essentially along the shelf, because it can be seen from Fig. 72 that from December 22 to December 26 the currents very closely paralleled the bottom contours. The finer silts that were still in suspension several days after the storm would have been carried another 40 to 60 km, but this time in a more northwesterly direction (N 28° W). In still water coarse silt takes about two days to settle 50 m; fine silt may take several weeks to settle the same distance.

As the actual volume of sediment transported depends not only on the instantaneous currents but also on the history of the time-varying velocity field, elementary statistical approaches to this problem are not very useful. For instance, several days of 30 cm sec⁻¹ currents with moderate offshore components occurring just after a storm having high current speeds such as in record O results in a completely different sediment-transport situation than if the 30 cm sec⁻¹ currents preceded the high currents and the latter were followed by a period of negligible flow. A statistical treatment of the problem such as given by Sternberg and McManus (1972) based on Hopkins' data assumes that these situations are the same. Also, it should be noted that the directional history of the currents immediately following a severe storm is at least as important as the directional history of the storm itself. Therefore, the proper way to calculate sediment transport under conditions such as found on the Washington continental shelf is to procure good velocity records, then find the times when erosion of shelf sediment is occurring, and trace the history of each size class until a negligible amount of material remains in suspension. This procedure is not necessarily any more work than the statistical method, because only a few significant events per year need be traced.

The net drift over the duration of record O is 320 km in the N 33° W direction; however, during storms the direction of drift is N 20° W or so.

The second storm during the record O period began on February 7, and although the currents were of greater peak intensity than those associated with the first storm, the net sediment transport by this storm was much less. This occurred for two reasons, both of which emphasize the importance of the above remarks on the nonlinear nature of the process of erosion and sediment transport. First, the currents exceeded 40 cm sec^{-1} for only 12 hr, so not a great deal of sediment was put in suspension, and, second, the currents changed direction on February 12, when a considerable amount of fine material was still in suspension.

SEDIMENT TRANSPORT CALCULATIONS

No completely satisfactory theory for the computation of suspended load transport is presently available. The sediment concentration profile can be found under a variety of conditions by solving Eq. (14), but in order to evaluate the total amount of material in suspension and the total transport of material the rate of erosion of material from the bed or the concentration at a level near the bed must be known. A proper theory from which these parameters can be obtained does not exist. However, there are several empirical equations available from which the average concentration of suspended material can be estimated, at least for steady-flow in rivers.

One of these, the Larras (1965) formula, appears to be valid over a limited range and only for fine quartz sands. This expression can be written as:

$$\langle \bar{\epsilon}_s \rangle = (2 \times 10^{-4}) \left(\frac{\tau_b}{\rho g D} \right)^5 \tag{29}$$

where $\langle \bar{\epsilon}_s \rangle$ is the concentration averaged in z over the depth of the flow. The total amount of suspended material is $\langle \bar{\epsilon}_s \rangle h$. Two other expressions for average concentration are the Laursen (1958) "total load equation":

$$\langle \bar{\epsilon}_s \rangle = \Sigma\, i \left(\frac{D_i}{h} \right)^{7/6} s\, f_1 \left(\frac{u_*}{w_s} \right) \tag{30}$$

where i is the fraction of the sample in size class D_i and $s = (\tau_b - \tau_c)/\tau_c$ is the excess shear stress, and the Bogardi (1965) equation:

$$\langle \bar{\epsilon}_s \rangle = \left(\frac{D_i}{h} \right)^{7/6} s\, f_2 \left(D, \frac{gD}{u_*^{\,2}} \right)$$

(31)

Calculations were made using all of the above equations and were found to agree in order of magnitude for fine sands; for silts the Laursen and Bogardi equations agree quite well with each other, but the Larras expression gives values that are much too high.

In the ocean, the apropriate depth to use in the calculation of $\langle \bar{\epsilon}_s \rangle h$ is not necessarily the water depth, because the stratification in the flow may serve to bound the velocity and sediment concentration fields. Fortunately, the depth enters into the Laursen and Bogardi expressions only to the one-seventh power and does not affect the final result very much when stated as total amount of suspended material, $\langle \bar{\epsilon}_s \rangle h$ instead of depth-averaged concentration. Table VIII gives the best estimates of $\langle \bar{\epsilon}_s \rangle h$ for various appropriate values of u_{300} and D. These data show much higher amounts of suspended silt relative to sand and a rather large increase in suspended material with increasing current speed.

Table VIII
Total amount of suspended sediment per unit area of the bed and total load transport of sediment for various u_{300} and D in cgs units

ϕ	3.0	3.5	4.0	5.0	u_{300}
D	0.0125	0.0088	0.0063	0.0031	
$\langle \bar{\epsilon}_s \rangle h$	0.0052	0.0083	0.019	0.087	42
	0.019	0.0281	0.070	0.37	48
	0.041	0.058	0.14	0.73	54
	0.067	0.098	0.23	1.10	60
	0.097	0.17	0.40	1.6	66
	0.13	0.24	0.61	2.2	72
	0.19	0.31	0.89	2.8	78
Q_s	0.22	0.35	0.8	3.6	42
	0.91	1.34	3.4	11.7	48
	2.2	3.1	7.6	39	54
	4.0	5.9	13.8	66	60
	6.5	11.3	26	106	66
	9.4	17.3	44	160	72
	14.8	24	69	220	78

Equation (18) shows that under conditions of increasing velocity and net erosion, less material will be put into suspension than under steady conditions at the same current speed. Therefore, the data presented in Table VIII overestimates the amount of sediment in suspension. However, the sediment concentration increases exponentially at first, and over the course of a one-day storm the error introduced by assuming the average concentration to be equivalent to the average concentration under steady conditions will be much less than the error introduced due to the general inapplicability of Eqs. (29), (30), and (31), to the type of flows that exist on the continental shelf.

Until actual measurements of suspended sediment concentration are made at some reference level through a storm, so that the solution to Eq. (18) can be used, our knowledge of the concentration field cannot be considered sufficient to warrant carrying out the detailed integrations necessary for an accurate evaluation of the total sediment discharge. However, an estimate of commensurate accuracy with the rest of the study can be obtained by multiplying $(31\ u_*)$ by $(<\bar{\epsilon}_s>)$ or (u_{300}) by $(<\bar{\epsilon}_s>)$. The results of this calculation for various u_{300} and D are also presented in Table VIII. As in the case of $<\bar{\epsilon}_s>h$ these results cannot be considered accurate to more than a factor of 2 or 3.

The data on total amount of suspended sediment presented in Table VIII can be converted to depth of erosion by dividing by the average concentration of sediment in the bed and by assuming uniform erosion over the entire shelf; that is, by assuming that the material goes into suspension, moves, and settles back to the bed without accumulating in any area or causing net erosion in any area. Although this assumption is not strictly valid, it is satisfactory for the central shelf region. This means that from a few millimeters to a few centimeters of sediment are eroded by each storm. The material is then redeposited, with the silt settling back on top of the sand to form a layered, possibly even graded bed.

Barnes and Gross (1966) estimated the rate at which a typical sediment particle of the shelf would travel using the radionuclide data shown in Fig. 74. On the central shelf where the Barnes and Gross data are applicable, the sediment is comprised of about 50% sand that moves as bed load and about 50% silt that moves as suspended load. During a storm the net movement by bedload is negligible compared to that of the suspended load, so the net transport of an average particle is about half the net movement of a silt particle. Assuming that sediment transport occurs during four storms per year and that the storms on record O are typical, yields a net distance moved per year of 80 km for a typical silt particle and a net distance moved per year of 40 km for an average sediment particle. This result can be compared to the 30 km yr^{-1} found by Barnes and Gross.

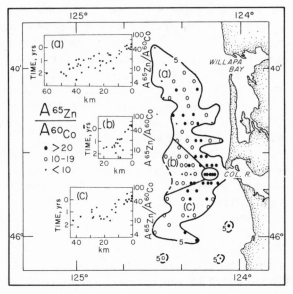

Figure 74.

Activity ratio of Zn-65 to Co-60 for surface sediments containing both radio nuclides. The insets show the time in years that it took for river derived sediments to reach a particular location. (After Barnes and Gross, 1966.)

CONCLUSION

It has been shown that the presently available current and sediment data for the Washington continental shelf present a consistent picture of sediment transport, one that might well also be coupled with the near-surface stratigraphy of the shelf. Detailed conclusions are as follows:

1. All of the presently available data indicate that silt from the Columbia River is transported northward and to a lesser extent westward.

2. Direct current measurements indicate that this transport occurs only during a few storms each winter.

3. Estimates of sediment transport suggest that an average particle on the shelf moves about 40 km yr⁻¹ in a longshore direction and about 7 km yr⁻¹ in an offshore direction. However, the sand fraction moves much more slowly, as bedload, and the silt fraction moves much more rapidly, as suspended load. Both the direction of transport and rate of average surface particle motion are in good agreement with the radionuclide data of Barnes and Gross (1966).

4. Estimates of the amount of material eroded from the central shelf by severe storms range from a few millimeters to over a centimeter and suggest that the surface sediments are layered and perhaps even graded by severe storms,

the sands being covered by silt that settles out of suspension after the storms. As silt layers on the order of a centimeter thick may be possible, short cores should be obtained if stratigraphic and sediment transport data are to be coupled in a meaningful way. Otherwise, a surface sample one centimeter thick may completely miss a sand layer and not be representative of the true dynamical situation. Moreover, if this model is correct then the amount of material suspended by a given storm may be measured by determining the thickness of a silt layer.

5. It is suggested that fine sand derived from the nearshore region is transported offshore only during severe storms, when it moves as suspended load, and that at other times this material, like the coarser sands, is trapped in the nearshore region by wave-driven bottom currents that have a net onshore direction.

The general picture presented in this paper is a tentative one in that it is based on an incomplete set of measurements and on estimates of suspended-sediment transport that will have to be related to actual field measurements of suspended sediment concentrations in the near future. Nevertheless, there is a very close correlation between the sedimentological observations and the general physical oceanography of the region.

ACKNOWLEDGMENTS

This work was supported by Atomic Energy Commission Contract AT(45-1)-2225-T25 (RLO-2225-T25-3). The authors would like to express their appreciation to Dr. C. A. Barnes and Dr. D. A. McManus for numerous very fruitful discussions and to Gael Welch and John Beck for their help in a multitude of ways during this study. This paper is Contribution No. 667, Department of Oceanography, University of Washington, Seattle.

LIST OF SYMBOLS

Symbol	Definitions	Dimensions
C_{100}	Drag coefficient for velocity measured 100 cm from the bed	
D	Grain diameter	L
g	Acceleration of gravity	L/T^2
h	Depth of flow	L
i	Fraction of sample in size class D_i	
i	$\sqrt{-1}$	
J	Bessel function of first kind	
K	Kinematic momentum diffusion coefficient	L^2/T
k	0.40 von Karman's constant	
k_s	Roughness height of the bed	L
N	Number of phases	
p	$w_s/(ku_*)$	

p_*	Order of Bessel function	
Q_s	Volume discharge of sediment per unit width of flow	L^2/T
s	$(\tau_b - \tau_c)/\tau_c$	
T	Function of t	
T_0	Constant of integration	
t	Time	T
u	Velocity component in x direction	L/T
u_*	Shear velocity or friction velocity	L/T
$(u_*)_c$	Critical shear velocity	L/T
w	Velocity component in z direction	L/T
w_s	Settling velocity of a sediment	L/T
Y	Bessel function of second kind	
Z	Function of z	
z	Elevation above boundary and vertical coordinate direction	L
z_0	Constant of integration	L
∇	Del	1/L
$(R_*)_c$	Critical roughness Reynolds number	
ln	Natural logarithm	
α_0	Constant of integration	
β	15.6	
δ	Thickness of viscous sublayer	L
λ	Separation constant	
λ_0	Deep water wave length	L
ν	Kinematic viscosity	L^2/T
ξ	z/h	
ρ	Fluid density	M/L^3
τ	Shear stress	M/LT^2
τ_b	Boundary shear stress	M/LT^2
τ_c	Critical shear stress	M/LT^2
τ_s	Skin friction on the boundary	M/LT^2
ϕ	phi size = $-\log_2$ (D in mm)	
ϵ	Instantaneous volume concentration of sediment at a point	
$<\epsilon>$	Concentration averaged in z over the depth of the flow	

Subscripts:

a	Reference level
b	Bed
i	Size class
n	Phase
s	Sediment
w	Water

Superscripts:

~	Designates vector
−	Time average
′	Fluctuating velocity

REFERENCES

Abramowitz, M., and Stegun, I. A., eds. (1964). "Handbook of Mathematical Functions." National Bureau of Standards Applied Mathematics Series. Vol. 55, Govt. Printing Office, Washington, D.C.

Ballard, R. L. (1964). Distribution of beach sediment near the mouth of the Columbia River. *Univ. Washington Depart. Oceanography Tech. Rept. (Seattle)* **98**, 81.

Barnes, C. A., Duxbury, A. C., and Morse, B. A. (1972). The circulation and selected properties of the Columbia River effluent at sea. *In* "Bioenvironmental Studies of the Columbia River Estuary and Adjacent Ocean Regions." (D. L. Alverson and A. T. Pruter, eds.). University of Washington Press, Seattle.

Barnes, C. A., and Gross, M. G. (1966). Distribution at sea of Columbia River water and its load of radionuclides. *In* "Disposal of Radioactive Wastes into Seas, Oceans and Surface Waters." International Atomic Energy Agency, Vienna.

Barnes, C. A., and Paquette, R. G. (1958). Circulation near the Washington coast. *Proc. Eighth Pacific Science Congress, Oceanography (Manila)* **3**, 585-608.

Bogardi, J. L. (1958). The total sediment load of streams: A discussion. *Proc. Am. Soc. Civil Engineers,* **84(HY6)**[1856], 74-79.

Bogardi, J. L. (1965). European concepts of sediment transportation. *Proc. Am. Soc. Civil Engineers* **91(HY1)**, 29-54.

Budinger, T. F., Coachman, L. F., and Barnes, C. A. (1964). Columbia River effluent in the Northeast Pacific Ocean, 1961, 1962: Selected aspects of physical oceanography. *Univ. Washington Dept. Oceanography Tech. Rept.* **99**(Ref M63-18), 78 pp.

Collins, C. A. (1968). "Description of Measurements of Current Velocity and Temperature over the Oregon Continental Shelf, July 1965-February 1966." Ph.D. Thesis, Oregon State Univ., Corvallis, Oregon.

Cutchin, D. L. (1972). "Low Frequency Variations in the Sea Level and Currents over the Oregon Continental Shelf." Ph.D. Thesis, Oregon State Univ., Corvallis, Oregon.

Duxbury, A. C., Morse, B. A., and McGary, N. (1966). "The Columbia River effluent and its distribution at sea, 1961-1963." *Dept. Oceanography Washington Tech. Rept.* **156**.

Einstein, H. A. (1950). The bedload function for sediment transportation in open channel flows. *U.S. Dept. Agriculture, Soil Conservation Serv. Tech. Bull.* **1026**.

Graf, W. H. (1971). "Hydraulics of Sediment Transport." McGraw-Hill, New York.

Gross, M. G., McManus, D. A., and Ling, H.-Y. (1967). Continental shelf sediment, Northwestern United States. *J. Sedimentary Petrology* **37**, 790-795.

Gross, M. G., Morse, B. A., and Barnes, C. A. (1969). Movement of near-bottom waters on the continental shelf off the Northwestern United States. *J. Geophysical Res.* **74**(28), 7044-7047.

Harman, R. A. (1972). The distribution of microbiogenic sediment near the mouth of the Columbia River. *In* "Bioenvironmental Studies of the Columbia River Estuary and Adjacent Ocean Regions (D. L. Alverson and A. T. Pruter, eds.). University of Washington Press, Seattle.

Hopkins, T. S. (1971a). "On the circulation over the continental shelf off Washington." Ph.D. Thesis, Univ. Washington, Seattle.

Hopkins, T. S. (1971b). Velocity, temperature, and pressure observations from moored meters on the shelf near the Columbia River mouth, 1967-1969. *Univ. Washington Dept. Oceanography Special Rept.* **45**, 143 pp.

Hopkins, T. S. (1971c). On the barotropic tide over the continental shelf off the Washington-Oregon coast. *Univ. Washington Dept. Oceanography* **46**, 22 pp.

Kelley, J. C., and McManus, D. A. (1969). Optimizing sediment sampling plans. *Marine Geology* **7**, 465-471.

Kelley, J. C., and McManus, D. A. (1970). Hierarchical analysis of variance of shelf sediment texture. *J. of Sedimentary Petrology* **40**, 1335-1339.

Larras, J. (1965). Hydraulique—Taux de concentration moyen des matériaux en suspension dans les écoulements rectilignes uniformes. *C.R. Acad. Sc. Paris,* **261**[2], 3525-3536.

Laursen, E. M. (1958). The total sediment load of streams, *Proc. Am. Soc. Civil Engineers,* **84(HYI),** 1530.

McManus, D. A. (1972). Bottom Topography and Sediment Texture near the Columbia River. *In* "Bioenvironmental Studies of the Columbia River Estuary and Adjacent Ocean Regions" (D. L. Alverson and A. T. Pruter, eds.). University of Washington Press, Seattle.

Mooers, C. N. K. (1970). The interaction of an internal tide with the frontal zone in a coastal upwelling region. Ph.D. Thesis, Oregon State Univ., Corvallis, Oregon.

Morse, B. A., Gross, M. G., and Barnes, C. A. (1968). Movement of seabed drifters near the Columbia River. *Proc. Am. Soc. Civil Engineers, J. Waterways and Harbors Div.* **94(WWI),** 93-103.

Nikuradse, J. (1933). Laws of flow in rough pipes. *Nat. Adv. Comm., Aeronautics Tech. Memo.* **1292,** 1-62 (transl. German, 1950).

Pillsbury, R. D. (1972). "A description of hydrography, winds and currents during the upwelling season near Newport, Oregon." Ph.D. Thesis, Oregon State Univ., Corvallis, Oregon.

Schlichting, H. (1962). "Boundary-Layer Theory." McGraw-Hill, New York.

Shepard, F. P. (1963). "Submarine geology." Harper and Row, New York.

Smith, J. D. (1969). Studies of non-uniform boundary-layer flows. *In* "Investigations of Turbulent Boundary Layer and Sediment-Transport Phenomena as Related to Shallow Marine Environments," Part 2. U.S. A.E.C., Contract AT(45-1)-1752. Ref: A69-7. Depart. Oceanography, Univ. Washington.

Smith, J. D. (1970). Stability of a sand bed subjected to a shear flow of low Froude number. *J. of Geophysical Res.* **75**(30), 5928-5940.

Sternberg, R. W. (1972). Predicting initial motion and bedload transport of sediment particles in the shallow marine environment. "Processes and Patterns of Sediment Dispersal on the Continental Shelf" (D. J. P. Swift, D. B. Duane, and O. H. Pilkey, eds.), Dowden, Hutchinson and Ross, Stroudsburg, Pennsylvania.

Sternberg, R. W., and McManus, D. A. (1972). Residual bottom currents on the continental shelf of Washington. "Processes and Patterns of Sediment Dispersal on the Continental Shelf" (D. J. P. Swift, D. B. Duane, and O. H. Pilkey, eds.). Dowden, Hutchinson and Ross, Stroudsburg, Pennsylvania.

CHAPTER 8

Implications of Sediment Dispersal from Long-Term, Bottom-Current Measurements on the Continental Shelf of Washington

Richard W. Sternberg* and Dean A. McManus*

ABSTRACT

Approximately 260 days of current speed and direction data collected 3 m off the seabed of the Washington continental shelf (Hopkins, 1971a) have been analyzed empirically for magnitude and directional response. Data have been separated according to current-speed class (0-10 cm sec^{-1}, 10-20 cm sec^{-1} . . . 90-100 cm sec^{-1}) and histograms have been made to determine the directional variability of the various speed classes. Data within speed classes have been averaged over monthly and annual periods to determine the net direction of movement of bottom water.

On the continental shelf of Washington, the bottom currents respond strongly to the surface wind stress. Currents of all speeds show a high degree of directional variability; however, all bottom currents greater than 10 cm sec^{-1} exhibit an overall trend to the northwest. Results indicate that sediment movement frequently occurs on the central continental shelf during the winter months. Lateral sediment dispersal (across the shelf) should occur as a result of the high variability of current direction while a net northward displacement of sediment is also suggested. The maximum annual displacement that could be experienced by a sedimentary particle under the observed conditions is estimated to be approximately 220 km to the northwest.

*Department of Oceanography, University of Washington, Seattle, Washington 98195

INTRODUCTION

Several recent works have dealt with an analysis of current regimes on the continental shelf and with the application of existing knowledge of continental shelf dynamics to the prediction of sediment dispersal patterns. These works are summarized by Swift (1969) and have been supplemented by those of Swift (1970) and Swift et al. (1971). In the latter work the energy sources available for sediment dispersal are used to resolve shelf currents into four classes (Fig. 75). According to Swift et al. (1971), ''The dominant categories of input, in order of importance for most shelves, are 'meteorological' (mainly direct wind and wind-wave input, but including barometric storm surge), 'tidal,' 'density' (seaward flow of brackish surface water with return bottom flow), and 'intruding oceanic current'.''

Swift et al. (1971) review some of the conclusions that have been reached regarding meteorologically induced currents. Some researchers contend that wave surge is not an important sediment dispersal mechanism in depths greater than approximately 15 m (e.g., King, 1959; Dietz, 1963). On the other hand, Curray (1960) concludes that in the Gulf of Mexico fine sands are stirred by hurricane waves approximately once every 5 yr on the outer shelf and more frequently than once every 2 yr on the inner portions of the shelf. Measurements in the Celtic Sea, which has a high average wave level, indicate that wave surge on the bottom strong enough to erode sand occurs approximately one day per year (Draper, 1967). Barometric storm surge and direct wind currents are also mentioned by Swift (1971) but no effort is made to estimate the bottom currents resulting from these conditions. In the summary by Swift et al. (1971) of the

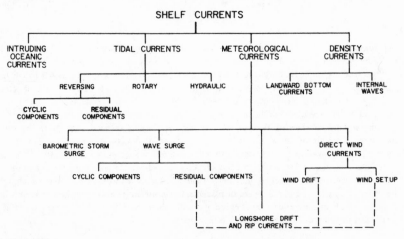

Figure 75.
Components of the shelf velocity field (after Swift et al., 1971).

hydraulic regime on the continental shelf two conclusions appear to be emphasized. First, bottom currents resulting from storm wave surge represent the dominant part of the meteorological currents; second, all of the currents proposed are dominantly oscillatory, although associated with a net transport referred to as "residual" currents. Thus, these authors concur that a long-term view of sediment dispersal mechanisms (on the order of years) is important because the residual currents associated with meteorological inputs could account for a systematic scheme of sediment dispersal on the continental shelf.

With respect to the nature of the net transport or "residual" currents, Swift (1970, p. 12) quotes Dunbar and Rogers (1957, p. 10): "Storm winds create local drift currents, but these are highly variable in direction and velocity. Heavy storms rarely cover an area more than 300 mi in diameter at one time, and seldom persist for more than a few days. Moreover, the wind normally changes velocity and direction as the storm passes. The net result of a passing storm, therefore, is to generate a temporary drift current that reaches maximum velocity after two or three days and then rapidly subsides, meanwhile changing or even reversing its direction; thus, the general effect of a series of passing storms on a shallow sea floor is to stir up bottom sediment and move it relatively short distances, first in one direction, then in the other."

The above statement, although referring to wind-drift currents instead of wave drift, appears to be the primary reference leading to the qualitative vectorial representation of shelf dispersal shown in Fig. 76 (from Swift, 1970). This diagram emphasizes the concept that randomly oriented bottom velocity vectors associated with wave-drift currents on the central and outer continental shelf play the dominant role in the mechanics of sediment dispersal. Under these conditions sediment movement is conceived as a large-scale diffusion of detritus

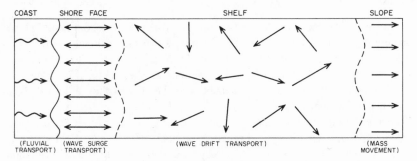

Figure 76.
Qualitative vectorial representation of shelf dispersal. Transport is on shore or offshore in the nearshore zone, as determined by resultants of oscillatory wave currents (null-point model). Transport is random across rest of shelf, determined by storm-generated wave-drift currents (wave-drift model). With sediment input at shoreline and output at shelf edge, net transport must be across shelf by a diffusion process (after Swift, 1970).

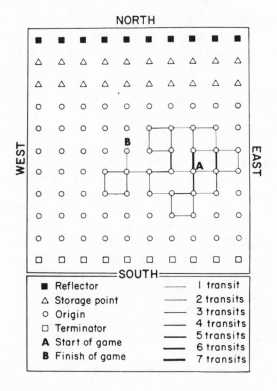

Figure 77.

Schematic model for random walk: solid squares represent the shoreline; open triangles, the zone of shoaling waves; open circles, the storage points for sediment in transit; and open squares, the shelf edge (after Swift et al., 1971).

from its source at the coast toward the deeper waters of the continental slope. Carrying the sense of Fig. 76 one step further, a two-dimensional random-walk model has been proposed (Swift, 1969, Swift et al., 1971) to account for the transport of sand over the outer continental shelf (Fig. 77). Thus the "residual" currents are considered to be associated with the random walk in some unstated manner.

The considerations illustrated in Figs. 76 and 77 are primarily the result of deduction and are severely limited by a lack of adequate field data. As stated by Swift et al. (1971), ". . . and we will be able to do little more than speculate about shelf sediment transport until geologists undertake to monitor shelf currents."

The purpose of this paper is to bring to light a rather long-time series of current-meter measurements recently carried out on the Washington continental shelf (Hopkins 1971a-c). In the past two years attempts to document the year-around weather and near-bottom currents on the central continental shelf of Washington have resulted in almost continuous coverage spanning nearly all months of the year. These data and their implications regarding the physical oceanography of the continental shelf have been recently reported by Hopkins (1971a-c). For our paper these data have been reevaluated to reveal the annual bottom current conditions on the Washington continental shelf. These are the most complete set of field data of which the authors are aware. Thus, they should serve a useful purpose in strengthening or amending the speculations discussed above.

DATA

Sampling

Current speed and direction were measured over a two-year period with Braincon type 316 and 381 Histogram current meters mounted on a taut-wire moor. Detailed descriptions of the sampling are given by Hopkins (1971a). Stations were positioned at several locations with current meters mounted at various depths, however, due to instrument failures, most of the data were collected from a single current meter mounted 3 m off the bottom at a location 80 m deep and 21 km off the Washington coast (latitude, 46° 25.0′N, and longitude 124° 20.0′W). Current speed and direction records consist of a continuous series of 20-min averages of the sensor outputs. Record lengths vary from several days to periods of 2 mo. By piecing together the records it is possible to obtain nearly a full year's coverage from the location of interest. Table IX summarizes the records used.

Analytical Techniques

Data have been sorted according to the month, speed, and direction within octants of a 360° circle. This technique allows the data to be summarized as "current roses" as shown in Fig. 78. Each current rose consists of eight histograms of current speed and duration arranged according to their magnetic headings as grouped into 45° azimuthal increments. The length of each vector represents the percent of time that a particular speed and direction occurred during the total record available for the month. The total of all vectors in each current rose is 100%. The number of days of data available for each month is also included in Fig. 78. Only partial coverage of data was available during February, April, May, August, and November and the complete month of March is missing from these computations.

The data in Fig. 78 show a considerable degree of directional variability (or randomness), so they were analyzed further to determine if any dominant directional orientations existed for the various speed increments. This analysis consisted of constructing progressive vector diagrams for all data during each month. Data within each speed increment or class were analyzed independently and vector sums were determined for current duration (vector length) and direction (vector orientation). The vector sum then, represents the net time and direction

Table IX
Summary of Current Speed and Direction Records Analyzed.[a]

Month	Hopkins record No.	Dates	Depth
January	0	Jan 1 - Jan 31	80 m
February	0	Feb 1 - Feb 16	80 m
March	Data not available		
April	D	Apr 2 - Apr 10	64 m
May	F	May 13 - May 15	59 m
	S	May 25 - May 31	80 m
June	S	Jun 1 - Jun 30	80 m
July	S	Jul 1 - Jul 16	80 m
		Jul 17 - Jul 31	80 m
August	A	Aug 1 - Aug 16	80 m
	X	Aug 21 - Aug 28	80 m
September	B	Sept 1 - Sept 23	82 m
	M	Sept 24 - Sept 30	80 m
October	M	Oct 1 - Oct 31	80 m
November	M	Nov 1 - Nov 14	80 m
	C	Nov 21 - Nov 30	50 m
December	C	Dec 1 - Dec 18	50 m
	O	Dec 19 - Dec 31	80 m

[a]The current meter is located 3 m above the bottom

of a particular speed class (0-10 cm sec^{-1}, 10-20 cm sec^{-1}, etc.) during a given month. If the length of the vector sum equals zero, then the directional variability of that speed class would be completely random during that period. The existence of a vector sum indicates that bottom currents of a certain speed flowed in one particular directional octant more often than in others during the month, and the length of the vector sum represents the number of hours that a preferred direction of movement occurred. An example of this analytical technique is shown in Fig. 79. The results for all monthly data are shown in Fig. 80. For the short-duration records it is assumed that the percentages calculated in Fig. 78 are representative of the total month. The data for Fig. 80 are tabulated in Table X.

A vector average for each speed class also has been determined for the total year. The results are shown in Table XI and Fig. 81 where vectors represent the net time during the year that a preferred direction occurred for each speed range. The annual vector sums are exactly analogous to the monthly vector sums as discussed above except for the time scale.

DISCUSSION

The most noteworthy aspect of this form of data analysis is the emphasis on current direction while retaining the integrity of the current speed measure-

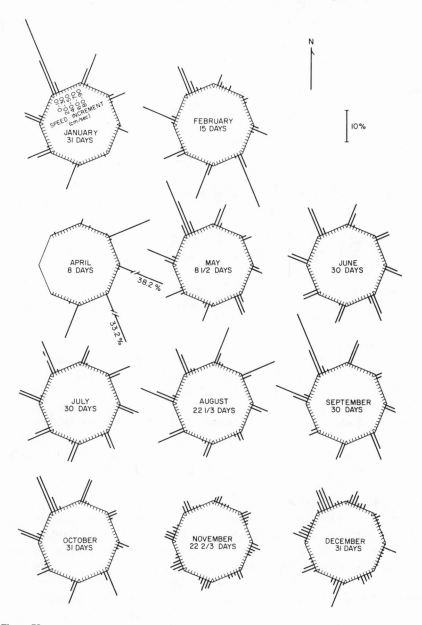

Figure 78.

Velocity data of Hopkins. Data are sorted according to month, speed class, direction (by octants), and time. The length of a vector represents the percent of the month that a current of a given speed increment and direction occurred. The sum of vectors for each month equals 100%.

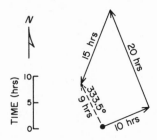

SPEED = 20-30 cm/sec

OCTANT (NOMINAL DIRECTION)	NUMBER OF 20 MINUTE AVERAGES	TOTAL TIME (HOURS)
2 (67.5°)	30	10
8 (337.5°)	60	20
5 (202.5°)	45	15
VECTOR SUM 333.5° = OCTANT 8		9

Figure 79.

Example of the construction of a progressive vector diagram for the speed class 20-30 cm sec[-1] where vector length equals time and vector orientation represents flow direction by octant (see Table IX for the numerical designation of octant).

ments. The vector sums may be considered as the mean directional components for the various speed classes after the random directional fluctuation has been averaged out. Although Fig. 78 indicates a strong directional variability for most speed classes, Figs. 80 and 81 show a significant northward trend especially for the stronger currents, those associated with sediment motion. These observations are discussed in detail by Hopkins (1971a) and summarized by Smith and Hopkins (this volume).

The northwest coast of the United States is characterized by northerly winds in the summer months and southerly winds throughout the winter season. Taking the winds and density stratification into account Hopkins (1971a) concludes that in the winter, with stratification low, the shelf bottom water moves primarily to the north and west in response to the barotropic pressure gradient. With the onset of summer conditions the northerly winds produce a barotropic flow to the south along the bottom on the central and outer shelf. After 1-2 days, however, a relatively strong baroclinic flow also develops which counteracts the barotropic pressure gradient and causes a net northward flow near the bottom on the central and outer shelf. Thus, even though the primary circulation results from the direct wind stress on the sea surface which changes direction seasonally, the resulting bottom flow is dominantly to the north.

Since the random fluctuations in direction have been averaged out of the data shown in Figs. 80 and 81, then the vector sums can be related, as a

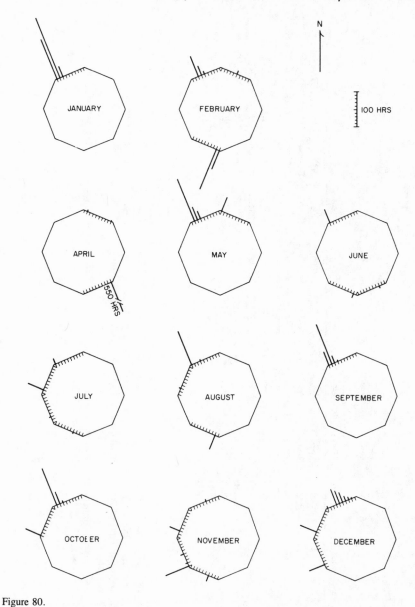

Figure 80.

Monthly vector sums of the speed classes. Data from Fig. 78 have been averaged according to time and direction so that the vectors represent the net time (in hours) that different current speeds (e.g., 0-10 cm sec^{-1}, 10-20 cm sec^{-1}) flowed in a preferred direction during each month. The speed classes are represented in similar fashion to Fig. 78.

Table X

Summary of Monthly Vector Sums for the Speed Classes[a]

Speed (cm sec⁻¹)	Jan Hours[b] (Octant)[c]	Feb Hours (Octant)	April Hours (Octant)	May Hours (Octant)	June Hours (Octant)	Jul Hours (Octant)	Aug Hours (Octant)	Sept Hours (Octant)	Oct Hours (Octant)	Nov Hours (Octant)	Dec Hours (Octant)
0-10	135(8)	61(5)	550(4)	47(1)	45(8)	19(8)	112(8)	45(8)	52(7)	79(6)	5(8)
10-20	204(8)	133(5)	3(2)	140(8)	13(5)	53(7)	46(5)	131(8)	121(8)	4(6)	52(6)
20-30	33(8)	68(8)		42(8)	5(4)	4(6)	6(7)	22(8)	39(8)	32(7)	57(7)
30-40		30(8)		23(8)		1(5)	3(8)	1(8)	5(8)	19(5)	50(8)
40-50		9(8)								10(6)	40(8)
50-60		10(1)								4(8)	31(8)
60-70											14(8)
70-80											7(8)
80-90											1(8)
90-100			1(1)								

[a] The number refers to the number of hours in a given month that a particular speed class flowed in a preferred direction. The numbers in parenthesis refer to the directional component (octant) of the net flow direction.

[b] To the nearest Hour.

[c] Numbers designate directional increment with respect to north: (1) = 0 - 45°; (2) = 45°- 90°; (3) = 90°- 135°; (4) = 135°- 180°; (5) = 180°- 225°; (6) = 225°- 270°; (7) = 270°- 315°; (8) = 315°- 360°.

Table XI
The Annual Vector Sums for the Speed Classes[a]

Speed cm sec^{-1}	Hours[b]	Octant[c]	Net annual displacement (km)
0-10	173	(4)	31
10-20	553	(7)	299
20-30	302	(8)	272
30-40	100	(8)	126
40-50	50	(8)	81
50-60	42	(8)	82
60-70	14	(8)	31
70-80	7	(8)	19
80-90	1	(8)	3
90-100	1	(1)	3

[a] See Table X for an explanation of the first three columns.
[b] to the nearest hour
[c] See Table X for explanation.

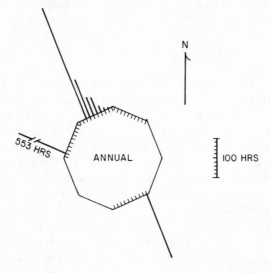

Figure 81.

Annual vector sums of the speed classes. Vectors represent the net time per year that currents of various speed classes flowed in a preferred direction. The speed increments are represented in similar fashion to Fig. 78.

first approximation, to the net monthly or annual transport of bottom water or sediment passing the station location on the continental shelf. A speed of 35 cm sec^{-1} one meter off the bottom is sufficient to erode and transport most sand-sized material. This corresponds to approximately 40 cm sec^{-1} 3 m off the bottom. Thus, it is seen from Figs. 78 and 81 that (a) high bottom velocities occurred during the winter months of November, December, and February, and (b) currents sufficiently strong to cause sediment movement had a net northward component for approximately 117 hr or almost 5 days during the year. A net northward component exceeding 70 cm sec^{-1} occurred for about one half day during the year. Considering that mass transport of sediment as bedload is proportional to $\overline{U}^3$, these higher velocities generated by the strong winds accompanying winter storms can account for significant quantities of mass transport.

The data summarized in the first two columns of Table XI also can be used for an order of magnitude estimate of the net annual displacement of bottom water on the continental shelf. These results are shown in column 4 of Table XI. Assuming that sedimentary particles move with the bottom water and that the threshold velocity is 40 cm sec^{-1} at 3 m above the bed, then column- 4 indicates that the maximum annual displacement of a sedimentary particle eroded from the bottom would be approximately 220 km in a northwesterly direction. If approximately six major storms sweep the Washington shelf per year then a sedimentary particle might be displaced as much as 35 km northward and 12 km westward per storm. These estimates are similar to those of Smith (1969, p. 4) who considered the net transport of suspended sediment across the continental shelf of Washington.

These results seem to point toward somewhat different generalities than those reached by workers on other shelves (e.g., Swift, 1969; Swift, 1970; Swift et al., 1971). The currents resulting from direct wind stress appear to be more important to the dynamics of the near-bottom-water transport on the central and outer continental shelf of Washington than those associated with wave surge (Hopkins, 1971a). Presumably this order of importance also applies to sediment transport. Furthermore, the motions observed, although highly variable, are not completely random and have a definite directional response to the dominant wind patterns in the area. The random fluctuations would produce significant lateral dispersal of sediment as proposed in Fig. 77, however, the net northward transport would generate a significant bias to the dispersal pattern that is not related to a random walk or other statistical assumption. It should be emphasized that the Washington shelf is a temperate latitude shelf exposed to the westerlies and the expanse of the Pacific Ocean. Thus, the observations do not necessarily apply to continental shelves in different oceanic settings. These results indicate, however, that the prediction of bottom currents or sediment transport on the Washington shelf by use of two-dimensional random-walk model would be misleading and significant modifications (such as those suggested by Swift et al., 1971) are warranted.

CONCLUSIONS

Considerations of results presented by Hopkins (1971a) and a specialized analysis of some of his data (1971b) reveal that:

1. The stronger bottom currents on the central continental shelf off the Washington coast are the result of meteorological conditions, most specifically wind stress associated with storm conditions rather than wave surge.

2. Measured current speeds at 3 m off the sea floor frequently exceed 40 cm sec⁻¹ during the winter months and may exceed 80 cm sec⁻¹ during severe storms.

3. Current direction vectors show a high variability throughout the year; however, when averaged over monthly and annual periods, bottom velocities exceeding 40 cm sec⁻¹ exhibit a net directional response, which on the continental shelf of Washington is dominantly northward.

4. The observed directional variability of the higher speed classes could account for significant sediment dispersal across the continental shelf; however, moving sediment would also experience a net northward migration as a result of the net transport of bottom water.

5. A sedimentary particle moving with the bottom water could have an annual displacement of as much as 220 km to the northwest along the continental shelf.

ACKNOWLEDGMENTS

The authors wish to thank Dr. Clifford Barnes who made Hopkins' data available for review. This work was supported by U.S. Atomic Energy Commission Contracts AT(45-1)-1725 and AT(45-1)-2200. University of Washington, Department of Oceanography; Contribution No. 668.

REFERENCES

Draper, L. (1967). Wave activity at the sea bed around northwestern Europe: *Marine Geol.* **5**, 133-140.

Dunbar, C. O., and Rodgers, J. (1957). "Principles of Stratigraphy." Wiley, New York.

Curray, J. R. (1960). Sediments and history of the Holocene transgression, continental shelf, northwest Gulf of Mexico. *In* "Recent Sediments Northwest Gulf of Mexico" (F. P. Shephard, F. B. Phleger, and T. H. van Andel, eds.), Am. Assoc. Petrol. Geologists, Tulsa, Oklahoma.

Hopkins, T. S. (1971a). On the circulation over the continental shelf off Washington. Ph.D. thesis, Univ. of Washington, Seattle.

Hopkins, T. S. (1971b). Velocity, temperature, and pressure observation from moored meters on the shelf near the Columbia River mouth, 1967-1969. Special Rept. No. 45, Dept. of Oceanography, Univ. of Washington, Seattle.

Hopkins, T. S. (1971c). On the barotropic tide over the continental shelf off the Washington-Oregon coast: Special Rept. No. 46, Dept. of Oceanography, Univ. of Washington, Seattle.

Smith, J. D. (1969). Investigations of turbulent boundary layers and sediment transport phenomena as related to shallow marine environments. Part II. Final report, U.S.A.E.C. Contract AT(45-1)-1752 (RLO-1752-13).

Swift, D. J. P. (1969). Evolution of the shelf surface, and the relevance of modern shelf studies to the rock record. *In* "The *New* Concepts of Continental Margin Sedimentation" (D. J. Stanley, ed.). American Geological Institute, Washington, D.C.

Swift, D. J. P. (1970). Quaternary shelves and the return to grade. *Marine Geol.* **8**, 5-30.

Swift, D. J. P., Stanley, D. J., and Curray, J. R. (1971). Relict sediments on continental shelves: A reconsideration. *J. Geol.* **79**, 322-346.

CHAPTER 9

Shelf Sediment Transport:
A Probability Model

Donald J. P. Swift,* John C. Ludwick,†
and W. Richard Boehmer‡

ABSTRACT

For many purposes the most useful conceptual framework for the study of detrital sediments is that of the sediment transport system. A system is comprised of a sediment source, and a dispersal zone which is at once a conduit and sink. Sediment transport systems impress granulometric and other petrographic gradients on their deposits by means of progressive sorting and similar mechanisms.

Markov process modeling is a versatile and illuminating method for testing assumptions concerning sediment-transport systems. Transition probabilities are employed as operational substitutes for incompletely understood physical mechanisms. In particular, the granulometric evolution of a detrital sediment stream through a dispersal zone may be examined stage by stage.

A simple, one-dimensional model consists of a Markov chain of nine transient states with nonzero transition probabilities between adjacent states only. Each transient state has associated with it an absorbing state from which the system (sand grain) cannot exit. For all trials, the transient state on the left side of the model (proximal state) has an initial state probability of unity. This fixes the place of sediment introduction. Ten transition probability matrices, one for each half-phi fraction of a hypothetical input sand, are used in ten successive trials. Calculation of the limiting state probability distribution permits the further calculation of simulated sediment size frequency distributions across a hypothetical shelf.

The model shows that progressive size sorting will occur along a sediment transport system

*Atlantic Oceanographic and Meteorological Laboratories, National Oceanic and Atmospheric Administration, 15 Rickenbacker Causeway, Miami, Florida 33149.

†Institute of Oceanography, Old Dominion University, Norfolk, Virginia 23508

‡Marine Affairs Program, University of Rhode Island, Kingston, Rhode Island 02892

if the transport is occurring over a depositional surface. If the input size distribution is not approximately normal, the distribution will evolve towards that configuration during the sorting process. In several variants of the model, standard deviation tends to decrease and kurtosis to increase. Skewness may increase, especially if the transport competence declines across the system.

While a one-dimensional Markov model of shelf sediment transport is useful for analyzing the textural or compositional evolution of sediment, two-dimensional models should permit evaluation of depositional topography, and the hydraulic mechanisms that produce shelf sand bodies.

INTRODUCTION

The study of sediments has witnessed an evolution of the central conceptual model of the discipline, from a stratigraphic product model, in which attention was primarily focused on the distribution of lithologies in three dimensions, to a stratigraphic process model based on the characteristics of depositional environments which produced the lithologies, and the changes of these characteristics through time. A further evolution of the model appears to be occurring as our accumulated knowledge of ancient sediments is increasingly supplemented by hydraulic analysis of modern sedimentary environments. This trend is concerned with dynamic systems of sediment transport. Its ultimate goal is not total resolution of the depositional environment, but instead analysis of sediment flux and quantification of the sediment budget.

The study described below is our initial attempt at development of a probability model of shelf sediment transport, as a tool in the analysis of shelf-transport systems and effects on grain-size distributions. However, because we have had to start at a very fundamental level, we find that our model is adaptable to almost any transport system.

SEDIMENT TRANSPORT MODELS

A basic conceptual model of a sediment transport system requires a planetary surface. In our own case, we are concerned with interfaces between the earth's several fluid envelopes and the lithosphere. The model is driven mainly by solar energy expended in the presence of a gravitational field. The surface is conveniently divided into a sediment-source area, a sediment conduit or gradient (Tanner, 1962), and a sediment sink. A slightly more advanced model recognizes that the second two elements are to a large extent synonymous; hence the model might be more realistically divided into a source area and a dispersal zone.

A real-world sediment-transport system may consist of any of the following: (a) an eroding headland, a zone of littoral drift, plus the expanding tip of a spit; (b) a drainage basin, its river net, plus a marine delta; (c) a shelf with relict sediment, a canyon and submarine fan, plus an aggrading abyssal plain; or (d) a desert deflation surface and adjacent dune belt.

Such sediment-transport systems may be identified in nature on the basis of circumstantial evidence, since the systems serve as filters whose sediment outputs differ systematically from their sediment inputs. The differences are usually observed as gradients in grain-size parameters, mineral composition, or other easily measured petrographic attributes within the sediment transport system. Most of these petrographic gradients have been attributed to "progressive sorting," whose effects have been described by Russell (1939) as:

> Shown by progressive changes in the characteristics of sediments, and can be detected by the analysis of a series of samples taken in the direction of transportation. . . . Progressive sorting may produce a progressive decrease in the mean grain size of sediments. There appear to be two possible causes of such a decrease; a progressive decrease in the competency of the transporting agent, and the lagging behind of the larger particles due to fluctuations in competency.

The latter idea has been expressed more succinctly as "the fines outrun the coarser particles." The concept of a sediment transport system identified by petrographic gradients appeared very early in sedimentological literature, and is ably summarized in the chapter on dispersal in Pettijohn's second edition (1957) of *Sedimentary Rocks*; but has received little notice in the interim. Since then attention has been focused on regional patterns of bedforms as indicators of sediment transport or "paleocurrent systems." More recently, examination of modern sediment transport systems has resulted in renewed interest in diagnostic petrographic gradients of sediment transport systems. Some recent studies of modern sediment transport systems revealed by petrographic gradients are the reports of Northrup (1951) on the southern New England shelf, of Stride (1963) on the sand streams of the western European shelf, Allen's (1965) analysis of prodelta sands on the Nigerian shelf, the studies by James and Stanley (1968) on sediment transport on the Nova Scotian shelf, and Beall's (1970) study of the fractionation of fine sand in the Mississippi River delta. These studies suggest that the "graded shelf" hypothesis of Johnson (1919) and the "relict shelf" hypothesis of Shepard (1932) and Emery (1952) are both partly correct. It appears that with time, continental margins tend to develop integrated cross-margin sediment transport systems. Systems of modern shelves have been disrupted by the Pleistocene sea-level fluctuations, but textural grade may be beginning to return (see Swift et al., 1971). We are not concerned here with larger implications of this premise, but merely note that textural gradients do locally exist in shelves, and propose a probabilistic model for their generation.

At this large scale of observation, the deterministic laws of sediment transport, as elucidated by hydraulic engineers and experimental sedimentologists, are not directly useful. Modeling of sediment-transport systems calls rather for a sort of statistical mechanics in which the element processed is seen as a population of particles whose behavior is best predicted by a probabilistic model. Deterministic laws of sedimentation have allowed us to estimate the behavior of a single-sediment particle entrained by a fluid on the basis of such parameters as grain size, shape, and density; and water temperature, density, and viscosity. Aggregate behavior is commonly inferred from single grain calculations. Probabilistic models postulate individual grain behavior, but permit direct evaluation of aggregate behavior, by focusing on a fundamental characteristic of entrained sediment; the propensity of individual particles to travel in a sequence of discrete, independent trajectories of varying orientation punctuated by periods of rest. These trajectories occur on at least two scales. At the smallest scale, saltating or rolling grains entrained in moving turbulent fluids undergo intermittent movement, with mean travel times on the order of seconds. Migrating bedforms impose residence times of their constituent grains of minutes to centuries. Grains suspended by fluids are subject to small-scale turbulence, and their trajectories may be viewed as a sequence of discrete steps separated by infinitely small residence times. A probabilistic model of sediment transport in random steps has recently been devised by Yang (1971a,b).

But sediment transport of the scope described on previous pages is intermittent on a larger scale. Rivers move most of their sediment during floods, and shelf surfaces are quiescent between storms, while hemipelagic environments are dependent on intermittent turbidity currents for sand input.

DIFFUSION MODEL

Thus a basic probabilistic model of sediment transport is a diffusion model. This is not to imply that the mean transport vector in such a system is zero. A case commonly studied in thermodynamics is a cylinder, partitioned into two chambers, one of which is evacuated. Air molecules in the remaining chamber are characterized by successive movements whose path lengths and orientations may be described as randomly distributed. If the partition is removed, the molecules redistribute themselves uniformly throughout the cylinder by means of these random movements. During the redistribution process, however, there is a net movement from one part of the cylinder to another. A simple geological analogy would be a hypothetical shelf with fluvial sediment input linearized by littoral drift. Sand escaping this linear source may then be distributed over the shelf by intermittent storm-generated currents of widely varying orientation and intensity (Swift, 1970). Net sand transport, however, must be directed seaward across the shelf in this hypothetical example.

But this example is unrealistic, since winds are systematically distributed within storms, and storm tracks are systematically distributed over the earth's surface. Therefore, a preferred sediment transport direction is to be expected. However, probabilistic methods are amendable to anisotropic situations, and are generally preferable to deterministic ones when the diffusive component (short term, variable movements) in the transport process is large with respect to the advective ("mean current") component.

Diffusion is a process by which matter is transported from one part of a system to another as a result of random motions (Crank, 1957). The simplest diffusion model would consist of a surface with polar coordinates and a point sediment source at the origin. If a coefficient of diffusion, D, is assigned to the sediment, then the concentration C of sediment over the surface at time T and distance x is given by

$$C = \frac{1}{2(\pi DT)^2} e^{\frac{-x^2}{4DT}}$$

Figure 82, showing concentration-distance curves for successive values of DT, is of particular interest to sedimentologists, for if different values of D are assigned to different size classes of sediment, and these classes are allowed to diffuse simultaneously, then the set of values C_{D1}, C_{D2}, C_{D3} . . . C_{Dn} for each successive value of x would correspond to the successive-size frequency distributions of a sediment-transport system in which progressive sorting is occurring.

As presented, the model is inadequate for the analysis of sediment transport. Preferred directions of movement cannot be programmed. A more subtle but equally serious problem is that the model makes no provisions for the withdrawal of sediment from the system into permanent storage, analogous to deposition. The result is that a sequence of ephemeral states may be observed, in which different geographic stations have different size frequency distributions. However, in the equilibrium state the model has been depleted of all sediment,

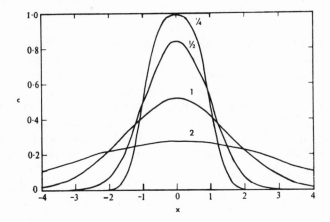

Figure 82.

Concentration-distance curves for a diffusion model. x = distance; c = concentration. Successively flatter curves are for successively larger values of DT.

or if it is a closed system with reflecting boundaries, the concentration for each size class is homogeneous throughout the model and size distributions at every point are equivalent to that of the input distribution. As noted by Pettijohn (1957, p. 541) this problem has disconcerting implications for Russell's (1939) concept of progressive sorting, in which the "fines outrun the coarse particles." The problem can be restated in graphic terms by considering an expressway at rush hour, whose traffic consists of cars and trucks. The onset of rush hour is signaled by a truck-free wave of the faster cars. However, after the arrival of the first slow-moving trucks at the point of observation, the mix of trucks and cars must remain constant, as long as the supply and speeds of each are constant. Clearly, for progressive sorting to occur in a sediment transport system, the sediment conduit must also be a sediment sink. Sediment particles must be selected for permanent deposition, and some sizes of particle must have higher probabilities of being so selected than others.

More advanced diffusion models overcome these difficulties. The simplistic geometry of Fig. 82 can be manipulated, and diffusion constants can be made to vary in time and space so as to simulate advective transport (Jost, 1960). Aspects of diffusion theory relating to heat sinks may be used to simulate selective deposition of sediment. However, the mathematics are abstruse, and it is difficult to intuitively grasp the evolution of particle populations modeled in this fashion.

MARKOV PROCESS MODEL

An alternative model using Markov processes permits ready visualization of aggregate-grain behavior during sediment transport. A Markov process model consists of states, which for the purposes of a sediment-transport model are locations on a Cartesian surface (Fig. 83). The term "state" is unfortunate in that it commonly has a time rather than a space connotation, but it is thoroughly established in Markov literature. Each state has associated with it a probability of transition to every other state in the chain. For a sediment transport model, these transition probabilities have nonzero values only between adjacent states. An important aspect of first-order Markov models is that they exhibit the Markov property: the probability p_j of the "system" (sand grain, for our purposes) occupying a given state, j, after having occupied the previous state i, is dependent on the probability $p_{i \rightarrow j}$, the probability of moving from state i to state j. First-order Markov models, therefore, have short "memories" extending for only a single transition. Higher order Markov processes "remember" more than one transition; since each state has associated with it probabilities for two or more successive steps.

Markov models, like diffusion models, may attain equilibrium. As well as transition probabilities, each state in a Markov chain has associated with it

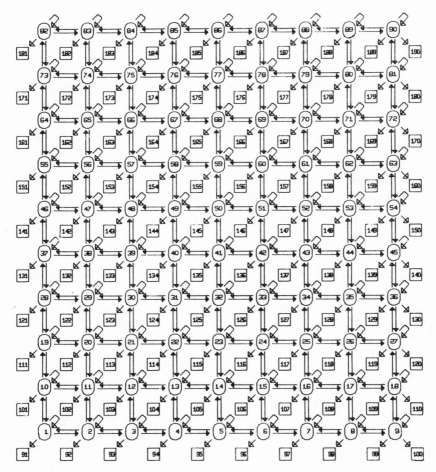

Figure 83.

A two-dimensional Markov model for simulating shelf sediment transport. Arrows are paths of transition between states. Circles are transient states which the system (a sand grain) can both enter and exit. Squares are absorbing states, with entrance paths but no exit paths. Transition probabilities for each path are associated with the source state. State probabilities, also associated with states, are the probabilities of the system occupying respective states after n transactions.

a state probability, or the probability of the system being that state after n transitions. With successive transitions, the state probability of a given state may asymtotically approach the limiting-state probability. For the Markov model of the kind shown in Fig. 83, the equilibrium distribution of probabilities is homogeneous only if all transition probabilities are equal. For a sediment transport model, the constraint that nonzero probabilities may occur only between adjacent states in the model means that strictly speaking, a uniform distribution of equilibrium probabilities will be impossible; there will always be boundary effects. Boundary states must have fewer permissible transitions; hence at least some of the associated transition probabilities must be larger than those of interior states.

This Markov model (Fig. 84) can be visualized as a sediment transport system by allowing a sand grain to work its way through it. Any state may be chosen as the starting state. The permissible exits from each state are given equal probabilities, and are assigned numbers one through four for north, east, west, and south transitions. The successive transitions may be selected by means of a random number table. If a large number of grains are introduced simultaneously, then "sediment" in a state of random diffusion is simulated. Preferred directions and pathways of sediment movement can be programmed into the model by specifying the appropriate initial state, and an appropriate matrix of transition probabilities.

A significant virtue of a Markov model is its ability to simulate an open system, whereby in the case of a sediment transport model, sediment is permanently lost at appropriate sites to deposition and burial. This is accomplished by adding states whose only nonzero transition probability is to itself: an autotransition. In the Markov models shown in Figs. 83 and 84, such absorbing states are represented as squares. For the sake of a sediment-transport model, they can be pictured as existing in a plane below the plane of the transient states.

The introduction of absorbing states into a Markov model renders it nonergodic; that is, the limiting state probability distribution is now dependent on the selection of the starting state. The "equilibrium" attained by a nonergodic Markov model simulates a static rather than a dynamic equilibrium. An input-sediment load, simulated by many trials, diffuses through such a chain until all grains are in absorbing states, rather like the conclusion of pin ball game. Thus, in a nonergodic Markov model of sediment transport, both deposits and sediment in transit are simulated.

PROGRESSIVE SORTING IN A MARKOV PROCESS MODEL

A Progressive Sorting Model

The Markov process model for sediment transport described in the previous section may be used to further explore Russell's concept of progressive sorting. For simplicity's sake, a section (Fig. 84) through the two-dimensional model of Fig. 83 has been selected to simulate a generalized bedload transport system. State 1 is the *initial state* or sediment source where successive "grains" may be introduced. States 1 through 9 are transient states representing a sediment pathway. The pathway is leaky since each transient state is associated with an absorbing state (states 10 through 18), and a final absorbing state, 19 represents

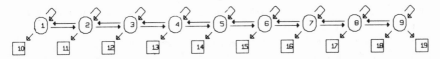

Figure 84.

A one-dimensional Markov model, equivalent to a cross section through Fig. 83. Circles are transient states; squares are absorbing states. Each set of one transient state and subjacent absorbing state comprises one geographical station in the Markov model. Farthest left station is shoreline, farthest right station, consisting only of an absorbing state, is the shelf edge.

Table XII
Size-Probability Matrix, Distribution Linear with Size and Invariant with Distance

	Grain size class									
	Coarsest ←									→ Finest
	1	2	3	4	5	6	7	8	9	10
Forward[a]	0.20	0.23	0.26	0.29	0.32	0.35	0.38	0.41	0.44	0.47
Backward[b]	0.20	0.23	0.26	0.29	0.32	0.35	0.38	0.41	0.44	0.47
Autotransition[c]	0.30	0.27	0.24	0.21	0.18	0.15	0.12	0.04	0.06	0.03
Trap[d]	0.30	0.27	0.24	0.21	0.18	0.15	0.12	0.04	0.06	0.03

[a] Doubled for first station, 0.000 for east station.
[b] 0.000 for first and last stations.
[c] 1.000 for last station.
[d] 0.000 for last station.

a sink at the end of the sediment pathway. Thus the 19 states combine to form a linear series of 10 geographical stations. This generalized system is for our purposes, a cross-shelf transect with a river mouth, aggrading shelf floor, and shelf edge; but it could be with suitable modifications a river, a littoral drift cell, a submarine fan valley, or any other conceivable transport system. In order to simulate progressive sorting in this Markov model it is necessary to consider a family of ten matrices of transition probabilities, one for each half Phi size class. The matrices are designed so that "very coarse sand" is most prone to being trapped, and "very fine sand" is the least prone. For simplicity, we have chosen a family of matrices with a linear relationship (Table XII) although in nature, the expression for the mobility of successive grain sizes must be a power function.

Procedure

The simplest method of determining the size frequency distributions of the deposits in absorbing states 10 through 19 would be to conduct, with the aid of a random number table, thousands of trials for each of the 10 matrices, then determine the relative proportions of trapped grains in equivalent states of each. This is an impractical procedure, and the same result can be obtained by calculating instead the limiting state probabilities of the states (Howard, 1960; pp. 3-7) for each of the 10 matrices. The initial state probability distribution for the Markov model of Fig. 84 can be expressed as a row vector; $\begin{bmatrix} 1 & 0 & 0 & 0 \\ \end{bmatrix}$ ... 0] indicating that when the system (sand grain) is in its initial state, there is a unit probability that it is in state 1. The state probability vector after the first transition S(1) can be found by post-multiplying the initial state probability vector S(0) by the transition probability matrix, P. Thus

$$S(1) = S(0)P$$

Likewise

$$S(2) = S(1)P = S(0)P^2$$

In general, the state probability vector of any transition S(n) can be found by post-multiplying the initial state probability vector S(0) by the nth power of the transition probability matrix P.

$$S(n) = S(0)P^n$$

In each state of the chain the successive state probabilities associated with successive transitions forms a sequence that asymptotically approaches the limiting state probability, thus

$$S(L) = S(0)P^{\infty}$$

Limiting state probability distributions can be calculated by recursion, by solving simultaneous equations, or by means of a zeta transform (Howard, 1960, pp. 3-12).

Sorting of a Rectangular Input Distribution

In order to model progressive sorting, ten transition probability matrices (see Tables XII and XIII) were designed to simulate the diffusion of 10 half phi sand classes through the Markov model of Fig. 84. Since the limiting-state probabilities values of the 10 chains were equally weighted a rectangular (uniform) input distribution has been assumed.

Limiting state probability distributions generated by these transition probability distributions are presented in Fig. 85a as percentages, indicating the distributions of different grades of sand across the "shelf." Figure 85a indicates that within each size class, the sediment is distributed according to a negative exponential function of the variety $Y = e^{-ax}$ where Y is the percent of sand present at distance x, e is the base of natural logarithms, and "a" is a coefficient determining the rate of sediment size decrease. This coefficient must become larger for successively finer-grained size classes.

Strong boundary effects are present at the ends of the geographic distributions.

Table XIII

Transition Probability Matrix for Size Class 1, Distribution Linear with Size and Invariant with Distance

From station	To station																		
	1	2	3	4	5	6	7	8	9	10	11	12	13	14	15	16	17	18	19
1	0.30	0.40								0.30									
2	0.20	0.30	0.20								0.30								
3		0.20	0.30	0.20								0.30							
4			0.20	0.30	0.20								0.30						
5				0.20	0.30	0.20								0.30					
6					0.20	0.30	0.20								0.30				
7						0.20	0.30	0.20								0.30			
8							0.20	0.30	0.20								0.30		
9								0.20	0.30									0.20	0.30
10										1.00									
11											1.00								
12												1.00							
13													1.00						
14														1.00					
15															1.00				
16																1.00			
17																	1.00		
18																		1.00	
19																			1.00

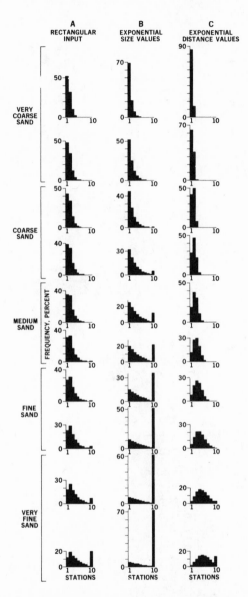

Figure 85.
Percent deposition of sand by geographic station for each of the 10 sizes classes, as calculated for the 4 variants of the model. Each distribution is a cross-section through the model, showing where most of the sand of that particular size class was deposited.

The "river mouth" (station 1) is an area of retarded deposition due to the fewer exit paths of its transient state and the necessarily higher probabilities associated with them, a not unrealistic feature in that real estuaries often have scour trenches associated with their mouths. The last station (10) corresponding to the upper continental slope, has an anomalously high deposition as a consequence of having no exit paths. The anomalies are the most pronounced for the finest, most mobile sediment class. They appear in plots of size parameters against distance (Fig. 87), as well in the geographic and grain size distributions of Figs. 85a and 86a.

Vertical sections through each of the size classes of the 10 geographical distributions are recalculated to 100% in Fig. 86a. These constitute the grain size frequency distributions of sand deposited at these stations.

In successive stations, the mode may be seen to progress through the sequence of size frequency distributions in a wavelike fashion, and the variation in mean size (Fig. 87) indicates that this mathematical model does indeed simulate progressive size sorting. The concavity in the decreasing curve for mean and the steady decrease in the standard deviation (Fig. 87) results from autotruncation of the distribution as the wavelike progression of the mode reaches the lower limit of the initial size distribution.

It is noteworthy that the initial "rectangular" distribution appears to evolve toward a normal distribution during the sorting process. The size distribution at station 1 consists of a linear decrease in relative sand abundance toward the fine end of the distribution. At stations 2 and 3 however, a mode or maximum may be seen, with frequencies decreasing away from it in both directions. By station 6, the mode has reached the finest class. However, continued evolution towards a more nearly normal distribution is apparent as a gain in modal frequency at the expense of the remaining classes, and as an increasingly steeper exponential form for the coarse side.

Sorting of a Normal Input Distribution

A major unrealistic aspect of the Markov model as described above is that it operates on an input with a rectangular size distribution. The distribution appears to evolve toward normality, but the sediment-transport path as defined by the model is too short and the number of size classes is too limited for the process to attain completion. This matter is remedied in post facto fashion by multiplying the size frequency distributions by weighting factors calculated from the normal distribution function (Hald, 1952). The resulting size frequency distributions (Fig. 86b) are the same as they would have been if the input distribution were normally distributed through 3 standard deviations on either side of the mean. Geographic distributions remain the same as for the rectangular input.

The effect of this transform is that the shift in mean and the variation in shape parameters is greatly muted (Fig. 87). The cyclic variation that does occur in the parameters of higher moment (standard deviation, skewness, kurtosis) is best understood by dissection of the size frequency curves into two components

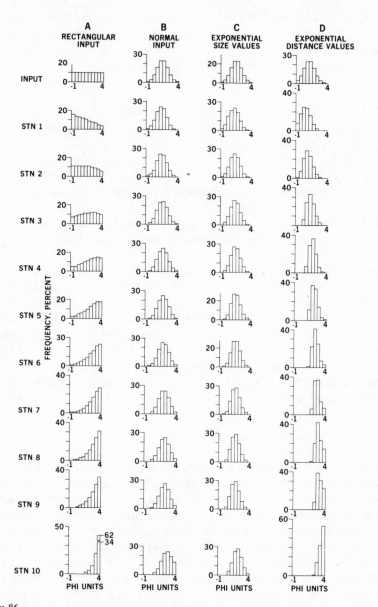

Figure 86.

Size frequency distributions at the 10 geographic stations of each of the 4 variants of the model described in this paper.

(Fig. 88): (a) An initially rectangular distribution through which a modal wave is seen to progress as successive frequency distributions are viewed, and (b) an invariant normal curve that has been imposed on each distribution by means of weighting factors. This curve represents the normally distributed input which the modal wave is modifying.

The modal wave may in turn be resolved into a central peak or zone of relative sand gain, and two flanking troughs, or zones of relative sand loss. In each zone, sand has been lost or gained relative to the baseline provided by the modal wave's initially rectangular shape.

The effect of the progression of the tripartite modal wave on the moment parameters of the normal distribution is a consequence of the mathematical and geometrical properties of these parameters. The normal curve may be viewed as the first member of a series of successive derivative curves, such that each curve represents a higher moment about the mean; the moment parameters are functions of these moments (Krumbein and Pettijohn, 1938, p. 25, p. 253). Skewness, for instance, is a function of the third moment about the mean, and of the second derivative of the normal distribution function. As such it is sensitive to changes in the four segments of the normal curve defined by its mode and inflection points, namely the distal coarse admixture, proximal coarse admixture, and proximal and distal fine admixtures (Fig. 88). In particular it is sensitive to changes in these segments that effect the symmetry or "tailedness" of the curve.

In station 1, for example, the forward zone of sand loss has developed over the fine admixture, while the distal zone of sand loss has developed over the coarse admixture (Fig. 88). These losses and gains are strongest over the distal portions of the admixtures. The result, barely discernable in the histogram of Fig. 86b, is that the coarse side of the curve has flattened (developed a tail) and the fine side has steepened (become truncated); skewness has become negative. In station 2, the central zone of sand gain now effects the proximal-fine admixture as well as the coarse admixtures; only the distal-fine admixture is now suppressed. The dominant effect is the differential modification of the two fine admixtures; with the proximal one amplified and the distal one suppressed, this side of the curve is rotated into a more nearly vertical position; skewness is more negative than in station 1. At station 3, however, the near zone of loss dominates the distal coarse admixture while the central zone of gain dominates the fine admixture (Fig. 88). The effect is to coarse truncate the distribution and to generate a tail on the fine side noticeable in the histogram of Fig. 86b as an additional fine class. Skewness is now positive. At station 4, the near zone of loss is now suppressing the proximal coarse admixture as well as the distal coarse admixture; steepening of this side is less effective, and skewness is not as positive as at the previous station.

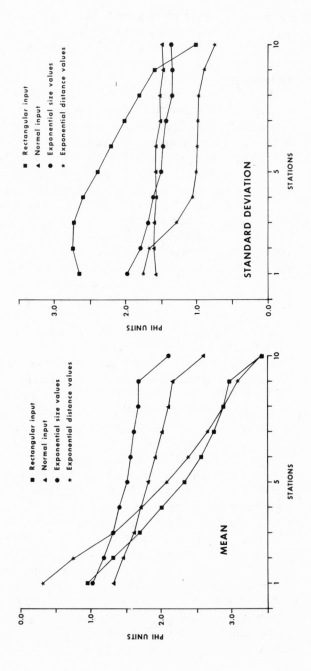

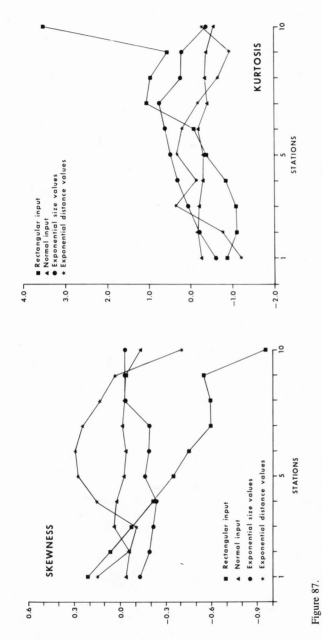

Figure 87.

Phi moment parameters for the 4 versions of the one-dimensional Markov process model described in this paper.

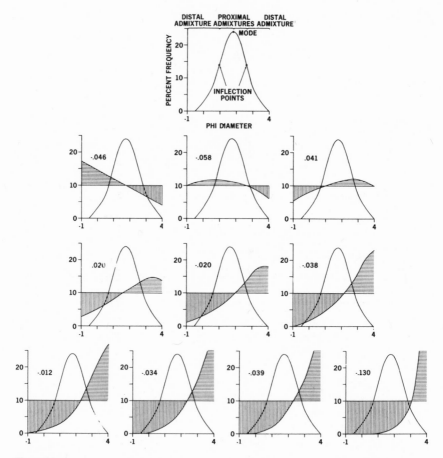

Figure 88.

Diagram illustrating interaction of zones of relative loss and gain of "modal wave" with distal and proximal admixtures of normal input curve. Ruled areas are defined by initial rectangular distribution and its subsequent modification by progressive sorting. Horizontally ruled sector is zone of relative sand gain; vertically ruled areas are forward and rear sectors of sand loss. The normal curve is the weighting distribution used in the experiment, and is invariant from station to station. Numbers are values of moment skewness for the resultant distribution.

Thus as the modal wave progresses through the normal input curve its 3 zones interact with the 4 curve segments so that the distribution's "tails" become more pronounced first on one side then on the other. Small-scale fluctuations must simultaneously occur in the other higher moment parameters, since the changes in curve shape also effect the spread of curve area about the mean (standard deviation), and the hollowness of the sides of the curve (kurtosis).

Table XIV
Size-Probability Matrix, Distribution Exponential with Size and Invariant with Distance

	Grain size class									
	Coarsest ←									→ Finest
	1	2	3	4	5	6	7	8	9	10
Forward [a]	0.196	0.321	0.362	0.432	0.459	0.475	0.485	0.491	0.494	0.496
Backward [b]	0.196	0.321	0.362	0.432	0.459	0.475	0.485	0.491	0.494	0.496
Auto-transition [c]	0.304	0.179	0.138	0.068	0.041	0.025	0.015	0.009	0.006	0.004
Trap [d]	0.304	0.179	0.138	0.068	0.041	0.025	0.015	0.009	0.006	0.004

[a] 0.000 for last station.
[b] 0.000 for first and last station.
[c] 1.00 for last station.
[d] 0.000 for last station.

Sorting with an Exponential Size Probability Distribution

A further unrealistic aspect of the Markov model is its family of transition matrices which have been assigned a linear relationship as a function of grain size. Experimental work has suggested that transport rates bear an exponential relationship to grain size (for instance Einstein, 1950). In this initial experiment we have not distributed the probabilities in accordance with any particular bedload function such as Einstein's, since the shelf regime that we are modeling is not one of continuous flow, but instead one of short periods of intensive movement, followed by long periods of quiescence. Thus the distribution of bed shear stress with time would be at least as important as the laws governing water-grain interaction under conditions of continuous flow in determining the transport rates of successive grain size classes. We are, therefore, free to adopt any size probability function of the general form $Y = 0.5 - e^{-axb}$ where Y is the probability of a grain moving forward or backward out of a station, x is here a number between 1 and 10 representing the 10 grain sizes in the model, and "a" and "b" are constants determining the shape of the curve. An "a" of 0.5 and a "b" of 1 were used to calculate the probabilities shown in Table XIV. The probability of a grain moving forward is $Y/2$, as is the probability of moving backward, and the probability of staying in place is $0.5 - Y/2$, as is the probability of a grain being trapped out.

With probability distributions thus exponentially distributed by size, the size distributions and geographic distributions of Figs. 85b and 86c result. Standard deviation decreases and kurtosis increases as the size range decreases with passage through the system. Skewness is negative in the near stations. Early losses in *both* the proximal and distal coarse admixtures of the size distribution result in a decrease in slope of the coarse side of the frequency curve. The skewness becomes less negative in the far stations. The greater difference in transport probabilities assigned to the coarser sand results in greater efficiency of sorting on that side of the curve, and progressive steepening of that side.

Sorting with an Exponential Distance Probability Distribution

Variations in size parameters in the sediment transport model now resemble those of actual sediment transport systems, with the principal exception of skewness. Many transport systems have a tendency to separate into two textural provinces defined by the skewness of the size frequency distributions produced; a proximal province of negatively skewed (coarse-tailed) lag deposits, and a distal province of positively skewed (fine-tailed) sediment. Beach sands tend to be negatively skewed, while derivative sands, in the back-beach dune on one side (Shepard and Young, 1961) and on the shore face on the other (Swift and others, 1971) are positively skewed. Crestal sands of shelf shoals are coarse-tailed while flank sands are fine-tailed (James and Stanley, 1970). Some notable exceptions occur. Sands of the Cretaceous Shelf of the Carolinas were initially fine-tailed, and trended toward symmetry with distance across the shelf (Swift and others, 1969). These sands were muddy, with 10% or more of silt and clay, hence the initial positive skewness. Since all sizes finer than gravel were present in the proximal deposits, progressive sorting could only reduce the skewness by autotruncation.

Transport systems whose deposits vary systematically from negative to positive skewness along the transport path are characterized by a downstream decline in competence. The proximal coarse-tailed lags are generated on high-energy shoals, while the distal fine-tailed deposits accumulate in deeper, quieter water. This characteristic can be simulated in our model by means of an exponential decrease in the transport probabilities for each grain size with distance through the system. The equation cited previously is used to generate probabilities, but in this case the distance-probability distribution rather than the size-probability distribution is obtained. The x is again a number between 1 and 10, this time representing the 10 stations. The ''a'' is now a grain-size coefficient and is itself exponentially distributed according to the formula $a = S^{(2-n)}$ where S is an arbitrary starting value, and n is the number in the series. To prepare the probability distributions of Tables XV and XVI, S was set equal to 2, resulting in the series, 2, 1, 1/2, 1/4. . . . For each of these values of the size constant ''a'' a distance-probability distribution was determined, using a curve-determining constant ''b'' of 2.

Application of the resulting distance-variant transition probability matrices to the model generates sediment size frequency distributions (Fig. 86d) in which skewness increases through the transport system (Fig. 87) until the trend is reversed at the last few stations by autotruncation of the distribution. We conclude that if a sediment-transport system, for instance a modern shore face (Fig. 89), is characterized by increasing positive skewness in its distal portion, then this trend is probably the consequence of a down-current decrease in energy level.

Other benefits accrue from the distance-variant model. Progressive size sorting is enhanced; the curve for mean-grain size in Fig. 87 is as steep as for the rectangular distribution. The trend of mean diameter with distance through the transport system (Fig. 87) is no longer nearly linear, but becomes an exponential decay curve of the sort $Y = e^{-ax}$, similar to the dispersal curves seen in nature (Pettijohn, 1957, p. 525-577). For the first time there is a real geographic separation of size classes (Fig. 85c).

An Alternative Characterization of Progressive Sorting

In preceding sections a model for the progressive size sorting of a sediment population has been examined. During the examination, the observer has in effect moved with the sediment, by viewing populations at successive downstream stations. In this section an alternative approach is used, wherein probability vectors and frequency distributions are considered from the vantage point of a single station.

For the purpose of the ensuing discussion, we propose the following terms and associated definitions:

system input frequency distribution denotes the grain-size distribution of sediment introduced into a model at one end as if from a source;

local input frequency distribution denotes the altered grain-size distribution that is available to enter a given local geographic location or site downstream from the point of system input;

admittance vector denotes an orderly arrangement of admittances each of which is a probability that a particle of stated characteristics and present in a local input can enter a local geographic site. Different admittances in the vector correspond to particles of different size, or shape, or density;

Table XV
Size-Probability Matrix, Distribution Exponential with Size and Distance, Station 1

	Grain size class									
	Coarsest ←									→ Finest
	1	2	3	4	5	6	7	8	9	10
Forward	0.034	0.184	0.303	0.389	0.444	0.470	0.485	0.492	0.492	0.498
Backward	0.000	0.000	0.000	0.000	0.000	0.000	0.000	0.000	0.000	0.000
Autotransition	0.500	0.500	0.500	0.500	0.500	0.500	0.500	0.500	0.500	0.500
Trap	0.432	0.316	0.197	0.011	0.059	0.030	0.015	0.008	0.004	0.002

Table XVI
Transition Probability Matrix for Size Class 5, Distribution Exponential with Size and Distance

From station	To station																		
	1	2	3	4	5	6	7	8	9	10	11	12	13	14	15	16	17	18	19
1	0.500	0.441							0.059										
2	0.303	0.197	0.303							0.197									
3		0.162	0.338	0.162							0.338								
4			0.068	0.432	0.068							0.432							
5				0.022	0.478	0.022							0.478						
6					0.006	0.494	0.006							0.494					
7						0.001	0.499	0.001							0.499				
8								0.500								0.500			
9									0.500								0.500		
10										0.100									
11											0.100								
12												0.100							
13													0.100						
14														0.100					
15															0.100				
16																0.100			
17																	0.100		
18																		0.100	
19																			0.100

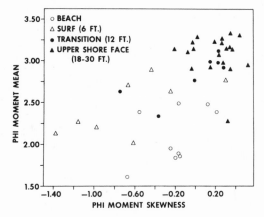

Figure 89.

Distribution of skewness on the shore face, Virginia-North Carolina coast. Data from Swift et al., 1971.

exit vector denotes an orderly arrangement of numbers each of which gives the probability that a particle of stated characteristics present at a local geographic site can exit, or leave, that site. Different probabilities corresponding as before to particles of different size, or shape, or density;

retention vector denotes an arrangement of probabilities each of which is obtained by subtracting the corresponding element in an exit vector from unity; and

retained distribution denotes a population of particles that were admitted by a geographic site and did not exit from that site.

If p_{jn} is an element in an admittance vector, where j denotes the jth station and n denotes one of 10 grain-size classes, and if p'_{jn} is a corresponding element in an exit vector for the same station, then the product of the two probabilities, $p_{jn}(1-p'_{jn})$, gives the probability that a particle in the local input enters and does not leave station j, and hence is present in the retained distribution at that site. The product of all corresponding elements in the admittance and retention vectors for a station gives a probability density function for that station (Fig. 90).

It is likely, but not necessary, that the admittance and exit vectors at a site are identical. Thus the probability of entering and not leaving a station can be rewritten as $p_{jn}(1-p_{jn})$. There is no requirement that either vector should be linear. It is to be noted that if transportability of a given particle size is so great that admittance at a station is certain, then $p_{jn} = 1$, and the probability of retention at that station is nil, and the particle moves through and on to the next station. Also note that if transportability of a given particle size is so small that admittance is zero, then particles of that size cannot occur in the retained distribution at that station.

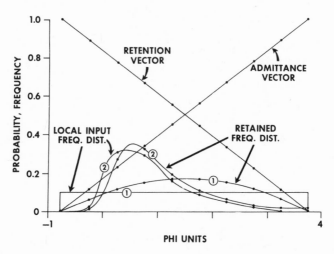

Figure 90.

Grain-size distribution at a station. For simplicity a local admittance vector and retention vector are linear and complementary. Local input distribution 1 is uniform; the shape of the corresponding retained distribution, 1, is solely dependent on the action of local admittance vector and retention vector. Local input distribution 2 is an arbitrary positively skewed, bell-shaped distribution; the corresponding retained distribution, 2, is altered in skewness by the action of the local admittance vector and retention vector.

The array of probabilities, $p_{jn} (1 - p_{jn})$, $n = 1, 2, \ldots, 10$, for a station does not necessarily indicate the form of a retained distribution, because what is retained is dependent also on shape of the local input distribution. If the local input is uniform, the array of probabilities referred to above gives the shape of the retained distribution. In general, the local input distribution is not uniform, and this causes a skewing of the form of the retained distribution. If I_{jn} is the frequency of occurrence of grains of size n in the local input to station j, then

$$R_{jn} = \frac{I_{jn} P_{jn} (1 - p_{jn})}{\sum_{n=1}^{n=10} I_{jn} P_{jn} (1 - p_{jn})}, \qquad n = 1, 2, \ldots, 10$$

is the frequency of grains of size n·in the retained distribution at station j. The local output from station j would usually be different than the local input to station j, because of the unequal partial removal of various grain-size classes.

The retained grain-size frequency distribution is, in this context, comprised of two significant parts: a coarser-grained part; and a finer-grained part. The extreme of the coarser-grained part is comprised of those available grains whose transportability is barely great enough for their entry into location j. The tail of the finer-grained part is comprised of those grains that although mobile were not quite mobile enough to escape location j. The central part of the retained distribution is less amenable to interpretation and reflects the crossing point of the admittance and retention vectors and biasing by the shape of the local input distribution. Grain-size sorting of the retained distribution might also better be conceived of as consisting of two measures: coarse-end sorting, or spread of the coarse tail of the retained grain-size distribution; and fine-end sorting, or spread of the fine tail of the retained grain-size distribution. These measures reflect the function of the admittance and retention vectors.

It is noteworthy that in the special case when the grain-size distribution of the local input matches the distribution, $p_{jn}(1-p_{jn})$, no change in grain-size distribution occurs with passage through location j. Ultimately all the sediment would be deposited along the transport path; however, the analysis described in the preceding few paragraphs does not deal with absolute amounts of sediment transported or deposited, but only with a progressive grain-size fractionization process.

Admittance and exit vectors are expressions for sediment transportability at station j, and relate to effects of fluid motions on sediment. It is the joint interaction of local input with these fluid motions that determines what is deposited. In the end the probability, p_{jn}, or admittance, is an alternative characterization of a sediment transport rate. A decrease in this rate with distance along a transport path will result in net deposition.

The evolution of a rectangular input frequency distribution towards a normal distribution, noted in a previous section, may now be seen as the consequence of the repeated application of complimentary admittance and retention vectors, or of a repeated "sorting event" (Middleton, 1968), in which a modal class is repeatedly selected for at the expense of distal classes. However, the phrase "[evolving] . . . towards a normal distribution" must be used with caution. The initially rectangular distribution of Fig. 86a is evolving toward a normal distribution to the extent that sorting is favoring the central portion (ultimately the fine distal admixture) at the expense of the distal portions. The application of successive admittance vectors to the coarse admixture causes its frequencies to become exponentially distributed, resulting in the sigmoidal shape characteristic of the sides of the normal distribution. No comparable process modifies the fine admixture, however, whose shape is determined primarily by the local retention vector. As a result, the retention distribution becomes more asymmetrical at successive stations, and, judged by the criterion of skewness, becomes increasingly less normally distributed. Thus, while the simplest version of the model modifies distributions toward more centralized patterns, a process inherent in the model inhibits the development of truly normal distributions.

CONCLUSIONS

Our Markov process model for simulation of a one-dimensional sediment-transport model has provided us with some fundamental insights into the generation and evolution of sediment size frequency distributions. We have watched rectangular input distributions evolve toward more centralized distributions. We have noted that progressive size sorting of normal size distributions will occur as long as the transport surface will accept grains for permanent deposition. If sediment mobility decreases exponentially with increasing size, there will be a concomitant down-current decrease in standard deviation and increase in kurtosis. If competence decreases with distance through the system, skewness will become markedly positive. These specific insights have been helpful, but the most valuable result of all has been a change in our conceptual approach to the interpretation of grain-size distributions from descriptive, statistical viewpoint to the probabilistic analysis of population dynamics.

We feel that we have by no means exhausted the possibilities of our model. To date we have attempted simulation of the simpler sort; we have changed our probability distributions in an arbitrary but reasonable way, until we obtained results that imitate those of nature. We have not yet attempted to work in the other direction; namely to obtain from the literature experimental and observational assessments of sediment movement probabilities, then to apply these to the model. A logical next step would be to mate our quantitative probabilistic model to the qualitative hydraulic approach of Inman (1944) to the nature of sediment sorting.

Real sediment transport systems may have more than one source (Cronan, 1972). Tributary sediment streams may join the trunk, resulting in hybrid size frequency distributions. This situation is easily modeled by attaching tributary sequences of states to the model of Fig. 84.

Two-dimensional models, such as the one in Fig. 83, may represent much more complex systems with several superimposed gradients. On open coasts the nearly coast-parallel, time-continuous surf-driven, littoral drift system is linked to the intermittent storm-driven system of coast-parallel currents seaward of the surf, by coast-normal processes such as rip currents, and the mid-depth return flow, and bottom wave-drift currents (Swift et al., 1971; Duane et al., 1972). Here longshore bars and troughs would be modeled by axes of convergence of transition probabilities; sand-circulation cells (Ludwick, 1972) by axes of convergence and shear. In two-dimensional models, matrices of limiting state probability distributions could be contoured for mean diameter or other textures. With some modifications isopleth maps of percent of total sediment deposited could be prepared, and interpreted as bathymetric maps of depositional topography. In these ways the dynamics of coastal and shelf morphology may be studied.

However, it is important to point out that Markov process models are subject to severe limitations. Like all wholly defined systems they constitute zero sum games. Considerable structural detail and precision is possible, but it is gained at the expense of dynamic realism, since it is difficult and perhaps ultimately impossible to translate real processes into probabilities. In the two-dimensional models stationarity promises to be an obstacle; bedload-substrate interactions occur as complex feedback systems, and in Markov terms, the transition probabilities must change with time. Higher order chains with multiple-dependence relationships must be explored. Ultimately, potential-flow theory must be added to the presently more limited forms of sediment transport modeling (Tanner, 1962, p. 114; Harbaugh and Bonham Carter, 1970, p. 205-254).

LIST OF SYMBOLS

Symbol	Definition
a	Shape constant for exponential curve
b	Shape constant for exponential curve
C	Concentration
C_{Dn}	Concentration for a given value n of D
D	Coefficient of diffusion
e	Base of natural logarithms
I_{jn}	Frequency of occurrence of grains of size at station j
i, j	Geographic stations in a Markov model
P	Transition probability matrix
p_{jn}	Probability of occurrence of grains of size at station j
R_{jn}	Frequency of occurrence of grains of size n in retained distribution at station j
S (L)	Limiting state probability vector
S (n)	State probability vector after n transitions
S (O)	Initial state probability vector
T	Time
x	Distance; grain size class; geographic stations
y	Transition probability between 2 transient states

REFERENCES

Allen, J. R. L. (1965). Late Quaternary Niger Delta, and adjacent areas: sedimentary environments and lithofacies. *Am. Assoc. Petroleum Geologists Bull.* **44**, 547-600.

Beall, A. O., Jr. (1970). Textural differentiation within the fine sand grade. *J. Geology.* **78**, 77-93.

Belderson, R. H., and Stride, D. H. (1966). Tidal current fashioning of a basal bed. *Marine Geology.* **4**, 237-257.

Crank, J. (1957). "The Mathematics of Diffusion." Oxford Univ. Press (Clarendon), London and New York.

Cronan, D. S. (1972). Skewness and kurtosis in polymodal sediments from the Irish Sea. *J. Sed. Petrology.* **42**, 1-2-106.

Duane, D. B., Field, M. E., Meisburger, E. P., Swift, D. J. P., and Williams, S. J. (1972). Linear shoals on the Atlantic inner continental shelf, Florida to Long Island. *In* "Shelf Sediment Transport: Process and Pattern" (D. J. P. Swift, D. B. Duane, and O. H. Pilkey, eds.). Dowden, Hutchinson and Ross, Stroudsburg, Pa.

Einstein, H. A. (1950). The bed-load function for sediment transportation in open channel flows: *U.S. Dept. of Agriculture, Soil Conservation Serv. Tech. Bull.* **1026**, 71 pp.

Emery, K. O. (1952). Continental shelf sediments off southern California. *Geol. Soc. America Bull.* **63**, 1105-1108.

Hails, J. R., and Hoyt, J. R. (1969). The significance and limitations of statistical parameters for distinguishing ancient and modern sedimentary environments of the lower Georgia Coastal Plain. *J. Sed. Petrology.* **34**, 559-580.

Hald, A. (1952). "Statistical Tables and Formulas." 97 pp, Wiley, New York.

Hjulstrom, F. (1939). Transportation of detritus by moving water. *In* "Recent Marine Sediments." (P. D. Trask, ed.), Am. Assoc. Petroleum Geologists, Tulsa.

Howard, R. A. (1960). *Dynamic programming and Markov processes.* Cambridge, M.I.T. Press, Cambridge, Massachusetts and Wiley, New York.

Inman, D. L. (1949). Sorting of sediments in the light of fluid mechanics. *J. Sed. Petrology.* **19**, 51-70.

James, N. P., and Stanley, D. J. (1968). Sable Island bank off Nova Scotia: sediment dispersal and recent history. *Am. Assoc. Petroleum Geologists Bull.* **52**, 2208-2230.

Johnson, D. W. (1919). *Shoreline processes and shoreline development.* (Facsimile, 1965), Hafner, New York.

Jost, W. (1960). *Diffusion in Solids, Liquids and Gases.* Academic Press, New York.

Kemeny, J. G., Snell, J. L., and Knapp, A. W. (1966). "Denumerable Markov Chains." Van Nostrand, Princeton, New Jersey.

Ludwick, J. C. (1972). Migration of tidal sand waves in Chesapeake Bay entrance. *In* "Shelf Sediment Transport: Process and Pattern" (D. J. P. Swift, D. B. Duane, and O. H. Pilkey, eds.). Dowden, Hutchinson and Ross, Stroudsburg, Pa.

Middleton, G. V. (1968). The generation of the log normal size frequency distribution in sediments. *In* "Problems of Mathematic Geology." Science Press, Leningrad.

Northrup, J. (1951). Ocean bottom photographs of the neritic and bathyal environmental south of Cape Cod, Massachusetts. *Geol. Soc. America Bull.* **62**, 1381-1383.

Pettijohn, (1957). "Sedimentary Rocks" (2nd ed.). Harper, New York.

Plumley, W. J. (1948). Black Hills terrace gravels: a study in sediment transport. *J. Geology.* **56**, 526-577.

Shepard, F. P. (1932). Sediments on the continental shelves. *Geol. Soc. America Bull.* **43**, 1017-1039.

Shepard, F. P. (1961). Distinguishing between beach and dune sands. *J. Sed. Petrology.* **31**, 196-214.

Spiegal, M. R. (1961). *Statistics.* Schaum, New York.

Swift, D. J. P. (1970). Quaternary shelves and the return to grade. *Marine Geology.* **8**, 5-30.

Swift, D. J. P. (1971). Relict sediments on continental shelves: a reconsideration. *J. Geology.* **79**, 322-346.

Swift, D. J. P., Heron, S. D., Jr., and Dill, C. E., Jr. (1969). The Carolina cretaceous: petrographic reconnaissance of a graded shelf. *J. Sed. Petrology.* **39**, 18-33.

Swift, D. J. P., Sanford, R. B., Dill, C. E., Jr., and Avignone, N. F. (1971). Textural differentiation of the shore face during erosional retreat of an unconsolidated coast, Cape Henry to Cape Hatteras, Western North Atlantic. *Sedimentology* **16**, 221-250.

Swift, D. J. P., Stanley, D. J., and Curray, J. R. (1971). Relict sediments on continental shelves: a reconsideration. *J. Geology* **79**, 322-346.

Tanner, W. F. (1962). Geomorphology and the sediment transport system. *Southeastern Geology* **4**, 113-126.

Yang, C. T., and Sayre, W. W. (1971). Longitudinal dispersion of bed-material particles. *J. Hydraulics Div. Proc. Am. Soc. Civil Engineers* **97**, 907-921.

Yang, C. T., and Sayre, W. W. (1971). Statistical model for sand dispersion. *J. Hydraulics Division, Am. Soc. Civil Engineers* **71**, 265-273.

II. Patterns of Fine Sediment Dispersal

CHAPTER 10

Transport and Escape of Fine-Grained Sediment from Shelf Areas

I. N. McCave

School of Environmental Sciences
University of East Anglia
Norwich, United Kingdom

ABSTRACT

Mud deposits on shelves occur as (a) muddy coasts, (b) nearshore, (c) mid-shelf, (d) outer shelf mud belts, and (e) as mud blankets most commonly found off major supply points. Mud in suspension from these supply points is moved partly straight out over the shelf and partly alongshore. This gives high concentrations nearshore with roughly exponential decrease seaward. In some areas such as plumes off deltas and areas of residual-current convergence, high-concentration zones are found trending across the shelf. Considering the scale of the shelf, waves and tidal currents contribute to a process of diffusive transport of sediment across the shelf. Quantitatively more important is advective transport in semipermanent currents exemplified by the plumes seen off deltas.

Recent work has shown that critical erosion friction velocities for cohesive sediment are related to yield strength. Deposition is controlled by the nearbed concentration, settling velocity, and a limiting shear stress τ_1 above which no sediment deposits. Critical erosion shear stresses greater than τ_1 are required after only a few hours compaction of a freshly deposited bed. Thus deposits formed nearshore from high-concentration flows during calm weather may develop sufficient cohesive strength to withstand storm conditions. This applies to the muddy coast and nearshore mud-belt accumulations. Mid-shelf mud belts arise from diffusive transport across the shelf to deposition sites where wave and tidal activity are relatively lower than on the inner or outer shelf. Outer-shelf and blanket mud deposits are caused by higher than average concentrations occurring under advective mud streams. A process of diffusion involving repeated deposition and resuspension provides a mechanism for the enlargement of mud depositional areas and gives the seaward progradation of the feather-edge of sedimentary regressions.

Consideration of reflection profiling work, ocean-basin morphology and material balance arguments indicates that most suspended sediment escaping from shelves must be deposited on the slope and rise. Of this the greater part must be in cones and fans off the major input points. As the concentrations are very low, settling does not give the required rates of deposition, thus high-level escape is precluded and escape from shelves must be mainly along the bottom. Outer shelf deposition and resuspension give the higher nearbed concentrations required for low-level escape. Possible transport mechanisms are cascading of cold water off the shelf in winter, water movement down submarine canyons, bottom Ekman layer transport downslope under major ocean boundary currents, and cascades of lutite flows (low concentration turbidity currents).

INTRODUCTION

Although much of the fill of the ocean basins and much of the modern cover of continental shelves consists of fine sediment, little consideration has been given to how this material is transported and deposited. This fine-grained suspended sediment—what students of fluvial processes would call "wash load"—is moved by a variety of processes in the sea. It is the purpose of this paper to examine some of these processes and to see how they may be reflected in the location of fine sediment deposits. The data base for a comprehensive appraisal of transport of fines at sea does not yet exist. This paper is, therefore, admittedly somewhat speculative. Nevertheless the modes of transport which are outlined here have a reasonable physical basis, and I hope that the suggestions which follow will stimulate thought and further work on these mechanisms.

The treatment in this paper proceeds from the point where suspended material escapes from rivers and estuaries. We are then confronted with the problems of how and where this sediment moves in shelf areas and where it is deposited. It is quite apparent from the distribution of terrigenous sediment in the oceans that much sediment escapes from shelf areas to be deposited in slope and rise areas and also to some extent on abyssal plains. How does this sediment escape from the shelf circulation system, and how does it get down to the bottom of the ocean? The latter point may be slightly outside the precise scope of this symposium volume, but one is loath to push the sediment out over the edge of the shelf without some consideration of the way in which it meets its fate!

MUD ON SHELVES AND IN SHELF WATERS

Distribution in Bottom Deposits

The distribution of sediments on shelves has recently been reviewed from different points of view by a number of workers (Emery, 1968; Hayes, 1967).

Salient features of the picture arising from this and other work are the presence of an inner-shelf modern sediment zone and an outer-shelf "relict" sediment zone. A more sophisticated treatment of the nature of these "relict" sediments has been given by Swift et al. (1971), but as these contain little mud, they do not concern us here. The modern sediment cover includes both sands and muds nearshore (mud is defined as sediment of quartz-equivalent sedimentation diameter $< 63\mu$). In most cases the coastal zone down to depths of about 10 m is covered by sand. In exceptional cases such as the coastal mud flats of Louisiana and the coastal areas of the Guianas and the Orinoco delta, an unprotected coast may contain a large proportion of muddy sediment (Morgan, van Lopik, and Nichols, 1953; van Andel, 1967; Allersma, 1968; Diephuis, 1966; Demerara Coastal Investigation, 1962). In areas with protection from open-ocean-wave attack, muddy bottom deposits may be found close to shore in water depths of 10-20 m, for example in the North Sea (Eisma, 1968) and the Adriatic (van Straaten, 1965; Brambati and Venzo, 1967). These are all shallow areas traditionally considered to be affected by waves and currents to such an extent that sandy deposits should prevail. Sand deposits generally occur seawards of this nearshore mud belt.

Off more exposed coasts, mud deposits are found at greater depths, bounded on the landward side by modern or reworked sands and to seaward by relict sands. Among such mid-shelf deposits are those found in the Celtic Sea (Belderson and Stride, 1966; McCave, 1971), and in the Gulf of Gascony in the Grande Vasiere (Andrieff et al., 1971; Rumeau and Vanney, 1968), at depths of 70 to 120 m. In the northwest Gulf of Mexico rather more muddy sediment is found on the outer shelf than in the central shelf region, in addition to a mud zone just off the nearshore sands (Curray, 1960; van Andel, 1960). Only off some deltaic areas do muds blanket the whole shelf, for example off the Mississippi (Shepard, 1960), the deltas of eastern India (Subba Rao, 1964) and the Irrawaddy (Rodolfo, 1969a). The east Asian rivers, the Chao Phya, Mekong, Huang Ho, and Yangtze, transport an enormous amount of material and have covered with mud the embayments into which they flow, yet the outer shelf beyond is still covered with relict sands. In a number of instances the covering of the shelf is mainly due to progradation of the delta whose fore-set slope may nearly coincide with the shelf edge (Shepard, 1960; Subba Rao, 1964).

Beyond the shelf on the continental slope and rise provinces the sediments are mainly fine-grained. Turbidites are important on the rise and in parts of the abyssal plains. In addition there is a considerable thickness of sediment commonly interpreted as pelagic. Some recent studies of basinal sediments have suggested, however, that not only the sands but also the fine sediments have been contributed by gravity-flow mechanisms (Gorsline et al., 1968; Bartolini and Gehin, 1970).

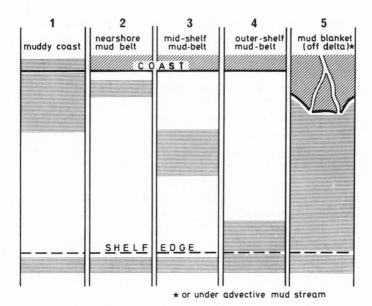

Figure 91.
Schematic representation of five cases of sites of shelf mud accumulation.

In summary; mud occurs on shelves (a) as coastal accumulations; (b) as nearshore mud belts bounded landward and seaward by sands; (c) as mid-shelf deposits similarly bounded; (d) as outer shelf accumulations; and (e) as blanket deposits across the shelf (mainly off deltas) (Fig. 91). In the oceans mud is abundant on slope and rise areas, and abyssal plain mud may be contributed by gravitational processes other than turbidity currents, and pelagic sedimentation as generally understood (Arrhenius, 1963).

Distribution in Suspension

Suspended sediment data are not as readily available as data on bottom sediment distribution. Certainly it will not be possible to map mean suspensate concentrations in the same way as other hydrographic parameters (e.g., temperature and salinity) for some years to come. Until that time any discussion of mud transport and deposition in the sea will be, to some extent, speculative.

The most pronounced feature of the distribution of suspended sediment is a broadly exponential decrease in concentration in a seaward direction. This pattern has been shown by a number of investigators using a variety of techniques including gravimetry, extinction measurements, Secchi disc, and others. These have included, e.g., the studies of Manheim et al. (1970) and Buss and Rodolfo (1972) on the eastern United States shelf; Joseph (1953), Hagmeier (1960), and Otto (1967) in the North Sea; Eisma (1967) and Allersma (1968) off the northeast coast of South America; Jerlov (1958) and Brambati and Venzo (1967) in the Adriatic; and Wageman et al. (1970) and Parke et al. (1971) on the east Asian continental margin. A number of these studies have dealt only with surface samples which do not represent the concentration in shelf waters as a whole. Other methods such as use of the Coulter Counter (Lee and Folkard, 1969) and visual inspection of space photographs (Wobber, 1967) have also been used. While the methods of light extinction, Secchi disc and space photography may not yield information on absolute suspended sediment concentration, the trends they show within the same water mass point to marked seaward concentration decrease. The order of magnitude of these concentration changes is from 10-100 mg/liter nearshore to 0.1-1 mg/liter over the outer shelf.

A second important feature of suspended sediment distribution is the existence of plumes of higher concentration extending across the shelf. Most commonly these are associated with major deltas, e.g., the Mississippi (Scruton and Moore, 1953), the Orinoco (probably mainly Amazon sediment) (Nota, 1958, p. 76), the Irrawaddy (Rodolfo, 1969b), the Po (Jerlov, 1958; Nelson, 1970) among others. Plumes of suspended sediment also occur where there is convergence of coastal currents, resulting in a current across the shelf. Such a case seems to occur in the southern North Sea (Joseph, 1953; Lee and Ramster, 1968) (Fig. 92). Although not confirmed by suspended sediment measurements another such plume may be inferred in the western Gulf of Mexico from Curray's (1960) grain-size modes and the current pattern shown by van Andel and Curray (1960) (Fig. 93). Another plume is indicated by Wobber (1967) on space photographs. This runs seawards from a point just southeast of Port Arthur, Texas. The Apollo 9 photograph given by Emery (1969) and the measurements of Buss and Rodolfo (1972, this volume) show a plume extending south south-east of Cape Hatteras. It seems likely that surveys of other areas with large inputs of suspended sediment (e.g., the Niger region and south-east Asia) would also show such plumes of suspended sediment.

PROCESSES AND MODES OF TRANSPORT AND DEPOSITION

Processes and Transport

Suspended sediment may be moved by processes of eddy diffusion and of advection. As we are considering the whole shelf (of an order of 10^2 km in width), then tidal motion with tidal excursions of the order of 1 to 10 km must be considered as an agent of diffusion, together with wind-generated waves. Both tides and waves mix shelf waters, and this leads to diffusive transport of sediment. The process of diffusion transports sediment from an area of higher

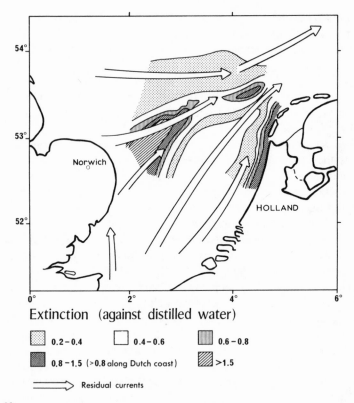

Extinction (against distilled water)

▨ 0.2 – 0.4   ☐ 0.4 – 0.6   ▥ 0.6 – 0.8

■ 0.8 – 1.5 (>0.8 along Dutch coast)   ▧ >1.5

⟹ Residual currents

Figure 92.

Turbidity (light extinction) given by Joseph (1953) with the residual current pattern (Lee and Ramster, 1968) in the southern North Sea. This illustrates two advective mud streams, one crossing from the English to the Dutch side of the area, the other running up the Dutch coast. Actual sediment concentrations are higher in the latter stream.

concentration to one of lower concentration, tending to give a uniform distribution. Given that the input of suspended sediment is at various points along the shoreline, there must be a seawards diffusive transport of material.

As we have seen, however, the distribution of concentration is not uniform but increases rapidly nearshore. This has two causes; the source is at the coast, and there may also be advective transport shorewards tending to reinforce the concentration gradient. The term "advective transport" means movement of suspended material resulting from net horizontal water movement. Shorewards advection may be due to intrusion of colder, more saline water from offshore, upwelling, (e.g., Hart and Currie, 1960; Blanton, 1971), or it may take the form of a quasi-estuarine circulation in which there is a net shoreward component in the bottom water movement in response to seaward flow of lighter surface

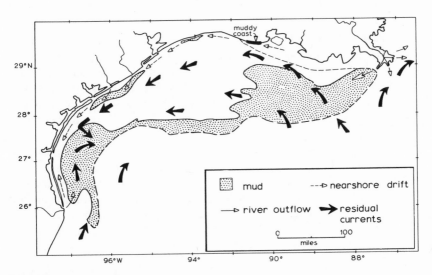

Figure 93.

Mud deposits and currents in the northwest Gulf of Mexico from Curray (1960) and van Andel and Curray (1960). The influence of Mississippi supply on the outer shelf west of the delta is marked; so is the relationship between mud and residual current pattern in the western Gulf.

water (e.g., Bumpus, 1965). In either case, settling of suspended sediment into the intruding water mass will tend to reinforce the concentration gradient.

There is an analogy here with the case of the vertical distribution of suspended load in a stream. Sediment is eroded from the bed (shoreline source) and is diffused upwards by turbulence (outwards by waves and tides) against the force of gravity tending to keep the sediment at the bed (advective water movements returning sediment shorewards). The net result in both cases is an approximately exponential decrease in concentration (upwards, stream; outwards, shelf).

Sediment can escape from the shelf, however, and the fact that the concentration in outer shelf waters generally is higher than that in adjacent oceanic waters indicates net seaward diffusive transport, and ultimately diffusive escape from the shelf system.

Far more important is the advective transport occurring in semi-permanent currents caused by wind, density, or inertia. Wind and waves are mainly responsible for the generation of longshore currents, and as these occur in the zone where concentration is highest, they are very important. Winds and inertial effects from adjacent oceanic current systems are responsible for the residual water movements in shelf seas and it is these residual currents which account for most mud transport. Examples of advective transport are the pronounced Rhine mud stream running northwards along the Dutch coast (Terwindt, 1967), the Po mud stream running south of the delta (Jerlov, 1958), and the Amazon mud stream running along the Guiana coast (Diephuis, 1966). Another important mud stream which is the result of the meeting of two coastal currents is that inferred from Curray (1960) and van Andel and Curray (1960) in the western Gulf of Mexico (Fig. 93). It is important because it moves directly across the shelf and so provides an escape route. Possibly the higher concentration plume off Cape Hatteras also represents a seaward transport path of sediment. The data of Stefánsson et al. (1971) suggests a fairly persistent offshore flow in this region.

The pronounced sediment plumes seen off deltas result from the inertia of the outflow from deltaic distributaries. In this important type of advective transport the flow is initially confined to the surface, but mixing occurs along the path of the plume. Settling of newly flocculated material also occurs along the transport path. These processes combined give a general concentration decrease downstream.

Processes and Deposits

Important advances have been made in the last few years in our understanding of the processes of erosion and deposition of fine-grained sediment. For the erosion of sediment Migniot (1968) has managed to collapse all the critical erosion friction velocities for a great variety of muds onto a single curve by using yield strength as the parameter defining the state of the bed. This gives

$$U_{*c} = 0.5 \, \tau_y^{\frac{1}{2}} \quad \text{for } \tau_y > 20 \text{ dynes cm}^{-2} \tag{1}$$

$$U_{*c} = \tau_y^{\frac{1}{4}} \quad \text{for } \tau_y < 10 \text{ dynes cm}^{-2} \tag{2}$$

where U_{*c} is the critical friction (shear) velocity in cm sec^{-1} and τ_y is the yield strength. The fact, well known since Hjulstrom's (1939) work, that the shear stresses required for erosion of mud are considerably higher than those allowing its deposition, is also reinforced by Migniot's data. For example, a critical $U_{*c} = 0.77$ cm sec^{-1} erodes mud of $\tau_y = 0.4$ dynes cm^{-2} and this corresponds to a concentration of 125 to 300 gm/liter (bulk density of 1.08 to 1.18 g cm^{-3}; porosity of 89% to 95%) for the muds used by Migniot. Such porosities (water contents) are achieved after a relatively short period of consolidation, ≈ 3 hr (Postma, 1967), 1 to 10 hr (Migniot, 1968), and are higher than those found in the top 25 cm of recently deposited muds (Skempton, 1970).

For the deposition of suspended sediment, Sundborg (1956, 1958) gives

$$C_t = C_0 \exp \left[-wtf(w)/D \right] \tag{3}$$

where C is concentration, w is settling velocity, t is time, D is flow depth, and $f(w)$ is some function of settling velocity. Essentially the same equation is given by Einstein and Krone (1962).

$$C_t = C_0 \exp\left(-wtp/D\right) \tag{4}$$

where p is the probability of deposition given by

$$p = 1 - \tau_0/\tau_1 \tag{5}$$

in which τ_0 is the fluid shear stress at the bed and τ_1 is the limiting shear stress above which no deposition takes place. Values of this limiting shear stress are in the region 0.41 to 0.81 dynes cm^{-2} (Einstein and Krone, 1962; Krone, 1962; Partheniades et al., 1969; Hydraulics Res. Sta., 1970). This corresponds to $U_{*_1} \left[= (\tau_1/\rho)_i^{1/2} \right]$ values of $0.64 - 0.09$ cm sec^{-1}. These equations have also been used by Owen and Odd (1970), in the form

$$\frac{dm}{dt} = -C_b w(1 - \tau_0/\tau_1) \tag{6}$$

which gives the mass rate of sediment deposition related to the nearbed concentration C_b. A simplified form was also used by McCave (1970) assuming $p = 1$, the case of complete entrapment in the viscous sublayer. However, as was argued there, this value should be <1 to allow for periodic ejection of sediment from the viscous sublayer, and thus the term $(1 - \tau_0/\tau_1)$ may be regarded as indicating the effective degree of sublayer instability. The U_{*1} values of 0.64 to 0.90 cm sec^{-1} correspond to velocities of 17 to 25 cm sec^{-1} at 1 m above the bed assuming smooth boundary turbulent flow. Deposition of mud occurs at velocities lower than these values which, at sea, may correspond to surface velocities only a little less than 50 cm sec^{-1} (1 knot). It will also be seen from these equations that the rate of deposition is proportional to the concentration near the bed and the settling velocity of the suspended sediment.

A further approach to the problem of mud deposition is also derived from the work of Einstein (1968) who introduced the notion of a "half-life" or half-residence time for particles in suspension. This half-life, the time taken for the concentration of particles of a given settling velocity to reduce to half the initial value, assuming uniform concentration throughout the depth, is given by

$$T_{1/2} = 0.693/(w/D) \tag{7}$$

This further assumes that each particle, once it gets very close to the bed deposits, and remains deposited. However, it has been suggested that some particles approach the bed closely but may be ejected from the viscous sublayer (McCave, 1970), and others may deposit but be eroded shortly afterwards by wave or turbulence-induced higher shear stresses. Einstein included a correction factor for the fraction of time during which deposition (assumed 100% efficient) occurred in his flume system. If we incorporate the overall probability of deposition (over a suitable time scale of perhaps a month) then the half-residence time is

$$T_{\frac{1}{2}} = 0.693/p(w/D) \tag{8}$$

Taking, e.g., $w = 10^{-3}$ cm sec^{-1}, $D = 100$ m, and $p = 0.5$, $T_{\frac{1}{2}} \approx 160$ days. However the whole depth is unlikely to participate in the exchange process with nearbed layers, particularly if there is thermohaline stratification, and the concentration generally is greater near the bed. On the other hand $p = 0.5$ is optimistic for a five-month period. Using $D = 10$ m and $p = 0.2$, then $T_{\frac{1}{2}} = 40$ days. The basic parameters are not known well enough for this yet to be of use, but it does suggest that concentration might decrease exponentially on a scale of a few weeks to months. This concentration decrease due to deposition on the bed with replenishment of the nearbed suspended sediment by turbulent mixing. So any tendency to deposition tends to concentrate the suspended material near the bed, and only an extreme storm is likely to diffuse the material back up through the whole water column.

Thus the occurrence of mud deposits should be regarded as an indication of the long-term balance between the factors C_b, w, and p which partly depends on the times for which $\tau_0 < \tau_1$ and $U_* > U_{*c}$. The latter factor is partly a function of the degree of consolidation of the deposit. On virtually no area of shelf will the velocity be always > 25 cm sec^{-1}. Every point on the shelf, therefore, is potentially a site of mud accumulation.

The velocity figures of 17 to 25 cm sec^{-1} are for steady uniform flow. Little is known about the effect of waves on a mud bed though Migniot (1968) has made some experiments. However, it is known that waves considerably augment the shear stress on the bed under flowing water, and this may result in a more than tenfold increase in sand transport rate (Bijker, 1967). A treatment of the wave–mud problem in terms of the "effectiveness" of waves (in doing work) on the seabed has been given by the present author (McCave, 1971). This was rather biased towards the mean conditions and showed, predictably, that the effectiveness increased rapidly nearshore. The weakness here is that both deposition and erosion of mud may be better related to extreme events, high and low, than to mean conditions.

With these points related to erosion and deposition, and also the advective and diffusive mechanisms of transport, in mind, let us now look at the present sites of mud accumulation given in the cases discussed in the section on bottom deposits and shown in Fig. 91, and explained schematically in Fig. 94. Both cases 1 and 2 are similar in that they are due to high-concentration longshore mud streams (advective transport). In the case of the muddy coast the balance at the coastline between concentration and wave and current activity is in favor

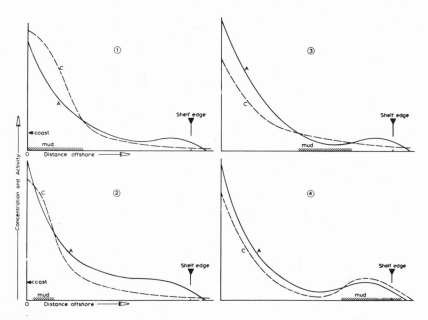

Figure 94.

Schematic explanation of cases 1 to 4 on Fig. 93: C = concentration and A = "activity" of water movement at the sea bed, a notional parameter combining wave and tidal effects.

of concentration. The concentration of mud in the waters off the Guiana coast damps the wave action so that the wave effect at the coast is less than would be expected from the normal Atlantic swell incident on the shelf. Off the Dutch coast the concentration of the Rhine mud stream is not so high and the balance allows a sandy coast with a zone of mud deposition nearshore. Case 5 is similar in that the mud blankets off deltas, and also under plumes such as that in the western Gulf of Mexico, result from a balance in favor of concentration in an advective mud stream. Under low-concentration plumes the balance may be in favor of wave and current activity and the bottom deposits are sand, the Cape Hatteras plume is a case in point.

It will be noted that the trends of the sedimentary deposits discussed so far are parallel to the direction of movement of the mud stream. This direction may be parallel or perpendicular to shelf contours.

The deposits of case 3 run parallel to shelf contours. These, together with the deposits of case 4 and the general lack of mud on the outer continental shelf, bring up the problem of the relationship between erosion, deposition, and diffusive transport produced by water movements. The deposits of case 3 are to some extent analogous to those of case 2 with the exception that they do not, for the most part, lie in the path of advective mud streams. The transport of mud to this mid-shelf depositional site is presumed to be by diffusive processes. On my diagram of wave effectiveness against depth for the Celtic Sea (McCave, 1971, Fig. 7) there is a low in the wave effectiveness between 40 and 70 fm (72-126 m). This parameter actually shows an increase seaward of this region. Tidal currents in a normally incident standing-wave system are argued by Fleming (1938) and Fleming and Revelle (1939) to be at a maximum at the shelf edge. While Fleming's assumptions do not hold for all shelves, the data of Korgen et al. (1970) from sites on the slope and rise west of Oregon show an increase in velocity with decreasing depth. It seems highly probable that there is some intensification in current velocities at the shelf edge. There may well also be a decrease in velocity over the midshelf region. In a progressive wave system running oblique to the shelf there will again be intensification in velocity nearshore. In some cases then, there may be a decrease in both tidal and wave effectiveness in the midshelf region. This decrease allows formation of a mud deposit even from water with a low suspended sediment concentration ($C < 1$ mg liter^{-1}).

This discussion really answers the problem of the lack of mud on the outer shelf. Permanent mud deposition is prevented either by wave action or by current activity, or both, as argued also by Kuenen (1939) and Nota (1958). There may also be other forces in operation such as breaking internal-waves and shelf-edge waves. The latter have been argued by Cartwright (1959) to be responsible for a sand-wave field on La Chapelle Bank just off the Western Approaches to the English Channel. Internal waves have been recorded in this region by Stride and Tucker (1960). It is instructive to consider the magnitude of deposit which has to be moved to ensure a mud-free outer shelf. Using the model of McCave (1970), which in any case overestimates deposition, the deposit from a flow maintained at $C_b = 0.5$ mg liter^{-1} with $w = 10^{-3}$ cm sec^{-1} would be 16 mg cm^{-2} in a year. This is only a little more than is required to provide a surface pore-filling for a sand bed! The data of Draper (1967) suggest that it is almost certain that waves alone could remove such a deposit on the outer shelf.

The balance is, however, tenuous as case 4 (based on the example of the outer shelf west of the Mississippi) demonstrates. Here there is an advective mud stream swept from the mouth of the Mississippi delta by the counterclockwise residual circulation of the northwestern Gulf of Mexico. This gives an increase in concentration with resulting deposition which tides and hurricane-generated waves can obviously not counteract. The preceding explanations are illustrated in Fig. 94.

The processes outlined above give a further dimension to the seaward diffusive transport of mud. Exchange between bed and suspension may play an important role. A deposit accumulated over a period of weeks or months may be resuspended during a storm giving a local increase in concentration. This material then diffuses out across the shelf and deposits during quiet weather conditions. It may go

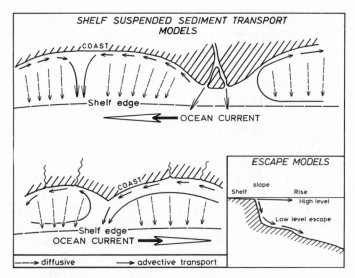

Figure 95.
Summary diagram illustrating shelf suspended sediment transport modes and patterns.

through a great many such cycles until it either escapes or is permanently deposited. This mechanism, incidentally, might explain how the feather edge of a sedimentary regression progresses seawards and how, eventually, the whole outer shelf may become covered with mud. In the region adjacent to an existing mud deposit there will be a greater chance of further deposition of mud. It will be apparent from the diagram of mud deposition sites (Fig. 91) that simple seawards movement may not be always the course of a regression. The mud deposit of case 3 could grow both seawards and shorewards. Models for shelf suspended sediment dispersal are given diagrammatically in Fig. 95.

ESCAPE FROM THE SHELF

Where to

The preceding discussion argues that mud is moved across the shelf advectively in streams and diffusively either continuously in suspension or through a process of bed/water exchange. Thus some of it arrives at the edge of the shelf. The question as to where it goes has an important bearing upon how one suggests that it gets there.

An average rate of sedimentation, on a carbonate-free basis, for the deep sea is about 0.4 g cm^{-2} 10^{-3} yr (Ku et al., 1968). With an area for basinal regions plus mid-ocean-ridge systems of 270 × 10^6/km^2 (Menard and Smith,

1966), the annual deposition in the deep sea is about 1.1×10^9 tonnes. This leaves a large part of the estimated fluvial suspended sediment supply of 18×10^9 tonnes (Holeman, 1968) unaccounted for. If one allows about 50% of the supply for deposition in estuaries and other coastal sites and on the shelf, then of the 9×10^9 tonnes going to the ocean basins 8×10^9 tonnes should be deposited on the continental slope and rise. Postglacial deposition rates for the continental rise off the eastern United States are given as greater than 3 g cm^{-2} 10^{-3} yr (Turekian, 1968), and if a value of say 5 g cm^{-2} 10^{-3} yr were valid for the world-rise system with a total area of $\approx 20 \times 10^6$/km^2 (Menard and Smith, 1968), still only a further 1×10^9 tonnes is accounted for.

It seems that the majority of suspended sediment must be deposited on the great fans and cones of sediment in front of the major supply points, e.g., the Ganges and Indus cones (Ewing et al., 1969). The foregoing figures may give some idea of the relative importance of diffusive and advective transport as the eastern United States has no major sediment supply points, and diffusive transport must be relatively important across this shelf. The bulk of suspended sediment must be moved in advective plumes off the major input points.

How

Settling

One possibility for escape is that shelf water bearing a uniform sediment concentration through a 150 to 200 m shelf-edge depth mixes out horizontally into the ocean from which sediment is then lost by settling. This sediment enters the ocean near the surface and the term "high-level escape" is used. In this way most sediment would be deposited on the slope, less on the rise, and least on the abyssal plains and ridge systems beyond. Such an argument fails on a number of counts. One may ask why this material does not settle on the outer shelf. If the sediment is in fact moving down in this region, then there will not be a uniform concentration with depth but a higher concentration near the bed. Morphology suggests that maximum deposition occurs on the rise rather than the slope. In many places surface concentrations are so low as to be undetectable (Jacobs and Ewing, 1969), but increase with depth. Sedimentation rates appear to be insufficient. A concentration of 0.05 mg liter^{-1} is reasonable for the western North Atlantic (Jacobs and Ewing, 1969; Folger, 1970) and a settling velocity of 10^{-4} cm sec^{-1} ($d = 1$ to 2 μm) is of the right order considering the data of Jerlov (1953), Arrhenius (1963) and Carder et al. (1971) on particle size. With these parameters the sedimentation rate by settling would be 0.16 g cm^{-2} 10^{-3} yr as opposed to the required >3 g cm^{-2} 10^{-3} yr.

In order to account for the distribution of certain sediment types, some authors have suggested an increase in settling velocity due to organic aggregation of particles (Rex and Goldberg, 1958; Osterberg et al., 1964; Calvert, 1966).

Calvert reports no trace of this aggregation in the sediments, nevertheless aggregation seems highly likely in the cases cited. Because productivity is higher in shelf waters this formation of organic aggregates should occur there too and tend to increase the suspended sediment concentration downwards. However, the question is not whether increase of settling velocity due to aggregation occurs, it almost certainly does, but whether this aggregation affects most of the sediment and so dominates the process and pattern of deposition. Limited size–distribution data suggest that it does not. Carder et al. (1971) find number mean diameters of about 3.35 μ for particles >2.22 μ, and Brun-Cottan (1971) shows a mode at 4 μ for particles >1 μ. As there is a significant amount of material <1 μ (Arrhenius, 1963) these are high values. The results of Brun-Cottan also allow calculation of a maximum mean particle density of 1.4 g cm^{-3} assuming primary particle density of 2.5 g cm^{-3}. These data do not suggest important enhancement of settling velocity due to aggregation. The same may also be said of flocculation, it occurs but is not dominant. Einstein and Krone (1961) review the kinetics of flocculation and show that the probability of particle collision due to Brownian motion depends directly on the particle number concentration and temperature, and inversely on viscosity. The probability of collision due to local shear is proportional to concentration and shear rate. Concentration, temperature, and shear rate will be higher and viscosity lower in nearshore and estuarine situations. Any important flocculation is, therefore, likely to have occurred before the sediment reaches the outer shelf. Whitehouse et al. (1960) show that flocculation of kaolinite, illite, and chlorite is essentially complete at chlorinity of 2⁰/₀₀, again making flocculation unlikely as a mechanism for significant settling-velocity distribution change over the outer shelf.

Water Movements

A mechanism which may be of some importance in temperate latitudes is cascading of water off the shelf (Cooper and Vaux, 1949). In winter the shelf waters may be cooled to such an extent that they are denser than bottom water on the adjacent continental slope. The mass of cold water then moves along the bottom and off the shelf. Cooper and Vaux demonstrate this process for the Celtic Sea and Stefánsson et al. (1971) show it on the North Carolina slope. Judging from the σ_t data given by these authors it seems doubtful that the transport would extend much below 1000 m depth. The mechanism is fairly attractive in that it occurs in winter, at which time wave action is most likely to resuspend bottom sediment. It is, however, unlikely to occur in tropical and subtropical latitudes.

A number of studies have suggested that submarine canyons act as sinks for suspended sediment (Moore, 1969; Beer and Gorsline, 1971; Felix and Gorsline, 1971). Both current measurements (Shepard and Marshall, 1969) and hydrographic properties of the water mass (Felix and Gorsline, 1971) in canyons off the California coast indicate net down-canyon water movement. This is one mechanism which may contribute the important quantities of mud found on these canyon fans. Current records from other sites are not all favorable for this process as is shown by Rowe's (1971) variable but dominantly up-canyon flow data from continental rise sites in the Atlantic.

A third possibility is of bottom Ekman-layer transport under major oceanic boundary currents. The Ekman layer is a boundary layer with Coriolis effect giving a spiral decrease in velocity. The net water transport in this layer will be at an angle to the flow of the main current (90° for the viscous case, some lesser angle in turbulent flow. The order of thickness of the layer is $(f/v)^{1/2}D$ where f is the Coriolis parameter, v the vertical eddy viscosity, and D the depth of flow. Formulation of the Ekman layer equations for viscous flow yields a transport perpendicular to the main current of $\frac{1}{2}(v/f)^{1/2} U$ per unit length of current, where U is the velocity in the main flow (Johnson, personal communication, 1971). At a latitude of 30° with $v = 10^{-2}$ m^2 sec^{-1} the thickness of the layer is of the order of 170 m and the transport 6 m^3 sec^{-1} m^{-1} length. This amounts to a downslope velocity of about 4 cm sec^{-1}. While this viscous solution gives transport values which may be too high, there must nevertheless be a bottom Ekman layer with a downslope component. What happens when the downslope transport beneath a current such as the Gulf Stream meets the required upslope transport from the base of the Western boundary undercurrent is not known. It must presumably move out at some mid-water level bearing its suspended sediment. These opposed transports may be responsible for the genesis of the nepheloid layer shown by Ewing and Thorndike (1965, Fig. 92) extending out from the base of the slope.

Lutite Flow

This is the name given to what in essence is a turbidity current of very low density and thus low velocity. Such a gravity driven flow has been proposed by Ewing and Thorndike (1965) and Jacobs and Ewing (1969) in connection with nepheloid layers and pelagic sediment distribution. Moore (1969) also has argued for a "turbid-layer" transport of material across the shelf initiated by wave-induced resuspension and driven by gravity. Although the concentration may increase downwards in an exponential fashion in some areas (Spencer and Sachs, 1970; d'Anglejan, 1970), nearbed concentrations do not tend to exceed 10 mg liter^{-1} on the outer shelf. This gives a density excess of 6×10^{-6} gm cm^{-3}, and, as Drake (1971) points out, a density contrast of 10^{-5} gm cm^{-3} is produced by a temperature change of 10.1°C to 10.0°C. Thus a lutite flow is unlikely to penetrate the thermal stratification of the oceans. Indeed, as Joseph (1957), Bouma et al. (1969), and Drake (1971) have shown there is an increase in turbidity at thermal interfaces suggesting that lutite flows are partially trapped there.

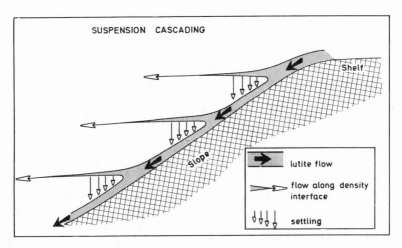

Figure 96.
Schematic representation of lutite flows with cascading. Decrease in width of arrows indicates decrease in concentration.

Table XVII
Velocities of Lutite Flows (in cm sec^{-1})

Slope	w cm sec^{-1}	C (mg liter^{-1})			
		1.0	2.0	5.0	10.0
1/50	$\leqslant 2 \times 10^{-3}$	2.8	4.0	6.3	9.0
1/500	$\left\{ \begin{array}{c} 10^{-3} \\ 10^{-4} \end{array} \right.$	0.68 0.88	1.07 1.25	1.8 2.0	2.7 2.8

A way around this problem has been suggested by Postma (1969) who postulates that as a lutite flow reaches a density interface and flows out over clearer water, it loses its particles by settling. As these reach the vicinity of the bed a new flow is formed. Thus the process, a suspension cascade, could transport fine sediment to considerable depths (Fig. 96). Limited support for this suggestion may be derived from Drake's (1971) data. He shows a number of suspensate layers held up by thermal stratification, each one of which decreases in concentration in a seaward direction. This decrease in concentration may be interpreted as due to settling from the layer and the number of layers as being due to a cascade of lutite flows. Concentration near the bed decreases downslope from 2.5 to 0.6 mg liter^{-1} due to entrapment in overlying layers of the cascade. When concentrations get this low the persistence of a lutite flow mechanism seems doubtful. Beer and Gorsline (1971), however, show the persistence of nearbed concentrations of 2 to 6 mg liter^{-1} down to 500 m depths in a canyon.

The sort of velocities which might be attained by lutite flows 10 m in thickness have been calculated using Komar's (1969) version of the autosuspension equation (Table XVII). On a slope of 1/50 velocities of 3 to 9 cm sec^{-1} give a considerable increase in the vertical descent rate over settling (60 to 180 × 10^{-3} as against 2 × 10^{-3} m sec^{-1}). Even on slopes of 1/500 velocities of 1.25 cm sec^{-1} give a vertical descent rate of 2.5 × 10^{-3} cm sec^{-1} which is about an order of magnitude greater than the settling velocity of 2 μ particles.

However, as a reviewer of this paper has pointed out, such low velocity flows on a plane slope will tend to move parallel to slope due to Coriolis force. A current 10 m thick moving at 3 cm sec^{-1} at $\lambda = 45°$ would move at an angle of only 2½° away from the contour, and a flow of 9 cm sec^{-1} would move downslope at 22° from contour. These paths, particularly the former, considerably reduce the slope actually traversed by the current, which makes likely the collapse of such a flow due to failure to meet the condition governing autosuspension $w < \alpha \bar{U}$, where α is the slope and $\bar{U}$ the current velocity. The faster flow could continue, but depends on abnormally high concentrations for its maintenance. An alternative to an open slope is a channelized flow and it may be that such a lutite flow could exist in a submarine canyon. Moore (1969) has observed slow (1 to 2 cm sec^{-1}) movement of turbid layers down valleys. In the case of much thinner turbid layers such as those less than half a meter thick recorded by Moore (1969, p. 87), the Coriolis effect becomes less than the frictional effect and the current would move dominantly downslope. If lutite flows and cascades are of any importance in moving fine sediment down the slope it seems that they should mainly be channelized. Observations suggest that they could be generated by wave activity (Moore, 1969).

CONCLUSIONS

Of the two modes of transport—diffusive and advective—outlined, the latter is quantitatively more important. This is suggested by the mud deposits under suspended-sediment plumes on the shelf and by the great accumulations on the rise off the major input points. In order that sediment should escape from shelves along the bottom and not be diffused out into the upper layers of the ocean, it seems that the material should be somehow concentrated in the lower part of the water column. Moore's (1969) observations of shelf-bottom turbid layers indicates that this does happen. The mechanism may be by deposition and resuspension over the outer shelf. Thus the fact that the outer shelf is an important site of mud deposition (but not accumulation) may exert some control over the mode of distribution of fine sediment to the ocean basins.

Two of the escape mechanisms, Ekman layer transport and lutite flow may in part be mutually exclusive. The more intense mixing found in high-velocity boundary currents such as the Gulf Stream may not allow the fine/turbidity structure of a suspension cascade to exist. This fine structure is recorded by Bouma et al. (1969) in the Gulf of Mexico and by Drake (1971) in Santa

Barbara Basin; neither location has strong currents. It must be admitted, however, that very little is known about either mechanism.

Studies on the transport, deposition, and erosion of fine sediments (which comprise more than 50% of the stratigraphic column) are in a rudimentary state. Compared with work on elucidation of the mechanics of sand transport, bed forms, etc., or the ecology of recent carbonates, studies on muds are few. Far better and more extensive data are required on the distribution and composition of suspended matter. In addition we need to know not only the particle size but also the settling velocity and mechanical properties of suspended and bottom sediments *in situ*. Don Swift put the case well when he wrote ". . . to further our knowledge of shelf processes we must start to 'shelf watch' on a large scale by correlating hydraulic regime with changing bottom sediment parameters over long periods of time." (Swift, 1970, p. 27.) That goes for suspended sediments too.

ACKNOWLEDGMENTS

My interest in the problems of suspended sediment transport stems from Professor H. Postma, director of the Netherlands Institute for Sea Research, and the extremely profitable time that I spent at his institute as a NATO research fellow. I thank both him and the Science Research Council (London) for this opportunity. John Johnson of the School of Mathematics and Physics in the University of East Anglia both suggested the Ekman layer transport mechanism and did the related sums. My colleague John Harvey critically reviewed the manuscript, and both he and Peter Liss made helpful comments.

LIST OF SYMBOLS

Symbol	Description	Dimensions
C	Concentration	ML^{-3}
D	Depth of flow	L
f	Coriolis parameter	
m	Mass	M
p	Probability of deposition	
t	Time	T
$T_{1/2}$	Half residence time	T
$\overline{U}$	Mean flow velocity	LT^{-1}
u	Shear velocity $= (\tau_0/\rho)^{1/2}$	LT^{-1}
w	Settling velocity	LT^{-1}
α	Slope	
ν	Vertical eddy viscosity	L^2T^{-1}

ρ	Fluid density	ML^{-3}
σ_t	$(\rho - 1) \times 10^3$	
τ	Fluid shear stress	$ML^{-1}T^{-2}$
τ_y	Yield strength	$ML^{-1}T^{-2}$
λ	Latitude	

Subscripts

b	Near-bed conditions
c	Critical conditions
l	Limiting conditions
o	At time = 0; at the bed
t	At time t

REFERENCES

Allersma, E. (1968). Mud on the oceanic shelf off Guiana. Paper given at CICAR conference, Curaçao, Nov. 1968, unpublished, Delft Hydraulics Laboratory.

Andel, Tj. H. Van (1960). Sources and dispersion of Holocene sediments, northern Gulf of Mexico. *In* "Recent sediments, northwest Gulf of Mexico" (F. P. Shepard, F. B. Phleger, and Tj. H. van Andel, eds.), pp. 34-55, Am. Assoc. Petrol. Geologists, Tulsa, Oklahoma.

Andel, Tj. H. Van (1967). The Orinoco Delta. *J. Sediment Petrol.* **37**, 297-310.

Andreieff, P., Bouysse, P., Chateauneuf, J. J., L'Homer, A., and Scolari, G. (1971). La couverture sédimentaire meuble du plateau continental externe de la Bretagne meridionale. *Cahiers Oceanogr.* **23**, 343-381.

d'Anglejan, B. F. (1970). Studies on particulate suspended matter in the Gulf of St. Lawrence. *McGill Univ. Marine Sci. Center Manuscript Rept.* **17**, 51 pp.

Arrehenius, G. (1963). Pelagic sediments. *In* "The Sea" (M. N. Hill, ed.), Vol. 3, pp. 655-727. Wiley, New York.

Bartolini, C., and Gehin, C. E. (1970). Evidence of sedimentation by gravity-assisted bottom currents in the Mediterranean Sea. *Marine Geol.* **9**, M1-M5.

Beer, R. M., and Gorsline, D. S. (1971). Distribution, composition and transport of suspended sediment in Redondo Submarine Canyon and vicinity (California). *Marine Geol.* **10**, 153-175.

Belderson, R. H., and Stride, A. H. (1966). Tidal current fashioning of a basal bed. *Marine Geol.* **4**, 237-257.

Bijker, E. W. (1967). Some considerations about scales for coastal models with movable bed. *Delft Hydraulics Laboratory* **50**, 142 pp.

Blanton, J. (1971). Exchange of Gulf Stream water with North Carolina Shelf water in Onslow Bay during stratified conditions. *Deep-Sea Res.* **18**, 167-178.

Bouma, A. H., Rezak, R., and Chmelik, F. B. (1969). Sediment transport along oceanic density interfaces. *In* "Geol. Soc. America Abstracts for 1969," Part 7, pp. 259-260., Geological Society of America, Boulder, Colo.

Brambati, A., and Venzo, G. A. (1967). Recent sedimentation in the northern Adriatic Sea between Venice and Trieste. *Studi Trentini di Scienze Naturali* **44**, 202-274.

Brun-Cottan, J.-C. (1961). Etude de la granulometrie des particules marines mesures effectuées avec un compteur Coulter. *Cahiers Oceanogr.* **23**, 193-205.

Bumpus, D. F. (1965). Residual drift along the bottom on the continental shelf in the Middle Atlantic Bight area. *Limnol. Oceanog. Suppl.* **10**, R50-R53.

Buss, B. A., and Rodolfo, K. S. (1972). Suspended sediments in continental shelf waters off Cape Hatteras, North Carolina. *In* "Shelf Sediment Transport: Process and Pattern" (D. J. P.

Swift, D. B. Duane, and O. H. Pilkey, eds.). Dowden, Hutchinson and Ross, Stroudsburg, Pa.

Calvert, S. E. (1966). Accumulation of diatomaceous silica in the sediments of the Gulf of California. *Geol. Soc. Amer. Bull.* **77**, 569-596.

Carder, K. L., Beardsley, G. F., and Pak, H. (1971). Particle size distributions in the eastern equatorial Pacific. *J. Geophys. Res.* **76**, 5070-5077.

Cartwright, D. E. (1959). On submarine sand-waves and tidal lee-waves. *Proc. Roy. Soc. Lond. Ser. A* **253**, 218-241.

Cooper, L. H. N., and Vaux, D. (1949). Cascading over the continental slope of water from the Celtic Sea. *J. Mar. Bio. Assoc. United Kingdom* **28**, 719-750.

Curray, J. R. (1960). Sediments and history of Holocene transgression, continental shelf, northwest Gulf of Mexico. *In* "Recent sediments, northwest Gulf of Mexico" (F. P. Shepard, F. B. Phleger and Tj. H. van Andel, eds.), pp. 221-266. Am. Assoc. Petrol. Geologists, Tulsa, Oklahoma.

Demerara Coastal Investigation (1962). Report on siltation of Demerara Bar Channel and coastal erosion in British Guiana. Delft Hydraulics Lab., 240 pp.

Diephuis, J. G. H. R. (1966). The Guiana Coast. *Tijdschr. Kon. Ned. Aardrijksk. Gen.* **83**, 145-152.

Drake, D. E. (1971). Suspended sediment and thermal stratification in Santa Barbara Channel, California. *Deep-Sea Res.* **18**, 763-769.

Draper, L. (1967). Wave activity at the sea bed around northwestern Europe. *Marine Geol.* **5**, 133-140.

Einstein, H. A. (1968). Deposition of suspended particles in a gravel bed. *J. Hydraul. Div., Proc. Am. Soc. Civ. Engrs.* **94**, 1197-1205.

Einstein, H. A., and Krone, R. B. (1962). Experiments to determine modes of cohesive sediment transport in salt water. *J. Geophys. Res.* **67**, 1451-1461.

Eisma, D. (1967). Oceanographic observations on the Surinam shelf. *Hydrographic Newsletter (The Hague), Special Issue* **7**, 21-53.

Eisma, D. (1968). Composition, origin and distribution of Dutch coastal sands between Hoek van Holland and the Island of Vlieland (Netherlands). *J. Sea Res.* **4**, 123-267.

Emery, K. O. (1968). Relict sediments on continental shelves of world. *Am. Assoc. Petrol. Geologists Bull.* **52**, 445-464.

Emery, K. O. (1969). The continental shelves. *Scientific Am.* **221**, 106-122.

Emery, K. O., and Niino, H. (1963). Sediments of the Gulf of Thailand and adjacent continental shelf. *Geol. Soc. Am. Bull.* **74**, 541-554.

Ewing, M., Eittreim, S., Truchan, M., and Ewing, J. I. (1969). Sediment distribution in the Indian Ocean. *Deep-Sea Res.* **16**, 231-248.

Ewing, M., and Thorndike, E. M. (1965). Suspended matter in deep ocean water. *Science* **147**, 1291-1294.

Felix, D. W., and Gorsline, D. S. (1971). Newport Submarine Canyon, California: an example of the effects of shifting loci of sand supply upon canyon position. *Marine Geol.* **10**, 177-198.

Fleming, R. H. (1938). Tides and tidal currents in the Gulf of Panama. *J. Marine Res.* **1**, 192-205.

Fleming, R. H., and Revelle, R. (1939). Physical processes in the ocean. *In* "Recent Marine Sediments" (P. D. Trask, ed.) pp. 48-141. Am. Assoc. Petroleum Geologists, Tulsa, Oklahoma.

Folger, D. W. (1970). Wind transport of land-derived mineral, biogenic and industrial matter over the North Atlantic. *Deep-Sea Res.* **17**, 337-352.

Gorsline, D. S., Drake, D. E., and Barnes, P. W. (1968). Holocene sedimentation in Tanner Basin, California continental borderland. *Geol. Soc. Am. Bull.* **79**, 659-674.

Hart, J. J., and Currie, R. I. (1960). The Benguela Current. *Discovery Reports* **31**, 123-298.

Hayes, M. O. (1967). Relationship between coastal climate and bottom sediment type on the inner continental shelf. *Marine Geol.* **5**, 111-132.

Holeman, J. N. (1968). The sediment yield of major rivers of the world. *Water Resources Res.* **4**, 737-747.

Hjulstrom, F. (1939). Transportation of detritus by moving water. In "Recent Marine Sediments" (P. D. Trask, ed.) pp. 5-31. Am. Assoc. Petrol. Geologists, Tulsa, Oklahoma.

Hydraulics Research Station (1970). Thames Estuary flood prevention investigation: mathematical silt model studies: the effect of a half tide barrier at either Woolwich or Blackwall on siltation in the estuary. Rept. EX 479, Hydraulics Res. Sta., Wallingford, England.

Jacobs, M. B., and Ewing, W. M. (1969). Suspended particulate matter: concentration in the major oceans. Science 163, 380-383.

Jerlov, N. G. (1953). Particle distribution in the ocean. Repts. Swedish Deep-Sea Expedit. 3, 73-97.

Jerlov, N. G. (1958). Distribution of suspended material in the Adriatic Sea. Arch. Oceanog. Limnol. 11, 227-250.

Joseph, J. (1953). Die Trubungsverhaltnisse in der Südwestlichen Nordsee während der "Gauss"-Fahrt im Februar/Marz 1952. Ber. Deutsch. Wiss. Komm. f. Meeresforsch. 13, 93-103.

Joseph, J. (1957). Extinction measurements to indicate distribution and transport of watermasses. Proc. UNESCO Symp. on Physical Oceanog. UNESCO, Paris, 59-75.

Komar, P. D. (1969). The channelized flow of turbidity currents with application to Monterey deep-sea fan channel. J. Geophys. Res. 74, 4544-4558.

Korgen, B. J., Bodvarsson, G., and Kulm, L. D. (1970). Current speeds near the ocean floor west of Oregon. Deep-Sea Res. 17, 353-357.

Krone, R. B. (1962). Flume studies of the transport of sediment in estuarial shoaling processes. Univ. of Calif. Hydraulic Eng. Lab. and Sanit. Res. Lab., Berkeley, Rept., 110 pp.

Ku, T-L., Broecker, W. S., and Opdyke, N. D. (1968). Comparison of sedimentation rates measured by paleomagnetic and the ionium methods of age determination. Earth and Planetary Sci. Lett. 4, 1-16.

Kuenen, Ph. H. (1939). The cause of coarse deposits at the outer edge of the shelf. Geol. Mijnb. 18, 36-39.

Lee, A. J., and Folkard, A. R. (1969). Factors affecting turbidity in the southern North Sea. J. Cons. Int. Explor. Mer. 32, 291-302.

Lee, A., and Ramster, J. (1968). The hydrology of the North Sea. A review of our knowledge in relation to pollution problems. Helgoländer Wiss. Meeresunters. 17, 44-63.

Manheim, F. T., Meade, R. H., and Bond, G. C. (1970). Suspended matter in surface waters of the Atlantic continental margin from Cape Cod to the Florida Keys. Science 167, 371-376.

McCave, I. N. (1970). Deposition of fine-grained suspended sediment from tidal currents. J. Geophys. Res. 75, 4151-4159.

McCave, I. N. (1971). Wave effectiveness at the sea bed and its relationship to bed-forms and deposition of mud. J. Sediment. Petrol. 41, 89-96.

Menard, H. W., and Smith, S. M. (1966). Hypsometry of ocean basin provinces. J. Geophys. Res. 71, 4305-4325.

Migniot, C. (1968). Etude des Propriétés physiques de differents sédiments très fins et de leur comportement sous des actions hydrodynamiques. La Houille Blanche 7, 591-620.

Moore, D. G. (1969). Reflection profiling studies of the California continental borderland: structure and quaternary turbidite basins. Geol. Soc. Amer. Special Paper 107, 142 pp.

Morgan, J. P., Lopik, J. R. van, and Nichols, L. G. (1953). Occurrence and development of mudflats along the western Louisiana coast. Coastal Studies Inst. Louisiana State Univ., Baton Rouge, Tech Rept. 2, 34 pp.

Nelson, B. W. (1970). Hydrography, sediment dispersal and recent historical development of the Po River Delta, Italy. In "Deltaic Sedimentation Modern and Ancient" (J. P. Morgan and R. H. Shaver, eds.), pp. 152-184, Soc. Economic Paleontologists and Mineralogists Spec. Publ No. 15, Tulsa, Oklahoma.

Niino, H. and Emery, K. O. (1961). Sediments of shallow portions of East China Sea and South China Sea. Geol. Soc. Amer. Bull. 72, 731-762.

Nota, D. J. G. (1958). Sediments of the western Guiana shelf. *Meded. Landbouwhogeschool Wageningen* **58**, 98 pp.

Osterberg, C., Carey, A. G., and Curl, H. (1964). Acceleration of sinking rates of radionuclides in the ocean. *Nature, London* **200**, 1276-1277.

Otto, L. (1967). Investigations on optical properties and watermasses of the southern North Sea. *Neth. J. Sea Res.* **3**, 532-552.

Owen, M. W., and Odd, N. V. M. (1970). A mathematical model of the effect of a tidal barrier on siltation in an estuary. Preprint Internat. Conf. on the Utilisation of Tidal Power, Halifax 'N.S. *Dept. of Energy, Mines and Resources, Ottawa*, 36 pp.

Parke, M. L., Emery, K. O., Szymankiewicz, and Reynolds, L. M. (1971). Structural framework of continental margin in South China Sea. *Am. Assoc. Petrol. Geologists Bull.* **55**, 723-751.

Partheniades, E., Cross, R. H., and Ayora, A. (1969). Further results on the deposition of cohesive sediments. *Proc. 11th Conf. on Coastal Eng., London (1967), Am. Soc. Civ. Engrs.* **1**, 723-742.

Postma, H. (1967). Sediment transport and sedimentation in the estuarine environment. *In* (G. H. Lauff, ed.), Estuaries, Publ. 83, pp. 158-179. American Association for Advancement of Science, New York.

Postma, H. (1969). Suspended matter in the marine environment. *In* "Morning Review Lectures of the Second International Oceanographic Congress, Moscow, 1966," pp. 213-219. UNESCO, Paris.

Rex, R. W., and Goldberg, E. D. (1958). Quartz contents of pelagic sediments of the Pacific Ocean. *Tellus* **10**, 153-159.

Rodolfo, K. S. (1969a). Sediments of the Andaman Basin, northeastern Indian Ocean. *Marine Geol.* **7**, 371-402.

Rodolfo, K. S. (1969b). Suspended sediments in surface Andaman Sea waters off the Irrawaddy Delta, northeastern Indian Ocean. *Geol. Soc. Am. Abst. 1969,* **7**, 190-191.

Rowe, G. T. (1971). Observations on bottom currents and epibenthic populations in Hatteras submarine canyon. *Deep-Sea Res.* **18**, 569-581.

Rumeau, J. L., and Vanney, J. R. (1968). Éléments-traces de vases marines du plateau continental Atlantique au sud-ouest du massif Armoricain (Grande Vasière). *Bull. Centre Rech. Pau - SNPA* **2**, 69-81.

Scruton, P. C., and Moore, D. G. (1953). Distribution of surface turbidity off Mississippi Delta. *Am. Assoc. Petrol. Geologists Bull.* **37**, 1067-1074.

Shepard, F. P. (1960). Mississippi Delta: marginal environments, sediments and growth. *In* "Recent Sediments, Northwest Gulf of Mexico" (F. P. Shepard, F. B. Phleger, and Tj. H. Van Andel, eds.) pp. 56-81. American Association Petroleum Geologists, Tulsa, Oklahoma.

Shepard, F. P. (1963). "Submarine Geology," Harper and Row, New York, 557 pp.

Shepard, F. P., and Marshall, N. F. (1969). Currents in La Jolla and Scripps Submarine Canyons. *Science* **165**, 177-178.

Skempton, A. W. (1970). The consolidation of clays by gravitational compaction. *Geol. Soc. London, Quart. J.* **125**, 373-411.

Spencer, D. W., and Sachs, P. L. (1970). Some aspects of the distribution, chemistry and mineralogy of suspended matter in the Gulf of Maine. *Marine Geol.* **9**, 117-136.

Stefansson, U., Atkinson, L., and Bumpus, D. F. (1971). Hydrographic properties and circulation of the North Carolina shelf and slope waters. *Deep-Sea Res.* **18**, 383-420.

Stride, A. H., and Tucker, M. J. (1960). Internal waves and waves of sand. *Nature, London* **188**, 933.

Subba Rao, M. (1964). Some aspects of continental shelf sediments off the east coast of India. *Marine Geol.* **1**, 59-87.

Sundborg, A. (1956). The River Klarälven: a study of fluvial processes. *Geogr. Annaler* **38**, 127-316.

Sundborg, A. (1958). A method for estimating the sedimentation of suspended material. *C. R. and Repts., Toronto, Intnatl Assoc. Sci. Hydrology* **43**, 249-259.

Swift, D. J. P. (1970). Quaternary shelves and the return to grade. *Marine Geol.* **8**, 5-30.

Swift, D. J. P., Stanley, D. J., and Curray, J. R. (1971). Relict sediments on continental shelves: a reconsideration. *J. Geol.* **79**, 322-346.

Straaten, L. M. J. U. Van (1965). Sedimentation in the northwestern part of the Adriatic Sea. "Proceedings 17th Symposium Colston Research Society," pp. 143-161. Butterworths, London.

Terwindt, J. H. J. (1967). Mud transport in the Dutch delta area and along the adjacent coastline. *Neth. J. Sea Res.* **3**, 505-531.

Turekian, K. K. (1968). "Oceans," 120 pp. Prentice-Hall, New Jersey.

Wageman, J. M., Hilde, T. W. C., and Emery, K. O. (1970). Structural framework of East China Sea and Yellow Sea. *Am. Assoc. Petrol. Geologists Bull.* **54**, 1611-1643.

Whitehouse, U. G., Jeffrey, L. M., and Debrecht, J. D. (1960). Differential settling tendencies of clay minerals in saline waters. *Proc. 7th Natl. Conf. on Clays and Clay Minerals,* 1-79.

Wobber, J. J. (1967). Space photography: a new analytical tool for the sedimentologist. *Sedimentology* **9**, 265-317.

CHAPTER 11

Sources and Sinks of Suspended Matter on Continental Shelves*

Robert H. Meade

U.S. Geological Survey
Woods Hole, Massachusetts 02543

ABSTRACT

While river sediments are the principal sources of suspended matter on many continental shelves, their importance on the Atlantic Continental Shelf of the United States is less than that of biogenic detritus produced by shelf organisms and of material resuspended from the shelf bottom. The major sites of accumulation of suspended matter from the Atlantic shelf are in the large estuaries and coastal marshlands rather than on the shelf itself or in the deep sea. Sources other than river sediments and sinks other than the sea floor should be given ample consideration in assessing the dispersal of suspended sediments on other continental margins of the world.

INTRODUCTION

The underlying assumption, explicit or implicit, of many studies of the suspended matter in waters of the continental shelves is that the material is supplied mainly by present-day input from rivers and that its destination is the floor of the open sea. This assumption, however, needs critical examination. Sources

*Publication authorized by the Director, U.S. Geological Survey. Contribution No. 2862 of the Woods Hole Oceanographic Institution.

other than present-day river input are demonstrably more important in some shelf areas. And except at the mouths of some of the world's major rivers, one can build convincing lines of evidence and argument to show that fine-grained sediment is not moving seaward across continental shelves.

The observations and inferences that follow are based mainly on experience on the Atlantic continental margin of the United States. Although the viewpoint is provincial, the Atlantic continental margin is one of the best studied marine areas of the world, and the problem needs to be considered in its depth as well as its breadth. With the ready admission that many of my comments are no less speculative than those in other papers in this volume, I will try to build a case for a somewhat different slant on the dispersal of suspended matter.

This paper grew partly out of comments and thoughts that were stimulated and developed while I was reviewing some of the other articles on suspended matter in this volume—those by Buss and Rodolfo, Pierce et al., Schubel and Okubo, and especially the extensive treatment by McCave—in all of which I found assumptions or attitudes with which I disagreed. I appreciate the opportunity to add my comments to this symposium, and I especially appreciate the advantage the other authors gave me by writing their papers first. This paper also owes a lot to discussions over the last several years with F. T. Manheim and J. S. Schlee of the U.S. Geological Survey; J. D. Milliman, D. W. Spencer, and J. M. Teal of Woods Hole Oceanographic Institution; and J. E. Sanders of Barnard College.

TYPES OF EVIDENCE AND
PROBLEMS OF INTERPRETATION

As of now, what we know about the sources, dispersal, and deposition of suspended sediment on continental shelves is based on two types of evidence. First, measurements of the instantaneous distribution of suspended matter, when combined with the present patterns of shelf-water circulation, give us snapshot views of what we take to be dispersal processes in action. Second, the distribution and characteristics of the bottom sediments that lie on the shelves give us an integrated picture from which we can make inferences about where sediments came from and how they got where they are. Taken by itself, neither of these types of evidence so far has given us a clear understanding of the dispersal of suspended sediment. When taken together in the same place, the two types of evidence often tell conflicting stories because either they represent processes that have occurred at different times or they represent the results of processes that operate within wholly disparate time scales.

Suspended Matter and Water Circulation

Concerning the first type of evidence: not only have too few studies been made so far, but too few of the data we do have are helpful in determining where the sediment comes from and where it goes. Even when measurements of suspended matter are tied to correlative data on the movement of water, we still have to reckon with the reality that suspended matter is not a conservative property of sea water. It settles to the bottom under the influence of gravity and it is resuspended by bottom currents. It is produced, consumed, deposited, and resuspended by organisms. So even if we compute the flux of suspended matter from measurements of suspended concentration and water movement at some point or series of points on the shelf, we still have no clear portrayal of the continuity of suspended-sediment flow.

Furthermore, we have too few data to show what happens during the rare events which our experience with rivers and beaches suggests are the times when most of the material is moved. This is why the studies of the dispersal of the sediments from the 1969 river floods in southern California (Drake et al., 1971; Drake et al., 1972) and the study of the effects of Hurricane Gerda off the North Carolina coast (Rodolfo et al., 1971) are so valuable. Studies of this type are rare, and we still have only a hazy conception of the frequency and magnitude of important sedimentary events on the continental shelf.

Bottom Sediments and the Relict-Sediment Problem

The problem with bottom sediments is not so much a lack of data: we have fairly accurate maps of the distribution of bottom sediments on many continental shelves of the world. The problem comes in the interpretation of what we know to be there. Sedimentologists began to investigate recent marine sediments in earnest in the 1930s with the confident hope that they could match modern sediments to modern conditions and develop good keys to the interpretation of sedimentary environments in ancient rocks. They discovered, however, that many surficial sediments on the continental shelves were not exclusively the products of modern marine conditions but were formed under earlier and different conditions and were now involved in the transitional processes of adjusting themselves to a continuously changing sea level. This absence of a clear key to the past in recent sediments has been described by a member of that hopeful generation of sedimentologists as "the great tragedy of the Pleistocene glaciation" (L. L. Sloss, oral communication, 1972).

The distinction between modern and relict sediments is not always a simple one because sediments adjust to the changing conditions that accompany rising sea level at rates that vary all the way from very rapid to very slow. Where large deltas extend onto or across the continental shelves, the bottom sediments are demonstrably the products of modern conditions. At the other extreme are areas of the continental shelves (especially the outer shelves) where bottom sediments may be altogether relict (Emery, 1968). Most shelf areas, however, are probably covered by bottom sediments that show a mixture of the influences of modern and preexisting conditions (Swift et al., 1971). On the Atlantic Continental Shelf of the United States, a useful generalization is that most of the sediments are relict in composition but modern in texture—that is, the material itself was brought to the shelf under conditions different from those existing today, but the present assortment and arrangement of particle sizes show the influence of present hydrodynamic conditions (Milliman et al., 1972). In any event, one cannot draw many unequivocal inferences about present-day processes by simply observing the distribution of different petrologic types of bottom sediments on continental shelves.

Examples of places to be wary of one's inferences are the mud deposits on the middle and outer shelves (cases 3 and 4 in McCave's Fig. 91, this volume). Such a deposit, 4 to 5 m thick, lies on the middle and outer shelf, south of New England. Although there has been a fair amount of disagreement on whether this mud is relict or modern (Garrison and McMaster, 1966), the mollusk remains that were found recently in the mud indicate that the deposit was formed when the shelf water was substantially colder than it is now (Schlee, in press). Judging from radiocarbon dates of shell material in sands that are thought to be continuous beneath this deposit, the mud accumulated some 9 to 10 thousand years ago when sea level was about 60 m below its present level. Why the mud accumulated at that particular time and place is not clear, but it does not appear to be a modern deposit. All deposits of this sort should be examined critically before they are assumed to be either modern or relict.

Disparate Time Scales

We add to the difficulties of interpreting the distributions of suspended matter in the water and deposited material on the bottom when we try to account for both types of evidence within the same time scale. The problem becomes even more difficult when we try to fit both types of evidence into geologic concepts that imply time scales measurable in millions of years. In thinking about sedimentation processes on continental shelves, we tend to look down the long span of geologic time and think in terms of inexorable seaward movement of sediment. However, this does not mean that we can go to the continental shelf at any randomly or even purposely selected instant and observe long-term processes in action. Furthermore, because of the drastic fluctuations in sea level during the Pleistocene ice ages, we cannot expect even the arrangement of bottom sediments on the shelves to reflect the long-term seaward movement of continental debris. Conversely, we cannot use models based entirely on long-term concepts to understand short-term problems such as the day-to-day disposition of suspended sediments in coastal waters.

SOURCES OF SUSPENDED MATTER

Our thinking about sources of suspended matter may be muddled by a lack of clear discrimination between ultimate and instantaneous sources. In the long run of geologic time, inorganic fine-grained sediment is ultimately derived from the continents. But what are the more immediate sources of sediments during shorter time spans such as the last 15 thousand years of rising sea level, or the last few millenia when man's activities have accelerated the rates of erosion and sedimentation, or the last few decades when man-made pollutants have been closely associated with fine-grained sediments?

Despite some lip service to the contrary, we tend to assume that most of the suspended matter in shelf waters is derived from present-day continents and that the large turbid plumes we see along the coastlines must represent the seaward flux of river sediment. Except for the areas near the mouths of some two dozen of the world's major rivers, however, other sources of suspended matter may well be more significant than river sediment. Two of these sources are the organic matter that is produced in waters of the shelf and the bottom sediments (relict or otherwise) whose particles can be resuspended. These sources complicate the analysis of suspended-matter data, but any model that ignores them is too simplified to be of much use in solving real problems.

Biogenic Detritus in Suspension

Organic matter makes up a large proportion of the suspended matter on continental shelves. It makes up essentially all the suspended matter in surface waters of most outer shelves. Data from the Atlantic coast of the United States (Manheim et al., 1970), and even from the Gulf of Mexico and the East China Sea where major rivers discharge their sediment loads (Manheim et al., 1972; Wageman et al., 1970), show that most of the suspended matter in surface waters of the outer shelves is combustible organic matter. Most of the remaining noncombustible ash, furthermore, is skeletal matter rather than terrigenous mineral debris. Our more recent measurements (unpublished) on the outer shelf between Cape Cod and Cape Hatteras show that even the near-bottom suspended matter contains 30 to 40% combustible organic matter plus another substantial fraction of noncombustible skeletal debris.

The large proportion of organic detritus on the outer shelves shows that a proportional decrease in the nonbiogenic component accompanies the seaward decrease in total suspended matter. That is, the concentration of terrigenous detritus decreases seaward across continental shelves at a greater rate than the concentration of total suspended matter. This decrease is shown in the hundredfold difference in the concentrations of mineral grains suspended in surface waters of the inner shelf and those of the middle and outer shelf off the southeastern United States (Manheim et al., 1970, Fig. 1C). Models of the diffusion or advection of terrigenous mineral matter, therefore, should be tested by their relations to the observed concentrations of nonbiogenic matter rather than to those of total suspended matter.

Resuspension of Shelf Bottom Sediments

Relict fine sediments can be an important source of suspended matter. Concentrations of suspended matter in near-bottom waters of the continental shelf are frequently greater above mud bottoms than those above adjacent sand or gravel bottoms. Detailed studies in the coastal waters of Massachusetts show that some of these higher concentrations can be related to bottom-dwelling animals that either resuspend the sediment themselves or loosen the bottom sediment so that it can be resuspended by tidal currents (Rhoads, 1963, 1970; Rhoads and Young, 1971; Young, 1971, Young and Rhoads, 1971). Equations based on strictly physical relations between cohesive strengths of inorganic sediment and erosional velocities of water, therefore, may have little relevance to actual resuspension mechanisms on the shelf.

In the Gulf of Maine, off the coast of New England, relict sediment is the principal source of suspended matter. When the last ice sheet withdrew from the Gulf of Maine, it left a topographically irregular area of the shelf covered with a poorly sorted mixture of particle sizes that ranged from boulders to clay. As the sea refilled the area, the fine-grained fraction of this material was winnowed from the elevated areas of the bottom and redeposited in the basins. The resulting pattern of distribution of the bottom sediments now shows gravels on top of the ridges and ledges, muds in the bottoms of the basins, and poorly sorted mud-sand-gravel mixtures (the original glacial deposit, largely unchanged) on the intermediate slopes (Hathaway et al., 1965; Schlee, in press). Suspended matter is presently distributed as normally low concentrations of mostly organic material in waters near the surface and several times greater concentrations of predominantly (about 90%) inorganic material in the waters near the bottoms of the basins (Spencer and Sachs, 1970). Judging from its spatial distribution and its chemical and mineral composition, the near-bottom material is resuspended bottom sediment. Its essential lack of relation to modern river input is shown by the large quantity in suspension. Spencer and Sachs calculated that the total amount of inorganic suspended matter in the gulf is about 37 million metric tons, more than half of which is suspended in the near-bottom waters of the deep basins. This total is about an order of magnitude greater than the annual contribution of sediment by the tributary rivers of the gulf.

SITES OF ACCUMULATION OF SUSPENDED MATTER

The long-term conception of the seaward movement of continental detritus is so much a part of basic geologic thinking that we sometimes forget that sediments can come to permanent resting places short of deep water. Although we now see how sediments can be trapped more or less permanently in estuaries (Meade, 1969a, 1972, and references cited therein; Schubel, 1971), we still have difficulty seeing how fine-grained sediments that reach the continental shelf can go anywhere but seaward. Where mud deposits are absent on continental shelves, we postulate "bypassing" mechanisms and support them by referring to postglacial rates of sedimentation on the continental slope and rise. However, sedimentation rates on the slope and rise are not the same as measures of contemporaneous input from the continental shelf. Furthermore, because we live in a transgressive era (that is, a period of rising sea level) and because the geologic record contains many examples of transgressive sedimentary deposits that accumulated at or near shorelines, perhaps we should look more closely at our present shorelines for significant accumulations of sediment.

Sedimentation Rates on the Continental Slope and Rise

While fine-grained sediments on the continental slope and rise may be demonstrably terrigenous, it does not follow that the measured rates of sedimentation on the slope and rise during the last 10 to 15 thousand years are a measure of the rate at which terrigenous sediments are now being supplied from continental sources. The material itself may well have reached the deeper areas at earlier times when sea level was lower and the shorelines were farther seaward than they are now, and the sedimentation rates may be a measure of local redistribution of the material since it was first delivered to the slope and rise.

The high rates of sedimentation measured on the continental rise off eastern North America (Turekian, 1968, p. 67), for example, are most easily explained as being due to the redeposition of older material by deep ocean currents. Ripple marks and other features on the continental rise suggest strongly that this is not an area of static vertical accumulation but one of active sediment movement parallel to the continental margin in the southerly direction of flow of the Western Boundary Undercurrent (Heezen et al., 1966; Heezen and Hollister, 1971, p. 390-416; Hollister and Heezen, 1972). That the principal component of sediment movement is parallel to rather than across the continental margin is also suggested by the presence of a band of suspended-matter concentration in the near-bottom waters on the continental rise (the "bottom nepheloid layer") which is overlain and bounded on the landward side by less turbid waters and shows little or no continuity with sources of suspended matter in overlying waters or on the continental shelf and slope (Eittreim et al., 1969). Likewise, the high rates of sediment accumulation in the western South Atlantic (Opdyke and Knowles, 1972) are probably more closely related to the northward flow of the Antarctic Bottom Water than to sediment input by any modern rivers (Heezen and Hollister, 1971, p. 365-381.).

Coastal Wetlands as Sites of Accumulation

A fair amount of evidence suggests that most of the fine-grained river sediments that reach the Atlantic Continental Shelf of the United States are transported back into the estuaries and coastal wetlands. South of Cape Cod, concentrations of suspended matter (and particularly of the terrigenous component) in shelf waters decrease rapidly in the seaward direction (Manheim et al., 1970; Meade et al., 1970), and no modern mud deposits are present on the middle or outer shelf to suggest the episodic seaward transport of fine material (Uchupi, 1963; Milliman et al., 1972). The prevalent drift of bottom water is landward, rather than seaward, on the inner continental shelf from the Gulf of St. Lawrence to Florida (Bumpus, 1965; Graham, 1970; Hathaway, 1972; Lauzier, 1967; Norcross and Stanley, 1967). Bottom sediments in most of the estuaries of the Atlantic seaboard differ significantly in their clay-mineral composition from sediments carried by their tributary rivers, and, therefore, they must contain either material that was derived from offshore and longshore sources or material that was discharged by other rivers and moved alongshore and into the mouths of the estuaries (Hathaway, 1972; Pevear, 1972; Windom et al., 1971).

Coastal marshlands seem to play an important role in trapping and accumulating fine-grained sediments along the Atlantic coast. Coastal marshes and their typical plant assemblages are sensitively adjusted to sea level. In the last 3500 yr, during which sea level has risen several meters, marshes along the Atlantic coast have been able to form and grow upward in response to the rising level of the sea (Bloom, 1967). As Sanders (1970) points out, this sustained upward growth requires a steady input of large quantities of sediment.

Using the available compilations of river-sediment input and coastal-wetland area, one can argue that the annual increment of river sediment to the Atlantic coast of the United States would add a layer to the coastal wetlands that is essentially what they require to keep up with rising sea level. The figures are as follows: Annual suspended-sediment input by rivers of the U.S. to the Atlantic coast during 1906 and 1907 was 25 million metric tons (Dole and Stabler, 1909). Area of coastal wetlands in the Atlantic states in 1954 was 9200 sq km (Spinner, 1969). Assuming a sediment particle density of 2.6 g cm^{-3} and a porosity of 60% (sediments will probably accumulate at porosities greater than 60%, but the underlying marsh deposits are being compacted, and the annual accumulation has to compensate for compaction as well as for the rising sea level), the 25 million tons would cover the 9200 km² to a depth of about 2.6 mm. This corresponds closely to the 2.5 to 3.5 mm of average annual rise of sea level along the Atlantic coast since 1930 (Meade and Emery, 1971). In order to survive, then, the coastal wetlands require an annual input equivalent to all the suspended sediment brought to the coast by the rivers.

Using these numbers involves some problems. The rate of input of river sediments that was estimated by Dole and Stabler was derived from only one year's data, and it represents perhaps five times the rate that prevailed before European colonists arrived (Meade, 1969b). The coastal marshes have been both enlarged and diminished by the activities of man: their areas have increased as a result of the increased input of river sediment (Gottschalk, 1945), and they have been decreased by dredging and filling (5.4% of Atlantic coastal wetlands were destroyed between 1954 and 1968, according to Spinner). The rate of sea-level rise since 1930 is somewhat greater than the rate from about 1900 to 1930 and substantially greater than the rate of 1 mm or less per year that seems to have been the average for the last 3000 years (Bloom and Stuiver, 1963; Redfield and Rubin, 1962; Scholl et al., 1969). Furthermore, the characteristics of the material accumulating in the marshlands may not be identical to those of the material being supplied by the rivers (although a few measurements made by Teal and Kanwisher, 1961, show that the particle-size distributions and organic-matter concentrations of the sediments in a Georgia salt marsh are essentially the same as those in suspended matter carried by rivers). So the meaning of the numbers is somewhat uncertain. But, for the sake of discussion, let us assume that (a) the texture and composition of the materials supplied by the rivers and those accumulating in the wetlands are roughly equivalent, (b) the area of the wetlands represents an equilibrium response to the sediment supply, and (c) the wetlands are rising as fast as the sea level; and then let us consider what might be happening to river sediment along two segments of the Atlantic coast.

For the coast between Cape Cod and Cape Lookout (North Carolina), Dole and Stabler estimated a total annual input of river sediment to the estuaries and coastal zone of about 9 million metric tons. Subsequent and more extensive measurements by the United States Geological Survey on the Delaware, Potomac, James and Tar Rivers suggest that Dole and Stabler's estimate may be slightly greater than one based on later data, but the two estimates would probably agree within 20%. The area of coastal wetlands in the same segment is about 4000 km². Spreading 9 million tons of river sediment evenly over the wetlands (assuming the same particle density and sediment porosity as above) would result in an annual accumulation of about 2 mm, which is within the range of accumulation rates measured in coastal marshes of Connecticut by Bloom (1967). Furthermore, as most of the coastal wetlands in this segment lie inside the large estuaries—Chesapeake and Delaware Bays, Pamlico, Albemarle, and Long Island Sounds—only a small fraction of the river sediment ever emerges onto the continental shelf. Biggs (1970), for example, estimates that of the sediment supplied to the upper half of Chesapeake Bay (which includes sediment eroded from the estuary shoreline as well as river sediment), only 9% moves seaward; the other 91% is retained in the upper half of the bay. The small amount of river sediment that does emerge from the estuaries probably is entrained by the southward-moving inshore circulation and the landward-moving bottom drift to eventually be deposited in the salt marshes that lie behind the barrier beaches of the outer coast. The only place where significant amounts of river sediment are likely to cross the shelf in this segment is south of Chesapeake Bay, where a virtually uninterrupted barrier beach prohibits access to the coastal marshes, and at Cape Hatteras, where the coastal currents are deflected seaward. But even at Cape Hatteras, despite the easy inferences we would like to make from the spectacular Apollo IX photographs (Emery, 1969; see also Mairs, 1970), the evidence for seaward transport of suspended sediment is not clear: our particle-size analyses of the suspended matter in this area show that most of the mineral matter in the shelf waters is not derived from the estuaries (Bond and Meade, 1966; Manheim et al., 1970), and there are no muds on the bottom beneath the Hatteras plume. My estimate of the fate of fine-grained river sediment between Cape Cod and Cape Lookout is that about 90% is eventually trapped in the large estuaries, and most of the material that escapes the estuaries goes eventually into the wetlands that line the outer coast. Only a few percent of the total river input crosses the shelf to the deep sea.

More pertinent to the discussion of material that escapes the rivers and estuaries is the coastal segment south of Cape Lookout, in which the main rivers discharge either into small estuaries or directly onto the continental shelf. Dole and Stabler estimated that the rivers between Cape Lookout and the southern tip of Florida discharged about 16 million metric tons per year. In the decades since Dole and Stabler's data were collected, soil-conservation measures and the construction of several large dams and many smaller ones on the major rivers has substantially diminished the flow of sediment to the sea. Probably no more than 10 million tons of sediment are now being delivered annually to the coast south of Cape Lookout. A hundredfold difference between the concentrations of suspended mineral grains in the surface waters of the inner and outer shelf (Manheim et al., 1970, Fig. 1C) suggests strongly that river sediment is not being moved seaward. The more likely fate of the river sediment discharged onto the shelf is that it is entrained southward by the prevailing longshore drift and worked back into the coastal wetlands by the net landward bottom drift. The area of wetland along this section of the coast is about 5000 km^2. The 10 million tons of river sediment would cover the wetlands to a depth of 2 mm per year which, as in the segment north of Cape Lookout, is slightly less than the rate at which sea level has risen in the last several decades.

The correspondence between the supply of river sediment and the growth requirements of the coastal wetlands should not be interpreted too literally because of the many uncertainties and assumptions involved in the calculations. However, the salt marshes do appear to be keeping up with rising sea level, and the necessary sediment must be coming from somewhere. Whether it comes from rivers, the shelf bottom, or the shoreline—or what proportions come from the different sources—is still an open question. In any event, the coastal wetlands are probably the major sink for fine-grained material suspended in waters of the Atlantic continental shelf.

Perhaps the processes we infer along the Atlantic coast are exceptional, and perhaps they are not. The coastal marshes of the Gulf coast of the United States, even with their large areas and rapid subsidence rates, can accommodate only a fraction of the sediment delivered by the Mississippi River. But what happens to river sediment in other areas? How much of the Rhine sediment is trapped in the Wadden Sea? How much of the Amazon sediment is permanently accreted to the Guiana coast and how much eventually makes its way to the deep sea? Does a significant fraction of the fine sediment from the Yellow and Yangtze Rivers ever reach the deep sea, considering the wide continental shelf in the Yellow and East China Seas, the outer half of which is floored with relict sands?

CONCLUSION

The present state of knowledge of sedimentary processes on continental shelves is exploratory rather than explanatory, and, therefore, we should keep our minds open to the multiplicity of possible sources and sinks of suspended matter. The models, both physical and conceptual, that we build at this point are necessary to the process of defining and evaluating the important variables, but we should keep in mind that they are not yet suitable for predicting the behavior of suspended sediment on the shelves. Too much of the critical field evidence is still lacking. The time scales of our thinking need to be coordinated so that we do not confuse processes that are observable from day to day or year to year with long-term geologic processes, which are mainly conceptual and are the integral sum of a number of individually observable short-term processes.

The remaining unanswered questions are the basic ones: Where does suspended matter come from? How far does it travel during a specific event or period of time? Where is it deposited? Once deposited, how long does it stay put?

REFERENCES

Biggs, R. B. (1970). Sources and distribution of suspended sediment in northern Chesapeake Bay. *Marine Geol.* **9**, 187-201.

Bloom, A. L. (1967). Coastal geomorphology of Connecticut. Cornell Univ. Dept. Geol. Sci. Final Rept. to Office Naval Res. (Contract Nonr-401(45), Task Number NR 388-065). Ithaca, New York.

Bloom, A. L., and Stuiver, M. (1963). Submergence of the Connecticut coast. *Science* **139**, 332-334.

Bond, G. C., and Meade, R. H. (1966). Size distributions of mineral grains suspended in Chesapeake Bay and nearby coastal waters. *Chesapeake Sci.* **7**, 208-212.

Bumpus, D. F. (1965). Residual drift along the bottom on the continental shelf in the Middle Atlantic Bight area. *Limnology and Oceanography* **10**, (Redfield Vol.), R50-R53.

Dole, R. B., and Stabler, H. (1909). Denudation. *In* "Papers on the Conservation of Water Resources. p. 78-93, U.S. Geol. Survey Water-Supply Paper 234, Washington, D.C.

Drake, D. E., Fleischer, P., and Kolpack, R. L. (1971). Transport and deposition of flood sediment, Santa Barbara Channel, California. *In* "Physical, Chemical, and Geological Studies" (R. L. Kolpack, ed.), p. 181–217; Vol. 2 of Biological and oceanographical survey of the Santa Barbara oil spill 1969–1970. Sea Grant Pub. 2, Univ. Southern California Allan Hancock Found., Los Angeles, Calif.

Drake, D. E., Kolpack, R. L., and Fischer, P. J., Sediment transport on the Santa Barbara-Oxnard shelf, Santa Barbara Channel, California. *In* "Shelf Sediment Transport: Process and Pattern" (D. J. P. Swift, D. B. Duane, and O. H. Pilkey, eds.). Dowden, Hutchinson and Ross, Stroudsburg, Pa.

Eittreim, S., Ewing, M., and Thorndike, E. M. (1969). Suspended matter along the continental margin of the North American Basin. *Deep-Sea Res.* **16**, 613-624.

Emery, K. O. (1968). Relict sediments on continental shelves of world. *Am. Assoc. Petroleum Geologists Bull.* **52**, 445-464.

Emery, K. O. (1969). The continental shelves. *Sci. American* **221**[3], 106-114, 116, 118, 120-122.

Garrison, L. E., and McMaster, R. L. (1966). Sediments and geomorphology of the continental shelf off southern New England. *Marine Geol.* **4**, 273-289.

Gottschalk, L. C. (1945). Effects of soil erosion on navigation in upper Chesapeake Bay. *Geog. Rev.* **35**, 219-238.

Graham, J. J. (1970). Coastal currents of the western Gulf of Maine. *Internatl. Comm. Northwest Atlantic Fisheries Res. Bull.* **7**, 19-31.

Hathaway, J. C. (1972). Regional clay mineral facies in the estuaries and continental margin of the United States east coast. *In* "Environmental framework of coastal-plain estuaries" (B. W. Nelson, ed.), p. 293-316. Geol. Soc. America Mem. 133.

Hathaway, J. C., Schlee, J. S., Trumbull, J. V. A., Hülsemann, J. (1965). Sediments of the Gulf of Maine (abs.). *Am. Assoc. Petroleum Geologists Bull.* **49**, 343-344.

Heezen, B. C., and Hollister, C. D. (1971). "The Face of the Deep." Oxford Univ. Press, New York.

Heezen, B. C., Hollister, C. D., and Ruddiman, W. F. (1966). Shaping of the continental rise by deep geostrophic contour currents. *Science* **152**, 502-508.

Hollister, C. D., and Heezen, B. C. (1972). Geologic effects of ocean bottom currents: western North Atlantic. *In* "Studies in Physical Oceanography" (A. L. Gordon, ed.), p. 37-66. New York, Gordon and Breach Sci. Pubs.

Lauzier, L. M. (1967). Bottom residual drift on the continental shelf area of the Canadian Atlantic coast. *Canada Fisheries Res. Board J.* **24**, 1845-1859.

Mairs, R. L. (1970). Oceanographic interpretation of Apollo photographs. *Photogramm. Eng.* **36**, 1045-1058.

Manheim, F. T., Hathaway, J. C., and Uchupi, E. (1972). Suspended matter in surface waters of the northern Gulf of Mexico. *Limnology and Oceanography* **17**, 17-27.

Manheim, F. T., Meade, R. H., and Bond, G. C. (1970). Suspended matter in surface waters of the Atlantic continental margin from Cape Cod to the Florida Keys. *Science* **167**, 371-376.

Meade, R. H. (1969a). Landward transport of bottom sediments in estuaries of the Atlantic Coastal Plain. *J. Sedimentary Petrology* **39**, 222-234.

Meade, R. H. (1969b). Errors in using modern stream-load data to estimate natural rates of denudation. *Geol. Soc. America Bull.* **80**, 1265-1274.

Meade, R. H. (1972). Transport and deposition of sediments in estuaries. *In* "Environmental Framework of Coastal-Plain Estuaries" (B. W. Nelson, ed.), p. 91-120. Geol. Soc. America Mem. 133.

Meade, R. H., and Emery, K. O. (1971). Sea level as affected by river runoff, eastern United States. *Science* **173**, 425-428.

Meade, R. H., Sachs, P. L., Manheim, F. T., and Spencer, D. W. (1970). Suspended matter between Cape Cod and Cape Hatteras. *In* "Summary of Investigations Conducted in 1969."Ref. 70-11, 47-49. Woods Hole Oceanographic Institution.

Milliman, J. D., Pilkey, O. H., and Ross, D. A. (1972). Sediments of the continental margin off the eastern United States. *Geol. Soc. America Bull.* **83**, 1315-1333.

Norcross, J. J., and Stanley, E. M. (1967). Inferred surface and bottom drift, June 1963 through October 1964. *In* W. Harrison et al., "Circulation of shelf waters off the Chesapeake Bight." Prof. Paper 3, p. 11-42. Environmental Science Services Adm.

Opdyke, N. D., and Knowles, R. (1972). Paleomagnetism of cores from the South Atlantic (abs.); *in* Continental drift emphasizing the history of the South Atlantic area: EOS. *Am. Geophys. Union Trans.* **53**, 170-171.

Pevear, D. R. (1972). Sources of recent nearshore marine clays, southeastern United States. *In* "Environmental Framework of Coastal-Plain Estuaries." (B. W. Nelson, ed.), p. 317-335. Geol. Soc. America Mem. 133.

Redfield, A. C., and Rubin, Meyer (1962). The age of salt marsh peat and its relation to sea level at Barnstable, Massachusetts. *Natl. Acad. Sci. Proc.* **48**, 1728-1735.

Rhoads, D. C. (1963). Rates of sediment reworking by *Yoldia limatula* in Buzzards Bay, Massachusetts, and Long Island Sound. *J. Sed. Petrology* **33**, 723-727.

Rhoads, D. C. (1970). Mass properties, stability, and ecology of marine muds related to burrowing activity; *in* Trace Fossils (T. P. Crimes, and J. C. Harper, eds.). *Geol. J. Spec.* **3**, 391-406.

Rhoads, D. C., and Young, D. K. (1971). Animal-sediment relations in Cape Cod Bay, Massachusetts, II; Reworking by *Molpadia oolitica* (Holothuroidea). *Marine Biol.* **11**, 255-261.

Rodolfo, K. S., Buss, B. A., and Pilkey, O. H. (1971). Suspended sediment increase due to Hurricane Gerda in continental shelf waters off Cape Lookout, North Carolina. *J. Sed. Petrology* **41**, 1121-1125.

Sanders, J. E. (1970). Coastal zone geology and its relationship to water pollution problems. *In* "Water Pollution in the Greater New York area" (A. A. Johnson, ed.), p. 23-35. Gordon & Breach, New York.

Scholl, D. W., Craighead, F. C., Sr., and Stuiver, M. (1969). Florida submergence curve revised: its relation to coastal sedimentation rates. *Science* **163**, 562-564.

Schlee, J. S. (in press). Atlantic continental shelf and slope of the United States—sediment texture of the northeastern part. *U.S. Geol. Survey Prof. Paper 529-L.*

Schubel, J. R. (1971). Estuarine circulation and sedimentation. *In* Schubel, J. R., et al., "The Estuarine Environment: Estuaries and Estuarine Sedimentation." Am. Geol. Inst. Short Course Lecture Notes, October 1971, p.VI-1 to VI-17. American Geological Institute, Washington,D.C.

Spencer, D. W., and Sachs, P. L. (1970). Some aspects of the distribution, chemistry, and mineralogy of suspended matter in the Gulf of Maine. *Marine Geol.* **9**, 117-136.

Spinner, G. P. (1969). The wildlife wetlands and shellfish areas of the Atlantic coastal zone. "Am. Geog. Soc. Serial Atlas of the Marine Environment" Folio 18.

Swift, D. J. P., Stanley, D. J., and Curray, J. R. (1971). Relict sediments on continental shelves: a reconsideration. *J. Geol.* **79**, 322-346.

Teal, J. M., and Kanwisher, J. (1961). Gas exchange in a Georgia salt marsh. *Limnology and Oceanography* **6**, 388-399.

Turekian, K. K. (1968). "Oceans." Prentice-Hall, Englewood Cliffs, New Jersey.

Uchupi, E. (1963). Sediments on the continental margin off eastern United States. *U.S. Geol. Survey Prof. Paper* **475-C**, C132-C137.

Wageman, J. M., Hilde, T. W. C., and Emery, K. O. (1970). Structural framework of East China Sea and Yellow Sea. *Am. Assoc. Petroleum Geologists Bull.* **54**, 1611-1643.

Windom, H. L., Neal, W. J., and Beck, K. C. (1971). Mineralogy of sediments in three Georgia estuaries. *J. Sedimentary Petrology* **41**, 497-504.

Young, D. K. (1971). Effects of infauna on the sediment and seston of a subtidal environment. *Vie et Milieu* **22** (*Supp.*), 557-571.

Young, D. K., and Rhoads, D. C. (1971). Animal-sediment relations in Cape Cod Bay, Massachusetts. I. A transect study. *Marine Biol.* **11**, 242-254.

CHAPTER 12

Suspended Sediments in Continental Shelf Waters Off Cape Hatteras, North Carolina

Barbara A. Buss and Kelvin S. Rodolfo

Department of Geology
College of Liberal Arts & Sciences
University of Illinois Chicago Circle
Chicago, Illinois 60680

ABSTRACT

Suspended sediment and salinity samples were taken at 46 stations over the Cape Hatteras shelf area from August 26 to 28, 1970. Analyses of these samples provide data on the Cape Hatteras plume seen in Apollo 9 photographs. Relatively high suspensate (> 2 mg liter^{-1}) and low salinity ($< 3.2^0/_{00}$) waters from Pamlico Sound are issuing onto the shelf through Okracoke and Hatteras Inlets. A portion of this outflow is moved southward by coastal currents. In the shoal waters off the cape itself, wave and current mixing causes suspensions of between 0.5 and 1.0 mg liter^{-1} Seaward, suspension are gradually diluted to less than 0.10 mg liter^{-1} on the outer shelf and slope with salinities generally between 3.5 and $3.6^0/_{00}$.

INTRODUCTION

Geographic Setting

The Neuse and Pamlico Rivers which drain the Coastal and Piedmont Provinces of North Carolina, deposit much of their loads in the quiet waters of Pamlico

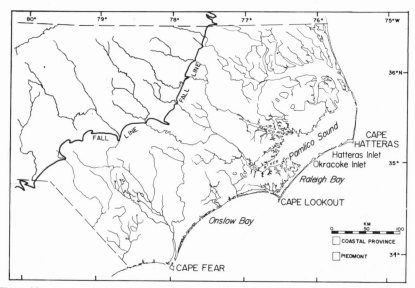

Figure 97.
Index map for the Cape Hatteras area.

Sound protected by the barrier islands. This chain of narrow islands less than 5.5 km wide backed by a sound between 2 and 55 km wide extends for hundreds of kilometers along the coast. The Sound is a depositional basin for the coarser river load but some of the fines are carried seaward (Pilkey, 1968), escaping through Okracoke Inlet, the nearest break in the barrier islands (Fig. 97). These sediments are partially deposited in a nearshore fan discernable on bathymetric charts and are also evident as sharp color transitions in Apollo 9 photographs (Mairs, 1970).

Cape Hatteras, one of the most intensively studied continental shelf areas (Newton and others, 1971) is important both sedimentologically and oceanographically. It is here that both the Gulf Stream and the continental slope make their closest approach to the Atlantic coast north of Florida. The cape is considered a sedimentological boundary (Stetson, 1939) and an area where water masses meet and mix (Stefansson and others, 1971). Currently the Carolina capes, believed to have formed as deltaic deposits off major rivers during Pleistocene low sea level stands (Hoyt, 1969), are being modified by erosion and deposition.

The purpose of this study was to explore the area's general sedimentary environment by sampling and examining the suspended sediments of the Cape Hatteras area, especially detailing the prominent plume located off the cape and nearby inlets. The hypothesis that all of the turbid plumes are subsurface, bottom hugging features (Rodolfo and others, 1971) was tested by analyses of the materials gathered. In broader context, the study and planned extensions will shed light on the sedimentation processes which transport fine materials to final depositional sites far from land.

Physical Oceanography

Seasonal circulation and water properties in the Cape Hatteras area have been detailed by Stefansson and others (1971). In this area, Virginia coastal water (a mixture of slope and river effluents) brought down from the north by an extension of the Labrador current and Caribbean water brought from the south by the Gulf Stream sweep onto the narrow shelf and mix with the Carolina water (a mixture of Gulf Stream water and Carolina river effluents). The major water circulation off the region is dominated by the Gulf Stream with a counterclockwise eddy present in Raleigh Bay during most of the year. During winter strong winds from the NE push cold, low salinity water around Cape Hatteras halfway across Raleigh Bay. Water has little if any vertical stratification but strong horizontal gradients. During summer winds are dominantly from the SE, minimizing Virginia coastal water inflow around the cape. Water shows strong vertical stratification. Salinities range between 30 and 36°/₀₀.

Tidal currents at the inlets may be especially important in contributing to suspension. Saline water occurs near bottom at the inlet mouth on the incoming tide below the fresher water outflow from the Sound. Turbidity maxima due to vertical mixing generally mark such tidal fronts in similar estuarine situations (Postma, 1967).

Bathymetry and Bottom Sediments

Detailed bathymetric charts of the Carolina shelf and slope have been compiled by Newton and others (1971). The shelf narrows to 30 km in width at Cape Hatteras sloping very gently (1 m : 900 m) out to a depth of 90 m (Fig. 98). Bottom sediment studies of the Cape Hatteras area include work by Stetson (1939), Uchupi (1963), Gorsline (1963), Emery (1965), Doyle et al. (1971), Pevear and Pilkey (1966), Lutenauer and Pilkey (1967), and Pierce (1969). The sediments off Cape Hatteras are largely relict sands (Milliman et al., 1968) with a large (5-50%) carbonate fraction. Silts and clays comprise less than 5% of the sediment and are present only near the coast and over the upper continental slope (Doyle et al., 1967). Cape Hatteras marks a sediment boundary, separating northern carbonate-poor sediments with high-silt and clay contents from silt-poor, well-sorted carbonate sands to the south. Doyle and others (1968) and Pierce (1969) believe that much of the nearshore fine sediments are brought onshore from reservoirs of unconsolidated sediments on the central and outer shelf. Although the Carolina coast as a whole may be experiencing onshore transport of fine sediments, the capes are localized areas along which fines appear to be migrating seaward. Aerial and satellite photography, notably the photographs taken by Apollo 9 in March, 1969, indicate that turbid plumes extend across the shelf from the vicinity of the capes. The Apollo 9 photographs discussed qualitatively by Mairs (1970) indicate turbid ebb-tide discharges from the inlets, noticeable across the shelf and entrained by the Gulf Stream. However, *in situ* sampling and analysis of the turbid waters has been conducted on only one other plume, that off Cape Lookout. Rodolfo et al., (1971) concluded that the Cape Lookout plume probably occurs near bottom where suspension concentrations are an order of magnitude greater than at surface.

PROCEDURE

Over the three day period from August 26 to 28, 1970, the R/V Eastward of Duke University made four sweeps paralleling the coast. From 46 stations located by Loran, 184 samples were taken across the Cape Hatteras shelf area from the shallow banks to the Gulf Stream and north from Okracoke Inlet to Gun Shoal (Fig. 99). Samples were taken at depth with a 30 liter Niskin water sampler, modified to close when a weighted trip line touched bottom.

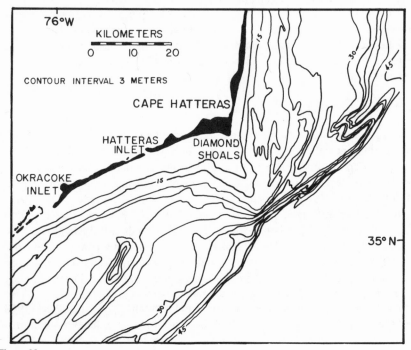

Figure 98.
Bathymetric chart for the study area (after Newton and others, 1971).

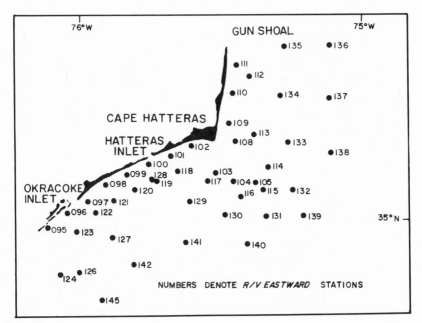

Figure 99.
Location of sampling stations for Cape Hatteras area.

Use of measured trip line lengths provided accurate vertical positioning of each sample with respect to the sea floor. At each station samples were taken 2 and 5 m off the bottom. Additional samples were taken at 5 and 10 m increments up to mid-depth. Portions of the 30 liter volume of each sample were used to rinse the polyethylene sample containers and handling equipment, leaving a final uncontaminated 25 liter sample. To minimize ship contamination, surface samples of 25 liter volume were collected underway from a position near the bow well forward of the Eastward's stack by casting a bucket away from the ship's hull. Bottom sediments were sampled with a Shipek sampler for comparison of bottom-grain size and composition with the suspended material at that station. Routinely, subsamples of the seawater were drawn for salinity determination on a Hytech salinometer. At each station the thermocline was located by bathythermograph.

To isolate particulate suspensates, water was filtered on shipboard by vacuum pump through 47 mm Millipore filters with 0.8 μ nominal pore diameters until the filters clogged. The volumes filtered ranged from 1 to 10 liters, depending on the suspensate content of the sample. All samples were filtered within four hours after they were taken. In the laboratory at Chicago, the clogged Millipore filters were dissolved and washed several times in filtered acetone, rinsed several times with filtered distilled water to remove residual salt, and the sediment transferred to preweighted 0.8 μ pore diameter Flotronic silver filters. Suspended sediment concentrations were calculated from the sediment weight and the original volume of filtered seawater.

The salinity and suspended sediment data for 2 m off bottom, 5 m off bottom, and surface samples are sufficient to construct computer plotted contour maps. The SYMAP computer program (Lab for Computer Graphics, Harvard University) was used to construct unbiased contour maps in desired intervals by numerical interpolation between values for a given level corresponding to grid located stations.

RESULTS

During the sampling period several fronts passed through the Cape Hatteras area causing winds between force 1 and 4 to vary in direction from SW to NE. Swells of up to a meter predominantly from the southeast, corresponded with a sea state no greater than 2. The complexities in tidal, current and weather variables, and the nonsynoptic occupancy of stations must be considered in interpreting shelf sedimentation and transport from the data. Weather can be excluded as a variable since it remained stable and typical of summer conditions. Samples were taken during several tidal cycles (Fig. 100) such that the central portion of the maps were sampled during ebb tide while the northern and western stations were occupied during flood tide. Tides ranged from 0.15 to 0.8 m above low mean sea level.

Salinity and Temperature

The surface salinity ranges between 29 and 36.29°/$_{oo}$ (Fig. 101), becoming more saline seaward. Two lobes of low salinity water (< 32°/$_{oo}$) are associated with Okracoke and Hatteras Inlets. Another lobe of low salinity (< 33°/$_{oo}$) water protrudes from the north boundary of the study area along the inner shelf. The northward dip of this front in vertical section reflects the influence of surface currents and bottom drag. The maps for 2 and 5 m off bottom show several high salinity lobes (> 36°/$_{oo}$) protruding up the slope onto the outer shelf from the eastern and southern boundaries of the study area (Fig. 101). Bathythermograph data agrees well with data collected by Stefansson and others in September of 1966.

Suspended Sediment

Surface

At all levels in the water column, suspensate concentrations decrease seaward from greater than 2 mg liter^{-1} to less than 0.10 mg liter^{-1} (Fig. 102). The surface

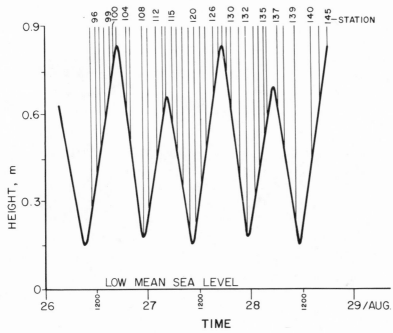

Figure 100.

Tidal curve for August 26 to 28, 1970, with stations located as to time sampled (from 1970 Tide Tables, East coast of North and South America).

map shows two lobes of high suspension concentration (> 2 mg liter^{-1}) associated with Okracoke and Hatteras Inlets although the Hatteras lobe is displaced southwestward. The highest concentration of 11.8 mg liter^{-1} occurred in surface waters immediately seaward of Okracoke Inlet. Concentrations of 0.50 to 1.0 mg liter^{-1} typify the inshore zone and also occur as a lobe over Diamond Shoals. One anomalous 0.50 to 1.0 mg liter^{-1} value occurs well onto the outer shelf but is mainly biogenic in origin. High concentrations of many surface samples are produced by local patchy concentrations of filamentous algae.

Mid-depth

The mid-depth (5 m off bottom) inlet lobes are similar in concentration to surface values but inshore and shoal concentrations increase to between 1.0 and 2.0 mg liter^{-1} in some areas.

Near Bottom

Near bottom contours most closely parallel the coast showing concentrations greater than 2.0 mg liter^{-1} in the lobe off Okracoke Inlet with 1.0 to 2.0 mg liter^{-1}

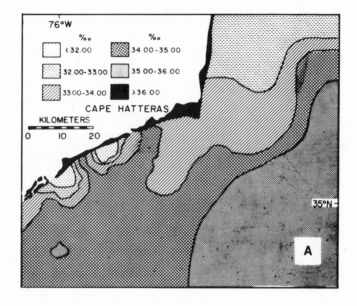

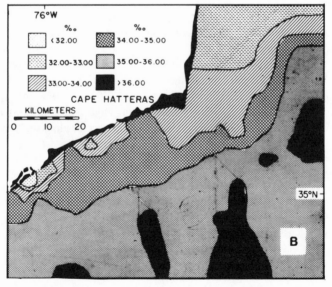

Figure 101.
Salinity distribution: (A) surface, (B) 5 m off bottom, and (C) 2 m off bottom.

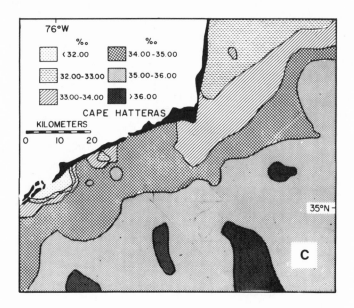

concentrations along the coast south of Hatteras Inlet. Nearshore values are uniformly 0.50 to 1.0 mg liter^{-1} encompassing the shoals area and are continuous with a deeper water patch 27 km south of the cape.

A graph of suspended sediment content against salinity displays the expected inverse relationship of fresh, turbid nearshore water to saline, clear Gulf Stream water (Fig. 103). A field of points outside this trend consisting of lower than expected salinity represents samples from stations north of Cape Hatteras. This less saline lobe from the north probably is a resultant of the increased and unobstructed runoff from rivers north of the study area which has been moved southward by coastal currents and has not yet been mixed enough to lose character. Five cross-sections normal to the coast and scattered across the study area (Fig. 104) agree closely with the summer regime conditions observed in 1966 (Stefansson et al., 1971). The summer regime is typified by a maximum vertical gradient caused by low runoff and wind conditions that minimize the influence of colder, fresher northern waters. Mid-shelf surface salinities are between 34 and 35$^0/_{00}$ with nearbottom salinities greater than 35$^0/_{00}$. Vertical distribution of salinity indicates upwelling of high salinity (> 36$^0/_{00}$) water along the slope during the summer. Geopotential topography and drift bottle studies during August, 1966, indicate that the Virginia Coastal waters split at Cape Hatteras. A portion of the water mass moves out across the shelf eventually to be moved north by the Gulf Stream while the other portion rounds Diamond Shoals heading south along the inner shelf at 13 cm sec^{-1}.

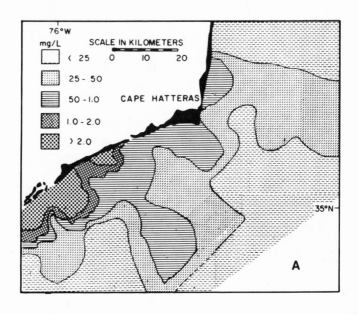

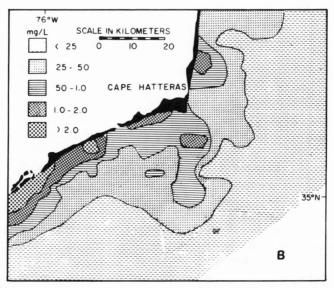

Figure 102.

Suspended sediment distribution: (A) surface, (B) 5 m off bottom, and (C) 2 m off bottom.

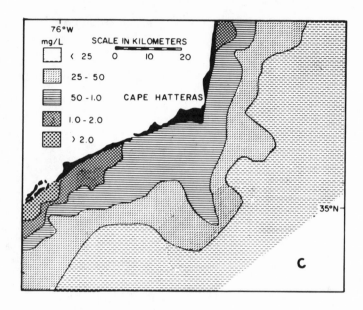

DISCUSSION

Data gathered during this late-August cruise agree very closely with the Stefansson et al. representation of typical summer conditions above. Vertical stratification (Fig. 104) and temperature and salinity variation (Fig. 101) are similar to those of September, 1966. Indications of a lobe of Virginia Coastal water reaching Cape Hatteras and a southwest current along the coast are present.

The isohaline contours depict a gradient between the fresher water from the inlets and barrier runoff to normal $35^0/_{00}$ salinities offshore. The outflow of relatively fresh Pamlico Sound water is evidenced by the low-salinity lobes ($< 32^0/_{00}$) associated with Okracoke and Hatteras Inlets. The southwestward displacement of the Hatteras Inlet lobe will be discussed with the suspended sediments. Subsurface waters 2 and 5 m off bottom contain lobes of high salinity ($< 36^0/_{00}$) protruding landward from the eastern and southern boundaries of the study area. These lobes probably represent near bottom patches of Gulf Stream water which are detached along the edge of the slope as the Gulf Stream moves northward, sweeping the outermost shelf at its nearest approach to shore. The Gulf Stream moves with velocities varying between 300 cm sec^{-1} at the surface and 10 cm sec^{-1} at 1500-2000 m depths (Von Arx, 1962). As the Gulf Stream moves across the shelf edge, it may erode fine-grained sediment and transport it northward. There is some evidence to support this since these patches are over areas where the bottom sediment consisted of coarse shell hash with little fine-grained material.

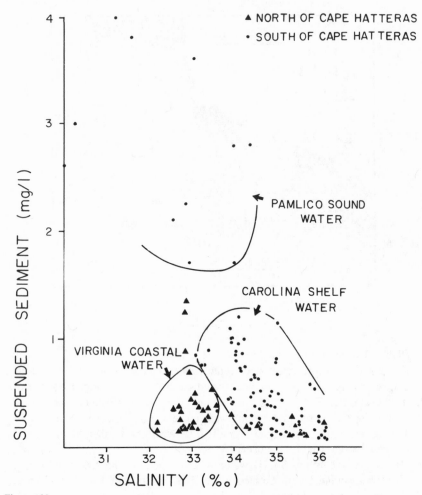

Figure 103.

Graph of suspended sediment content versus salinity for all samples.

Shallow waters (7 to 15 m depths) nearshore and further offshore over Diamond Shoals were isothermal. The thermocline occurred about 10 m below the surface over the middle shelf and between 30 and 40 m on the outer shelf and slope (77 to 337 m).

Analyses of the suspensates (Fig. 102) for the three sampling levels (2 m) and 5 m off bottom, and surface) show a seaward decrease in concentration roughly paralleling depth contours. The highest concentrations (> 2 mg liter^{-1}) were located near the inlets with high concentrations (1.0 to 2.0 mg liter^{-1}) found all along the coast. Over the continental slope, clear water with less

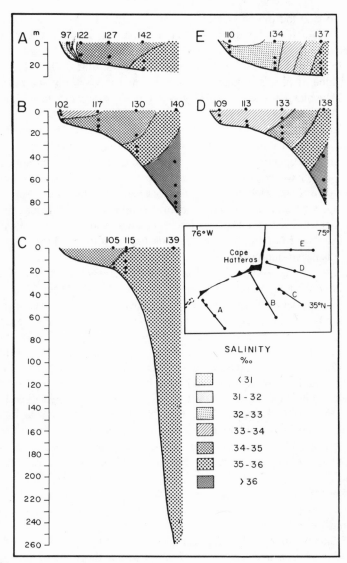

Figure 104.
Salinity cross-sections showing summer regime conditions (August, 1970).

than 0.10 mg liter^{-1} of suspensate suggests the edge of the Gulf Stream. These concentrations agree closely with previous reconnaissance surface data (Manheim et al., 1970). Concentrations vary with distance from shore and sediment sources, the water depths of effective wave and current action, and tidal influences. The maximum concentration value found of 11.8 mg liter^{-1} was located in the surface lobe at the mouth of Okracoke Inlet. Although photos and maps show a similar though smaller lobe at Hatteras Inlet, north of Okracoke Inlet and further from the major river outlets, salinity and suspended sediment contour maps do not depict a fan at the mouth of Hatteras Inlet. They do, however, show a lobe of high sediment concentration and low salinity southwest of the inlet. Okracoke Inlet (Station 96) was sampled from 1115 to 1153 EDT on August 26th during the period of maximum tidal velocity between high and low tides (Fig. 102). The high concentration was caused by surface outflow of relatively fresh Pamlico Sound water and roiling of near bottom water as more saline water moved in below on the incoming tide. At this time a similar outflow must have occurred at Hatteras Inlet. When the Eastward had moved northward to Station 99, between Okracoke and Hatteras Inlets, samples with high suspended contents and low salinity similar to the Okracoke Inlet lobe were collected. Since the Hatteras Inlet lobe alone shows any kind of displacement, the data from this station suggests that a portion of the inlet outflow is being displaced southwestward by known nearshore currents (Gray and Cerame-Vivas, 1963). High tide occurred at 1618 EDT, near occupancy of Hatteras Inlet (Station 100). At this time ideally predicted tidal currents are minimal and no fresh water would issue from the inlet. Although waves and coastal currents must still have been active, the lack of tidal current must have influenced mixing and suspension even in such shallow depth. When Hatteras Inlet was sampled (Station 100), no significant sediment or salinity anomalies were found. The samples were obtained between 1547 and 1559 EDT well within the several hour lag period before a strong tidal outflow could again develop.

Another lobe of high concentration at all depths is located over Diamond Shoals (Fig. 102) 15 km off Cape Hatteras. In these shallow depths current induced suspension of sediment results in the turbid plume seen in the Apollo 9 photographs. There is some evidence that bottom sediments are being moved landward in the bay areas on the continental shelf and into estuaries along the coast (Pevear and Pilkey, 1966; Lutenauer and Pilkey, 1967; Pierce, 1969). This, however, in no way precludes localized seaward movement of suspended fines. There is no question that relatively fresh water heavy in suspended material is issuing from the barrier inlets and is contributing to the shelf's suspended load. Although some of this suspension may be deposited rapidly on the inner shelf off the inlets, and much continues out across the shelf in diluted pulses (Mairs, 1970), this study in agreement with Manheim et al. (1970) indicates that a portion of the sediment outflow is being transported along the littoral zone. As littoral currents moving southward along the coast reach the inflexion point of the cape, some of the water is deflected continuing outward across the shelf. Work by Lutenauer and Pilkey (1967) show sediment transport is limited to slight spill-over between embayments by the Carolina capes. The plume at Cape Hatteras extending well over the shelf represents a sediment fraction temporarily lost from the littoral system which may be permanently lost if the sediment is entrained by the Gulf Stream and carried onto the slope. Aerial photographs over the past 25 yr of the North Carolina capes demonstrate considerable erosion of the coast by storms (El-Ashry and Wanless, 1968) and increased suspension over the shoals occurs during storms (Rodolfo et al., 1971). This sediment, along with material already suspended by normal agencies probably is carried downslope across the shelf at these localities.

A minor lobe of relatively high concentration (0.62 mg liter^{-1}) in near bottom samples was located unexpectedly at Station 130 in water more than 37 m deep (Fig. 102). This is inconsistent with the explanation for generalized distribution of suspensate with depth and distance from shore which explains the grading and major lobes. At 49 m depths, 27 km offshore, the suspended particle content is similar to that in 13 m depth, 6 km offshore. It is neither close to a shore source like Okracoke Inlet nor in a shallow area of mixing like Diamond Shoals. Local bottom stirring by benthic organisms in muddy areas is a possible cause of such suspensions. The sediment at Station 130 consisted of a fine silty sand. This lobe is continuous with the high concentration lobe 2 m off the bottom over the shoals and might also be an out and downward movement of some of that suspensate from the shoals through a small gulley found on bathymetric maps at the same location.

The color plates of the Apollo 9 photograph of the shelf area suggest that plume suspensates exist in many overlapping layers located at different depths (Mairs, 1970). This is reflected in the vertical suspensate distributions. Of the 46 stations samples, maxima were found at surface, near bottom or at intermediate depths. Surface maxima were found at 21 stations, of which 9 also had secondary maxima near bottom. Fourteen stations possessed near bottom maxima, of which 3 also had a secondary maxima at the surface. Only 6 stations showed intermediate depth maxima mostly associated with a layer in the southwest edge of the Diamond Shoals plume. The high number of surface maxima is in part due to high organic content. Soft-bodied algae, diatoms radiolarians and silicoflagellates were observed under the microscope. Nearshore surface maxima also result from the upward and outward flow of fresher Pamlico Sound water during the tidal cycle at Okracoke Inlet. Primary or secondary bottom maxima at half the stations sampled reinforce the belief that near bottom currents are important in the transport of shelf sediments.

CONCLUSIONS

A plume with high (1.0 to 2.0 mg liter^{-1}) concentrations of suspended sediment within the shoal area of Cape Hatteras is probably caused by wave rolling and littoral currents. Plumes of sediment laden (> 2 mg liter^{-1}), fresher

($< 32^0/_{00}$) waters are also being issued onto the inner shelf from Pamlico Sound and its adjacent rivers through Okracoke and Hatteras Inlets. Some of this material is displaced southwestward by coastal currents. Fresher water ($< 33^0/_{00}$) with less suspended sediment is also being brought down from north of the study area by coastal currents. These late-August results add support to the seasonal nature of the Carolina shelf waters noted by Stefansson and others in 1966.

ACKNOWLEDGMENTS

Work at the University of Illinois, Chicago Circle, was aided by National Science Foundation Grants GA-1072 and GA-12783. Duke University's R/V Eastward is funded by the National Science Foundation. Work aboard the Eastward was under the aegis of the Duke University Oceanographic Program. This paper is one result of the joint Duke University-U.S. Geological Survey marine geology program. The author wishes to thank the crew and scientific staff of the Eastward for shipboard assistance; Richard Kolb for his advice and help with the SYMAP program; and Thomas Nardin for suggestions concerning the manuscript.

REFERENCES

Bumpus, D. F. (1969). Reversals in the surface drift in the middle Atlantic bight area. *Deep-Sea Res.* [Suppl.] **16**, 17-23.

Doyle, L. J., Cleary, W. J., and Pilkey, O. H. (1968). Use of mica in determining shelf depositional regimes. *Marine Geol.* **6**, 381-389.

El-Ashry, M. T., and Wanless, H. R. (1968). Photo interpretation of shoreline changes between Capes Hatteras and Fear (North Carolina). *Marine Geol.* **6**, 347-379.

Emery, K. O. (1965). Geology of the continental margin off eastern United States: Submarine Geol. and Geoph. Colston papers, p. 1-20. Butterworths, London.

Gorsline, D. S. (1963). Bottom sediments of the Atlantic shelf and slope off the southern United States. *J. Geol.* **71**, 422-440.

Gray, I. E., and Cerame-Vivas, M. J. (1963). The circulation of surface waters in Raleigh Bay, North Carolina. *Limnol. Oceanogr.* **8**, 330-337.

Hoyt, J. H. (1969). Origin of capes and shoals along the southeastern United States coast (abs.); Abstracts with Programs for 1969, p. 109, Geological Society America.

Lutenauer, J. L., and Pilkey, O. H. (1967). Phosporite grains: their application to the interpretation of North Carolina shelf sedimentation. *Marine Geol.* **5**, 315-320.

Mairs, R. L. (1970). Oceanographic interpretation of Apollo photographs. *Photogrammetric Engineering* **36**, 1045-1058.

Manheim, F. T., Meade, R. H., and Bond, G. C. (1970). Suspended matter in surface waters of the Atlantic continental margin from Cape Cod to the Florida Keys. *Science* **167**, 371.

Milliman, J. D., Pilkey, O. H., and Blackwelder, B. W. (1968). Carbonate sediments on the continental shelf, Cape Hatteras to Cape Romain. *Southeastern Geology* **9**, 245-267.

Newton, J. G., Pilkey, O. H., and Blanton, J. O. (1971). An Oceanographic Atlas of the Caroline Continental Margin. 56 p. North Carolina Dept. of Conservation & Development, Raleigh, North Carolina.

Pevear, D., and Pilkey, O. H. (1966). Phosphorite in Georgia shelf sediments. *Geol. Soc. Am. Bull.* **77**, 379-858.

Pierce, J. W. (1969). Sediment budget along a barrier island chain. *Sed. Geol.* **3**, 15-16.

Pilkey, O. H. (1968). Sedimentation processes on the Atlantic southeastern United States continental shelf. *Maritime Sediments* **4**, 49-51.

Postma, H. (1967). Sediment transport and sedimentation in the estuarine environment. *In* "Estuaries," (G. Lauff, ed.), publ. #83, p. 158-179, Amer. Soc. Adv. Sci.

Rodolfo, K. S., Buss, B. A., and Pilkey, O. H. (1971). Suspended sediment increase due to Hurricane Gerda in continental shelf waters off Cape Lookout, North Carolina. *J. Sed. Pet.* **41**, 1121-1125.

Stefansson, U., Atkinson, L. P., and Bumpus, D. F. (1971). Hydrographic properties and circulation of the North Carolina shelf and slope waters. *Deep-Sea Res.* **18**, 383-420.

Stetson, H. D. (1939). The Sediments of the continental shelf off the eastern coast of the United States. *In* Recent Marine Sediments. *Am. Assoc. Petrol. Spec. Publ.,* 230-244.

SYMAP Program, "Lab for computer graphics," Harvard University, Cambridge, Massachusetts.

Uchupi, E. (1963). Sediments on the continental margin off eastern United States. *U.S. Geol. Survey Prof.* **475-C**, C132-C137.

Von Arx, W. S. (1962). "An Introduction to Physical Oceanography," 422 p. Addison-Wesley, Reading, Massachusetts.

CHAPTER 13

Mineralogy of Suspended Sediment Off the Southeastern United States

J. W. Pierce,* D. D. Nelson,*
and D. J. Colquhoun†

ABSTRACT

Sampling of material suspended in the waters on the continental shelf from Chesapeake Bay to Savannah, Georgia, for mineralogical analysis has demonstrated that four water masses have distinct mineral suites suspended in the water. The mineralogy associated with each watermass reflects the contribution of continental sediment made in the area where the watermasses originate. This is readily shown for two coastal watermasses. Minerals suspended in Carolinian coastal water apparently are the result of river runoff in the Carolinas and Georgia. The suite in the Virginian coastal water has northern affinities and probably originates north of Chesapeake Bay. Gulf stream and Carolinian slope water also have diagnostic suites but their source is beyond the scope of this study. Mineralogy is a parameter that can be used to identify watermasses in the way that temperature and salinity have been used previously.

INTRODUCTION

The distribution patterns of bottom sediments on the continental shelves of the world are reasonably well known (Emery, 1968). Some of this sediment owes its distribution to primary emplacement during lower stands of sea level. Much of the sediment now blanketing these shelves consists of particles whose size indicates that they must have been emplaced by bedload or saltation processes.

*Division of Sedimentology, Smithsonian Institution, Washington, D.C. 20560.
†University of South Carolina, Columbia, South Carolina 29208.

The transport and dispersion of fine-grained material is less well studied. This is surprising because fine particles (shale and clay) make up the bulk of the rocks in the geological column (Pettijohn, 1957, p. 11). Until the dispersal patterns of fine particles across or on the present continental shelves are known, little progress can be made toward unravelling the seemingly complex patterns of the lateral and vertical distribution of clay minerals in shales.

Particulate material, both organic and mineral, contributes to the turbidity of water masses. Data on turbidity measurements, either transmission or scattering, are much more readily available than data on actual concentrations or type of particulate. The reason for this, obviously, is the greater ease and rapidity of turbidity measurements over actual concentrations and particulate type. The difficulties of relating transmission measurements to particulate concentration have been pointed out by Visser (1970). Nevertheless, measurements of turbidity have been used to delineate and trace water masses with some success (Jerlov, 1968; Beer and Gorsline, 1971; Drake, 1971). Concentration of particulate material, determined by various methods, have also been used for tracing of water masses.

This article reports concentrations and mineralogy of the suspended material and their use to characterize water masses with the additional parameters of salinity and temperature used as supplementary data.

METHODS AND MATERIALS

Thirty-six stations between Chesapeake Bay and Savannah, Georgia, have been occupied during five cruises (Fig. 105). Cruise dates were August-September, 1970; October, 1970; and two cruises during March, 1971. An additional cruise was made in September, 1971. Data from this last cruise will only be used to supplement data from the other cruises. Navigational control was by Loran A. Reoccupation of stations during the October, 1970, cruise and on later cruises plus sampling at more than one depth provided us with 201 samples.

Most of the stations are located along transects oriented nearly perpendicular to the coast. Normally, three stations are on each transect: one, nearshore; another, on the outer shelf; the third, intermediate between the other two. The water depths for the nearshore stations range from 8 m to 19 m; the depths of the stations on the outer shelf, from 26 m to 76 m. Most of the deeper samples from stations on the outer shelf are from depths of about 40 m.

At each station, one-hundred liters of water was pumped from predetermined depths. A submersible water-well pump, with flow lines and electrical cable, was attached to the hydrographic wire and lowered to the appropriate depth; the flow lines were completely flushed, and the water pumped into and held

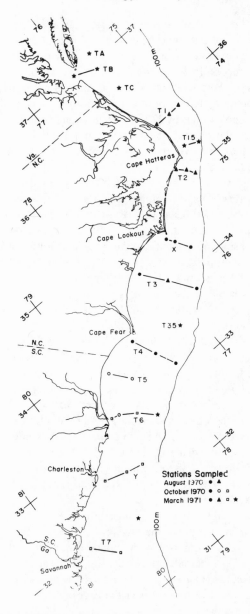

Figure 105.

Station locations between Chesapeake Bay and Savannah, Georgia, occupied in August-September, 1970, October, 1970, and March, 1971. Dates of sampling of the stations shown by symbols. Letters and numbers associated with the lines give number of cross-shelf transects.

in polyethylene bottles until filtration. In no case was the water held longer than three hours in the bottles. After the first few trials, sampling depths were standardized at 3 m below the surface and 3 m above the bottom. Near-bottom positioning was accomplished by use of a 12 kHz pinger, if available, or the amount of lead-weighted wire out, if the pinger was not available.

The present pump system is limited to working in depths of less than 80 m. Larger electrical cable or use of 220-V current would permit extension to greater depths. However, the present depth limitation is sufficient to sample most of the continental shelf area. Where deeper samples were desired, 30-liter Niskin bottles were used.

Concurrent Nansen casts, for temperature and salinity data, were made during the last four cruises. The temperature-salinity data permit collation with the results of Stefannson, Atkinson, and Bumpus (1971) and Blanton (1971) on the hydrographic properties of the waters of the shelf and slope.

The 100 liter of sea water were filtered through a Millipore MF® filter, 142 mm diameter, using a Millipore transfer pump. Either all of the 100 liter was filtered or filtering was continued until the filter plugged.

All but four of the water samples were passed through filters having $8.0 \pm 1.4 \ \mu m$ pores. The filtrate from 114 samples was then refiltered through pores of $0.45 \pm 0.02 \ \mu m$ size.

Each filter was dissolved in 30 ml of acetone and the liquid clarified by centrifugation. To assure complete solution of the filter, two additional acetone washes were used, followed by one of an ether-acetone mixture and one ethanol wash. The dry-weight of the remaining material was used to calculate the concentrations of total particulate matter. It is realized that some organic compounds are soluble in acetone, ether, and alcohol. What we refer to as total particulate matter lacks these compounds, which in some cases may aggregate a significant percentage of the organic material. The use of 142-mm diameter filters precludes preweighing the filters and use of weight gain as a measure of concentrations. An O-ring seats directly on the rim of the filter, sometimes causing the filter to tear at the seat upon removal.

Organic material was removed by digestion in 30% hydrogen peroxide for at least 16 hr. The dry weight of the residue was used to calculate the concentrations and percentages of oxidizable and nonoxidizable material.

The nonoxidizable material, after weighing, was saturated with magnesium ions by five washings, and subsequent centrifugation, with 1 N $MgCl_2$. Two washings with distilled water removed the excess salts. A slide was prepared by allowing approximately 5 mg of material of 0.5 ml of water to dry onto a 2.5 cm square glass slide under an infrared heat lamp.

X-ray diffraction patterns were made of these slides using Ni-filtered, Cu-K_α radiation, at scan speeds of $1° \ 2\theta$ min^{-1}. Successive patterns were made of the original slide, after exposure to an atmosphere saturated with ethylene glycol

(Brunton, 1955). Selected slides were heated to 300°C and 500°C for one hour and diffractometer scans made after each heat treatment. Slow-scan (1/4° 2θ min^{-1}) diffraction patterns were made between 24.5° and 25.75° 2θ to assist in differentiating chlorite and kaolinite (Biscaye, 1964). Positions of the peaks and changes, which occur as a result of different treatments, were used to determine the minerals that were present.

No attempt was made to obtain the relative amount of the different minerals in either a quantitative or semiquantitative way. All references to changes are strictly qualitative and should be so construed.

We chose to do this for several reasons, some imposed by the size of the samples or by the presence of certain minerals. Techniques for determination of relative amounts of minerals present in a sample by X-ray diffraction are far from standardized, are beset by many uncontrollable variables, and reporting of the results in numerical form often gives a false sense of accuracy (Pierce and Siegel, 1969). Recommendations for slide preparation suggest either the smear technique or suction onto ceramic plates (Gibbs, 1965). Because of the presence of feldspars and quartz, use of ceramic slides was not permissible. The small amount of material did not permit use of the smear technique. We desired oriented aggregates. Settling onto a glass slide tends to bias the upper layers of the slide toward the finer particles (Gibbs, 1965) although use of a thick slurry with fast drying decreases this tendency somewhat over that obtained by settling from a disperse suspension.

EFFICIENCY OF FILTERS

In the initial phase of the work, the water was filtered through 8.0 ± 1.4 μm pores in order to collect the particulate material. It was realized that this pore size is considerably larger than is generally used in investigations of suspended sediment, with a range of from 0.8 μm down to 0.45 μm reported in the literature. The primary purpose of the project was the mineralogy of the suspended particulate matter, not concentration with which most investigators are concerned. We did not feel that the large pore size presented serious obstacles to the determination of the mineralogy of the suspended sediment, being greatly impressed by the data presented by Sheldon and Sutcliffe (1969) and finding little rationale for use of any specific pore size. Use of the smallest size available can be argued on intuitive grounds. This must be compromised with the logistical problems of filtering large volumes of waters, in order to obtain enough material for analysis, over a relatively short time period, with a limited number of people, and limited equipment.

It has been pointed out that a filter cake is rapidly built up, effectively reducing the pore size and consequently the size of the particles trapped on the filters (Sheldon and Sutcliffe, 1969; Manheim, Meade, and Bond, 1970). Sheldon and Sutcliffe present numerical data on the capability of stated pore sizes to retain particles smaller than the pores. This capability is dependent upon concentration and the size distribution of the suspended particulate matter.

Very few experiments have been reported on the efficiency of filters to trap particles smaller than the stated pore size. Filter manufacturers readily give the size of particles that will be retained but provide no information on the efficiency to retain (or pass) smaller particles. Smaller particles are retained by van der Waals' forces, by random entrapment in the tortuous pores, and by buildup on previously retained particles (Millipore Corporation, 1966). Because of the dearth of reported data on filter efficiency and because we anticipated some questions regarding the use of large pore size, we began refiltering the filtrate from the 8.0 μm pores through 0.45 μm pores. To date, data is available for 114 samples that have been refiltered (Tables XVIII and XIX).

Table XVIII
Average Results of Passing 8 Micrometer Filtrate through 0.45 Micrometer Pores[a]

	Total particulate material		Nonoxidizable material	
	Average (%)	Range (%)	Average (%)	Range (%)
All stations	67	12-98	68	15-98
Nearshore	75	31-98	74	30-98
Mid-shelf	70	48-95	72	51-96
Outer shelf	58	12-96	60	15-96

[a]Average and range of total particulate material and nonoxidizable matter caught by 8 μm pores as a percentage of that caught by both filters.

Table XIX
Grouping of Samples into Percentage Classes Showing Efficiency of 8 Micrometer Pores to retain Specified Percentage of the Material Caught by Both 8 Micrometer and 0.45 Micrometer Pores[a]

	No. of samples in class	
Percentage	Total particulate material	Nonoxidizable material
10.0-19.9	2	2
20.0-29.9	4	3
30.0-39.9	2	5
40.0-49.9	9	4
50.0-59.9	11	15
60.0-69.9	23	26
70.0-79.9	12	15
80.0-89.9	17	21
90.0-100	16	15
	96	106

[a]Laboratory errors prevent reporting of the results of dual filtering of all 114 samples.

Of all the suspended material trapped by both the 8.0 μm pores and the 0.45 μm pores, an average of 67% of total particulate material and 68% of the nonoxidizable matter were caught by the 8.0 μm pores, only 33% and 32% passing through these pores. The range was from 12% to 98% for total particulate matter and 15% to 98% for the nonoxidizable material. Higher concentrations rapidly clog the filter, permitting only a small percentage of the suspended material to pass. The relationship between particle passage in lower concentrations is less clear. In some cases, it is relative to the amount of organic material present. In other cases, it may be related to the size distribution of the particles.

The 8μm pores trapped 50% or more of the total particulate material from 81 of 96 samples; the 8 μm pores trapped 50% or more of the nonoxidizable particles from 92 of 106 samples (Table XIX). The distribution of samples in percentage classes is skewed toward the higher classes. On an intuitive basis, we did not expect such a good recovery by such large pores.

Even though the filters may be reasonably efficient at trapping particles smaller than the stated pore size, one must determine if there is discrimination against certain minerals and, if so, how drastic the discrimination is. The various diffractograms of material trapped by the 8.0 μm pores and by the 0.45 μm pores were carefully scrutinized in an attempt to determine if there were any minerals present in the 0.45 μm material that could not be detected in the material trapped by the larger pores. In no case was a mineral present on the filter with the smaller pores that could not be detected with some combination of diffractograms made from the material trapped by the larger pores. This in no way implies that peaks from some of the minerals were not much more prominent in the material from the finer filter (Fig. 110b,c). Some minerals could be considered to be present only in trace amounts on the coarser filter but may appear as the major constituent on the finer. This, in part, supports our decision not to report the mineralogy in a quantitative or semiquantitative manner.

Several minerals apparently are completely trapped, or nearly so, by the coarse filter. Peaks indicative of these minerals do not appear on the diffractograms of the water refiltered through 0.45 μm pores.

Very few of the common clay minerals, as well as other silicates and the carbonates, passed through the coarser pores. Traces of illite, kaolinite, and chlorite were found on a few of the finer filters. The most common mineral passing through the 8.0 μm pores was talc. Most often this was the only mineral present on the finer filter.

CONCENTRATION OF PARTICULATE MATERIAL

The major purpose of this paper is not concerned with concentration values but, even with the limitations of coarse filters in determining concentrations, some points can be brought out. In the case of dual filtration, 85% or more of the material was trapped by the 8.0 μm pores when concentrations exceeded 1.00 mg liter[1]. We have no reason to expect that this would not hold for the samples illustrated and believed that the values given are reasonable for the time the samples were taken.

Concentration of total particulate material, trapped by the 8.0 μm pores, ranged from 0.05 mg liter[1] to 8.44 mg liter[1] (Table XX). All samples that were dually filtered fell within this range except for one taken on maximum ebb flow at the mouth of an estuary. Concentrations as reported here are also in the same range as has previously been reported for this area Manheim, Meade, and Bond, 1970). Concentrations of nonoxidizable material averaged higher in the near-bottom waters than for near-surface waters (Table XX). Exceptions do occur where near-surface concentrations exceed those near the bottom (Figs. 106 and 107).

Two of the stations (T2S2, TXS3, Fig. 107) are located near the shoals associated with Cape Hatteras and Cape Lookout. Higher concentrations in the near-surface water may be the result of turbulence over the shoals and subsequent transport to the sampling location (Buss and Rodolfo, 1971).

The remainder of the illustrated stations are located nearshore. Some of these also illustrate higher concentrations in the near-surface layers (T3S1, all cruises; T5S1, October; T2S1, March; T4S1, March). In all of these cases except for T5S1, sampling occurred during some phase of ebb tide. Temperature-salinity data taken at the same time indicated a stratified condition at the station. The concentrations undoubtedly represent effluent, from estuaries, more concentrated near the surface Mairs, 1970; Buss and Rodolfo, 1971). The remaining station, off Myrtle Beach, S.C., was taken near mid-flood tide. Near-surface concentrations may be the result of effluent from Little River Inlet, about 15 mi to the north, carried south alongshore under the stress of strong northeast winds blowing at the time.

Concentrations of total material and nonoxidizable matter were found to change, at several stations, by nearly an order of magnitude between successive cruises. An increase in concentration by nearly a factor of two can occur over a four-day period (Table XXI; Fig. 106 and 107). The sea state dratically affected concentrations and, for stations near estuaries, the tidal stage has a pronounced affect.

COMPOSITION OF SUSPENSATE

Oxidizable organic material makes up from 4 to 87% of the total particulate matter. Changes in concentration of oxidizable organic material is due to biologic activity and is generally lowest at stations on the outer shelf where the water

Table XX
Summary of Concentrations of Particulate Matter by Type, Station Location, and Sampling Level[a]

		All stations			Nearshore			Middle shelf			Outer shelf		
		$\bar{X}$	Range	N	$\bar{X}$	Range	N	$\bar{X}$	Range	N	$\bar{X}$	Range	N
Total particulate	U	1.35	5.65–0.05	61	2.38	5.65–0.36	25	0.67	1.28–0.05	18	0.61	1.45–0.10	18
	L	1.72	8.44–0.06	60	3.18	8.44–0.35	24	0.84	3.03–0.10	19	0.66	1.46–0.06	17
Nonoxidizable	U	1.09	5.09–0.01	56	1.93	5.09–0.39	23	0.47	1.07–0.01	17	0.41	1.16–0.02	16
	L	1.53	6.91–0.02	52	2.75	6.91–0.18	22	0.75	2.70–0.08	16	0.51	1.12–0.02	14
Oxidizable	U	0.29	0.85–0.02	56	0.42	0.85–0.10	23	0.22	0.41–0.04	17	0.17	0.68–0.02	16
	L	0.32	1.53–0.03	52	0.53	1.53–0.13	22	0.20	0.54–0.09	16	0.12	0.24–0.03	14

[a]Data from August-September, 1970; October, 1970, and March, 1971 cruises.

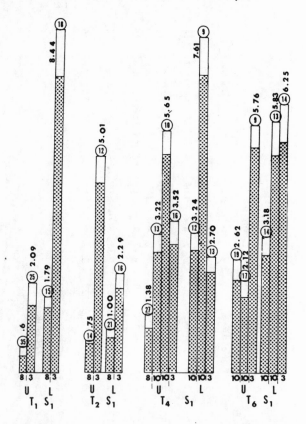

Figure 106.

Bar graph of concentrations at four nearshore stations showing changes in concentration at different sampling times. Total amount of particulate material at top of graph, in mg liter[-1], percentage of oxidizable material by circled number. Stippled part of bar, nonoxidizable material; unstippled part, oxidizable. Cruises: (8) August-September, 1970; (10), October, 1970; and (3) March, 1970. Some stations occupied twice in October, 1970.

is low in nutrients (Bumpus and Pierce, 1955; Stefansson, Atkinson and Bumpus, 1971). Organic matter constitutes the smallest percentage of the total suspensate at nearshore stations. Although the concentration of organic material remains high in the nearshore waters, the percentage of the total material is low because of dilution by high concentrations of terrigenous sediments. At any given station, including mid and outer shelf, the concentration of oxidizable material is less subject to change than that of the nonoxidizable phase.

The nonoxidizable phase is composed of various minerals originally derived from the land plus those minerals that make up the hard parts of the organisms living in the water. The mineralogy of the biogenic contribution is limited, being composed of the various calcium carbonate minerals and amorphous silica.

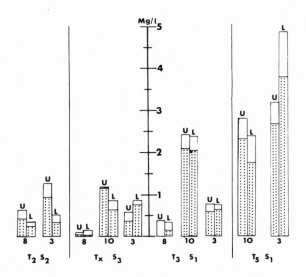

Figure 107.

Bar graph of concentrations at selected stations. Length of bar equals total concentration in mg liter^{-1}; stippled, nonoxidizable material; un-stippled oxidizable.

One of the difficulties, as yet unsolved, is determination of the contribution of these organically precipitated minerals relative to the amount of these same minerals in the terrigenous fraction. This is especially critical where large amounts of "amorphous" material are present. The mineralogy of the terrigenous fraction has a much broader spectrum.

The only organically precipitated carbonate that could be considered common in the area is low-Mg calcite—not surprising, as this is the primary carbonate of planktonic organisms. Neither aragonite nor high-Mg calcite are in sufficient quantities to register on the X-ray diffractograms. Low-Mg calcite is in nearly every sample south of Cape Fear, common at the stations between Cape Fear and Cape Lookout, and rare north of Cape Lookout, indicating decreasing plankton productivity to the north at the time of sampling.

Dolomite was present in 71% of the samples. There seems to be no correlation between the presence of dolomite and low-Mg calcite ($r = -0.0005$).

Many silicate minerals are present; most, if not all, are of a detrital origin. Quartz and illite are ubiquitous, appearing in 97% of the samples. Obviously, these two minerals have little value as tracers for water masses or as indicators of the source of the sediment.

Clay minerals other than illite have more limited lateral distributions and some appear to be useful in discriminating water masses. Well-crystallized, thermally stable (500°C) chlorite is present in all samples north of Cape Hatteras, some samples between Cape Hatteras and Cape Fear, and absent south of Cape

TABLE XXI

Changes in Concentration at Selected Stations at Different Sampling Periods[a]

| | | | Concentration | |
| | | | Total (mg/liter) | Nonoxidizable (mg/liter) |
Station	Date	Pore size		
T1S1 L	30/VIII/70	8	1.79	1.51
	17/III/71	8	8.44	6.91
	23/IX/71	⎧ 8	0.38 ⎰ 0.31	0.36 ⎰ 0.30
		⎩ 0.45	⎱ 0.07	⎱ 0.06
T2S1 U	16/VIII/70	8	0.75	0.65
	16/III/71	8	5.01	4.40
	22/IX/71	⎧ 8	0.49 ⎰ 0.39	0.44 ⎰ 0.34
		⎩ 0.45	⎱ 0.10	⎱ 0.10
TXS3 U	3/IX/70	8	0.10	0.03
	19/X/70	⎧ 8	1.46 ⎰ 1.18	1.35 ⎰ 1.14
		⎩ 0.45	⎱ 0.28	⎱ 0.21
	16/III/71	8	0.58	0.36
	22/IX/71	⎧ 8	0.49 ⎰ 0.22	0.32 ⎰ 0.16
		⎩ 0.45	⎱ 0.27	⎱ 0.16
T5S1	20/X/70	8	2.39	1.73
	20/X/70	8	4.31	3.75
T6S1	21/X/70	8	3.18	2.72
	24/X/70	8	5.83	5.06
	24/III/71	8	6.25	5.37
	20/IX/71	⎧ 8	0.42 ⎰ 0.28	0.42 ⎰ 0.28
		⎩ 0.45	⎱ 0.14	⎱ 0.14

[a]Where available data given for amount of material caught by both 9 μm and 0.45 μm pores.

Fear. Kaolinite, often with poor crystallinity or extremely small crystallite size, has the most widespread distribution of the clay minerals other than illite. It is the predominant mineral at mid-shelf and nearshore stations south of Cape Fear. Between Cape Fear and Cape Lookout, it is a common constituent of the sediment suspended in the water over the middle and inner parts of the shelf. North of Cape Lookout, kaolinite is present only in small quantities or absent. Montmorillonite has a very limited distribution. It is generally restricted to nearshore waters and is found at all nearshore stations south of Cape Fear. The abundance decreases rapidly to the north and is absent north of Cape Hatteras (Figs. 108 and 109).

Other silicates are also present, although with much more limited lateral distributions. An amphibole, probably hornblende, is present as part of the

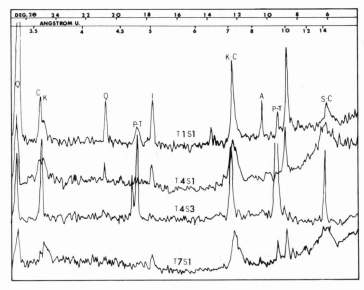

Figure 108.

X-ray diffractograms of samples from different water masses. (a) top, Virginian coastal water; (b) Carolinian coastal water; (c) Gulf Stream; and (d) Bottom, effluent from Savannah, Georgia harbor showing close relationship to Carolinian coastal water buth with addition of talc to suite.

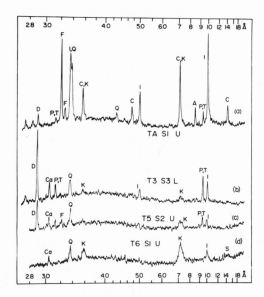

Figure 109.

X-ray diffractogram of samples. (a) Top, typical Virginian coastal water suite taken from north of Chesapeake Bay; (b) Gulf Stream water with addition of illite; (c) Mixing of Gulf Stream water and Carolinian coastal water; and (d) Carolinian coastal water.

mineral suite at all stations north of Cape Lookout. South of Cape Fear, this mineral is absent. Between these two capes, the amphibole has spotty occurrences being present at some stations during one sampling and absent at other times.

A series of integral order reflections at 9.37, 4.69, and 3.13A occur in slightly over one-half of the samples. These peaks correspond to the basal spacings of the crystal lattices of pyrophyllite and talc. Because of our small sample size and interference from other silicates, which are present, it is impossible to determine, by use of the 060 spacing, whether the mineral is pyrophyllite or talc. Both may be present in some instances while, in other cases, only one or the other may be present (Pierce, Nelson and Colquhoun, 1971). The ratio of the intensities of the first three reflections gives some indication of which of these minerals is present; in our case, these ratios suggest that the mineral is talc. This mineral is present at most of the stations north of Cape Lookout and only at the outer shelf stations south of here.

Feldspar is common north of Cape Lookout and in all samples north of Cape Hatteras. It is rare south of Cape Lookout.

WATER-MASSES

Discrimination of the water masses present in this area has been accomplished previously by use of hydrographic properties, such as temperature, salinity, dissolved oxygen, and nutrients (Bumpus, 1955; Bumpus and Pierce, 1955; Blanton, 1971; and Stefansson, Atkinson, and Bumpus, 1971). In addition, water masses have been outlined and circulation patterns postulated from Apollo IX photographs (Mairs, 1970; Stevenson and Uchupi, 1969) as well as the correlation of the space imagery with physical oceanographic data (Nichols, 1969; Rao et al., 1971).

Four water masses, covering parts of the North Carolina shelf, have been defined on the basis of hydrographic properties: Virginian coastal water, Carolinian coastal water, Gulf Stream water, and Carolinian slope water. The mineral suite suspended in each of the four different water masses is distinctive and characteristic of that water mass. The minerals suspended in the water are indicative of different source areas.

Virginian Coastal Water

Virginian coastal water was originally defined by Bumpus and Pierce (1955). As defined, this water mass covers the shelf north of Cape Hatteras and is composed of slope water and effluent from rivers north of Cape Hatteras. Bumpus (1955) postulated that the shelf south of Cape Hatteras did not receive regular contributions of Virginian coastal water and that this water mass generally was confined to the north by the oceanographic barrier at Cape Hatteras. Bumpus and Pierce (1955), and Gray and Cerame-Vivas (1963) noted that northeast winds could push Virginian coastal water past Cape Hatteras. Chase (1959) believed that changes found in temperature and salinity at Cape Fear were due to the influx of Virginian coastal water that had broken through the oceanographic barrier. Harrison et al. (1967) report the recovery of drift bottles, which were released north of Cape Hatteras, north of Cape Fear.

Virginian coastal water has salinities of less than $35^0/_{00}$ (Bumpus and Pierce, 1955; Harrison et al., 1967). Temperatures are seasonally dependent but are generally low.

Concentrations of suspended particulate material in the Virginian Coastal water are generally high, greater than 0.5 mg liter^{-1}. The minerals of the terrigenous phase are extremely well cyrstallized as shown by the sharp peaks on the diffractograms (Figs. 108 and 109). Minerals suspended in the water are quartz, amphibole, illite, chlorite, talc, dolomite, feldspar, and rare occurrences of slight amounts of kaolinite (Table XXII).

Carolinian Coastal Water

Bumpus and Pierce (1955) found that a different water mass occupied most of the North Carolina shelf south of Cape Hatteras. They named this Carolinian coastal water. According to them, this water is composed of an admixture of Gulf Stream water and effluents from the rivers south of Cape Lookout. This water mass is warmer and more saline than the Virginian coastal water.

Concentrations of suspended particulate matter are approximately the same as for the Virginian Coastal waters. The minerals of the terrigenous phase of the Carolinian Coastal water have poorer crystallinity, as shown by lower and more diffuse diffraction peaks, than those in samples from in Virginian coastal waters. This can be attributed to smaller size of the crystallites or poorer crystal structure. In the case of many samples, the X-ray diffraction peaks were extremely low even though the amount of sediment on the slide appeared to be in sufficient quantity to diffract X-rays if the material was of a crystalline nature. It is possible that much of the nonoxidizable material is highly degraded and some, if not a large amount, is approaching an amorphous state.

Table XXII
Summary of Mineralogy of Different Water Masses

Water mass	Mineralogy
Virginian coastal	Chlorite, illite, talc, amphibole, quartz, feldspar, dolomite, rare traces of kaolinite
Carolinian coastal	Montmorillonite, illite, kaolinite, quartz, calcite
Gulf stream	Minor chlorite, talc, kaolinite, quartz
Carolinian slope	Illite, talc, kaolinite, quartz, feldspar, calcite, dolomite

There is also a difference in the contained minerals (Figs. 108b,d and 109d, Table XXII). The Carolinian coastal water has, in addition to the ubiquitous quartz and illite (the latter with lower and broader diffraction peaks), kaolinite and some montmorillonite. Chlorite is not present in detectable quantities. Talc is believed to be absent except in the effluent plume of the Savannah (Georgia) estuary (Fig. 108d), or when mixing occurs with Gulf Stream water or Carolinian slope water (Fig. 109c).

Gulf Stream Water

The Gulf Stream water mass occupies the outer edge of the shelf and the surface water layers east of the shelf break. High temperatures, high salinities and low nutrient concentrations characterize this water mass.

Nonoxidizable suspended particulate material is generally low, much lower than that of the two coastal waters. The minerals are apparently well crystallized (Fig. 108c, 109b, 110b and d). Quartz, kaolinite, dolomite, and talc, with very minor amounts of chlorite, characterize this water mass. The kaolinite exhibits no better crystallinity than that in the Carolinian coastal water. The chlorite is nonexpandable, noncollapsible, but is completely destroyed upon heating to 600°C.

Carolinian Slope Water

Lying below the Gulf Stream water mass on the slope and, encroaching at times on the shelf, is the Carolinian slope water (Stefannson, Atkinson, and Bumpus, 1971). Blanton (1971) did not differentiate this as a separate water mass from the Gulf Stream water. Stefannson, Atkinson, and Bumpus, based on temperature-salinity diagrams, believed this water mass to be composed primarily of Caribbean water.

These authors did not observe intrusions of colder water onto the shelf from the slope during winter or spring months due primarily to the increased density of the coastal waters by lowered temperatures. In fact, cascading can occur during the winter months because of this temperature drop.

During the late summer and early fall, on the other hand, intrusions of Carolinian slope water onto the shelf were present for four consecutive years. Meteorological and climatological conditions, south to southwest winds and low amounts of fresh-water run-off, leading to this encroachment, are most common during the summer and early fall. A south to southwest wind may lead to an "upwelling" of the slope water over the edge of the shelf while the thermal stratification, generally present at this time, prevents mixing. Northeast winds, prevalent during the winter months, plus colder denser shelf waters are not conducive to upwelling.

If Carolinian slope water was present during any of our cruises, the late summer or early fall cruises would have been the most likely ones to sample

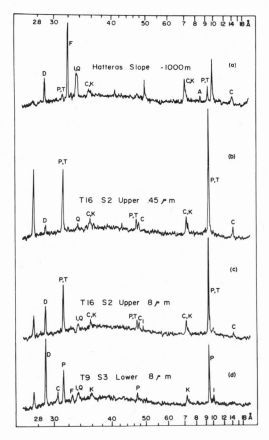

Figure 110.

X-ray diffractogram of samples taken in September, 1971. (a) Sample from continental slope off Cape Hatteras; (b) typical Gulf Stream water taken off Miami, Florida, material passing through 8 μm pores; (c) material of same sample as (b) caught on 8 μm filter; and (d) Carolinian slope water.

this water mass. Complementary temperature-salinity data were not taken in September of 1970. Temperature-salinity data from the October, 1970, cruise did not clearly indicate the presence of slope water on the shelf.

Temperature-salinity values of near-bottom water of a few stations of the fall, 1971, cruise plot within the envelope indicative of Carolinian slope water (Stefansson, Atkinson, and Bumpus, 1971, Figs. 8 and 10). The area of encroachment of this deep water would be around Cape Fear and south. Many of the values plot near the edge of the slope water envelope, suggesting some mixing. Only one sample occurred well within the envelope (T9, 30°25.ON, 80°17.O'W, 37 m depth). X-ray diffractograms of the nonoxidizable particulate of this sample showed the presence of talc, dolomite, kaolinite calcite, feldspar, quartz, and illite (Fig. 110). This mineral suite is different than that of the Gulf Stream or Carolinian coastal water. The other samples have, in some cases, admixtures of what could be Carolinian coastal water. Mixtures of Gulf Stream water are very difficult to determine.

SOURCE OF SUSPENDED SEDIMENTS

Any discussion of concerning the source of the suspended sediments must account for the suites present in four water masses. Because these suites are distinctive, some difference in source must exist.

Gulf Stream and Carolinian Slope Water

These water masses originate well outside the area of investigation. Material can be picked up or lost during transit from the original source. Because of this, comments on the source of the minerals suspended in these water masses are purely speculative.

Jacobs and Ewing (1969) show that waters in the Caribbean and Gulf of Mexico contain significant amount of chlorite, mica, talc, kaolinite, quartz, and feldspar with very minor amounts to no montmorillonite. Minor differences do exist between samples. This suite of minerals is not greatly different than that of the Carolinian slope water, which contains illite (mica), talc, kaolinite quartz, feldspar, calcite, and dolomite.

Subtraction of the feldspar and addition of chlorite would produce the suite of minerals present in the Gulf Stream water.

Carolinian Coastal Water

The suite of minerals suspended in the Carolinian coastal water is similar to that found in the sediments on the floors of adjoining estuaries and in the beds of the rivers feeding into the estuaries (Heron et al, 1964; Neiheisal and Weaver, 1967; Pevear, 1968, Windom et al., 1971; Hathaway, in press). Kaolinite is the predominant mineral derived from the erosion of a Piedmont source while montmorillonite is dominant in Coastal Plain soils.

There is a general paucity of clay-sized material in the bottom sediments covering the shelf off the southeastern United States (Hathaway, 1971). Of 337 samples collected from the shelf between Chesapeake Bay and Georgia by the USGS-Woods Hole Oceanographic Institution Joint Program, only four had more than 1% clay-sized material (< 4 μm) and only 63 had more than 1% silt-sized material. The four samples averaged 4.5% clay-sized material while the 63 samples averaged 3.6% silt.

Reworking of bottom sediments and resuspension of the finer particles could provide some of the material now in suspension, despite the lack of the finer particles in the shelf sediments. In addition to reworking, some material must be carried out of the estuaries and become part of the sediment suspended in the shelf waters. This sediment may have been originally deposited on the shelf during periods of lower sea level and transported landward and into the estuaries during rising sea level (Pevear, 1968; Hathaway, in press). Resuspension in the estuaries by tidal currents or wind-agitated waves could then permit this material to be carried back into the sea.

Virginian Coastal Water

The source of the sediments suspended in the Virginian coastal water is less clear. The minerals in the estuaries and rivers between Cape Lookout and Chesapeake Bay are more closely akin to those found farther south and in the Carolinian coastal water (Brown and Ingram, 1954; Griffin and Ingram, 1955; Nelson, 1960; Duane, 1962; Hathaway, in press). There is a dominance of kaolinite, dioctahedral vermiculite, and montmorillonite. Near the inlets and mouths of estuaries, a mineral suite similar to that in Virginia coastal water is encountered and probably indicates sediment contributions from the ocean as has been suggested to occur farther south (Heron et al, 1964, Windom et al., 1971).

Mineralogy of the sediment in the Virginian coastal water resembles that of the sediments of Boston Harbor in that both contain illite and chlorite (Mencher, et al., 1968). A closer correlation is exhibited with a northern mineral assemblage as defined by Hathaway (in press). This northern assemblage consists of illite, chlorite, feldspar, and hornblende.

This northern mineral facies is present in samples from the continental slope and in the estuaries as far south as Chesapeake Bay. Lack of clay-sized material on the shelf prevented definition of the assemblage in samples from the shelf. Hathaway (in press) believes that the relatively unweathered minerals of this suite were deposited in the estuaries, on the shelf, and on the slope by melt waters from the glaciers which eroded rocks of the northern Appalachians.

During the period of rising sea level in the Holocene, the shelf sediments were winnowed of clays and the net movement of the sediment was landward (Hathaway, in press). This resulted in a northern suite of minerals in parts of Chesapeake Bay and the extreme eastern parts of the northern North Carolina sounds that is dissimilar to the material now being brought in by the rivers. No more than traces of illite are found in rivers south of the Hudson and chlorite is not detectable in river samples south of the Potomac.

Thus, the source for the mineral suite suspended in Virginian coastal water has northern affinities. The ultimate source seems to be crystalline rocks of the northern Applachians via glacial erosion and distribution by melt water. The present source is either stream contributions from north of New York or introduction of sediment from the lower reaches of the Chesapeake and Delaware Bays into the ocean. This sediment would have been carried into the estuaries during rising sea level.

LATERAL MOVEMENTS OF WATER MASSES

Each water mass occupies a distinct part of the shelf at any given time. Due to changes in meteorological and oceanographic conditions, the position of the water masses change, there is mixing with adjacent waters, or a pod of one water mass may become detached from the main body.

The boundaries between the different water masses are transitional, more so in a lateral sense than in a vertical. Where one water mass overlies another, the boundary is often sharp with little mixing across the boundary. As yet, we have been unable to derive a satisfactory technique for determining quantitative dilution factors of one water mass by another, although a heterogenous mineral suite can indicate that mixing has occurred.

In early fall (September) of 1970, Virginia coastal water was present, as a distinct entity, as a nearshore wedge north and south of Cape Hatteras (Fig. 111). Carolinian coastal water could be defined, in an undiluted state, only nearshore at Cape Fear. The Gulf Stream was relatively close to shore, covering the outer shelf in nearly an undiluted state. Onslow Bay had a mixture of Carolinian coastal water and Gulf Stream water. Raleigh Bay had mixtures of Gulf Stream, Virginian, and Carolinian waters with some admixtures of Carolinian coastal water as far north as Cape Hatteras.

One month later (October) the southern half of the area was resampled. The Gulf Stream was much farther offshore and contributed less to the water present in Onslow Bay.

The entire area, from Chesapeake Bay south, was sampled in the spring (March) 1971. The Gulf Stream was far offshore, in an undiluted state, beyond the outermost stations (Fig. 112). Virginian coastal water had been pushed south into Onslow Bay by the strong northeast winds blowing at the time. Onslow Bay thus had a mixture of the two coastal waters and some Gulf Stream water brought in by a counterclockwise eddy. Part of the plume of Virginia coastal water turned east at Cape Hatteras and became mixed with Gulf Stream water.

SUMMARY AND CONCLUSIONS

The amount of material trapped by filters with 8.0 μm pores ranged from 0.05 mg liter^{-1} to 8.44 mg liter^{-1}; changes occur and can increase by a factor

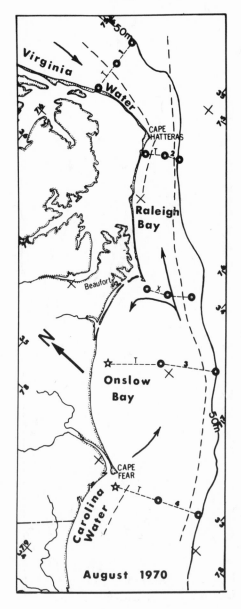

Figure 111.
Position of undiluted water masses and areas of mixing during August-September, 1970.

of 2 in a matter of days. Concentrations were highest at nearshore stations and near the bottom, although with a well mixed water column, concentrations near the bottom and near the surface did not differ greatly.

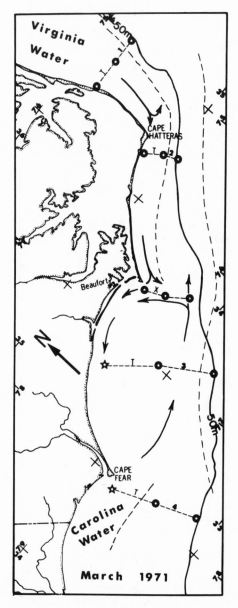

Figure 112.

Position of undiluted water masses and areas of mixing during March, 1971.

Four water masses may be defined on the basis of the mineral suite suspended in the water. These water masses are Virginian coastal, Carolinian coastal, Gulf Stream, and Carolinian slope. They are analogous to water masses defined by hydrographic properties.

Carolinian coastal water has a suspended mineral suite that apparently is the result of river runoff in the Carolinas and Georgia. Virginia coastal water originates north of Chesapeake Bay. The Gulf Stream water and Carolinian slope water probably originate in the Caribbean or Gulf of Mexico.

The location of the water masses are relatively constant in a general sense. Shifts in the boundaries and large areas of mixing between the water masses occur in response to meteorological conditions.

ACKNOWLEDGMENTS

The authors acknowledge the support of the Oceanographic Program of Duke University Marine Laboratory for use of the R/V Eastward. The Oceanographic Program is supported by National Science Foundation Grants GB17545 and GA27725. We also acknowledge the support provided by the National Oceanographic and Atmospheric Administration for providing time on the NOAA ship Whiting. Assistance in sampling was provided by the officers and men of the Whiting. Funds for this research were made available through the Smithsonian Research Foundation, RA427224, and through the Smithsonian Office of Academic Programs.

D. F. Bumpus, J. C. Hathaway, and R. H. Meade critically read the manuscript. Interpretations presented in the paper, however, are the responsibility of the authors.

REFERENCES

Beer, R. M., and Gorsline, D. S. (1971). Distribution, composition and transport of suspended sediment in Redondo Submarine Canyon and vicinity, California. *Marine Geol.* **10**, 153-175.

Biscaye, P. E. (1964). Distinction between kaolinite and chlorite in recent sediments by X-ray diffraction *Am. Mineralogist* **49**, 1281-1289.

Biscaye, P. E. (1965). Mineralogy and sedimentation of Recent deep-sea clay in the Atlantic Ocean and adjacent seas and oceans. *Bull. Geol. Soc. Am.* **76**, 803-831.

Blanton, J. (1971). Exchange of Gulf Stream water with North Carolina shelf water in Onslow Bay during stratified conditions. *Deep-Sea Res.* **18**, 167-178.

Brown, C. Q., and Ingram, R. L. (1954). The clay minerals of the Neuse River sediments. *J. Sediment. Petrol.* **24**, 196-199.

Brunton, G. (1955). Vapor-pressure glycolation of oriented clay minerals. *Am. Mineralogist* **40**, 124-126.

Bumpus, D. F. (1955). The circulation over the continental shelf south of Cape Hatteras. *Trans. Am. Geophys. Union* **36**, 601-611.

Bumpus, D. F., and Pierce, E. L. (1955). The hydrography and distribution of Chaetognaths over the continental shelf off North Carolina. *Deep-Sea Res.* **3** [Suppl.], 92-109.

Buss, B. A., and Rodolfo, K. S. (1971). Suspended sediments off Cape Hatteras, North Carolina. *Geol. Soc. Amer. Abst.* **3**, 521.

Chase, J. (1959). Wind-induced changes in the water-column along the East Coast of the United States. *J. Geophys. Res.* **64**, 1013-1022.

Drake, D. E. (1971). Suspended sediment and thermal stratification in Santa Barbara Channel, California. *Deep-Sea Res.* **18**, 763-769.

Duane, D. B. (1962). "Petrology and Recent Bottom Sediments of the Western Pamlico Sound Region, North Carolina." Ph.D. dissertation, 155 pp. Kansas Univ., Lawrence, Kansas (unpublished).

Emery, K. O. (1968). Relict sediments on the continental shelves of the world. *Bull. Am. Assoc. Petr. Geologists* **53**, 445-468.

Gibbs, R. J. (1965). Error due to segregation in quantitative clay mineral X-ray diffraction mounting techniques. *Am. Mineralogist* **50**, 741-751.

Gray, I. E., and Cerame-Vevas, J. J. (1963). The circulation of surface waters in Raleigh Bay, North Carolina. *Limnol. and Oceanog.* **8**, 330-337.

Griffin, G. M., and Ingram, R. L. (1955). Clay minerals of the Neuse River estuary. *J. Sediment. Petrol.* **25**, 194-200.

Harrison, W., Norcross, J. J., Pore, N. A., and Stanley, E. M. (1967). "Circulation of Shelf Waters off the Chesapeake Bight." Prof. Paper 3, 82 pp. U.S. Dept. Commerce, ESSA., Washington, D.C.

Hathaway, J. C. (1971). "Data file, Continental Margin Program, Atlantic Coast of the United States, vol. 2, Sample Collection and Analytical Data" Ref. 71-15, 496 pp., Woods Hole Oceanographic Institution (unpublished).

Hathaway, J. C., in press. Regional clay mineral facies in the estuaries and contiental margin of the United States East Coast. "Environmental framework of Coastal Plain Estuaries" (B. W. Nelson, ed.) Geol. Soc. Amer. Mem. 133.

Heron, S. D., Jr., Johnson, H. S., Jr., Wilson, P. G., and Michael, G. E. (1964). Clay mineral assemblages in a South Caroline lake-river-estuary complex. *Southeastern Geol.* **6**, 1-9.

Jacobs, M. B., and Ewing, M. (1969). Mineral source and transport in waters of the Gulf of Mexico and Caribbean Sea. *Science* **163**, 805-809.

Jerlov, N. G. (1968). "Optical Oceanography." 194 pp. Elsevier Publ. Co., New York.

Mairs, R. L. (1970). Oceanographic interpretation of Apollo photographs. *Photogrammetric Engineering* **36**, 1045-1058.

Manheim, F. T., Meade, R. H. and Bond, G. C. (1970). Suspended matter in surface waters of the Atlantic continental margin from Cape Cod to the Florida Keys. *Science* **167**, 371-376.

Mencher, E., Copeland, R. A., and Payson, H. J. (1968). Surficial sediments of Boston Harbor, Massachusetts. *J. Sediment. Petrol.* **38**, 79-86.

Millipore Filter Corporation (1966). *Detection and Analysis of Particulate Contamination.* Millipore Filter Corporation Application Manual, ADM 30.

Neiheisal, J., and Weaver, C. E. (1967). Transport and deposition of clay minerals, southeastern United States. *J. Sediment. Petrol.,* **37**, 1084-1116.

Nelson, B. W. (1960). Clay mineralogy of the bottom sediments of the Rappahannock River, Virginia. "Clays and Clay Minerals, Proc. 7th Natl. Conf." p. 135-147. Macmillan (Pergamon), New York.

Nichols, M. M. (1969). Aspects of coastal oceanography from space photography. *Geol. Soc. Amer. Abst.* **1**, 160.

Pettijohn, F. J. (1957). "Sedimentary Rocks." 2nd Ed. 718 pp. Harper, New York.

Pevear, D. R. (1968). "Clay Mineral Relationships in Recent River, Nearshore Marine, Continental Shelf, and Slope Sediments off the Southeastern United States." Ph.D. dissertation, 164 pp. Univ. Montana, Missoula, Montana (unpublished).

Pierce, J. W., Nelson, D. D., and Colquhoun,D. J. (1971). Pyrophyllite and talc in waters off southeastern United States. *Marine Geol.* **11**, M9-M15.

Pierce, J. W., and Siegel, F. R. (1969). Quantification in clay mineral studies of sediments and sedimentary rocks. *J. Sediment. Petrol.* **39**, 187-193.

Rao, P. K., Strong, A. E., and Koffler, R. (1971). Gulf Stream and Middle Atlantic Bight: Complex thermal structures as seen from an environmental satellite. *Science* **173**, 529-530.

Stefansson, U., Atkinson, L. P., and Bumpus, D. F. (1971). Hydrographic properties and circulation of the North Carolina shelf and slope waters. *Deep-Sea Res.* **18**, 383-420.

Sheldon, R. W., and Sutcliffe, W. H., Jr. (1969). Retention of marine particles by screens and filters. *Limnol. and Oceanogr.* **14**, 141-144.

Stevenson, R. E., and Uchupi,E. (1969). The dispersion of suspended sediments off southeastern United States. *Geol. Soc. Amer. Abst. with Programs,* **1**, 216.

Visser, M. P. (1970). The turbidity of the southern North Sea. *Deutsche Hydrog. Zeit.* **23**, 97-117.

Windom, H. L., Neal, W. J., and Beck, K. C. (1971). Mineralogy of sediments in three Georgia estuaries. *J. Sediment. Petrol.* **41**, 497-504.

CHAPTER 14

Sediment Transport on the Santa Barbara-Oxnard Shelf, Santa Barbara Channel, California

David E. Drake,* Ronald L. Kolpack,*
and Peter J. Fischer†

ABSTRACT

The floods of January and February, 1969, caused a record discharge of more than 50×10^6 metric tons of flood sediment into the waters over the Santa Barbara-Oxnard Shelf. The initial deposition of >70% of this material was at depths of less than 30 m on the inner shelf. Following the floods, suspended sediment concentrations were highest over the inner shelf and exhibited an approximately exponential decrease across the shelf at all water depths. Concentrations were generally highest near the bottom and ranged from approximately 50 mg liter^1 over the inner shelf during the floods to less than 1.0 mg liter^1 at the shelf break late in 1969 and early in 1970.

Redistribution of the flood sediment involved rapid erosion over the inner shelf (at depths of <30 m) by wave associated currents and transfer of this material to the middle shelf and beyond. Deposition of resuspended detritus on the middle and outer shelf occurred within protected topographic depressions and in response to the seaward decline in wave-induced current energies. Seaward transport of material resuspended by wave associated, near-bottom currents occurred within a north-westward and westward flowing geostrophic current and within a current convergence trending across the shelf from Ventura. Although a mean velocity of approximately 25 cm sec^{-1} is typical

*Department of Geological Sciences, University of Southern California, Los Angeles, California 90007

†Department of Geology, San Fernando Valley State College, Northridge, California 91324

of the northwestward current sweeping the middle and outer shelf, this current was not sufficient to prevent deposition of fine silt and clay.

Transport of suspended particles across the shelf in the presence of thermal stratification within the water column resulted in detachment of turbid water from the near-bottom nepheloid layer and the formation of mid-water particle maxima. Initial deposition of flood material was rapid and concentrations of suspended matter during and following the record floods never reached levels required to penetrate the water column density stratification. Therefore, the development of mid-water particle maxima, the strong seaward gradient in particle concentrations at all levels and the deposition of fine silt and clay over the middle shelf support a conclusion that the mechanism of turbid-layer (density-excess) flow downslope was insignificant.

INTRODUCTION

In January and February of 1969 southern California experienced two intense rainstorms which resulted in a record flood discharge of terrigenous detritus by the coastal watershed. Owing to the semiarid climate of the area the stream-borne sediment was a distinctive red-brown in contrast to the generally gray and olive-gray marine sediments off southern California (Emery, 1960). This fact, combined with the availability of excellent data on the discharge of suspended fluvial material by the major drainage systems, provided an opportunity to study the fate of the flood material on a portion of the mainland shelf. Furthermore, detailed information on the preflood distribution of sediments on the continental shelf off southern California is available from previous investigations by the Allan Hancock Foundation (1965) and Wimberly (1964). The purpose of this research was to determine the initial pattern of deposition of the flood material, the changes which occurred with time and, in particular, the pathways and mechanisms of fine sediment transport.

The segment of mainland shelf selected for this research lies between Santa Barbara and Oxnard (Fig. 113) and is a natural subdivision of the mainland shelf off southern California since it is a relatively wide shelf separating markedly narrower shelves to the west and east.

METHODS

The distribution of oxidized flood material on the shelf was determined by measuring layer thickness recovered in more than 175 Phleger cores and modified Reineck box cores. Loss of surface material by the box corer is minimal, therefore, primary consideration was given to these data in the preparation of flood sediment isopachs. Bottom sediment textures were determined by settling procedures (Krumbein and Pettijohn, 1938; Cook, 1969). The distribution and concentrations of suspended particulate matter over the shelf were determined by membrane filtration of more than 250 discrete water samples and more than 150 continuous

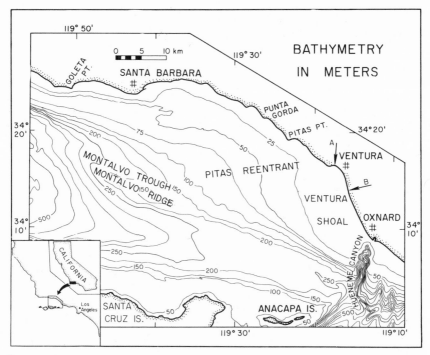

Figure 113.

Location map and generalized bathymetry of eastern Santa Barbara Channel. Bathymetry is in meters and is based on Environmental Sciences Services Administration charts. The Ventura River is denoted by the arrow labeled "A" and the Santa Clara River by the arrow labeled "B."

vertical profiles of light-beam transmission. Water samples were recovered at the surface with a polyethylene bucket and at depth with 7-liter Van Dorn and 30-liter Niskin bottles. The 30-liter Niskin bottle was modified to close one meter above the bottom when a weighted trigger line struck the sea floor. In order to minimize the effects of the poor flushing rate of the Niskin sampler with 7.4 cm end openings (Weiss, 1971), the sample bottle was lowered at the slowest possible rate (approximately 15 m min⁻¹) in the bottom one-half of the water column and held for a short time at 5 to 10 m above the bottom before final lowering.

Particulate matter and water were separated using 47 mm Millipore® HA filters with a nominal pore diameter of 0.45 μ. Studies at our laboratory show that recovery of particles from a standard sample of Pacific red clay (0.5 mg liter⁻¹ concentration) is better than 95% with the 0.45 m pore-size filters whereas as much as 15% of the sediment passed through 0.8 μ pore-size filters.

Water samples of 4 liters were filtered through two numbered, preweighed filters with the lower of the pair acting as a blank. Salts were washed from

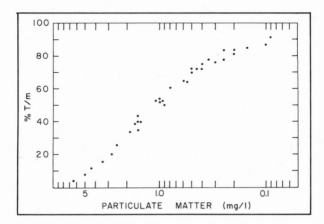

Figure 114.
Scatter diagram of % transmission meter[-1] and the total suspended particles concentration in mg liter[-1]. Data points are based on samples collected during February 6-17, 1970. Suspended particle concentrations were determined by filtration with Millipore HA filters.

the filters with a minimum of 100 ml prefiltered distilled water and the filters were then dried in the shore laboratory at 60°C for 24 hr and weighed to the nearest one-tenth of a milligram. Repeated analyses of test samples indicate a reproducibility of ±20% for this method.

A continuously recording beam transmissometer and thermistor developed by the Visibility Laboratory of Scripps Institution of Oceanography (Petzold and Austin, 1968) was used to determine the distribution of suspended particles and the thermal structure of the water column. The transmissometer measures the attenuation of a 2 cm diameter beam of white light over a one meter water path and the light is filtered for peak sensitivity at 470 nm. Approximate concentrations of suspended particles can be obtained with the transmissometer over the range of 6 mg liter[-1] to less than 0.1 mg liter[-1] (Fig. 114). However, the accuracy of this method is about ±30% (Drake, 1971) due to the optical effects of dissolved organic substances.

Surface-water current patterns over the shelf (Fig. 115) were determined from monthly drops of 30 weighted drift cards at 10 stations (Kolpack, 1971). The drift cards were designed to float with approximately 25 cm of their plastic envelopes below the sea surface and only 2-3 cm above the surface. Card recovery averaged about 10% during a period of one year.

Current-meter data obtained at 10 stations during July and August of 1967 were made available to the authors through the courtesy of Shell Oil Company. All values were obtained by a Hydro Products Model 501 current meter positioned 6 m above the bottom for periods of 12-24 hr.

PHYSIOGRAPHY

The Santa Barbara-Oxnard Shelf is a northwest-southeast trending topographic feature of about 40 km in length and has a maximum width of 25 km normal to that trend (Fig. 113). The total area shallower than 100 m is approximately 800 km².

Montalvo Ridge, Montalvo Trough, Pitas Reentrant, and Ventura Shoal are four structurally controlled topographic features on the shelf (Fig. 113) which are pertinent to the oceanographic and sedimentologic regime of the area. Montalvo Ridge is a relatively steep-sided, linear and fault-bounded anticlinial feature located approximately 20 km south of Santa Barbara. Montalvo Trough slopes at 10 m km^{-1} to the northwest along the north flank of Montalvo Ridge. Pitas Reentrant slopes southwestward at about 4 m km^{-1} from the coast between Ventura and Pitas Point and connects with Montalvo Trough. The structural grain was initiated in late Oligocene to early Miocene time (Fischer, 1972) and seismic activity as well as surface ruptures along fault traces demonstrates structural modification is still in progress (Sylvester, Smith, and Scholz, 1970; Hamilton et al., 1969; Fischer, 1972).

OCEANOGRAPHY

The surface-water circulation in Santa Barbara Channel is characterized by two strong currents which converge in the area between Santa Barbara and Santa Rosa Island. Surface currents over the Santa Barbara-Oxnard shelf (Fig. 115) are dominated by the Anacapa Current which is a northwestward flowing current that enters the Channel through the passage between Anacapa Island and the mainland coast south of Oxnard. However, circulation over the inner shelf between Santa Barbara and Ventura is characterized by a series of eddies which break off the Anacapa Current. These eddies result from the impingement of the Anacapa Current on the coastline near Santa Barbara, local bottom topographic effects and also the predominant northwest-wind regime.

The northern Channel Islands Ridge and mainland coast effectively shield the shelf from swell approaching from all directions except the western quadrant. Westerly wave approach and the orientation of the shoreline combine to produce longshore currents which range from 15 to 60 cm sec^{-1} to the southeast (Kolpack, 1971). The surface swell typically has a period of 9 to 20 sec with heights of 1 to 2 m (Table XXIII) but winter and spring storms plus prevailing onshore winds increase the wave heights during that part of the year. Consequently, the Santa Barbara-Oxnard shoreline is a moderate energy coast (Ingle, 1966; Kolpack, 1971) and transport of nearshore sediments is an important component of the shelf transport system. Beach width and sediment textural changes demon-

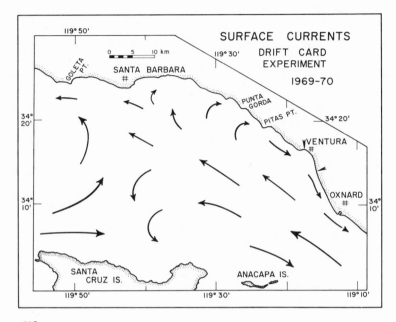

Figure 115.
Summary of surface current pattern based on monthly drops of drift cards in 1969 and 1970. The arrow lengths are a qualitative indication of the persistence of the surface currents.

Table XXIII
Mean Wave and Wind Conditions during Survey Periods[a]

Survey	Wave ht. (m)	Wave period (sec)	Wind speed (km/hr)
San Buenaventura Beach: 1.4 km north of Santa Clara River			
January 22-28, 1969	1.0	11.0	7.2
May 1-7, 1969	0.9	12.8	9.6
July 16-20, 1969	0.9	14.0	10.9
August 23-30, 1969	1.0	12.0	12.8
November 4-7, 1969	0.6	6.4	8.0
December 8-14, 1969	2.1	8.3	6.9
McGrath Beach: 0.9 km south of Santa Clara River			
January 22-28, 1969	1.7	11.5	8.0
May 1-7, 1969	1.1	11.9	12.3
July 16-20, 1969	1.1	10.6	10.9
August 23-30, 1969	1.0	12.8	13.4
November 4-7, 1969	1.2	7.2	11.0
December 8-14, 1969	3.0	9.3	8.0

* [a]Unpublished data courtesy of the U. S. Army Corps of Engineers Coastal Engineering Research Center.

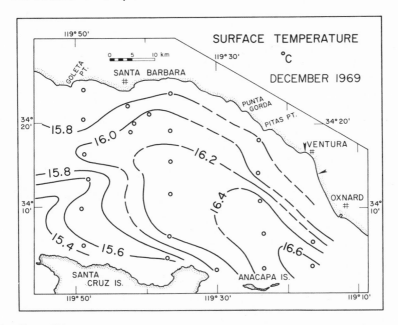

Figure 116.
Distribution of surface water temperature in December, 1969.

strate that much of the nearshore sand transported past Oxnard is intercepted and routed offshore by Hueneme Submarine Canyon which cuts across the southern margin of the shelf and heads within several hundred meters of the shore (U.S. Army Corps of Engineers, unpublished data).

The three hydrographic seasons in the area (CALCOFI, 1955-1970; Emery, 1960; Kolpack, 1971) are characterized by; a spring season of wind driven upwelling, a late summer season of vertical stability, and a fall and early winter season of vertical mixing and broad northwestward flow.

The warm water that enters the area from the southeast during the winter months is bordered by a cooler water zone toward the mainland and island coasts (Fig. 116). This simple pattern becomes more complex in localized areas of the coastal zone during February-May when there is vertical water movement associated with a period of upwelling. During the summer months of June through September the surface water temperature pattern is distinguished by the development of a lens of warm water over the northern half of the shelf (Fig. 117) as a result of solar insolation. In addition, the surface water isotherms for the summer months define a convergence zone southwest of Pitas Point. This feature is also evident from light transmission information obtained in December, 1968 (Fig. 118) and November, 1969. Thus, the westward transport of inner shelf water off Pitas Point that is implied by these patterns suggests that the convergence zone is an important feature of the shelf circulation system.

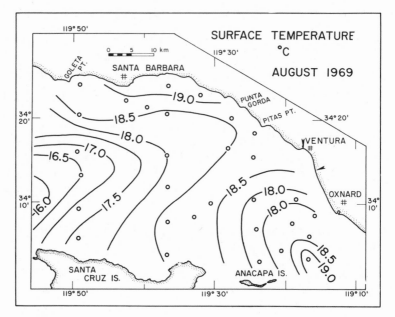

Figure 117.

Distribution of surface water temperature in August, 1969.

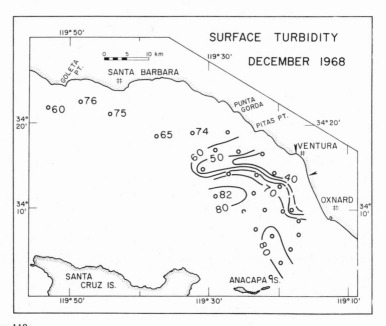

Figure 118.

Surface water transparency values ($\% \, T/m$) for December 18-21, 1968. Small circles represent station control.

Table XXIV
Mean Bottom Current Velocities and Direction, Related to Tidal Cycle

	Flood tide		Ebb tide	
Station No.	Mean velocity (cm sec^{-1})	Mean direction (° true)	Mean velocity (cm sec^{-1})	Mean direction (° true)
1	21	243	21	230
2	36	307	28	282
3	31	270	25	250
4	27	325	27	270
5	23	252[a]	27	289
6	26	270	31	220
7	26	240	29	275
8	27	015	31	265
9	23	263	27	307
10	22	246	25	253
Mean	26	280	27	260

[a]Mean current direction was 155° true for 2 hours prior to stabilization at 252° true.

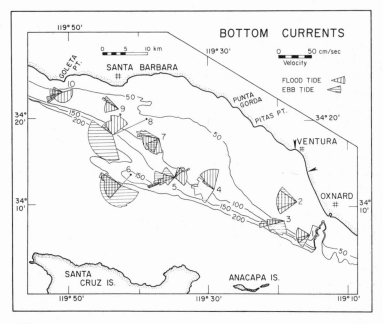

Figure 119.
Current velocities 6 m off the bottom measured during July and August, 1967. Bathymetry is in meters.

Bottom current measurements made during the summer of 1967 (Table XXIV and Fig. 119) show a northwestward (280°) flow during flood tide and a slightly more southerly (260°) flow during ebb tide. Current speeds ranged from 10 cm sec^{-1} to greater than 50 cm sec^{-1} with a computed mean velocity of approximately 25 cm sec^{-1} to the west.

Although no direct information on the magnitude of wave-induced, near-bottom currents on the Santa Barbara-Oxnard shelf presently is available, the oscillatory currents can be estimated by solving the following equation:

$$v = (2\pi/T)\,ae^{-2\pi Z/L}$$

where T = period, a = amplitude, L = wavelength, and Z = depth. A typical channel wave with a period of 9 sec, a length of 200 m and an amplitude of 1 m would produce surge velocities of approximately 13 cm sec^{-1} at 50 m, 25 cm sec^{-1} at 30 m, and 35 cm sec^{-1} at 20 m. The recent work by Southard, Young, and Hollister (1971) indicates that wave-induced bottom currents on the Santa Barbara-Oxnard shelf are capable of eroding deposited sediments only at depths shallower than 20 m. However, these currents probably help maintain the suspension of fine-grained detritus to depths of at least 50 m.

FLOOD SEDIMENT

Hourly information on water discharge and suspended load (U.S. Geological Survey Water Resources Division, unpublished data) shows that approximately 50×10^6 metric tons of sediments were transported to the coast by the Santa Clara River during 1969. More than 70% of the total flood discharge occurred on two days, January 25 and February 25, 1969, and a maximum suspended sediment discharge of 22×10^6 metric tons was recorded on February 25, 1969.

The Ventura River system has an uncontrolled drainage of approximately one-eighth the area of Santa Clara River. Although the U.S. Geological Survey gauging station on the Ventura River was destroyed, discharge was estimated and suspended sediment samples were recovered near the river mouth on 22 days during the one month flood period. These data indicate a suspended sediment contribution of about 6.5×10^6 metric tons, which is equivalent to the amount calculated from direct comparison to the drainage area and well documented suspended load of Santa Clara River.

The size distribution of suspended sediments was determined for 10 samples recovered from Santa Clara and Ventura Rivers when flow rates exceeded 1000 m^3 sec^{-1}. Ranges of 15-20% clay, 40-50% silt, and 30-40% sand were determined for these samples but the values are subject to considerable variation owing to the difficulty of obtaining representative samples from flooding rivers. Calculations based on the previously mentioned suspended load data indicate that approximately $35\text{-}45 \times 10^6$ metric tons of silt and clay and $12\text{-}20 \times 10^6$ metric tons of sand were introduced by the Santa Clara and Ventura Rivers during the floods.

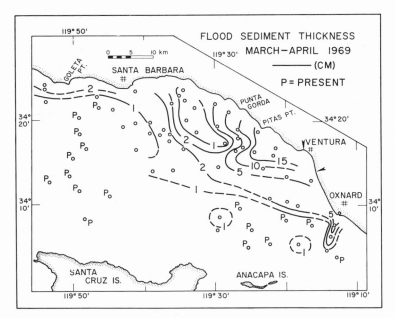

Figure 120.
Initial depositional pattern of 1969 flood sediment, March and April, 1969.

Textural analyses of flood sediments deposited on the shelf west of Ventura show that during the floods, sand-size particles were confined to a coastal strip extending from 1 to 1.5 km seaward of the beach. The finer fractions were transported farther offshore as suspended load and were deposited at a distance of more than 2 km off the coast. Initial deposition of measurable thicknesses of flood detritus from Santa Clara and Ventura Rivers occurred within 20 km of the coast (Fig. 120).

Bathymetric data (U.S. Army Corps of Engineers, unpublished data) show that a significant sand delta formed off the Santa Clara River mouth as a result of flood sediment deposition. The subaerial portion of this delta extended 0.7 km offshore, however, submarine shoaling was detected at a distance of about 2 km from the river mouth. If one uses a density of 1.5 g cm^{-3} for the freshly deposited material then, approximately 16×10^6 metric tons of coarse silt, sand and gravel were present in this delta two months after peak flooding. The similarity of estimates of the amount of sand transported in suspension and the amount of material forming the river mouth delta suggests that the influence of traction load was relatively minor. Although the particle sizes and transport rates of traction load sediment were not measured, it is likely that the bulk of this material was coarser than sand and was deposited before reaching the shoreline.

The initial depositional pattern of the flood sediments resulted in a westward-thinning triangular blanket of sediment (Fig. 120). Estimates of the volume

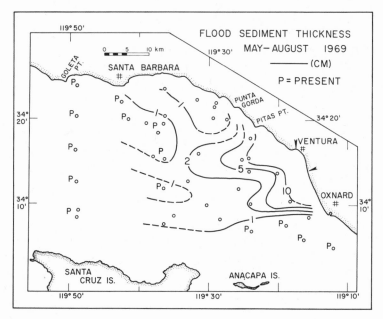

Figure 121.
Flood layer thickness pattern based on bottom samples of May-August, 1969.

of flood sediment present on the shelf during March and April, 1969, are subject to some uncertainty because of natural variations in thickness and the lack of complete thickness data between the river delta and the fine-grained sediment wedge to the west. However, if one assumes that there was no change in the thickness of the flood sediment layer between the 15 cm thick offshore layer and the delta, then approximately $30\text{-}35 \times 10^6$ metric tons of flood sediment were deposited seaward of 2 km from the coast. Combining the offshore ($30\text{-}35 \times 10^6$ metric tons) and the delta (16×10^6 metric tons) values yields a total of about $45\text{-}50 \times 10^6$ metric tons of flood sediment. Thus, more than 70% of the detritus was retained on the shelf at the end of April, 1969.

The pattern of oxidized sediment deposition during the summer of 1969 was similar to the pattern of initial deposition except the layer was about 20% thinner and a second lobe of material developed to the south of the primary lobe (Fig. 121). In addition, the subaerial sand deposit off the mouth of the Santa Clara River had been reduced to a river mouth bar by August, 1969.

The secondary lobe of flood sediment off Oxnard must have been derived from material eroded from the primary lobe, transported south by nearshore currents and entrained by the Anacapa Current because the Anacapa Current transports very little suspended matter into the Santa Barbara Channel (Fig. 122). This interpretation is substantiated by the suspended sediment information presented later.

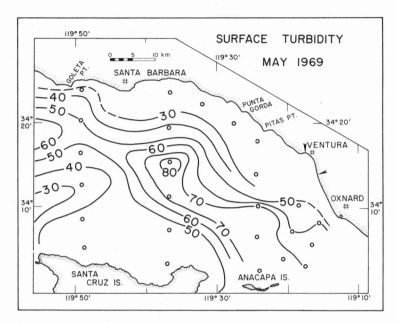

Figure 122.
Surface water transparency values (% *T/m*) for May 1-7, 1969.

Eighteen months after the floods (Fig. 123), the surface layer was still readily detected on the shelf although considerable bioturbation, scour, and redistribution had occurred in the area south of Ventura, whereas there was little indication of erosion off Pitas Point and over the northern shelf. The stability of the 10 cm thick deposit off Pitas Point, at depths of less than 20 m, indicates that the current convergence in this area is a stable feature of the shelf circulation system.

Textural analyses of surface samples show that the sediments initially deposited in the Oxnard lobe graded from moderately well-sorted fine sand near the coast to poorly sorted silt and clay at a distance of 15 km west of Oxnard. Subsequently, there was a progressive westward movement of silt and clay along the southern shelf beneath the Anacapa Current. This redistribution of sediments involved the transfer of fine silt and clay from the inner shelf to form northwest trending elongated lenses of fine silt and clay on the middle and outer shelf and within Montalvo Trough (Fig. 124).

A series of box core samples were collected from shelf depths of 30 to 100 m in July, 1971 in order to determine the stability of this middle shelf deposit. These samples showed that the thickest deposits of oxidized sediment occurred at a depth of 30-60 m but the middle shelf lens had been shifted to the north and northwest parallel to the shelf break. Mean diameters of the sediment within the thickest middle-shelf deposits ranged from 6-8 μ, but the

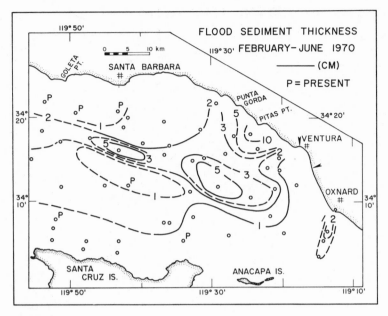

Figure 123.
Flood layer thickness pattern based on bottom samples of February-June, 1970.

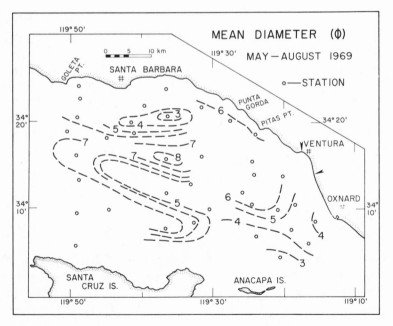

Figure 124.
Mean diameters of flood sediment in May-August, 1969.

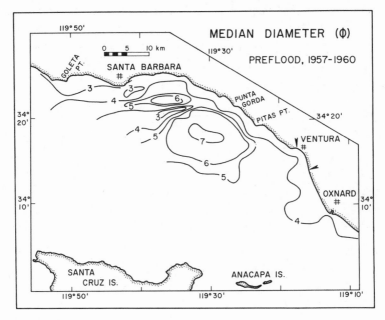

Figure 125.
Preflood distribution of median sediment diameters (modified after Wimberly, 1964).

thickness of the layer was not significantly less than the thickness measured one year earlier. This suggests that the erosion of fine silt and clay indicated by the slight increase in diameter was offset by continued seaward movement of material from the inner shelf. The texture of the sediments collected in July, 1971, was about one grade size finer than the underlying preflood middle- and outer-shelf sediments (Fig. 125; Wimberly, 1963). Consequently, a period of at least 3 yr is required for sediments on the Santa Barbara-Oxnard Shelf to return to ''normal'' following major introductions of river-borne material.

SUSPENDED PARTICULATE MATTER

Surface

The distribution of particulate matter at the surface and within the water column over the Santa Barbara-Oxnard shelf was determined during December 1968; January, April, May, July, August, November, and December 1969; and February, 1970. These cruises started prior to a major flood in southern California and the concentrated effort in the area during and after the flood permitted us to follow the sequential history of the flood sediments after they entered the marine environment.

Vertical transparency profiles obtained after 4 days of flooding, and introduction

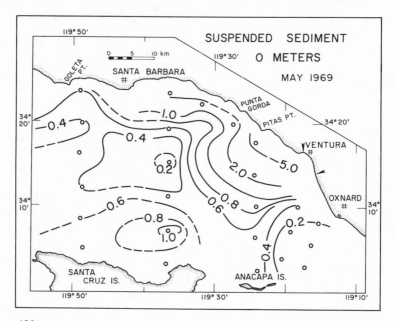

Figure 126.

Distribution of suspended particulate matter at the surface for May 1-7, 1969. Values are in mg liter[-1] and were determined by filtration with Millipore HA filters.

of about 5×10^6 metric tons of suspended detritus by the Santa Clara and Ventura Rivers, showed that most of the suspended matter was contained within a 10-20 m thick surface layer. Particulate concentrations in this layer ranged from 7-10 mg liter[-1] at a distance of 3 km off the Santa Clara River mouth to <3 mg liter[-1] over the outer shelf. Whereas sand and coarser particles were deposited in the nearshore zones, finer grained fractions were transported through the nearshore and inner-shelf zones to form a westward-trending surface plume within the current convergence off Ventura and Pitas Point. This pattern is essentially identical to the distributional patterns of surface suspended matter recorded prior to the floods and also late in 1969. In addition this pattern correlates well with the depositional pattern of flood sediments deposited before April, 1969, indicating that surface and subsurface transport paths were similar.

Concentrations of suspended particles in the surface water remained high through May 7, 1969 (Fig. 126) even though the contribution of terrigenous detritus declined to <60 metric tons day[-1] during the last week of April, 1969. During May these concentrations exceeded 5 mg liter[-1] over the inner shelf and decreased to values of <0.5 mg liter[-1] within the Anacapa Current over the outer shelf. Transmissometer profiles obtained in April and May, 1969, show that the turbid surface layer was approximately 15 m thick and had at least

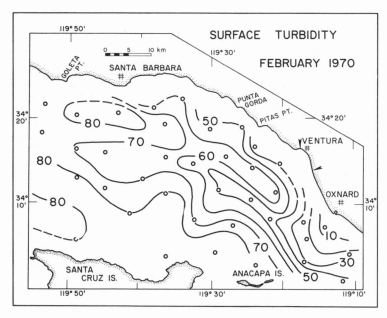

Figure 127.

Surface water transparency values (% *T/m*) for February 6-17, 1970. This survey followed three days of moderate river discharge and introduction of approximately 2×10^4 metric tons of suspended detritus by Santa Clara River.

2 mg liter^{-1} of suspended particulate matter. Thus, the surface layer present in May, 1969, contained between 10 and 20×10^4 metric tons of sediment. Most of this material was probably derived from resuspension of nearshore flood deposits since the range of values is equivalent to the river-borne suspended sediment discharge for the entire month of April. By July, surface water concentrations of particulate matter had decreased markedly to <1.0 mg liter^{-1} 4 km west of Ventura and <0.5 mg liter^{-1} over the outer shelf. The greatest decrease, exceeding one order of magnitude, occurred over the inner shelf.

Moderate rainfall and discharge of approximately 2×10^4 metric tons of suspended particulate matter by Santa Clara River occurred on February 9-13, 1970, and followed a period of no river discharge. The resulting pattern of suspended detritus at the surface (Fig. 127) defines the transport path followed by terrigenous sediment during periods of moderate river flow. In contrast to the large turbid plume developed during the 1969 floods, the February discharge was not sufficient to break through the nearshore zone and, therefore, high particle concentrations were limited to the innermost shelf. This material was transported south of Oxnard where much of it was entrained by the Anacapa Current and delivered to the outer shelf. Although the initial transport paths of terrigenous suspended matter are, in part, related to the magnitude of river discharge, the circulation system over Santa Barbara-Oxnard shelf results in a high retention of fine sediment within Santa Barbara Channel.

The surface transparency data for February provide an independent verification of the mean velocity of Anacapa Current. Westward transport of the turbid plume within this current was first detected 70 km from the Santa Clara River three days after the initial introduction of fluvial sediment. Assuming longshore currents to Oxnard averaged 40 cm sec^{-1} (Kolpack, 1971), the mean flow of Anacapa Current was approximately 30 cm sec^{-1}.

Intermediate Depths

Bouma, Rezak, and Chemlik (1969) and Drake (1971) have shown that the subsurface distribution of particulate matter is influenced strongly by the vertical density stratification resulting from the relatively abrupt temperature decreases associated with the macrostructure and microstructure of the water column. In the Santa Barbara Channel, transmissometer and temperature profiles show that transparency decreases exceeding 10% always are associated with more or less abrupt thermal discontinuities of $< 0.1°C$ and greater. Nevertheless, thermal "steps" are not always accompanied by particle maxima nor is there any significant correlation of the magnitudes of particle maxima and the thermal discontinuities. The latter implies that the subsurface turbid layers are not supplied principally by particle settling, but through the lateral spread of material from topographic highs (Fig. 128). In particular, transmissometer cross-sections through Pitas Reentrant and Montalvo Trough in November, 1969 (Fig. 129), and in February, 1970 (Drake, 1971), show that a series of stacked mid-water particle maxima are the result of the detachment of turbid water from the bottom "nepheloid" layer at each increase in the shelf gradient and the westward current flow. Particulate concentrations within these layers are generally at least one order of magnitude higher than those at the sea surface and within intervening clear layers. Thus, while much of the suspended particulate matter near the surface is restricted to the inner shelf, lateral supply from the bottom turbid zone maintains high concentrations within subsurface layers to the shelf edge and beyond.

Near Bottom

During all surveys, the highest concentrations near the bottom were over the inner shelf and ranged from approximately 50 mg liter^{-1} during the major-flood period (January and February) to approximately 4-6 mg liter^{-1} in November and December of 1969 (Fig. 130). Concentrations always decreased seaward with values along the outer shelf ranging from 10 mg liter^{-1} early in 1969 to <2.0 mg liter^{-1} in November and December of 1969 and February, 1970. Average

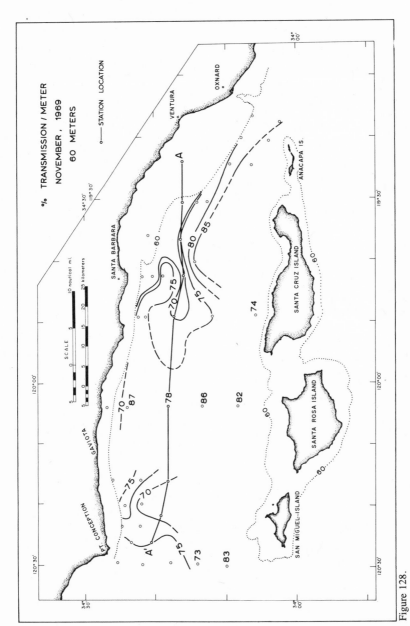

Figure 128.

Distribution of water transparency values (*% T/m*) at 60 m for November 4-7, 1969. Refer to Fig. 114 for approximate particle concentrations. The solid line (A-A') refers to the cross-section shown in Fig. 129.

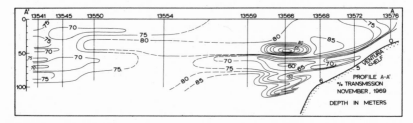

Figure 129.

East-west cross-section (A-A′) showing the vertical distribution of light-attenuating substances over Santa Barbara-Oxnard shelf. For clarity, the bottom 20 m of the water column is not contoured, but the % *T/m* value at the bottom is noted. Refer to Fig. 128 for profile location.

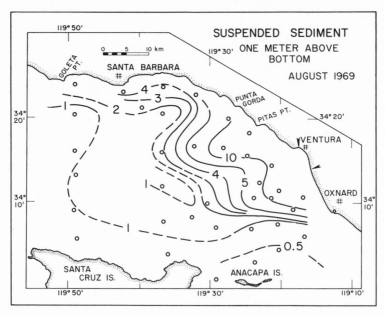

Figure 130.

Distribution of suspended particulate matter (mg liter⁻¹) one meter above the bottom for August 23-30, 1969.

concentrations near the bottom were relatively high (approximately 5-10 mg liter⁻¹) through August, 1969, but decreased to about 3 mg liter⁻¹ in November and remained relatively low through February, 1970. This near-bottom trend mirrors the trend of concentrations at the sea surface which occurred three to five months earlier. The concentration gradient across the shelf demonstrates the major importance of bottom currents associated with surface waves. Therefore, the general decrease in particle concentrations near the seabed after August is ascribed to prior erosion of fine sediment from depths of less than about 30 m and transport of this material to the lower-energy middle and outer portions of the shelf.

Analysis of the vertical distribution of particles near the bottom in more than 100 transmissometer and temperature profiles obtained in 1969 and 1970 shows that the material was contained within more or less distinct bottom nepheloid layers averaging 14 m in thickness with a range of from 35 m to less than 5 m. No trend in the thickness versus depth was detected. In approximately 72% of the records, the transition from the relatively clear water above the bottom nepheloid layer occurred within 7 m and was associated with a distinct increase in the thermal gradient. With the exception of those records at shelf depths of less than 20 m where the water temperature did not change with depth, *all* near-bottom turbid layers were associated with a cooler water layer which on 64% of the records was isothermal within the limits of thermistor resolution ($\pm 0.05°C$). The association of the bottom nepheloid layer and isothermal water is the resultant of near-bottom turbulence over the shelf. In those cases in which the upper limit of the bottom nepheloid layer is relatively distinct and is associated with an equally abrupt temperature discontinuity (72% of the records), the beam transmission values change only slightly within the turbid layer. The small transparency range within these layers and their association with isothermal water demonstrates sufficient turbulence to produce a uniform distribution of particles. The thickness of the nepheloid layer is a function of the magnitude of such turbulence and the restriction to vertical diffusion imposed by the thermal stratification of the water column (Drake, 1972).

DISCUSSION

Moore (1969) presented a thorough discussion of the hypothesis of density excess flow of suspended matter over the continental shelf and slope off southern California. He proposed that the bulk of the fine-grained detritus delivered by the coastal watershed and deposited in the basins of the borderland is transported by gravity flow of near-bottom turbid layers. His development of this model for suspended-sediment transport suffered from a lack of data on typical particle concentrations in shelf waters. The 1969 floods provided an excellent opportunity to monitor suspended-matter concentrations during a period of extremely high terrigenous-sediment influx and test this potentially important transport mechanism.

The particle concentrations measured during and after the floods at the surface, within intermediate depth particle maxima, and near the bottom never exceeded 50 mg liter[1] at distances of 3 km or more from the coast. Furthermore, the bulk of even the finest flood sediment was deposited rapidly to form a blanket over the inner and middle shelf and initial transport of this detritus is completely explained by the known advective currents over Santa Barbara-Oxnard shelf.

Once deposited, current energies were incapable of maintaining particle concentrations in excess of approximately 30 mg liter^{-1} within the bottom turbid layer over the inner shelf and concentrations ranged from about 1 to 5 mg liter^{-1} along the shelf break.

In order to drive a density flow downslope across the Santa Barbara-Oxnard shelf, temperature and salinity data collected during May, August, and December of 1969 (Kolpack, 1971) show that the turbid flow must be of sufficient concentration to overcome a minimum vertical water density increase of about 2×10^{-4} g cc^{-1}. This is equivalent to a particle concentration exceeding 100 mg liter^{-1}. It is evident that the actual concentrations during and following the major 1969 floods were too low by a factor of from 2 to > 10. The initial and subsequent patterns of flood sediment deposition, and the approximately exponential offshore concentration decrease of particles suspended near the bottom and at intermediate depths support the conclusion that downslope transport by density excess flow was unimportant. This interpretation is supported by the vertical distribution of particles over the middle and outer shelf. Transmissometer and temperature profiles show that detritus suspended near the bottom becomes detached from the seabed at increases in the shelf gradient; resulting in the development of a number of distinct particle maxima at intermediate depths (Fig. 129). The vertical distribution of these maxima is correlated closely with the density stratification of the water column (Drake, 1971).

In an attempt to reconcile the observed formation of particle maxima at intermediate depths with the hypothesis of turbid-layer flow, Postma (1968) proposed a model involving downslope density-excess transport between the points of turbid-layer detachment from the bottom. This model requires that particle settling from the mid-water layers to the bottom is sufficiently rapid and complete to generate a second turbid flow.

Textural analyses of suspended particles recovered one meter above the bottom along the shelf profile shown in Fig. 129 were obtained with a light microscope. Mean diameters ranged from 16 μ over the inner shelf to approximately 8 μ at the shelf break within Montalvo Trough. Seaward of station 13572 the samples contained no grains larger than 31 μ and the bulk of the material was clay and fine silt. In order for particles within detached mid-water turbid layers to return to the shelf through settling, the following simple equation must be satisfied:

$$\omega \geqslant (u) \sin \beta$$

where ω is the settling velocity, u is the mean flow of the prevailing advective current and $\sin \beta$ is the shelf slope (for small β). Assuming a mean westward flow of 10 cm sec^{-1} for Anacapa Current (which is less than one half the actual velocity) and a shelf slope of 1/100, the suspended particles must be larger than 40 μ (using Stoke's Law settling rate for particles with densities of 2.65

in 15°C water). It is clear that the detachment of suspended fine silt and clay from the bottom nepheloid layer within an advective current trending across the shelf precludes the possibility that these particles will significantly influence near-bottom concentrations over deeper portions of the shelf. In fact, in cases in which the vertical and horizontal particle distributions within subsurface plumes are reasonably well defined, the seaward decrease in concentrations is adequately explained by lateral eddy diffusion. In any case, downslope flow should displace relatively warm and less saline water to greater shelf depths (off southern California). Yet, no measurable temperature or salinity inversions near the seabed or within recently detached turbid plumes are present in more than 150 transmissometer-temperature profiles and STD profiles recorded in 1969 and 1970 (Drake, 1972).

CONCLUSIONS

1. Initial deposition of fine detritus from the January-February, 1969, floods was controlled by a wedge-shaped current convergence trending across the shelf off Ventura and Pitas Point, whereas sand and larger particles were confined to a 1 to 2 km wide nearshore zone. This pattern of deposition is compatible with shelf current patterns during nonflood periods and suggests that the fresh water discharge did not change the circulation system significantly.

2. The concentration of suspended particulate matter at the surface over the shelf is controlled principally by the sediment texture and wave energy in the nearshore zone. Diffusion and particle settling result in a steep seaward concentration gradient at the surface.

3. Vertical particle distributions over the middle and outer shelf are characterized by sharply bounded particle maxima which are closely associated with the thermal microstructure of the water column and are principally supplied by lateral transport of detritus in suspension near the bottom. Sufficiently high levels of turbulence were present near the bottom at all shelf depths to produce isothermal bottom-water layers and maintain sediment particles in suspension.

4. Concentrations of suspended detritus are invariably highest within the near-bottom water over the inner shelf, but there is a tendency for the highest concentrations to occur within mid-water layers over the deeper portions of the shelf. Concentrations one meter above the bottom ranged from 50 mg liter^{-1} near the rivers during the flood period to less than 2 mg liter^{-1} at the shelf edge during the latter part of 1969. Concentrations of suspended particles decreased with increasing depth on all surveys. In agreement with the texture and redistribution of the flood layer, the negative seaward gradient shows that supply of particles to the near bottom water over Santa Barbara-Oxnard shelf comes principally from detritus resuspended by wave associated currents over the inner and middle shelf.

5. Although the shelf surveys followed a record discharge of vast amounts of fine terrigenous sediment, suspended particle concentrations near the bottom were not high enough to overcome the water density increase due to temperature and salinity, and, thus, were not capable of driving downslope turbid layer flows. Data collected during the latter half of 1969 and in 1970 suggest that near bottom concentrations over the shelf typically are well below 10 mg liter^{-1} and attain higher values only in the nearshore zone.

The temporal changes of the 1969 flood layer and the textures and depositional patterns of sediments on Santa Barbara-Oxnard shelf are in good agreement with the available information on advective water circulation and the expected current and wave-energy variations over the shelf due to topography.

ACKNOWLEDGMENTS

We wish to thank Richard Loudermilk and the staff of the Visibility Laboratory, Scripps Institution of Oceanography; Captain Fred Ziesenhenne and the crew of RV Velero IV, University of Southern California; Shell Oil Company for releasing bottom-current data obtained in 1967; and the many students of Fullerton Junior College, Fullerton, California, who assisted with sample collections. Special thanks are extended to Ken Robinson, Robert Young, Ron Zakrzewski, and Mike Carruth for data reduction; Elise Kahn and Susie Dimitriou for manuscript typing.

Support for this research was provided under National Science Foundation grants GA-13083, GA-22842, and GB-8206; the Western Oil and Gas Association; and NOAA Office of Sea Grant, Department of Commerce grants GH-89 and 2-35227.

REFERENCES

Allan Hancock Foundation (1965). An oceanographic and biological survey of the southern California mainland shelf. Publ. No. 27. Calif. State Water Quality Control Board, Sacramento, California.

Bouma, A. H., Rezak, R., and Chmelik, F. B. (1969). Sediment transport along oceanic density interfaces. *Geol. Soc. America Abs. 1969*, part 7, 259-260.

CALCOFI (California Cooperative Oceanic Fisheries Investigations), 1955-1970. Progress reports. Calif. Dept. Fish and Game, Sacramento, California.

Cook, D. O. (1969). Calibration of University of Southern California automatically recording settling tube. *J. Sed. Pet.* **39**, 781-786.

Drake, D. E. (1972). Distribution and transport of suspended matter, Santa Barbara Channel, California. *Deep-Sea Res.* **18**, 763-769.

Drake, D. E. (1972). Distribution and ttransport of suspended matter, Santa Barbara Channel, California. *Thesis, Univ. Southern California, Los Angeles, Calif.*, 357 pp., (unpublished).

Emery, K. O. (1960). "The Sea Off Southern California." Wiley, New York.

Fischer, P. J. (1972). Geologic evolution and Quaternary geology of the Santa Barbara basin, southern California. *Ph.D. thesis, Univ. Southern California, Los Angeles. California,* 337 pp. (unpublished).

Hamilton, R. M., Yerkes, R. F., Brown, R. D., Jr., Burford, R. O., and DeNoyer, J. M. (1969). Seismicity and associated effects, Santa Barbara Region. *In* "Geology, Petroleum Development, and Seismicity of the Santa Barbara Channel Region, California." p. 1-13, Prof. Paper 679, U.S. Geol. Survey.

Ingle, J. C., Jr. (1966). "The Movement of Beach Sand." Elsevier, Amsterdam.

Kolpack, R. L. (ed.) (1971). "Biological and Oceanographical Survey of the Santa Barbara Channel Oil Spill." Vol. II: Physical, Chemical and Geological Studies, Allan Hancock Foundation, 477 pp., Univ. Southern California, Los Angeles, California.

Krumbein, W. C., and Pettijohn, F. P. (1938). "A Manual of Sedimentary Petrography." Appleton, New York.

Moore, D. G. (1969). Reflection profiling studies of the California continental borderland: structure and Quaternary Turbidite Basins, *Geol. Soc. Am.* Spec. Paper 107, 142 pp.

Petzold, T. H., and Austin, R. W. (1968). An underwater transmissometer for ocean survey work, *Tech Rep. Scripps Inst. Oceanogr.*, Ref. 68-9, 5 pp.

Postma, H. (1968). Suspended matter in the marine environment. *In* "Some Problems of Oceanology." Int. Oceanographic Congress 1966. pp. 258-265. State Publish. House, Moscow.

Southard, J. B., Young, R. A., Hollister, C. D. (1971). Experimental erosion of calcareous ooze. *J. Geophys. Res.* **76** 5903-5909.

Sylvester, A. G., Smith, S. W., and Scholz, C. H. (1970). Earthquake swarm in the Santa Barbara Channel, California, 1968. *Bull. Seismol. Soc. Amer.* **60**, 1047-1060.

Weiss, R. F. (1971). Flushing characteristics of oceanographic sampling bottles. *Deep-Sea Res.* **18** 653-656.

Wimberly, S. (1963). Sediments of the mainland shelf near Santa Barbara, California. *In* "Essays in Marine Geology in Honor of K. O. Emery" T. Clements (Editor), p. 191-201.Univ. Southern California Press.

Wimberly, S. (1964). Sediments of the southern California mainland shelf. *Thesis, Univ. Southern California, Los Angeles, California,* 207 pp. (unpublished).

CHAPTER 15

Comments on the Dispersal

of Suspended Sediment

Across the Continental Shelves

J. R. Schubel and Akira Okubo

Chesapeake Bay Institute
The Johns Hopkins University
Baltimore, Maryland 21218

ABSTRACT

Since most of the fine-grained inorganic deep-sea sediment is derived from the continents, it must have crossed the continental shelves to reach the ocean basins. Most of it crossed the shelves in suspension, but the routes and rates of transport are obscure. The bulk of the sediment now being discharged by rivers is trapped in the estuaries, but nearly all of the fraction that escapes apparently crosses the shelf. A simple steady-state, two-dimensional mathematical model applied to a section off the mouth of Chesapeake Bay supports the idea of bypassing of the shelf by fine-grained terrigenous sediment, and suggests that much of the fine-grained sediment that escapes the estuaries may be accumulating on the continental slope and rise. The results from the model also suggest that the cross-shelf flux of sediment associated with the mean (residual) flow can account for most of the mass of sediment that is estimated from erosion rates to reach the ocean basins.

INTRODUCTION

Since the bulk of inorganic deep sea sediment is derived from the continents, it must have crossed the continental shelf to reach the ocean basins. Most of this sediment is fine-grained clay which probably crossed the shelf while in suspension. Sediment suspended in the waters overlying the shelf is dispersed by the circulation of the shelf waters—both advection and diffusion.

Studies of the dispersal of suspended sediment require determinations both of the suspended sediment population—the concentration and particle size (settling velocity) distribution—and of the transporting agents. There is a dearth of information on both of these topics for the continental shelves of the world. There are scattered observations of the distributions of suspended sediment in the waters overlying some segments of the shelf (see articles and references in this volume), but the spatial and temporal variations of the suspended sediment populations are virtually unknown for all shelf areas of the World Ocean. The circulation of the shelf waters is just as poorly known. An indication of the level of our understanding can be gained from the highly speculative nature of McCave's article (this volume), or directly from Carter's (1969) description of the present state of knowledge of coastal circulation. The paucity of pertinent geological and physical data, and the lack of any sediment dispersal models, have precluded realistic estimates of the long-term average flux of suspended sediment across any segment of the continental shelf.

The objectives of this paper are to describe in very general terms some aspects of suspended sediment dispersal on the continental shelves, to formulate a simple mathematical model to estimate the distribution of suspended sediment in the waters overlying a section of the continental shelf, and to estimate the flux of suspended sediment across the shelf segment.

SUSPENDED SEDIMENT DISPERSAL
AND THE CONTINENTAL SHELVES

The sediment suspended in the waters overlying the continental shelves is derived in part directly from the continents by rivers, glaciers, wind, and shore erosion; in part from biological activity; and in part from the resuspension of bottom sediments. The principal ultimate source of the inorganic-suspended matter is the continents. The coast forms essentially a line source of terrigenous sediment which varies markedly in intensity, being greatest where the coastline is interrupted by large rivers such as the Mississippi and the Ganges-Brahmaputra. Each year the world's rivers discharge an estimated 18×10^3 metric tons of sediment (Holeman, 1968). Much of this sediment is entrapped in the estuaries of these rivers (Schubel, 1971), but substantial amounts of it escape to the sea. Part of the sediment that initially escapes the estuaries is carried back into them by the net upstream flow of the lower layer (Meade, 1969). Most of the fine-grained sediment that ultimately escapes the estuaries must be transported across the shelves and be deposited on the continental slope, continental rise, and deep sea floor since most of the shelves are covered with coarse-relict sediments. As McCave points out in his paper, muddy sediments do occur on some shelves, but the deposits are not extensive. Some of these deposits are relict, but other represent modern deposits of terrigenous sediment.

The general paucity of fine-grained sediments on the continental shelves is the result of a combination of factors—the trapping of fine-grained sediment in estuaries, the bypassing of the shelf by the fine-grained sediment that escapes the estuaries, and the relatively short time that sea level has been near its present position.

Fluctuations in sea level have played a major role not only in determining the distribution of bottom sediments on the shelf, but also in determining the flux of fine-grained terrigenous sediment to the ocean basins. During the glacial periods of the Pleistocene Epoch, sea level was lower and much of the continental shelf was exposed to the atmosphere. The rivers flowed across the shelf and debouched into the sea through narrow valleys incised into the outer shelf and upper slope. These features are preserved today as the heads of submarine canyons. During these periods of lowered sea level, large amounts of sediment were discharged directly into the ocean basins.

Once in the sea the fine sediment was dispersed by the prevailing coastal current systems that lay seaward of the edge of the shelf. In the northern hemisphere the average surface flow near shore would, as today, have been roughly parallel to the coast with the land to the right. The mechanism that drives this surface circulation is the riverflow. Because the water discharged by the estuaries is fresher and less dense than the offshore waters, a seaward density gradient is established which produces a circulation pattern with outflow at the surface and inflow in the deeper layers. Because of the Coriolis force the seaward-flowing surface current is deflected to the right in the northern hemisphere producing the observed surface currents parallel to the coast with the land on the right-hand side. In the southern hemisphere, of course, the surface currents are parallel to the coast with the land on the left. The coastal currents during periods of glacial advance were located over the outer shelf and upper slope and were in areas subject to runoff from glaciated regions, probably less intense than today because of the decreased river flow. Large amounts of sediment probably accumulated on the continental slope and rise during the glacial periods.

During the interglacial periods, physical conditions on the shelves were similar to those of today. The shelves were covered by sea water, and the coastal current system was displaced landward. Because of the increased fresh water inflow, the coastal circulation was probably intensified. As sea level rose the estuaries were progressively displaced landward to positions probably approximating those of today. Much of the fluvial sediment which had been discharged directly into the ocean basins during periods of lowered sea level was entrapped in the estuaries now located well up on the continental blocks. With the next fall in sea level the estuaries were flushed out, and much of the sdeiment which had accumulated in their basins was discharged into the ocean basins.

Sea level has been near its present position for the past several thousand years and probably most of the sediment discharged by rivers has been entrapped in the estuaries. The trapping of sediment by estuaries has been discussed by a number of investigators (Postma, 1967; Schubel, 1968, 1971; Meade, 1969). The trapping efficiency of an estuary is determined largely by its geometry and its net nontidal circulation pattern. Reliable estimates of the trapping efficiency, and of the net sediment discharge, are unavailable for any of the world's estuaries.

During glacial periods there was, because of decreased river flow, a tendency for estuaries to shift in their circulation toward the well-mixed estuarine type, and probably to be less efficient as sediment traps. Changes in the geometry of the estuarine basins, however, may have compensated for the changes in river flow.

An increase in depth or decrease in width would tend to move an estuary away from the well-mixed estuarine type, and in the direction of a river-dominated estuary. With the data available it is not possible to determine whether the decreased role of the river or changes in geometry had a greater effect on the estuarine circulation patterns during glacial periods. Changes in the position of the estuaries relative to the edge of the shelf were of greater significance in determining the contributions of fluvial sediment to the ocean basins.

The bypassing of the shelf by sediment has received only qualitative consideration (Curray, 1965). Curray suggested two mechanisms by which terrigenous sediments are bypassed across the shelf. According to Curray, near the shelf edge excessive turbulence from waves and tides prevents the deposition of fine sediments on the outer shelf, and leads to the formation of slope-facies muds. In other areas, sediments are temporarily stored in submarine canyons and then funneled to the deep sea floor. Longshore drifted sand is temporarily accumulated in the heads of submarine canyons which extend close to shore, while finer suspended sediment is presumably trapped by settling into the canyon farther out on the shelf. The transportation of these sands and muds to the continental rise and abyssal plain takes place, according to Curry, by turbidity currents, slow glacierlike flowage, and by current action.

Most of the fine-grained sediment that reaches the deep sea floor is probably carried in suspension across the shelf without being temporarily stored in submarine canyons. There are, however, no estimates or models for estimating the masses of fine-grained sediment that ''bypass'' any segment of the continental shelves of the world.

A MODEL FOR SUSPENDED SEDIMENT DISPERSAL

We have constructed a simple steady-state, two-dimensional model for predicting the long-term average distribution of suspended sediment in shelf waters.

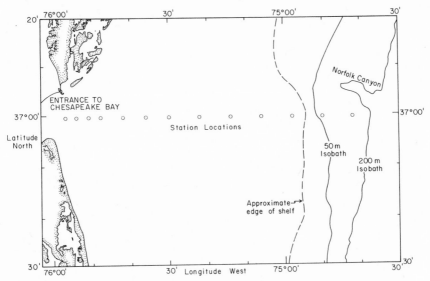

Figure 131.
Map of shelf area off Chesapeake Bay showing the station locations.

We have applied our model to a segment of the shelf off the mouth of Chesapeake Bay, Fig. 131. The model requires a knowledge of the mean (residual) flow, the horizontal and vertical eddy diffusivities, and the mean settling velocity of the suspended particles. The model is not designed to describe temporal variations in the distribution of suspended sediment, local features such as plumes of turbid water off promontories of the coast, or near-shore events.

Take the x-axis seaward from the coast and the z-axis vertically downward from the sea surface or naviface (Montgomery, 1969). Let u, w, w_s, A, and K be, respectively, the horizontal water velocity, the vertical water velocity, the mean settling velocity of the suspended particles, the horizontal eddy diffusivity and the vertical eddy diffusivity, Fig. 132. The transport-diffusion equation for suspended material with a concentration, S, can then be expressed as

$$\frac{\partial uS}{\partial x} + \frac{\partial wS}{\partial z} = \frac{\partial}{\partial x}\left(A\frac{\partial S}{\partial z}\right) + \frac{\partial}{\partial z}\left(K\frac{\partial S}{\partial z}\right) - \frac{\partial}{\partial z}(w_s S) \qquad (1)$$

(advection) (horizontal (vertical (settling)
              diffusion)   diffusion)

The boundary conditions are:

at $x = 0$ (shore), $S = F(z)$ (given source) $\qquad (2)$

at $z = 0$ (naviface), $K\dfrac{\partial S}{\partial z} - (w + w_s)S = 0$ (i.e., no sediment flux) $\qquad (3)$

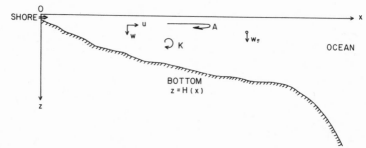

Figure 132.

Pictorial representation of the parameters used in calculating the distributions shown in Figs. 133 and 134.

at $z = H(x)$ (bottom), $K\dfrac{\partial S}{\partial z} - (w + w_s)S = G(x,t)$ (4)

where $G(x,t)$ represents the rate of deposition or resuspension of material at the bottom.

For the sake of analysis some simplifications are necessary. Since Eq. (1) is of second order with respect to x, a second boundary condition for S at $x = 0$, for example $\partial s/\partial x$ is required for solution of Eq. (1). The assignment of two boundary conditions for S is undesirable for modeling because of the increased mathematical complexity, and because of the uncertainty of our knowledge of the source terms. To avoid these difficulties, we define a virtual horizontal advection velocity, $\hat{u}$, for the flux of suspended sediment by

$$\frac{\partial}{\partial x}(\hat{u}S) \equiv \frac{\partial}{\partial x}(uS) - \frac{\partial}{\partial x}\left(A\frac{\partial s}{\partial x}\right)$$ (5)

The virtual advection velocity, $\hat{u}$, defined by Eq. (5) is then a composite velocity including both advection and diffusion.

We will ignore the vertical motion of water over the shelf except in the nearshore region where a two-layer circulation persists. This assumption appears to be justified because of the very gentle slope of the continental shelf. In the waters overlying the continental slope, however, there appears to be a significant vertical motion which may play an important role in slope and rise sedimentation.

We assume that $\hat{u}$, K, and w_s are constant, and that the water depth, H, is constant across the shelf.

After making the above assumptions Eq. (1) becomes

$$\hat{u}\frac{\partial S}{\partial x} = K\frac{\partial^2 S}{\partial z^2} - w_s\frac{\partial S}{\partial z}$$ (1′)

The exchange rate of sediment at the bottom $G(x,t)$ is critical in predicting the flux of suspended sediment across the shelf. This rate is poorly known, but since most of the shelf is covered with coarse-grained relict sediment and since very little fine-grained material is accumulating on most areas of the shelf, $G(x,t)$ must be very small. We will assume that $G = 0$, i.e., that there is neither net deposition nor net resuspension.

The solution of Eq. (1′), subject to the boundary conditions given by Eqs. (2), (3), (4), with $F(z) = 2\ m\ \delta(z)*$, $H = $ constant, and $G = 0$ is obtained by

$$\frac{S}{S_0} = \frac{D}{eD-1}\, e^{Dz'} + 4\pi e^{\frac{1}{2}Dz'} \sum_{j=1}^{\infty} \frac{j}{D^2 + 4j^2\pi^2}\, (D\,\sin j\pi z' + 2j\pi + \cos j\pi z').$$

$$e^{-(0.25D\,+\,j^2\pi^2 D^{-1})x'} \tag{6}$$

where $x' \equiv x/L\ \left(L \equiv \dfrac{\hat{u}H}{w_s}\right)$, $z' \equiv z/H$, $D \equiv \dfrac{Hw_s}{K}$. and $S_0 \equiv \dfrac{m}{H}$

D is a nondimensional parameter essential to the solution. It is a measure of the relative importance of the settling velocity, w_s, to a diffusion velocity K/H, in determining the vertical distribution of suspended sediment. S_0 is a measure of the "source concentration" of suspended sediment. It is the total mass of suspended sediment discharged per unit time divided by a volume equivalent to the volume of water swept out by a plane 1 cm wide and depth H moving with a speed $\hat{u}$. The x' and z' are nondimensional distance and depth, respectively, L is a measure of the horizontal distance from the source at which the characteristic settling time H/w_s equals the travel time, $L/\hat{u}$.

In the shelf waters between Cape Cod and Chesapeake Bay there is a small net transport of water from the shore to the ocean. This net seaward flow must exist because the supply of fresh water from rivers is approximately five times greater than the excess of evaporation over precipitation along this stretch of shelf (Ketchum and Keen, 1955). There is little direct information concerning the residual flow pattern. The few investigations that have been made indicate that there is a persistent offshore component of the surface flow (Ketchum and Keen, 1955; Bumpus, 1965; Bumpus and Lanzier, 1965). The data also indicate that near the bottom there appears to be a landward flow on approximately the inner third of the shelf, and a seaward flow across the remainder of the shelf. Estimates of the rate of the bottom flow, based on seabed drifters, range from about 0.4 to 1.0 cm sec⁻¹. Currents inferred from the movement of seabed drifters include both advection and diffusion. On the average throughout a column of water we may expect a long-term offshore advective flow of very small velocity, probably of the order of 0.1 cm sec⁻¹, or less, except on the inner shelf where there is a net landward flow.

*m is a measure of the total amount of suspended sediment discharged from land and has dimensions of $[M\,L^{-1}]$. The $\delta(z)$ is a Dirac delta function mathematically representing the situation that the source is located at $x = z = 0$. Note that $\int_0^H \delta(z)\, dz = 1/2$.

The appropriate value to assign to $\hat{u}$ is obscure. According to Carter (1969), the flushing time of fresh water from the continental shelf between Cape Cod and Chesapeake Bay can be interpreted in terms of a composite velocity, a velocity which includes both advection and diffusion. Carter's "diffusion velocity" had a net offshore value of 1-2 cm sec^{-1}. While $\hat{u}$ and the "diffusion velocity" are not equivalent in form, both are measures of the combined effects of advection and diffusion, and must at least be of the same order of magnitude.

The horizontal eddy diffusivity in the waters overlying the shelf between Cape Cod and Chesapeake Bay has been estimated from the seasonal changes in salinity by Ketchum and Keen (1955). Their values ranged from 0.58×10^6 to 4.96×10^6 cm^2 sec^{-1}.

The value of the vertical eddy diffusivity in shelf waters was estimated by Carter and Okubo (1965) from dye diffusion studies off Cape Kennedy, Florida. The vertical eddy diffusivity, unlike the horizontal eddy diffusivity, is strongly dependent on the stability of the water column, and Carter and Okubo (1965) found that the value ranged from 1.3 to 19 cm^2 sec^{-1}. A value of 10 cm^2 sec^{-1} appears to be a reasonable estimate of the vertical eddy diffusivity for shelf waters.

We use the following numerical values to compute S/S_0 from Eq. (6), $H = 50$ m, $\hat{u} = 2$ cm sec^{-1}, K = 10 cm^2 sec^{-1}, $w_s = 10^{-3}$ cm sec^{-1} and $w_s = 10^{-4}$ cm sec^{-1}. These values give two sets of values of D and L: $D = 1/2$, $L = 100$ km, and $D = 1/20$, $L = 1000$ km. The computation extends seaward from $x = 0$ to $x = 100$ km. The computations result in two nondimensional distribution patterns of the concentration of suspended matter, Figs. 133 and 134.

An actual distribution observed on May 15, 1971, along a section extending from the mouth of Chesapeake Bay to the edge of the shelf along 37°00'N is presented in Fig. 135. The agreement between the theoretical distribution patterns and the observed distribution pattern, Fig. 135, is generally quite good, particularly for the case where $D = 1/2$ which corresponds to an average particle settling velocity of about 10^{-3} cm sec^{-1}—a reasonable value for w_s according

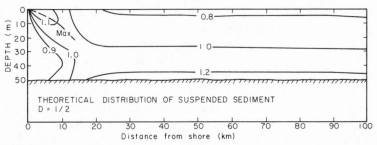

Figure 133.

Theoretical nondimensional distribution pattern of suspended matter along the section off the mouth of Chesapeake Bay for $D =$ one half and $L = 100$ km.

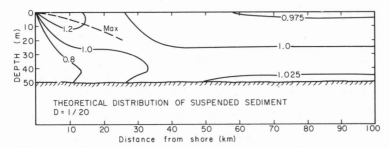

Figure 134.

Theoretical nondimensional distribution pattern of suspended matter along the section off the mouth of Chesapeake Bay for D = one twentieth and L = 1000 km.

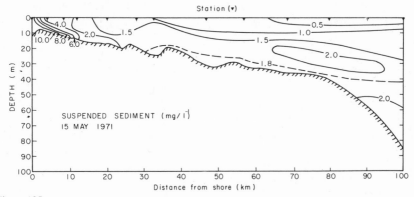

Figure 135.

Observed distribution of suspended matter (mg liter[-1]) along the section off the mouth of Chesapeake Bay on May 15, 1971 (see Fig. 131 for station locations).

to our investigations. To facilitate comparison, the nondimensional theoretical values of Fig. 133 have been converted to actual concentrations, Fig. 136. To make the conversion we chose an S_0 such that the computed values of the concentration of suspended sediment, averaged over depth, matched the observed values presented in Fig. 135, at the outer three stations on the shelf. The appropriate value of S_0 was 1.5×10^{-6} g cm^{-3}. The choice of S_0 does not, of course, alter the calculated distribution pattern.

Both the model and the observed distribution show that at some distance from the coast the suspended matter is distributed relatively uniformly with depth. Farther seaward the concentrations increase with depth as expected. The distribution pattern observed on May 15, 1971, is representative of this segment of the shelf, but the concentrations on the outer shelf are somewhat higher than those normally observed. The maximum centered at approximately 80 km from shore at a depth of about 20-30 m is unexplained. According to Meade (personal communication) during a period in 1969 when high concentrations of suspended inorganic sediment were expected, the values observed on the inner shelf were similar to those reported in Fig. 135, but the concentrations on the outer shelf were only one third to one fifth those depicted in Fig. 135. There is no reason to believe that these do not represent real differences. Five other cruises have been made over the shelf from about 37°30'N to 35°25'N, nearly to Cape Hatteras. The density (σ_t) distribution on May 15, 1971, is shown in Fig. 137.

The observed values of the vertically averaged concentration of suspended sediment on May 15, 1971, were 7.8, 2.0, 1.7, 1.4, and 1.6 mg liter[-1] at 5, 20, 35, 68, and 90 km from the coast. Seaward of about 30 km from the coast the vertically averaged value of the concentration of suspended matter remains remarkably constant, ranging only from 1.4 to 1.7 mg liter[-1]. This indicates that there is essentially no net deposition or resuspension of fine-grained material

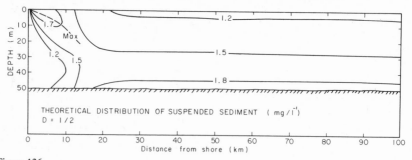

Figure 136.

Calculated distribution of suspended matter (mg liter[-1]/ along the section off the mouth of Chesapeake Bay for D = one half and L = 100 km.

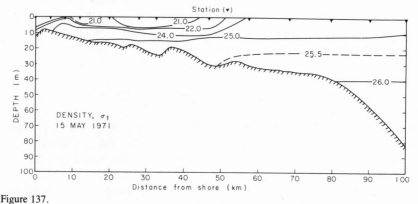

Figure 137.

Distribution of σ_t along the section off the mouth of Chesapeake Bay on May 15, 1971 (see Fig. 131 for station locations).

on the shelf seaward of this distance. This supports our assumption that $G(x,t) \approx 0$, is consistent with the observed distribution of bottom sediments, and supports the concept of bypassing. Near the coast, the persistent two-layered circulation pattern results in a transport of fine-grained suspended sediment landward in the lower layer. Some of this sediment is carried back into the estuaries (Meade, 1969; Schubel, 1971).

We can use the data from Fig. 135 to estimate the flux of sediment across the shelf. Since we assumed that there is no flux of sediment either at $z = 0$ or at $z = H$, the seaward transport of suspended material q, is constant through any vertical section (unit width times depth H) and is equal to

$$q = \int_0^H S\hat{u}\,dz = m\hat{u} \tag{7}$$

where q has the dimensions of $[MT^{-1}L^{-1}]$. The data from Figs. 135 and 136 suggest that the seaward transport of suspended matter through vertical sections seaward of about 30 km from shore is essentially the same. The flux of suspended matter per unit width is given by $q = S\hat{u}H$. Based on the data from May 15, 1971, we have: $q = S\hat{u}H = (1.5 \times 10^{-6}\ \text{g cm}^{-3})\,(2\ \text{cm sec}^{-1})\,(5 \times 10^3\ \text{cm}) = 1.5 \times 10^{-2}\ \text{g sec}^{-1}\ \text{cm}^{-1}$.

It is an amusing, and perhaps informative, exercise to assume that the distribution of suspended sediment observed on May 15, 1971, at $37°00'$N is representative of the waters overlying the world's continental shelves and to use these data to estimate the total net transport of suspended material from the continents to the ocean basins. This estimate can be made by multipying q by the total length of shelf edge, and yields a value of about 2×10^{10} metric tons year[-1] for the net transport of fine-grained suspended matter across the shelves from the continents to the ocean basins. Probably at least 75% of this is combustible organic matter. Our value is not far from estimates of the amount of material assumed to reach the ocean based on worldwide erosion rates. Holeman (1968), e.g., estimated that approximately 1.8×10^{10} tons of suspended sediment reach the ocean each year, while Judson (1968) put the value at about 0.9×10^{10} tons year[-1].

Our model gives a value of about 1.2×10^8 tons year[-1], at least 75% of which is combustible organic matter for the mass of fine-grained suspended matter transported across the continental shelf between Georges Bank and Miami. This value appears to be too high. Dole and Stabler (1909) estimated that approximately 25×10^6 tons of sediment are discharged each year by the rivers along the Atlantic coast of the United States. Probably a large fraction of this is entrapped in the estuaries of these rivers. Our estimated value for the flux of inorganic sediment across the shelf is probably 5-8 times that which might be expected from Dole and Stabler's data. Since our observed concentrations were perhaps 3-5 times greater than the "average" concentrations on this section of shelf (Robert Meade, personal communication, 1972), our estimated sediment flux may be high by this same amount. Some of the material in suspension may also be due to resuspensions which could account for our higher values. The choice of $\hat{u}$ is also important in estimating the flux of suspended sediment across the shelf. Our value of $\hat{u}$ may be high by a factor of 2, but it is unlikely in view of the present evidence that it is off by more than this.

Much of the sediment that bypasses the shelf is probably deposited on the continental slope and rise. We can investigate the sedimentation rate in these provinces by integrating the basic Eq. (1), after using Eq. (5), over z from $z = 0$ to $z = H(x)$. Assume that there is no water flux through $z = 0$, the naviface, or through $z = H(x)$, the bottom, but that there is an unknown flux of sediment at the bottom $H(x)$. We obtain

$$\hat{u}\frac{d}{dx}\int_0^H S dz = \left.\left(K\frac{\partial S}{\partial z} - w_s S\right)\right|_{z = H(x)} \tag{8}$$

In order for the deposition of suspended sediment to occur, $w_s S$ must be larger than $K\,\partial S/\partial z$. The upper limit corresponds to the case where the diffusive flux of sediment vanishes, owing probably to the ineffectiveness of the wind and tidal actions as stirring mechanisms on the slope. If this is the case, then

$$\hat{u}\frac{d}{dx}\int_0^H S dz = -w_s(S)_H \tag{8'}$$

Let

$$\left.\begin{aligned}\int_0^H S dz \equiv H\bar{S} \\[2em] (S)_H \equiv \beta\,\bar{S}\end{aligned}\right\} \tag{9}$$

where the overbar denotes the depth averaging and β is a numerical parameter depending upon the vertical profiles of $\hat{u}$ and S. Substituting Eq. (9) into Eq. (8') we obtain

$$\hat{u}\frac{d}{dx}(H\bar{S}) = -\beta w_s\bar{S} = -\frac{\beta w_s}{H}(H\bar{S}) \tag{10}$$

The integration of Eq. (10) over x from $x = x_0$ (the outer-edge of the continental shelf) to $x = x$ results in

$$M = M_0 \exp\left\{-\int_0^x \frac{\beta w_s}{\hat{u}H} dx\right\} \tag{11}$$

where M is the total amount of suspended sediment in a vertical column of water ($M = \overline{S}H$) and the subscript zero denotes the value at the edge of the shelf. The variation of water depth over the continental slope can be represented by an exponential function, $H = H_0 \exp(x/l)$, where l, a measure of the horizontal separation of the depth contours, is a relative measure of the slope. Equation (11) is then expressed by

$$M \sim M_0 \exp \left\{ -\frac{w_s l}{\hat{u} H_0} (1 - e^{-x/l}) \right\} \tag{12}$$

assuming that $\beta = 0(1)$. Taking $H_0 = 50$ m and $1 = 100$ km, we can see that an appreciable deposition of particles with a settling velocity of the order of 10^{-3} cm sec^{-1}, or larger, takes place within a distance of the order of 100 km from the shelf edge. Finer particles with a settling velocity of the order of 10^{-4} cm sec^{-1} travel much greater distances and may require a distance of the order of 1000 km before appreciable deposition occurs.

CONCLUSIONS

The results of the simple steady-state, two-dimensional mathematical model described in this paper support the concept of the bypassing of the shelf by fine-grained terrigenous sediment. The results also suggest that the cross-shelf flux of sediment associated with the mean (residual) flow can account for most of the mass of sediment that is estimated from erosion rates to reach the ocean basins. Refinements of this model or development of more sophisticated ones must await more detailed information on the mean flow of the shelf waters, the suspended sediment source terms, the settling velocities and concentration of the suspended particles, and the exchange rate $G(x,t)$ at the bottom. The occasional surveys of the distribution of temperature and salinity in coastal waters are of limited value when applied to the classical theories of physical oceanography because of periodic variations in temperature and salinity produced by tidal currents and internal waves, and variations produced by wind-driven currents. Continuous observations of the important parameters over periods of greater than a year are badly needed at a number of fixed points across the shelf. These physical data should be supplemented with detailed observations of the suspended matter to delimit the spatial and temporal variations of the important characteristic properties, particularly the settling velocities of the particles and their concentration. In addition, the exchange rate of sediment at the bottom, $G(x,t)$, must be determined both on the shelf and on the slope for refinement of any model of sediment dispersal. This could be determined from time series observations of the concentration of suspended sediment at several levels very close to and fixed with reference to the bottom.

ACKNOWLEDGMENTS

Contribution 175 from the Chesapeake Bay Institute of The Johns Hopkins University. We are indebted to W. B. Cronin, T. W. Kana, and C. H. Morrow for collection and analysis of the samples of suspended sediment. This work was supported in part by the Oceanography Section, National Science Foundation, NSF Grant GA 28276; in part by the Fish and Wildlife Administration of the State of Maryland and the Fish and Wildlife Service, Bureau of Sport Fisheries and Wildlife, Department of the Interior through Dingell-Johnson Funds, Project F-21-1; and in part by National Science Foundation, NSF Grant GA 16603 (Geophysical Fluid Mechanics).

REFERENCES

Bumpus, D. F. (1965). Residual drift along the bottom on the continental shelf in the Middle Atlantic Bight Area. *Limnology and Oceanography* **10**, [Suppl.] R50-R53.

Bumpus, D. F., and Lauzior, L. M. (1965). Surface circulation on the continental shelf off eastern North America between Newfoundland and Florida. *In* "Serial Atlas of the Marine Environment." Folio 7. 4 P. 8 pl. (Appendix iii), Am. Geograph. Soc., New York.

Carter, H. H. (1969). Physical processes in coastal waters. *In* "Background Papers on Coastal Wastes Management," Vol. 1, p VI-V16, Natl. Academy of Engineering, Washington, D.C.

Carter, H. H., and Okubo, A. (1965). A study of the physical processes of movement and dispersion in the Cape Kennedy area. "Final Report under the U.S. Atomic Energy Commission," Report No. NYO-2973-1, Chesapeake Bay Institute, Johns Hopkins Press, Baltimore, Maryland.

Curray, J. R. (1965). Late Quaternary history, continental shelves of the United States. *In* "The Quaternary of the United States" (H. E. Wright and D. G. Frey, eds.), p. 723-735, Princeton Univ. Press, Princeton, New Jersey.

Dole, R. B., and Stabler, H. (1909). Denudation. *In* "Papers on the Conservation of Water Resources," p. 78-93. U.S. Geol. Survey Water Paper 234.

Holeman, J. N. (1968). The sediment yield of major rivers of the world. *Water Resources Res.* **4**(4), 737-747.

Judson, S. (1968). Erosion of the land—or what's happening to our continents. *Am. Scientist* **56**, 356-374.

Ketchum, B. H., and Keen, D. J. (1955). The accumulation of river water over the continental shelf between Cape Cod and Chesapeake Bay. Papers in Marine Biology and Oceanography. *Deep-Sea Research* **3** [Suppl.], 346-357.

Meade, R. H. (1969). Landward transport of bottom sediment in estuaries of the Atlantic Coastal Plain. *Jour. Sed. Pet.* **39**[1], 222-234.

Montgomery, R. B. (1969). The words naviface and oxyty. *J. Mar. Res.* **27**, 161-162.

Postma, H. (1967). Sediment transport and sedimentation in the estuarine environment. *In* "Estuaries" (G. Lauff, ed.), p. 158-179. Am. Assoc. Adv. Sci., Washington, D.C.

Schubel, J. R. (1968). Turbidity maximum of northern Chesapeake Bay. *Science* **161**, 1013-1015.

Schubel, J. R. (1971). Estuarine circulation and sedimentation. *In* "The Estuarine Environment: Estuaries and Estuarine Sedimentation" (J. R. Schubel, ed.), Short Course Lecture Notes, p. VI1-VI17. Am. Geol. Inst., Washington.

III. Patterns of Coarse Sediment Dispersal

CHAPTER 16

Some Specific Problems in Understanding Bottom Sediment Distribution and Dispersal on the Continental Shelf*

Joe S. Creager and Richard W. Sternberg

Department of Oceanography
University of Washington
Seattle, Washington 98195

ABSTRACT

Geologists involved in research into the problems of continental-shelf-sediment classification, distribution, and dispersal mechanisms must disengage themselves from the simplistic study of sediment distribution alone as a desired goal. The dispersal mechanisms which have heretofore been intuitively deduced must now be observed and measured. Study of the processes rather than the results must be our immediate goal.

To reinforce this suggestion an example is presented to point out how intuition has led to a general misunderstanding of the sediment types present on Arctic continental shelves which has led to a second-order misinterpretation of dispersal mechanisms deduced from the generality that Arctic continental shelves are covered by glacial sediment. A second example is given of the problems encountered in attempting to apply a wave-surge dominated, shelf sediment dispersal model to the Washington continental shelf with no success. The results of this attempt suggest that the model is not only locally limited in usefulness but may be basically wrong.

*Contribution No. 669, Department of Oceanography, University of Washington, Seattle.

INTRODUCTION

Dispersal is defined in Webster's Third New International Dictionary as "the process *or* result of spreading by migration." Actually, marine geologists, in their quest for the sedimentary history, present condition, and future development of continental shelves, must concern themselves with both processes and results of "sediment spreading." Our knowledge of bottom sediment dispersal mechanisms comes primarily from inference as a by-product of studies of shelf sediment distribution. Thus, it is virtually impossible to separate the processes from the results while attempting to unravel the sedimentary framework and history of a continental shelf area.

This paper is not meant to be an exhaustive exposition on continental shelf bottom sediment dispersal but rather, an overview of some present difficult problems that are encountered when trying to understand continental-shelf sedimentation.

Historically it is common among geologists to deal with classifications and distribution of sediments first and then interpret dispersal mechanisms from their findings. For many reasons this is how it had to be. Geologists are presently in a phase of evolving working hypotheses of shelf sediment dispersal mechanisms based almost exclusively upon reasoning. Furthermore, the proposed mechanisms have not been directly measured. Instead, generalities about sediment distribution are used to test the hypothetical dispersal models. This approach may be establishing a set of guidelines that lead nowhere. Enough weaknesses have slowly crept into our views to produce a euphoria which has led to our acceptance of working hypotheses as real and applicable to the world's shelves in general when in fact they may be only locally correct or insignificant. We should no longer accept overly generalized or ambiguous classification schemes, data on sediment distribution from a minority of mid-latitude shelves as an indication of worldwide and latitudinally significant variations, and reasoning alone instead of field measurements as our sole knowledge of dispersal mechanisms.

From our own personal experience we have found that the recent intensity of study of continental-shelf sediments has led to a number of significant overstatements of what is really known. We would like to bring to your attention a few examples of the interrelated difficulties which inherently arise when adequate data are not available before working hypotheses are established. Space dictates that we confine ourselves to a few of the more comprehensive and widely read papers on distributional variation with latitude and the sedimentary grading of continental shelves. These problems recently have been best articulated in the United States by J. R. Curray, K. O. Emery, M. O. Hayes, D. G. Moore, and D. J. P. Swift, and in the Soviet Union by A. P. Lisitsyn.

CLASSIFICATION OF SHELF SEDIMENTS

At the center of many studies into the nature of shelf sediments and sedimentation are problems of categorization. Basic to all problems of dispersal are the difficulties and ambiguities of size analysis including both the methods and size classes chosen. We do not mean to digress into these problems but merely to restate that they exist and that intercomparisons with non-United States workers should be approached with caution. Our ideas of sediment classification by origin are materially influenced by size. We are all familiar with the idea that grain size should decrease offshore, and when it does not we infer something about the origin of the anomalous size.

Swift et al. (1971) give a good review of the workings of the classification scheme proposed by Emery (1952) which is used extensively in the United States. This classification categorizes sediments as authigenic (precipitated *in-situ*), biogenic (mainly calcareous), residual (produced by *in-situ* weathering of bedrock), detrital (modern sediments transported by water, ice, or wind), and relict (remanent from an earlier, different environment). In use this scheme has been difficult to apply for it is a mixture of time and nontime related terms. A biogenic sediment, for instance, is classified on the basis of origin (from its composition) and yet conceivably it could be either modern or relict. We then should have a modern or relict modifier to each of the above five sediment classes rather than a relict category as a class in itself. Additionally, there has been a significant class of sediments which cannot be easily placed in this scheme. Where do we class a basal transgressive sand (relict) that is being mixed today with silts and fine sands (modern)? We have been faced with this problem in interpreting the shelf sediments of the Chukchi and Bering Seas (McManus et al., 1969). Swift et al. (1971, p. 343) suggest the term "palimpsest" for a sediment which "exhibits petrographic attributes of a later environment" (i.e., a mixed sediment). We recommend this term's acceptance.

The term palimpsest, because it refers to a mixed sediment, requires knowing something of the dispersal mechanisms associated with emplacement of one component or lack of removal of the other component. All too often, however, the dispersal mechanism is interpreted entirely on the basis of the sediment nature after its identifiable characteristics have been ground into the classification scheme. A better system would involve feedback through some knowledge of the transport mechanics available either from theory or field observation. This unfortunately is seldom done with the result being that we are creating a body of deceiving information. This will be discussed in more detail later.

SHELF SEDIMENT DISTRIBUTION

Knowledge of the nature and distribution of shelf sediments is required before we can understand the historical, environmental, and process aspects of shelf sedimentation. Many researchers have recognized the most serious problem which Hayes (1967, p. 130) states as "whatever the precise relative importance of Pleistocene and Holocene processes in continental shelf sedimentation may be, it is obvious that relict Pleistocene sediments present an ominous problem for workers trying to understand present-day sedimentation problems."

As an example, the problems of latitudinal (or climate) variation of shelf sediment have recently been discussed by Lisitsyn (1966), Niino and Emery (1966), Hayes (1967), Emery (1968) and McManus (1970) among others. Basic to this problem is the sorting out of relict, modern, and palimpsest sediments. Ideally, through appropriate observation, geologists could identify modern and palimpsest sediments leaving the remainder to be relict. This will take many years of " 'shelf watch' on a large scale, by correlating hydraulic regime with changing bottom sediment parameters over long periods of time" (Swift, 1970, p. 27). Alternatively, geologists have applied petrographic, structural, and biogenic criteria (Emery, 1968; Swift et al., 1971) to recognize the various kinds of sediment. From this approach, generalizations have been made about the history of shelf morphology and sediments.

Emery (1968) summarized the distribution of relict sediments on the world's shelves with the stated intention of providing "a note of caution against the uncritical relating of sediments of the shelves to present-day water environments" (p. 445). In his attempt to provide a strong note of caution it appears that he may have overemphasized the distribution of exposed relict sediment and misinterpreted some sediment as relict. In numerous places throughout the paper, Emery (1968) states that relict sediments "exposed" and "unburied" at the shelf surface are the ones to which he is referring when he states that 70% of the area of the world's shelves consist of exposed relict sediment. This is based upon a study of some 21% of all shelf surfaces, the vast majority of which are located in temperate northern lattudes. Two important points stand out. First, throughout the paper, relict sediment is identified as exposed or thinly covered (up to 50 cm in the Barents Sea) by modern sediments. Second, high latitude sediments are referred to as glacial and, therefore, relict. From a very rough estimation of the areas covered by modern and relict sediment from the figures in Emery (1968), we estimate a different ratio (Table XXV). These ratios do not suggest to us that 70% of the world's shelves are covered by relict sediment although we have not planimetered the areas involved to normalize the coverage.

Before we began our studies of Arctic shelf sediments in 1959, we considered that all Arctic shelves had been glaciated and, hence, were covered with relict glacial debris. We have found this not to be the case. We know of no documented source which confirms that sizable areas of the total continental shelf of the Arctic (excluding possibly the Canadian Archipelago) were glaciated during the Wisconsin. This misconception possibly stems from the reporting of what appear to be ice-rafted pebbles in Arctic shelf sediments. Lisitsyn (1966) emphasizes sea-ice rafting along the northern and western margins of the Bering Sea. He interprets the gravel fraction as modern and being transported today. This may be, but comparison with shallow subbottom reflection profiles (Grim and McManus, 1970; Kummer and Creager, 1971) indicates that the Lisitsyn

Table XXV
Area of Shelves Covered by Modern and Relict Sediment[a]

	Percent	
Area	Modern	Relict
East Asiatic shelf	50	50
Bering Sea[b]	40	60
Southern California	80	20
U. S. Gulf of Mexico	40	60
U. S. Atlantic	20	80
Northeast South America	50	50
Nigeria	70	30
Barents and Kara Seas[c]	100	0
Western Europe	10	90

[a] As estimated by the authors from Emery (1968).

[b] Based upon radiocarbon dates, we have found that the modern belt of sediments around St. Matthew Island is much more extensive.

[c] Returning the 50 cm of modern sediment discounted by Emery. The modern nature of the entire Kara Sea sediment is supported by Gorshkova (1957).

gravel distribution correlates with the interpreted seaward extent of glacial ice suggesting a relict interpretation. Even if the relict interpretation is correct, this gravel does not cover a large area.

Another example of misconception is more obscure. Hayes (1967, p. 124) states: "The discovery of numerous local concentrations of pebbles in Holocene shelf sediments of the southeastern Chukchi Sea by Moore (1964) and others, which are attributed to ice-rafting, stresses the importance of this sediment-transporting mechanism in polar climates." Quoting Moore (1964, p. 339) we see that "local concentrations of ice-rafted pebbles and coarse sand deposits" . . . are "quite prevalent in the southeastern Chukchi Sea and suggests that a high level of sedimentation by ice-rafting in this region has been more or less continuous since the last sea-level rise. The surface sediments are also known to be cobble- and gravel-spattered and to have local concentrations of gravel and cobbles within the normal shelf sediments, which results in very poor size sorting" (Dietz et al., 1964). Dietz et al. (1964, p. 248) state: "The transport of detritus by ice rafting is clearly apparent. Many of the muds and sandy muds contain ice-rafted erratics in the granule, pebble, and cobble sizes." They also state that " . . . mud and mud-and-sand are predominant bottom covering on the Chukchi shelf." The importance of ice-rafting has increased with each successive quote above. Inspection of Fig. 12.2 in Dietz et al. and their text indicates that all but one of their reports of significant rock or gravel in the bottom sediments between Bering Strait and the Cape Lisburne-Herald Shoal Sill are directly associated with presumed bedrock outcrops or relict gravel beaches in the Strait, off cliffs, or on shoals (McManus et al., 1969). Emery (1968, p. 447) states that "At high latitudes are broad areas of ice-contributed detrital sediments having the same characteristically varied composition as glacial till on land." Figure 2 of Emery (1968) classes these sediments as glacial and in equilibrium with their environment. All of the above ideas taken together are incongruous and confusing to us for glacial refers to land ice which is not and probably has not been significant over the Arctic shelves since Illinoian time. Therefore, glacial sediment could not be the equilibrium sediment in the Arctic.

The cumulative end point seems to lead to the generality that gravel and pebbles (sediments with a "tilllike" composition) are a significant part of Arctic shelf sediments, hence sea-ice rafting and/or glaciation are important processes, hence the sediments are relict if glacial and modern if sea-ice rafted.

Our studies in the Laptev, East Siberian, and Chukchi Seas indicate that ice-rafting has not been a significant dispersal mechanism and the age of the sediment indicates its modern origin. Unpublished radiocarbon dates on surface sediments at eight locations in the Chukchi Sea range in age from 100 to 3400 yr BP. The sediments dated were the upper 25 cm of cores so the dates represent some span of time, but can certainly be classed as modern. Two dates from the Laptev Sea, 6000 and 8000 years BP, may suggest relict. Four dates, 2300-2800 yr BP, suggest modern and one date, 7100 yr BP, suggests relict for the surface sediments of the Bering Sea.

Over 600 bottom samples from the southern Chukchi Sea have been studied. Repeatedly we have reported the paucity of gravel in this area (Creager, 1963; Creager and McManus, 1966, 1967; McManus et al., 1969). We have examined more than 1000 samples from over 100 cores from the Chukchi Sea including a number along the profile collected by Moore (1964). The oldest samples date well before the last maximum sea level lowering. Excluding the cores near cliffs on submerged exposed bedrock, only 13% of the samples contain any material coarser than sand size. Further, only 4% of the samples contain material coarser than -1ϕ or any gravel-size particles in quantities as great as 1% by weight. Based on these percentages it would be questionable to conclude that the gravel fractions had been glacially transported and even if they were sea-ice rafted, we would not consider this mode of dispersal as anywhere near dominant. Similar lack of significant ice rafting has been reported for the Kara Sea (Gorshkova, 1957), the Laptev Sea (Holmes, 1967; Holmes and Creager, in press), and the East Siberian Sea (Naugler, 1967; Naugler et al., in press). The same series of papers all interpret the dominant sediment present as modern over this large combined shelf area. The importance of sea ice in transporting the silt which is predominant in the Arctic is still an unanswered and difficult question.

We suggest that considerably less than half of the known shelf areas are covered by exposed relict sediment. We agree completely with Emery's (1968, p. 260) statement that, "The present is the key to the past, but sedimentologists must be sure that the sediments of the shelves that they wish to relate to known environmental conditions really are present-day sediments that were deposited in the existing conditions." We further suggest that we all more critically review our own and others' works before we accept them or generalize from generalizations.

A step beyond the above discussion involves interpreting latitudinal or climatic variations in shelf sediments. We find that in northern polar regions the shelf sediment does not coarsen to gravel, become glacial and relict as variously presented by Niino and Emery (1966), Hayes (1967), and Emery (1968). McManus (1970) summarizes the criteria of climatic changes in the inorganic components of marine sediments and corrects a number of the problems which have disturbed us here. Niino and Emery (1966), Emery (1968), and McManus (1970) all reference Lisitsyn (1966) but fail in their discussions of latitudinal variation to include a significant discussion on pages 341-345. Figure 138 is from Lisitsyn (1966) and summarizes his interpretation of latitudinal variations throughout the oceans. It is strikingly applicable to shelves and is similar to other schemes except that it extends over a greater range of latitudes and appears to be more correct for the higher latitudes. When perusing this figure, keep in mind that by coarse fragmental a Soviet scientist means 0ϕ and coarser, sand is 1ϕ to 3ϕ inclusive, and that pelitic material includes 8ϕ and finer. Lisitsyn also suggests the significant importance of coarse fragmental ice-transported material in glacial (polar) zones. He is greatly influenced by his studies in the western Bering Sea where there is considerable coarse material. There is a great deal of coincidence between his map of rock outcrops (Fig. 72, p. 221) and his map of rock material (Fig. 92, p. 275) and the extent of disturbed subbottom reflectors reported by Kummer and Creager (1971). Lisitsyn's ice-rafted material may be relict glacial deposits. In spite of this bias, he still gives coarse fragmental and pelitic material equal importance and shows sand and silt as being twice as important.

SHELF SEDIMENT DISPERSAL

Without adequate detailed studies of both sediment distribution and water circulation over an entire shelf, we are left to intuition and deduction about sediment dispersal.

The most recent models for continental shelf sedimentation have been provided by Dietz (1963), Moore and Curray (1964), Curray (1965), Swift (1970), and Swift et al. (1971). In common they each include a basal transgressive sequence or outcropping rock as the beginning of the sedimentation model. Simplistically,

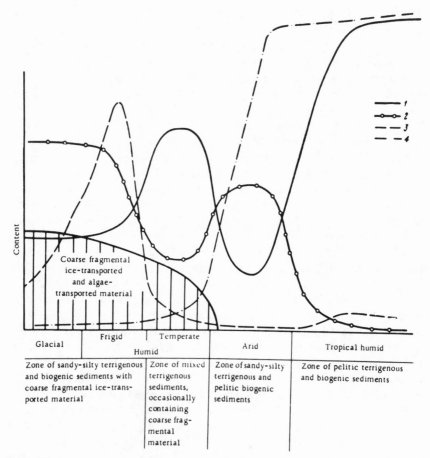

Figure 138.
Granulometric composition of terrigenous and biogenic material as a function of climatic zonality. Influx of terrigenous material: (1) yield of pelitic fraction from the weathering crust in different climatic zones and (2) yield of sand and silt fractions from the weathering crust. Influx of biogenic material: (3) influx of pelitomorphic $CaCO_3$ into abyssal sediments and (4) influx of silt size opal into the deepwater sediments related to development of diatomaceous phytoplankton. The hachured area represents the region of occurrence of coarse fragmental material in deepwater sediments in the zone affected by ice and large algae. Reproduced from Lisitsyn (1966, p. 343; in translation).

the basal sequence is a nearshore sand that has been spread across the shelf as sea level rose during the Holocene transgression. After sea level reached its present level approximately 5000 to 7000 years ago, a modern nearshore sand facies is produced and the finer-grained suspended fraction is transported seaward in sufficient quantity to result in a modern seaward prograding mud blanket or shelf facies. The exposed basal sand beyond the shelf facies will

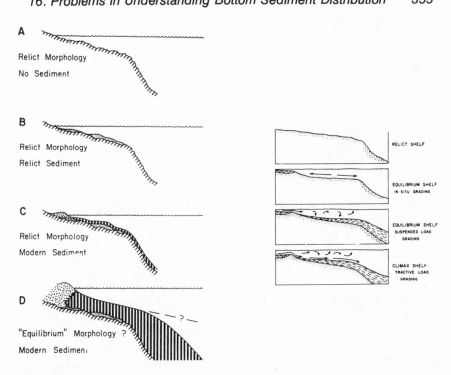

Figure 139.

Left: A is a diagrammatic representation of relict (inherited from previous environmental conditions) versus equilibrium sediment and shelf morphology; B, C, and D show a hypothetical sequence, with relict topography and basal transgressive sediments being covered with sediments in equilibrium with the new environment. Eustatic stability and gradual regional subsidence assumed. Reproduced from Curray (1965, p. 726). Right: Hypothetical evolution of a graded shelf. Reproduced from Swift (1970, p. 26).

either be relict or palimpsest. With time, estuaries and lagoons are filled and sands begin to prograde seaward. The Curray (1965) and Swift (1970) models (Fig. 139) differ only in the extent and amount of seaward progradation of this sand layer.

Curray (1965) discusses dispersal mechanisms for the suspended sediments in only general terms and considers bed load transport as being insignificant in water depths greater than about 10 m. He suggests (p. 727) that variations in bottom current intensity (associated with waves, tides, and permanent oceanic currents) with depth of water and distance from shore should control the transport and deposition of the suspended sediments producing the shelf facies.

Swift (1970) and Swift et al. (1971) discuss dispersal mechanisms in considerably more detail recognizing a wide variety of shelf currents (Fig. 140). Early in each of these papers the relative importance of the various classes of shelf

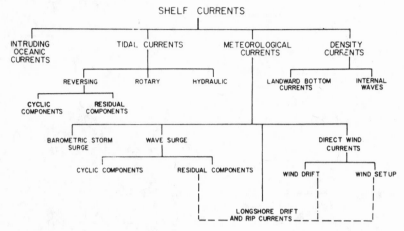

Figure 140.

Components of shelf velocity field. Reproduced from Swift et al. (1971, p. 324).

currents is weighed with the conclusion that important differences exist between the shore face or nearshore zone as opposed to the shelf zone. Oscillatory and mass transport currents related to wind waves; compensatory currents (i.e., longshore currents, rip currents, mid-depth return flow); and fresh-water currents and associated compensatory currents (shoreward bottom flow in response to seaward surface flow of land runoff) are recognized as important components of the nearshore dispersal system. This is generally accepted and differences of opinion are centered on effective offshore distance of this zone.

Dispersal systems on the shelf seaward of this nearshore zone are the ones least understood and are, as both papers point out, the ones which should receive most attention by interested investigators. We would like, with the indulgence of D. J. P. Swift, D. J. Stanley, and J. R. Curray, to point out what we think are problems in their discussions of relative importance of various currents in the shelf zone in considering sediment dispersal. Some of the problems may in part be generated by our lack of understanding of parts of their papers; if so, clarification would be extremely useful. We wish to make it clear that we agree that it is the so-called "rare" event that is probably the significant dispersal agent on many shelves and that these " 'rare' events summed over long periods of time result in a steady-state process; thus, many apparently gradual geological processes have a catastrophic fine structure." (Swift et al, 1971, p. 325). This, as all such generalities, should be applied with caution. For example, the predominant sediment dispersal mechanism in the southern half of the Chukchi Sea is a permanent current system driven by a difference in elevation between the Bering Sea and Arctic Ocean (Creager and McManus,

1967; Coachman and Tripp, 1970) whereas in numerous other areas the predominant dispersal mechanism is the tide current (e.g., Belderson and Stride, 1966). These examples do not exclude the possibility that some "rare" event might be of significance in dispersing sediment; however, they emphasize the point that in some shelf areas tides and currents play an important role as dispersal mechanisms.

Referring to Fig. 140, Swift (1970) and Swift et al. (1971) separate shelf currents into four major categories of which Meteorological Currents is considered as dominant. Meteorological Currents include barometric storm surge, wave surge, and direct wind currents. Wave surge consists of an oscillatory component and a residual or "wave-drift" component. Direct wind currents consist of wind set up and wind drift.

There is need for clarification as to what is meant by the terms "wind-drift" current and "wave-drift" current because they appear to represent a source of confusion in the literature and because recent studies suggest that direct wind currents are more important than these papers recognize. Wave-drift currents are the residual (or net) currents associated with the passage of an oscillatory water wave that has mass transport in the direction of wave propagation. A wind-drift current is one produced by direct wind stress on the sea surface. This produces an integrated transport, in the Ekman Layer, directed 90° to the right of the direction in which the wind blows (in the Northern Hemisphere) if the water is deep enough to permit the layer to develop fully. In water depths less than that required for full development of the Ekman Layer, the integrated transport direction shifts toward the direction that the wind is blowing. Over the continental shelf wind-drift transport may cause surface water to be driven onshore or offshore depending upon the orientation of the wind and coast, producing convergence or divergence zones and leading to barotropic or baroclinic compensatory currents below the surface. These should also be part of the wind-drift current system but are not mentioned in either Swift (1970) or Swift et al. (1971). Instead, in Swift's discussion of shelf currents (1970, p. 12), the only mention of wind-drift currents is to state that, "Direct wind drag would also be important," and to quote at length from an elementary text whose authors present a very confusing discussion of drift currents.

In Swift et al. (1971) the main discussion on the shelf hydraulic regime (p. 324-325) considers oscillatory waves (both wind waves and tide) as the dominant dispersal agents. They also conclude that both 'meteorological' tides (barometric storm surge) and astronomical tides would have shoreward residual currents in the direction of propagation. Shallow-water waves and landward bottom flows associated with density currents (Fig. 140) also would produce shoreward residual currents. Wind-drift currents are mentioned as "associated with wave surge" and are not discussed again (except as illustrated in Fig.

140). Either wind-drift currents and associated barotrophic- or baroclinic-compensatory currents are not produced or they are not recognized over the Atlantic continental shelf.

We understand Swift et al. (1971) to be saying that the United States Atlantic continental shelf is a wave dominated shelf, that all significant sediment dispersal mechanisms have a residual onshore component of flow, and that bed load transport occurs only during "rare" storms which also are the sole resuspenders of sediment. This leads to the statement that "sediment transport on continental shelves must, therefore, be a diffusion process" (Swift et al., 1971, p. 326). The original articulation of this idea was given by Swift (1970, p. 13) where he shows the wave-drift transport in the shelf zone to be random leading to dispersal by diffusion. The statements in one paper (1970) that shelf transport is random and in the other (1971) that all significant components of the sediment transport scheme have onshore residuals is confusing to us. Further, if the net bottom currents are onshore, then how does continentally derived bed load transported sediment find its way across a continental shelf that is not incised by submarine canyons?

In reviewing the rather comprehensive articles discussed above, it is important to remember that only one (to our knowledge) relatively long-term *in-situ* measurement of waves or currents has been recorded over the floor of the continental shelf. Thus, the authors cited above are summarizing other scientific works and compiling their own ideas without the aid of environmental observations and under these conditions one would expect some conflict of interpretation.

This dilemma, at least partially, has been solved for the Washington continental shelf through the efforts of Hopkins (1971). Over a period of two years he collected measurements of current velocity, pressure, and temperature at various shelf locations but mostly from a station 3 m off the seabed located 21 km offshore in a water depth of 80 m. Continuous measurements of up to two-months duration for all seasons including storm conditions were obtained. Results indicated net north and offshore water movement with speeds up to 85 cm sec^{-1}. Bottom-current speeds great enough to move the silty-sand found at the station (> 40 cm sec^{-1}) occurred approximately 3½% of the time. These higher speeds were associated with winter storm conditions.

A theoretical solution which will account for this flow suggests that both the southwest winter winds and northwest summer winds produce a wind-driven surface current, a surface slope current, and a resulting Ekman bottom frictional flow toward the north and offshore (offshore 5°-15° from the bottom contours). Direct observation suggests that 2-3 days of wind will result in the bottom flows noted. Superimposed on this general system is a bottom-water drift toward the Columbia River near its mouth as compensation for fresh-water discharge. The tidally produced currents measured were strong but produced little net circulation.

More detailed discussions of Hopkins' results and interpretations regarding the dynamics of the shelf circulation system are presented in this volume by Smith and Hopkins, and Sternberg and McManus. Hopkins' measurements are significant in that (a) they represent the most complete documentation of the physical environment near the floor of the continental shelf to date and (b) the dominant dispersal mechanisms on the central- and outer-Washington continental shelf appear to be associated with the wind-drift current system rather than wave motion.

Considering the long-term physical, chemical, biological, and geological studies that have been carried out on the Washington continental shelf, it represents one of the most completely studied shelf areas in the world and serves as an excellent focal point for this discussion. Comparisons of bottom currents (Hopkins, 1971; Smith and Hopkins, this volume), shelf morphology (McManus, in press), movements of seabed drifters (Morse et al., 1968; Gross et al., 1969), distribution of radionuclides (Gross and Nelson, 1966), three-year hindcasts of surface wind waves (O'Brien, 1951), and mineral and chemical composition of sediments (White, 1968, 1970) suggest the existence of three environmentally distinct zones over the shelf. These zones have very diffuse boundaries and extend approximately parallel to the Washington coast between the depths < 50 m, 50-145 m, > 145 m.

Characteristic features of the shelf zone landward of the 50-m contour include both sedimentological and hydrodynamic properties. Seabed drifters tend to move northward along the Washington shelf; however, landward of the 40-55 m depth contour the drifters migrate toward the beach (Morse et al., 1968; Gross et al., 1969). A short direct current measurement in 50 m of water during northwest winds contained southerly onshore motions which are related to wind waves and not to wind-drift currents (Hopkins, 1971). The 50-m contour also coincides with the ½-wavelength depths of monthly wave statistics (O'Brien, 1951) which further suggests the importance of wave surge in this zone. The 50-m contour marks a change in shelf inclination while the bottom sediments landward are characteristically 2ϕ to 3ϕ median diameter and contain less than 10% silt.

The central region of the continental shelf is characterized by a dominance of silt (40%-70% by weight). As discussed in detail by Smith and Hopkins (this volume), this silt is thought to emanate from the Columbia River mouth and is transported to the north-northwest along the central shelf. Bottom currents in this region are primarily wind-drift currents which have a dominant flow direction towards the north-northwest.

Seaward of approximately the 145-m depth contour the median size increases and the silt content decreases relative to the central shelf. This region is also characterized by a rougher topography and in some areas the formation of glauconite (White, 1968, 1970).

These many interlocking lines of evidence permit us to suggest that the Washington continental shelf exhibits a sediment distributional pattern that is normal in the sense of having a nearshore wave controlled zone (< 50 cm), a central shelf zone in which wind-drift current systems control sediment dispersal (50 m-145 m), and an outer zone showing characteristics of relict topography and relict or palimpsest sediments.

Wind-drift current systems also appear to dominate the outer zone; however, the decreasing sediment supply accounts for the decreased silt concentrations. If waves from severe storms on rare occasions influence the sediment distribution on the central and outer shelf, this is not detectable in the parameters studied.

The general sedimentological distributions on the Washington continental shelf appear to be in good agreement with the models of shelf grading proposed by Curray (1965), Swift (1970), and Swift et al. (1971). It is noteworthy, however, that the mechanisms by which this end point is reached appear to be different than proposed. The above summary represents the combined results of both sedimentological and physical oceanographic data collected from the same area. Insofar as these results stem from *in-situ* measurements, and indicate other important sediment dispersal mechanisms than those suggested in recent literature, they illustrate the effectiveness of studying both processes and results concurrently.

Extrapolation of the above results is questionable without attendant measurements in other areas. We urge caution in the application of any narrowly tested model of shelf sedimentation if it involves acceptance of dispersal mechanisms where they are unobserved. This is not meant to discount the models of shelf grading proposed by Curray (1965) and Swift (1970) for they seem to adequately describe the distributions of sediment observed on many shelves. However, the dispersal mechanisms presented for attaining the end point of these models is unacceptable to us as a generality. The idea of wave-drift dominance may be correct for the Atlantic coast of the United States but before this is generally accepted there must be observational evidence. The *a priori* extrapolation of shelf sedimentation schemes from one area to another without knowledge of the dynamics of the environment could lead to serious scientific misconceptions.

CONCLUSIONS

In retrospect the major implication of this paper is that none of us can escape from provincialism and an inability to sort through and read all that is written on any subject. Although we implied that generalizations are bad, this is relative. Without generalization no progress beyond the known is made. With too far reaching generalizations and insufficient data, we can be distracted down a blind alley with an attendant overly great waste of time. Unless controversy exists, complacency sets in.

Fundamental to our problems on shelf sedimentation is lack of measurement of the environment in which the sediment exists. We can no longer interpret the environment from a vague knowledge of the nature and distribution of the sediments. If we are not doing this correlation for the modern environment, how can we expect to interpret older environments?

ACKNOWLEDGMENTS

During this compilation of material, we probably have overlooked many good papers on pertinent subjects; for this we apologize and we would appreciate their being brought to our attention. We have benefited from numerous conversations with D. A. McManus, R. J. Echols, M. L. Holmes, and J. D. Smith. This paper, however, is our responsibility, particularly if we have warped some of our colleagues' ideas.

Appreciation is expressed to the National Science Foundation for support for many years under Grants GA-2457, GA-808, GA-11126, and GA-28002 and to the Atomic Energy Commission under contracts AT-45-1-540, AT-45-1-225TA21-1, and AT-45-1-2225-TA24. Mr. R. W. Roberts has supervised the analyses of most of the sediments from the Arctic.

REFERENCES

Belderson, R. H., and Stride, A. H. (1966). Tidal current fashioning of a basal bed. *Marine Geol.* **4**, 237-257.

Coachman, L. K., and Tripp, R. B. (1970). Currents north of Bering Strait in Winter. *Limnol. and Oceanogr.* **15**, 625-632.

Creager, J. S. (1963). Sedimentation in a high energy, embayed, continental shelf environment. *J. Sediment. Petrol.* **33**, 815-830.

Creager, J. S., and McManus, D. A. (1966). Geology of the southeastern Chukchi Sea. *In* "Environment of the Cape Thompson Region, Alaska" (N. J. Wilimovsky, ed.), pp. 755-786. U.S. Atomic Energy Commssion, Washington, D.C.

Creager, J. S., and McManus, D. A. (1967). Geology of the floor of the Bering and Chukchi Seas—American studies. *In* "The Bering Land Bridge" (D. M. Hopkins, ed.), pp. 7-31. Stanford University Press, California.

Curray, J. R. (1965). Late Quaternary history, continental shelves of the United States. *In* "The Quaternary of the United States" (H. E. Wright and David G. Frey, eds.), pp. 723-735. Princeton University Press, New Jersey.

Dietz, R. S, (1963). Wave base, marine profile of equilibrium, and wave-built terraces: a critical appraisal. *Bull. Geol. Soc. Am.* **74**, 971-990.

Dietz, R. S., Carsola, A. J., Buffington, E. C., and Shipek, C. J. (1964). Sediments and topography of the Alaskan shelves. *In* "Papers in Marine Geology" (R. L. Miller, ed.), pp. 241-256. Macmillan, New York.

Emery, K. O. (1952). Continental shelf sediments off southern California. *Bull. Geol. Soc. Am.* **63**, 1105-1108.

Emery, K. O. (1968). Relict sediments on continental shelves of world. *Bull. Am. Assoc. Petrol. Geologists* **52**, 445-464.

Gorshkova, T. I. (1957). Sediments of the Kara Sea, *Trudy Vsesoyuznogo, Gidrobiologicheskogo Obshchestva* **8**, 68-99 (Translated, U.S. Naval Oceanographic Office, No. 333, 1967).

Grim, M. S., and McManus, D. A. (1970). A shallow seismic-profiling survey of the northern Bering Sea. *Marine Geol.* **8**, 293-320.

Gross, M. G., and Nelson, J. L. (1966). Sediment movement on the continental shelf near Washington and Oregon. *Science* **154**, 879-885.

Gross, M. G., Morse, B. A., and Barnes, C. A. (1969). Movement of near-bottom waters on the continental shelf off the northwestern United States. *J. Geophys. Res.* **74**, 7044-7047.

Hayes, M. O. (1967). Relationship between coastal climate and bottom sediment type on the inner continental shelf. *Marine Geol.* **5**, 111-132.

Holmes, M. L. (1967). Late Pleistocene and Holocene History of the Laptev Sea. *Thesis, Univ. Washington, Seattle, Wash.*, 176 pp. (unpublished).

Holmes, M. L., and Creager, J. S. (in press). Holocene history of the Laptev Sea continental shelf. *In* "Arctic Geology and Oceanography" (Yvonne Herman, ed.).

Hopkins, T. S. (1971). On the circulation over the continental shelf off Washington. 204 pp. *Ph.D. thesis, Univ. Washington, Seattle, Washington* (unpublished).

Kummer, J. T., and Creager, J. S. (1971). Marine geology and Cenozoic history of the Gulf of Anadyr. *Marine Geol.* **10**, 257-280.

Lisitsyn, A,P. (1966). Recent sedimentation in the Bering Sea. *Akad. Nauk SSSR., Izdatelstvo, Moskva* (English Ed., translation, 1969). 614 pp.

McManus, D. A. (1970). Criteria of climatic change in the inorganic components of marine sediments. *Quaternary Res.* **1**, 72-102.

McManus, D. A. (1972). Bottom topography and sediment texture near the Columbia River. *In* "Bioenvironmental Studies of the Columbia River Estuary and Adjacent Ocean Waters" (A. T. Pruter and D. L. Alverson, eds.), pp. 241-253. U.S. Atomic Energy Commission, Washington, D.C.

McManus, D. A., Kelley, J. C., and Creager, J. S. (1969). Continental shelf sedimentation in an Arctic environment. *Bull. Geol. Soc. Am.* **80**, 1961-1983.

Moore, D. G. (1964). Acoustic-reflection reconnaissance of continental shelves: Eastern Bering and Chukchi Seas. *In* "Papers in Marine Geology" (R. L. Miller, ed.), pp. 319-362. Macmillan, New York.

Moore, D. G., and Curray, J. R. (1964). Wave-base, marine profile of equilibrium, and wave-built terraces: Discussion. *Bull. Geol. Soc. Am.* **75**, 1267-1274.

Morse, B. A., Gross, M. G., and Barnes, C. A. (1968). Movement of seabed drifters near the Columbia River. *Proc. Amer. Soc. Civil Eng., J. Waterways and Harbors Division* **94(WWI)**, 93-103.

Naugler, F. P. (1967). Recent sediments of the East Siberian Sea. *Ph.D. thesis, Univ. Washington, Seattle, Washington,* 71 pp. (unpublished).

Naugler, F. P., Silverberg, N., and Creager, J. S. (in press). Recent sediments of the East Siberian Sea. *In* "Arctic Geology and Oceanography" (Yvonne Herman, ed.).

Niino, H., and Emery, K. O. (1966). Continental shelf sediments off northern Asia. *J. Sediment. Petrol.* **36**, 152-161.

O'Brien, M. P. (1951). Wave measurements at the Columbia River light vessel, 1933, 1936. *Trans. Amer. Geophys. Union* **32**, 875-877.

Swift, D. J. P. (1970). Quaternary shelves and the return to grade. *Marine Geol.* **8**, 5-30.

Swift, D. J. P., Stanley, D. J., and Curray, J. R. (1971). Relict sediments on continental shelves: a reconsideration. *J. Geol.* **79**, 322-346.

White, S. M. (1968). The mineralogy and geochemistry of continental shelf sediments off the Washington-Oregon coast. *Ph.D. thesis, Univ. Washington, Seattle, Washington,* 210 pp. (unpublished).

White, S. M. (1970). Mineralogy and geochemistry of continental shelf sediments off the Washington-Oregon coast. *J. Sediment. Petrol.* **40**, 38-54.

CHAPTER 17

Implications of Sediment Dispersal from Bottom Current Measurements; Some Specific Problems in Understanding Bottom Sediment Distribution and Dispersal on the Continental Shelf— A Discussion of Two Papers

Donald J. P. Swift

National Oceanic & Atmospheric Administration
Atlantic Oceanographic & Meteorological Laboratories
15 Rickenbacker Causeway
Miami, Florida 33149

ABSTRACT

The emphasis of Creager and Sternberg (1972) and also Sternberg and McManus (1972) on the wind-drift component of shelf storm currents is valid, since this component is a major, if not dominant, cause of unidirectional currents over large portions of most nontidal shelves. In view of the genetic complexity of the current velocity field over the shelf, however, an operational distinction is better made between storm-generated bottom currents in which a unidirectional component of compound origin is dominant, and the surge-dominated flow, which characterizes but is not confined to the fair-weather hydraulic regime. This division recognizes that shelf currents are frequently of compound origin, and is keyed to the interpretation of substrate response elements. Such an approach may encourage examination of the very real role of waves as a contributing factor in shelf sediment transport.

A simple random walk model for sediment transport across the shelf does not, as noted by Sternberg and McManus (1972) and by Creager and Sternberg (1972), accurately describe the advective component of sediment transport. It does, however, accommodate the diffusive component. The more flexible Markov-process model is valid for both components.

CONCEPTUAL MODELS

Creager and Sternberg (1972) have found an ideal key with which to review the broad field of the study of bottom sediment distribution and dispersal on continental shelves. They seek the paradigm, or conceptual framework with which their colleagues order reality. Unfortunately, the more successful conceptual models, like other valuable tools, are easily misused. They can be treated as closed systems, which cease to create new order and reason, but merely recirculate the old; that is, they lead to circular reasoning whereby relationships predicted in the model are too hastily "found" in nature, then cited as the model's verification. Creager and Sternberg are concerned with such use and misuse of conceptual models for shelf sediment transport.

Probably the best defense for those of us brought to task for our complicity in thus detaching reasoning from reality is Griffiths' (1967) "stochastic convergence" model for the advancement of knowledge. Griffiths notes that major scientific advances are based on preceding concepts (Fig. 141) and that while a given concept may be erroneous, it is generally less erroneous than its predecessor, portions of which it tends to incorporate. In his example, historical estimates for the age of the earth range from approximately zero to infinity, but converge towards a "true value" (which itself is time variant). The fine structure of the process is significant. A "vertical breakthrough" often contains many erroneous concepts, yet gives impetus to its field. Those who follow in the pioneer's footsteps spend most of their efforts in tidying up his structure of thought; eventually the reordering is thoroughgoing enough to constitute a new breakthrough which, despite its freshness, contains elements of the old. Our own field of shelf sediment transport affords a classic example. The turn of the century students, of whom D. W. Johnson (1919) was a noted example, conceived of the equilibrium shelf profile and its corollary, the size-graded shelf. Shepard (1932) sharply questioned the corollary on the basis of map notations, and Emery's (1952) classification of shelf sediments restructured the Johnson model in such a systematic fashion as to constitute a new breakthrough. The concept

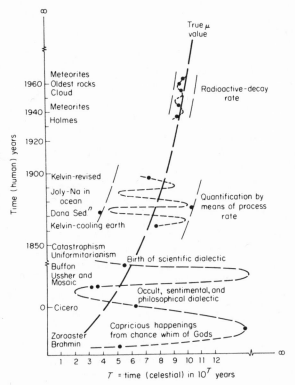

Figure 141.

Estimation of the age of the earth as an example of stochastic convergence toward some true value (from Griffiths, 1967).

of relict sediments is one which I and others in the phase of consolidation have attempted to qualify; yet without the conceptual platform provided by both Johnson's and Emery's models, we could only move painfully toward these points of departure.

SHELF VELOCITY FIELD: WAVE-DRIFT AND WIND-DRIFT COMPONENTS

With this perspective I would like to take advantage of my early acquaintance with both the Sternberg-McManus and the Creager-Sternberg manuscripts and their request for clarification of some of my earlier statements. Clarification is necessary, for several of us (Swift, 1970; Swift, Stanley, and Curray, 1971) seem (with some provocation) to have become straw men in a debate on the relative merits of wind-drift and wave-drift velocity components as agents of shelf sediment dispersal. Both papers were written at a time when a major issue had been defined by the exchange between Dietz (1963) and Moore and Curray (1964) as to whether or not any significant movement at all occurred on the shelf. Consequently, no major attempt to assess relative competence of wave- and wind-drift velocity components was undertaken. Instead, we invoked evidence wherever we found it to indicate that sediment is moving on shelves. The lengthy discussion of oscillatory wave, tidal, and storm surge movements, and their residual components (Swift et al., 1971a, p. 324-325) was introduced to help prove that movement occurred. I note that we place "direct wind input" first among energy inputs into the shelf velocity field, although the effect is somewhat marred by later unfortunate references to "wave-dominated" shelves. Weather-dominated shelves is a preferable term. We clearly distinguished between tide-dominated shelves and weather-dominated shelves (Swift, 1970, p. 17; Swift et al. 1971a, p. 324) before discussing the catastrophic fine structure of storm-driven sedimentation on the latter (p. 324), hence we were presumably applying this generality with the caution required by Creager and Sternberg.

Since completion of the manuscript, "Relict Sediments: a Reconsideration," I have been investigating hydraulic process and substrate response on portions of the Atlantic continental shelf. The latter studies have progressed more rapidly than the former; although I have acquired only a small store of hydraulic data, I now have a great awareness of the appalling logistics of *in-situ* recording of the hydraulic regime. My substrate studies (Swift et al., 1971b, 1972a, 1972b; Duane et al., 1972) have provided considerable circumstantial evidence for southward sediment transport on the inner central Atlantic shelf. I have attributed these to storm-generated currents rather than to the fair-weather wave regime, and have been considerably influenced in my thinking by theoretical considerations set for by Harrison et al. (1964) in their study of circulation in the Chesapeake bight of the Atlantic shelf. These authors postulated south-flowing currents on the inner shelf as a consequence of a barotropic response to wind set-up during northeasters. Consequently when, during the oral presentations of this symposium, a wind- versus wave-current controversy was referred to and it became apparent that I was expected to defend the latter end of it, I was rather startled. However, after reviewing my 1970 and 1971 papers I note that while a clear distinction between wind- and wave-drift velocity components was made, wave-drift components were stressed, and that wind-drift components were not discussed in detail. The papers in this volume by Sternberg and McManus and Creager and Sternberg are justifiably concerned with this imbalance. Their review of the problem is a service to our readers and has sharpened my own thinking. However, my thoughts, thus sharpened, have led to doubts concerning a simple division between wave-drift and wind-drift currents.

One concern is in part semantic. The total instantaneous shelf velocity field at a point of measurement, and the succession of these fields through time are the only realities. Velocity components are fictional though useful abstractions. My own work on the central Atlantic shelf (Swift et al., 1972b; Duane et al., 1972) has led me to conclude that brief, intense, storm-generated pulses are the most significant factor associated with observed forms of substrate response such as large and small bedforms and sand size-distributions. The wind-drift velocity component presumably dominates these currents, but they are probably never purely wind-drift currents in the sense of being solely due to direct wind stress on the water surface. They are most nearly such on the outer shelf during storms, when in the generating area, the waves are high and steep but also of wavelength too short to reach to the outer shelf floor. The surface currents thus generated, theoretically a pure wind-drift component for deep water gravity waves, may eventually reach the shelf floor; either directly through he downward transfer of momentum by turbulent diffusion, or indirectly, in the form of compensatory currents. However, complex wind waves are the least amenable to the linear approximations of classical wave theory, and the wave-drift component of the surface current cannot be accurately assessed. An additional significant wave-drift component, likewise mathematically intractable, occurs as a consequence of the breaking of the crests of the steep storm waves.

Other velocity components are present in this situation. Galt (1971) has devised a model for pressure-induced storm surges over the continental shelf that indicates a velocity component form this cause on the order of 10 cm sec^{-1} near the shelf break, with the possibility of considerably larger components in the vicinity of submarine canyons. Finally, if long period waves unrelated to the present storm are arriving, the bottom may be directly affected by wave surge and an associated residual velocity component. During this time, the central and inner shelves, if within the storm area, are experiencing a similar velocity field, with the oscillatory and residual wave components intensified on the bottom in proportion to the decreasing depth.

It may be more helpful for the interpretation of shelf-floor response elements to distinguish not between such genetic concepts as wind-drift and wave-drift currents, but instead between two operational bottom regimes; namely a unidirectional, storm-generated current in which the wind-drift component is a major residual and dominates over oscillatory wave surge; and surge-dominated flow, which characterizes (but is by no means limited to) the fair-weather hydraulic regime. The response element corresponding to surge-dominated bottom flow is, of course, an oscillation ripple field (Komar et al., 1972). The response elements associated with dominantly unidirectional storm currents have been little investigated, but on the central Atlantic shelf they appear to include asymmetric transverse ripples up to 20 cm high, asymmetric transverse forms up to 3 m high, and longitudinal sand ridges up to 10 m high (Swift et al., 1972).

Of concern also is the problem of over-reaction. The emphasis of Sternberg

and McManus and Creager-Sternberg on the importance of the wind-drift component in high-competence currents on nontidal shelves seems reasonable, but
the advent of this ray of light on one aspect of the hydraulic climate is all
the more reason to get a better fix on the significance of waves in shelf sediment
transport. Part of the present emphasis on unidirectional shelf currents stems
from our mode of observation; generally by means of Savonius rotor meters
which are poorly designed for resolving wave surge. As noted by Creager and
Sternberg, "If waves from severe storms on rare occasions influence the sediment
distributions on the central and outer shelf, this is not detectable in the parameters
studied." However, the problem is basically one of conceptual approach, since
other workers in the same general area, with access to similar equipment, have
come to very different conclusions concerning the role of wave action in sediment
transport (Komar et al., 1972).

This role is thus at present ambiguous. Creager and Sternberg (1972)
detect a wave-dominated zone landward of the 50-m isobath on the Washington
shelf, while Komar et al. conclude that there is a significant wave-driven component of sediment transport as far seaward as rippling occurs on the adjacent
Oregon shelf, or to depths in excess of 200 m. However, Komar et al., (1972),
following the model of Bagnold (1963) stress that wave surge is an efficient mover
of sediment mainly in conjunction with a superimposed unidirectional current.
Not only does wave surge serve to entrain sediment, but as noted by Keulegan
(1948), Scott (1952), and Bagnold (1963), complex synergistic effects occur
whereby the movement of various size fractions of the pulsating sediment stream
across a wave-rippled surface tends to be rectified, or even reversed. Here
is a further potentially misleading aspect of the wave-drift versus wind-drift
controversy. We are justly proud of our present concern with the hydraulic
regime, but the outlines and dimensions of shelf sediment transport will continue
to elude us until we undertake to directly monitor the sediment flux itself by
sediment traps, tracers, time lapse photography, microbathymetric time series,
or other means.

In the middle Atlantic bight, unidirectional, storm-generated currents appear
to control constructional shelf topography well landward of Creager and Sternberg's 50-m boundary. The characteristic response element, the ridge and swale
topography, appears as far inshore as the 10-m isobath, where it gives way
to the wave-smoothed shore face. It appears to result in part from a helicoidal
flow structure of storm-generated currents in the shallow-water generating zone.
In the seaward sectors, the topography itself probably impresses this flow pattern
on storm currents and consequently would tend to be maintained by it. Here
surge-dominated flow serves primarily to degrade wave crests between storms,
although Creager (personal communication) has made the interesting observation
that the mid-shelf divergence of bottom drift noted by Bumpus (1962) in the
Middle Atlantic Bight may reflect to some extent the shoreward residual compo-

nent of bottom wave surge on the inner shelf, as well as the salinity-driven circulation.

But any comparison of the east- and west-coast shelves must consider their respective regimes and substrates. The east-coast shelf is in the lee of the continent with respect to the prevailing westerlies, and has a very different and less intense wave climate, whereas the west-coast shelf has a high, rocky coast, whose shore face may be less suitable for the generation of a constructional ridge topography. Thus the ridge-building propensity of the east-coast shelf may tend to overshadow its response to wave-driven components of the hydraulic regime, just as the apparent inability of the west coast inner shelf to build ridges may tend to disguise the importance of the unidirectional component of storm-generated currents in that area.

PROBABILITY MODELS: ADVECTIVE VERSUS DIFFUSIVE SEDIMENT TRANSPORT

Creager and Sternberg object to Dunbar and Rodgers' (1957, p. 11) vivid, intuitive picture of drift currents generated by storms (quoted by Sternberg and McManus, 1972), presumably because the wave-versus-wind dichotomy is not considered. However, Sternberg and McManus' Fig. 4 (1972) is basically this quotation given flesh. It has been impressively quantified, but we still know only that sediment must have been moved first "one way, then another," and very little about the channels of energy transfer. Here, in our respective attitudes to the Dunbar and Rodgers quotation, lies a more basic difference in viewpoint than any emphasis on wave-drift or wind-drift velocity components. Flume studies show that bedload sand grains travel in a series of short trajectories punctuated by periods of rest. Dunbar and Rodgers (1957) inferred, and Sternberg and McManus (1972, Fig. 78) have demonstrated that on weather-dominated shelves, there is a higher order of intermittent storm-driven trajectories punctuated by rest intervals. In this hierarchy of intermittent movements, the role of diffusive sediment transport is large relative to that of advective transport, particularly in the large-scale movements. As noted by Sternberg and McManus, there is a net transport direction, but the variation among directions of successive trajectories is high. In consequence, the deterministic relationships expressed by the laws of classical sediment hydraulics can never fully resolve the regional sediment budget, and probabilistic analysis must be an important aspect of any such study.

In my 1970 paper, "Quaternary shelves and the return to grade," I considered a simple, qualitative, diffusion model for shelf sediment transport. I wanted to show that in the simplest case, with random movement of sediment on the shelf, there ensued a net cross-shelf transport to the shelf-edge sink. However, the random walk scheme that I described is a framework for conceptualization and ultimately for calculation, and is not to be confused with a real process "associated" with residual currents "in some unstated manner" (Sternberg and McManus, 1972). As the model does not have the ability to simulate advective transport and preferred directions of movement, its virtue lies in its ability to accommodate the diffusive component of sediment transport. In our 1971 paper (Swift et al., 1971), we developed, on a qualitative basis, the more flexible general case, a Markov process model which permits preferred directions and paths of sediment movement. A first step in quantifying this model has been taken in this volume (Chapter 9). Thus when Creager and Sternberg represent me as saying on one hand that shelf transport is random and on the other hand that all significant components of the sediment transport scheme have onshore residuals, I can only reply that I hold neither of these opinions. I would suggest instead that the direction of residual currents depends on the climate and configuration of the shelf in question, that the role of diffusive processes is a strong one on some shelves, and that a probabilistic analysis of these processes is rewarding.

Advective shelf sediment transport is in fact probably shelf-parallel or nearly so in most cases (Swift et al.; Smith and Hopkins, 1972; McCave, 1972), due to the tendency of oceanic water movements to parallel pressure gradients. Cross-shelf diffusion may be volumetrically insignificant on some modern disequilibrium shelves. Ancient shelves, however, must frequently have had higher coastal sediment input, and the rate of cross-shelf diffusion of the finer grades of sand and silt, which travel largely by suspension, may have been much higher, as is probably the case today off large deltas.

There once transpired a debate between Nils Bohr and Albert Einstein concerning the role of probability in the universe (Cline, 1965). Einstein rejected a universe in which probabilistic laws played a significant role as being unworthy of its Creator, and considered probabilistic laws as merely expedient shortcuts to deterministic reality. Bohr noted that "blind chance" followed laws of precision and symmetry. He felt that these laws correspond to a deep level of reality, and considered them a credit to the Creator. Shelf geologists, faced with the complexities of diffusive sediment transport, must look thankfully to Bohr as a spokesman for a probabilistic approach.

REFERENCES

Bumpus, D. F. (1965). Residual drift along the bottom on the continental shelf in the middle Atlantic area. *Limnology and Oceanography* 10[Suppl.], R50-R53.
Bagnold, R. A. (1963). The mechanics of marine sedimentation. *In* "The Sea Vol. 3 of The Earth Beneath the Sea." (M. N. Hill, ed.), p. 507-525. Wiley, Interscience, New York.
Cline, B. L. (1965). *Men who made the new physics.* 223 pp., Signet Science, New York.
Creager, J. S., and Sternberg, R. W. (1972). Some specific problems in understanding bottom sediment dispersal and distribution on the continental shelf. *In* "Shelf Sediment Transport:

Process and Pattern" (D. J. P. Swift, D. B. Duane, and O. H. Pilkey, eds.). Dowden, Hutchinson and Ross, Stroudsburg, Pa.

Dietz, R. S. (1963). Wave-base, marine profile of equilibrium and wave built terraces, a critical appraisal. *Geol. Soc. America Bull.* **74**, 971-990

Duane, D. B., Field, M. E., Meisburger, E. P., Swift, D. J. P., and Williams, S. J. (1972). Linear shoals on the Atlantic inner continental shelf, Florida to Long Island. *In* "Shelf Sediment Transport: Process and Pattern" (D. J. P. Swift, D. B. Duane, and O. H. Pilkey, eds.). Dowden, Hutchinson and Ross, Stroudsburg, Pa.

Dunbar, C. O., and Rodgers, J. (1957). "Principles of Stratigraphy." 356 pp., Wiley, New York.

Emery, K. O. (1952). Continental shelf sediments off southern California. *Geol. Soc. America Bull.* **63**, 1105-1108.

Galt, J. A. (1971). A numerical investigation of pressure induced storm surges over the continental shelf. *J. Physical Oceanography* **1**, 82-91.

Galt, J. A. (1971). A numerical investigation of pressure induced storm surges over the continental shelf. *J. Physical Oceanography* **1**, 82-91.

Griffiths, J. C. (1967). "Scientific Method in Analysis of Sediments." 508 pp. McGraw-Hill, New York.

Harrison, W., Norcross, J. J., Pore, N. A., and Stanley, E. M. (1967). Circulation of shelf waters off the Chesapeake Bight. *Environmental Science Services Administration Prof. Paper* **3**, 82 pp.

Johnson, D. W. (1919). "Shore Processes and Shoreline Development." 584 pp. Hafner, New York (1965 facsimile).

Keulegan, G. H. (1948). An experimental study of submarine sand bars. *Beach Erosion Board Tech. Rept.* **3**, 42 pp.

Komar, P. D., Neudeck, R. H., and Kulm, L. D. (1972). Observations of deep water oscillatory ripple marks on the Oregon continental shelf. *In* "Shelf Sediment Transport: Process and Pattern" (D. J. P. Swift, D. B. Duane, and O. H. Pilkey, eds.). Dowden, Hutchinson and Ross, Stroudsburg, Pa.

Moore, D. J., and Curray, J. R. (1964). Wave base, marine profiles of equilibrium, and wave built terraces: discussion. *Geol. Soc. America Bull.* **75**, 1267-1274.

Scott, T. (1954). Sand movement by waves. *Beach Erosion Bd. Tech. Memo.* **48**, 37 pp.

Shepard, F. P. (1932). Sediments on the continental shelves. *Geol. Soc. America Bull.* **43**, 1017-1039.

Sternberg, R. W., and McManus, D. A. (1972). Implications of sediment dispersal from long term, bottom-current measurements on the continental shelf of Washington. *In* "Shelf Sediment Transport: Process and Pattern" (D. J. P. Swift, D. B. Duane, and O. H. Pilkey, eds.). Dowden, Hutchinson and Ross, Stroudsburg, Pa.

Swift, D. J. P. (1970). Quaternary shelves and the return to grade. *Marine Geology* **8**, 5-30.

Swift, D. J. P., Holliday, B. W., Avignone, N., and Shideler, G. (1972). Anatomy of a shoreface ridge system, False Cape, Virginia. *Marine Geology* **12**, 59-84.

Swift, D. J. P., Stanley, D. J., and Curray, J. R. (1971a). Relict sediments on continental shelves: a reconsideration. *J. Geol.* **79**, 322-346.

Swift, D. J. P., Sanford, R. B., Dill, C. E., Jr., and Avignone, N. F. (1971b). Textural differentiation of the shore face during erosional retreat of an unconsolidated coast, Cape Henry to Cape Hatteras, western North Atlantic shelf. *Sedimentology* **16**, 221-250.

CHAPTER 18

Sediment Transport
Around the British Isles

A. H. Stride

National Institute of Oceanography
Wormley, Surrey, Great Britain

ABSTRACT

Observations, inferences, and predictions show that present-day water movements are at work on the continental shelf around the British Isles. The effects are obvious where the tidal currents are strong but can be discerned even where the water movements are only occasionally strong enough to move sand.

INTRODUCTION

The sea floor around the British Isles is a good region for studying sedimentation processes. First, so much is known about the environment that tentative correlations for one area can be tested out at numerous other similar sites, so that the main process can be recognized despite the inevitable geological and oceanographical noise that obscures it in more restricted studies. Second, the strong currents occurring in some regions produce a suite of characteristic bed forms, indicative of sediment transport, each of which can now be used singly where full supporting evidence is not yet available.

373

THE ENVIRONMENT

There are extensive areas of floor at most depths down to about 200 m, with enclosed seas, narrow straits, and long open reaches of ocean-facing continental shelf. Some floors were heavily glaciated while others lay beyond the ice sheets.

The main currents are tidal, with strength near the sea surface reaching between one half and 4 knots over large areas; other currents are intermittent, more variable in direction and generally weak. Storm waves can be large and are expected to rile much of the floor for an appreciable number of days per year (Draper, 1967). The 50-year wave is predicted to reach almost as high as 33 m (110 ft) on the open continental shelf and to be not much less in the northern North Sea (Mr. L. Draper, personal communication), so that there is little floor that is not subject to some disturbance.

SAND AND GRAVEL TRANSPORT

Proof of sediment transport is provided by bed forms, characteristic of mobile gravels and sands, on extensive areas at all depths on the continental shelf where current strength is adequate. The known bed forms include longitudinal furrows and gravel waves (Stride et al., 1972), sand ribbons (Kenyon, 1970a), and sand waves and sand patches in zones of progressively decreasing current strength (Belderson et al., 1971), the variety of which is shown elsewhere (Belderson et al., 1972).

Bed-transport directions are indicated by the bed forms and their zonal distribution, by the usual progressive decrease in grade and current strength and by the direction of greatest competence of the tidal currents (Kenyon and Stride, 1970; Stride, in press). Bed-transport directions are mostly parallel with the coasts. They are interrupted by bed-load partings and convergences, typical of the tidal current environment. Local transport paths are associated with bays, longitudinal sand banks and with the edge of the continental shelf, where it is transverse to the tidal flow. The direction of net transport of sand and gravel seems to be determined by the peak tidal current asymmetry (for example, Belderson and Stride, 1969), while storm-waves can increase the transport more than twentyfold in winter (Johnson and Stride, 1970). The mobile material is derived partly from earlier Quaternary deposits, partly from present-day faunas, and some from coastal erosion.

THE DEPOSITS

Proof of the operation of the transport paths is also provided by the new deposits, which have their own grade, structure, petrology, and fauna in keeping with the present environment (Belderson et al., 1971). There are extensive sheetlike deposits of gravel, sand, and silt where grade is in equilibrium with current strength. Sand deposits being laid on a gravel floor can be represented by isolated patches when sand is in short supply (Kenyon, 1970b). Elsewhere, there are isolated deposits of sand tied to estuaries, bays, headlands, straits and to the edge of the continental shelf (Stride et al., 1972). Deposits associated with bed-load convergences need not be in equilibrium with current strength (Stride, in press).

REFERENCES

Belderson, R. H., Kenyon, N. H., and Stride, A. H. (1971). Holocene sediments on the continental shelf west of the British Isles. The ICSU/SCOR Working Party 31 Symposium, Cambridge, 1970: "The Geology of the Eastern Atlantic Continental Margin." (Miss F. M. Delany, ed.), Vol. 2, Europe. Institute of Geological Sciences Report, No. 70/14.

Belderson, R. H., Kenyon, N. H., Stride, A. H., and Stubbs, A. R. (1972). "Sonographs of the Sea Floor: A Picture Atlas," 185 pp. Elsevier, Amsterdam.

Belderson, R. H., and Stride, A. H. (1969). Tidal currents and sand wave profiles in the north-eastern Irish Sea. *Nature* **222**, 74-75.

Draper, L. (1967). Wave activity at the sea bed around northwestern Europe. *Marine Geol.* **5**, 133-140.

Kenyon, N. H. (1970a). Sand ribbons of European tidal seas. *Marine Geol.* **9**, 25-39.

Kenyon, N. H. (1970b). The origin of some transverse sand patches in the Celtic Sea. *Geol. Mag.* **107**, 389-394.

Kenyon, N. H., and Stride, A. H. (1970). The tide-swept continental shelf sediments between the Shetland Isles and France. *Sedimentology* **14**, 159-173.

Johnson, M. A., and Stride, A. H. (1970). Geological significance of North Sea sand transport rates. *Nature* **224**, 1016-1017.

Stride, A. H. (in press). Sediment transport by the North Sea. *North Sea Science* (E. D. Goldberg, ed.). MIT Press, Cambridge, Mass.

Stride, A. H., Belderson, R. H., and Kenyon, N. H. (1972). Longitudinal furrows and depositional sand bodies of the English Channel. Mémoire Bureau Recherches Géologique et Minières, No. 79, 233-244.

CHAPTER 19

Migration of Tidal Sand Waves
in Chesapeake Bay Entrance

John C. Ludwick

Institute of Oceanography
Old Dominion University
Norfolk, Virginia

ABSTRACT

Subtidal sand waves occur atop shoals and on shoal margins in the northern part of the tidal entrance to Chesapeake Bay, Virginia. Wavelength ranges from 60-245 m; height ranges from 1.5-3.4 m. All major slopes are 2-3°, far less than the angle of repose of the constituent sediment. In flood-dominated tidal channels, sand waves are asymmetrical and face landwards; in ebb-dominated tidal channels and atop most banks, sand waves face seawards. Where near-bottom flood and ebb tidal currents are equal in time-velocity impulse, sand waves of symmetrical-trochoidal profile are developed.

Data from 21 successive echo-sounding profiles taken over a 17-mo period show that the seaward-facing asymmetrical sand waves are migrating seaward. Rates of migration of these waves range from 35 to 150 m yr^{-1}. Symmetrical sand waves did not show significant migration. Sand wave height changes seasonally, symmetrical-trochoidal waves experiencing a twofold height change. Small heights occur from October to late April when surface water waves are frequently higher than 1.5 m; large sand wave heights occur from May to September, particularly during the latter month when surface water waves are usually lower than 1.5 m in height.

Prominent shoals in the entrance area characteristically have ebb-dominated near-bottom tidal currents on one side and flood-dominated near-bottom tidal currents on the other side. This pattern is suggestive of a sand circulation conjoint with the shoal. Geomorphic evidence, including that from sand wave facing direction, is consistent with the existence of sand-circulation cells. The circulation mechanism, because it creates lengthened sediment residence time in the cells, may explain how the shoals maintain their positions in the face of strong tidal currents and heavy wave action.

INTRODUCTION

At present, the migration rate of subtidal sand waves is most readily obtained from comparisons made between at least two successive bathymetric surveys in which corresponding sand wave crests can be confidently identified. In this study a series of 21 successive profiles taken over a period of 17 mo are examined for migration rate and for evidence of the effects of tidal phase, run-off, and wave action. The ubiquity of sand waves in tidal inlets and entrances, port approaches, navigation channels, as well as the probable role of sand waves in shoal evolution, sediment movement, and burial of minefields would appear to warrant an analysis of travel rates.

Even in the very earliest studies of undulose bedforms, migration rate was seen to vary with fluid flow velocity. Before 1871 an empirical equation had been developed in which sand bank travel rate was related to the square of the surface flow velocity (Leliavsky, 1955, p. 12). This expression (Fig. 142) was deduced by the French engineer, Sainjon, from observations in the Loire River. Chang (1939) and Salsman et al. (1966) have presented other migration data from flume and field that are fitted by fifth power functions of flow velocity. Allen (1963) reduced the Loire River migration data of Ballade (1953) to a power function in which the fluid flow velocity bears the exponent, 2.41. It is obvious that a simple power-law relationship cannot be generally valid since in such a law no provision is made for the special occurrences of upper regime flow including antidune motion. It is also true that although flow velocity is probably the single factor that produces the largest effect on sand-wave migration rate, other factors are of considerable significance.

Sand-wave migration is perhaps more purposively associated with the movement of bed sediment than with fluid flow velocity. Unfortunately, here one is immediately confronted with all the presently unsolved problems of initiation of sediment motion, turbulent transport, and non-Newtonian behavior of sediment-ladened suspensions. These unsolved problems are serious obstacles to a full physical understanding of sand wave travel rate. It can be deduced readily that among the relevant factors are: (a) particle properties, including density, size, shape, and cohesion; (b) fluid properties, including density, viscosity, and, therefore, temperature; (c) boundary properties, including roughness, sand wave height, wave length, and bed slope; and finally (d) water depth or depth below an internal fluid flow boundary. The shear stress exerted by the moving fluid on the bed is a key factor in sediment transport and in sand-wave migration.

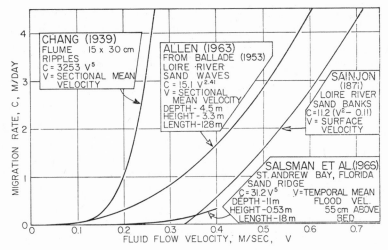

Figure 142.
Various proposed relations between the migration rate of undulose bed forms and unidirectional fluid flow velocity. Note the differences in size of feature, water depth, and definition of velocity used. In each equation, C is in units of m day^{-1}, and V is in units of m sec^{-1}.

In most studies of sand wave migration rate, detailed consideration of cause of migration is usually preceded by a determination of flow regime (Deacon, 1894; Gilbert, 1914; Simons et al., 1961, 1965), since in the upper regime, the direction of motion of the bed forms may even be opposite to the direction of the fluid flow. Kennedy (1963, 1969) gives the following expression for the minimum Froude number at which antidunes form: $F^2 = (1/kd)$ tanh kd. The symbols are defined below. This lower limit is seen to depend on kd, water depth relative to wavelength.

The migration rate equation that follows was also developed by Kennedy (1969) and illustrates for the two-dimensional case, the plurality of factors relevant to travel rate of fully-developed bedforms:

$$U_b = (\overline{T}_b/\beta)nk \left[U/(U - U_c) \right]\left[(1 - F^2 kd \tanh kd)/(\tanh kd - F^2 kd) \right]$$

where U_b is bed form migration velocity; $\overline{T}_b$ is the average volume rate, per unit width of channel, of downstream sediment transport due only to the migration of bed sediment; β (taken as unity by Kennedy in a sample computation) is the ratio of bed load to total sediment transport; n (taken as 2.64 in a sample computation for sediment 0.93 mm in mean size) is a dimensionless exponent in an assumed sediment-transport law; k is the wave number, $2\pi/L$, where L is wavelength; U is unidirectional mean fluid flow velocity; U_c (taken as 1.30 ft sec^{-1} in a sample computation for sediment 0.93 mm in mean size) is the critical velocity for initiation of motion; and F is the Froude number, $U/\sqrt{gd}$, where g is gravitational acceleration, and d is water depth. The relationship has not been confirmed by comparison with field data. Such comparisons are seriously hampered by a lack of reliable measurement of $\overline{T_b}$.

Under strong, but subcritical, unidirectional flow, profiles of actively moving undulose bed forms are asymmetrical, and often exhibit steep downstream faces which slope at the underwater angle of repose of the constituent sediment. This angle ranges from 24° to 34°. Under reversing flow, as in tidal channels, if ebb and flood currents are of equal strength and duration, symmetrical bed form profiles of trochoidal shape are developed. Forel (1883) regarded profiles of this type as composites of two asymmetrical opposed forms. Profiles of intermediate symmetry, produced when ebb and flood currents are not equal in strength and duration, are aptly termed asymmetrical-trochoidal (Van Veen, 1935).

Migration of strongly asymmetrical forms with slip faces occurs by bed erosion on the upstream slope of the wave and by deposition on the downstream face which builds forward by intermittent or continuous avalanching of sediment (Allen, 1962, 1965; Jopling, 1965; Harms, 1969). Small symmetrical forms on a level bottom may oscillate under reversing flow, but do not undergo net translation. Bed waves that are asymmetrical-trochoidal in profile migrate in the direction of the net, or residual, current and are steeper downstream of the dominating current.

Increasingly evidence is coming to light that maximum slopes of many large sand waves are considerably less than the angle of repose of the constituent sediment (Bucher, 1919; Lane and Eden, 1940; Carey and Keller, 1957; Cartwright, 1959; Allen, 1963; Visher, 1965; Imbrie and Buchanan, 1965). It seems likely that there is no boundary layer separation of the flow at the subdued crests of such features, no persistent lee eddy, and no lee counter current. Under a competent unidirectional flow, sediment at the bed would be transported unidirectionally along the entire form profile but, of course, not necessarily at constant speed. The origin and maintenance of low-slope sand waves are not yet completely understood.

Smith (1970) has recently made an important addition to the long succession of theories proposed to account for sand-wave origin, shape, and motion. According to a part of his comprehensive theory, during each growth and migration of an initial perturbation of the bed sediment, side slopes of the features would be less than the angle of repose; however, Smith indicates (p. 5938, 5939) that if maximum shear stress lies upstream of a sand-wave crest, eventually the downstream slope increases to the angle of repose. It seems unwarranted, however, to accept without reservation the inference that sand waves of low slope are simply features in an early stage of growth or simply features on which maximum shear stress does not lie upstream of the crest. As described in a later section of the present paper, the height of sand waves in shallow water may be limited by wave or current action.

Sand waves of large amplitude advance more slowly than small sand waves of identical profile if the bed sediment transport rate per unit width of channel is the same for both features (Cornish, 1901; Cloet, 1954a, b: Simons and Richardson, 1960; Allen, 1965; Ashida and Tanaka, 1967; Crickmore, 1970). Smith (1968) and Nordin (1968) concluded that sand waves of short wavelength migrate faster than those of longer wavelength. The preferred explanation is that large sand waves have a large sediment storage volume and thus require more sediment to move a unit distance. A consequence of the rapid movement of small bedforms is that they overtake larger features in the same wave train, merge with them, increase the size of the larger forms, and increase the incidence of long wavelengths by decreasing the incidence of short wavelengths. Some investigators have argued that sand wave height should, therefore, increase among the forms in a wave train with distance in the direction of migration. It is a corollary of this idea that larger sand waves are older waves.

Sand wave migration rates vary seasonally in some estuaries and tidal rivers. In the Loire River, Ballade (1953) found that during the season of low run-off, sand waves at Ile de Bois, 32 km from the mouth of the river, experienced only back and forth excursions of 3 to 8 m and only the tops of the bed features were affected. During the season of high runoff, the ebb current was aided, sand wave asymmetry developed, and the entire train migrated downstream at a rate of 2.5 m day^{-1} under peak discharge.

The migration of sand waves and the evolution of some sediment shoals are apparently related. Using the then newly developed recording echo-sounder, Ven Veen (1935, 1936) showed that the great current-parallel sand ridges of the southern North Sea were surmounted by sand waves. Earlier, Cornish (1901) had noted in his study of intertidal shoals and sand waves of English estuaries, that there was a clear correspondence between the facing direction of asymmetrical sand waves and the asymmetry of the shoals that they surmounted.

Following the firm establishment by Cornish (1901) of the finding that ebb-dominant and flood-dominant currents tend to be at least partly confined to different, mutually evasive channels among tidal shoals, and the related finding by Van Veen that parabolic-shaped shoals were a fundamental form conjoint with this distribution of currents, Van Straaten (1950, 1953) showed that sand waves found in the the tidal channels between limbs of parabolas faced the closed or dead-ends of the channels. He also showed that in the tidal channels on either side of a linear shoal, sand waves faced with the dominant current and in opposite directions.

It would now appear likely that many sand shoals subject to mutually evasive tidal flows are coexistent with sand-circulation cells, the presence of which is revealed by the mapped pattern of sand wave facing directions (Houbolt, 1968; Smith, 1968, 1969; James and Stanley, 1968; Klein, 1970). Other sand banks subject to tidal currents appear to owe their existence to the convergent motion of sand waves from opposite sides of the banks (Jordan, 1962; Dingle,

1965; Jones, Kain, and Stride, 1965). Alternate exposure and shielding of the two sides of the bank to ebb and flood currents account for the observed disposition of the asymmetrical sand waves.

Certain fields of large asymmetrical sand waves in the North Sea off the Netherlands did not experience migration within detectable limits of 60 m over a survey period of 2.5 yr (Langeraar, 1966; Anonymous, 1967); however, this is possibly an indication of a slow rate of migration. Other workers in the seas surrounding the British Isles have mapped in considerable detail probable regional sand distribution pathways from the facing directions of sand waves (Stride and Cartwright, 1958; Stride, 1963; Belderson and Stride, 1966; Harvey, 1966; Kenyon and Stride, 1968; Terwindt, 1971). Dominant tidal currents and sand-wave facing direction are strongly correlated.

FIELD AREA AND METHODS

The present study was performed in the entrance area of Chesapeake Bay, the largest estuary on the Atlantic Coast of the United States (Fig. 143). Physiographically the water body is a coastal plain estuary formed by drowning of a Pleistocene-incised river valley. The estuary is approximately 314 km long and 24 km in average width in the lower reaches. Water depth in the lower part of the bay is 9.1 m and in the entrance area averages 11.1 m. Fresh water river inflow averages 1580 m^3 sec^{-1}. Salinity in the entrance area ranges from $18^0/_{00}$ to $33^0/_{00}$ seasonally.

The distance between the capes at the mouth of the estuary is 18 km. Two principal shipping channels pass through this entrance: Thimble Shoal Channel, the approach to which is 25.3 m deep; and Chesapeake Channel which is 15.8 m deep. In the northern half of the entrance area there is a complex array of subtidal sand banks and intervening tidal channels. Water depth to the top of these banks ranges from 1 to 5 m relative to MLW. Water depth in the deepest parts of the channels ranges from 10 to 20 m. By following a zigzag path it is possible to move from the north side of the entrance to within 4 km of the south headland without entering water deeper than 7.3 m. This is possible because of the serpentine shape of the shoals and the position of some shoals in a blocking position athwart main channels.

Chesapeake Bay is unique in that it is sufficiently long to accommodate one semidiurnal tidal wave at all times. Owing to frictional losses there is very nearly no reflected wave felt in the entrance area and hence the tide there is of the progressive type: maximum tidal currents are nearly synchronous with high water and low water. Mean tidal range at the north side of the entrance is 0.89 m. At the south side it is 0.86 m. At spring tide the corresponding ranges are 1.10 m and 1.04 m. Duration of fall exceeds duration of rise on the northern side of the entrance in contrast to the south side.

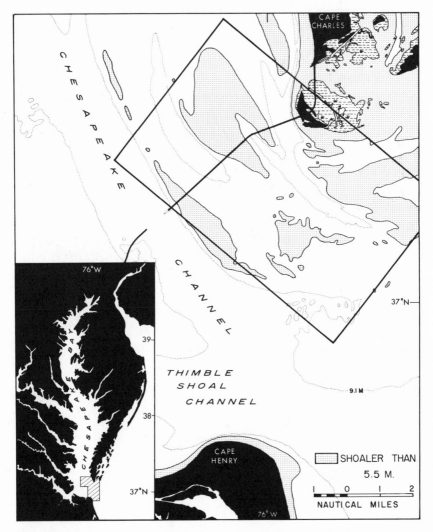

Figure 143.
Chesapeake Bay, the tidal entrance, and the study area of shoals and tidal channels.

Surface tidal currents in the entrance area of the bay, range from 50 to 100 cm sec^{-1} at ebb and flood maxima on most days. The greatest annual forecasted surface current at any place in the entrance is 175 cm sec^{-1}. This is in the channel west of Fisherman Island (Fig. 144) and occurs on a flood current. Annual maximum ebb current for the entrance occurs in the same channel but is only 140 cm sec^{-1}.

There are pronounced differences in ebb current speed and flood current speed

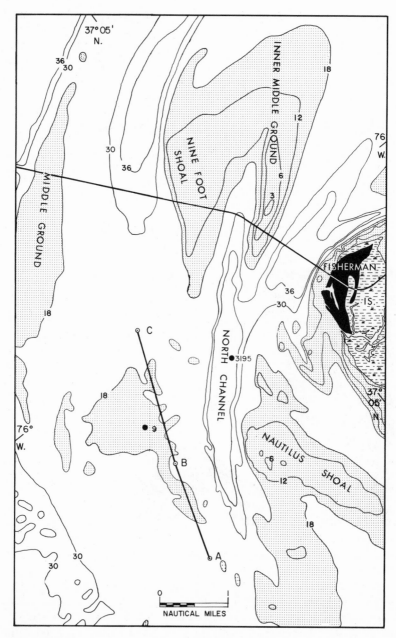

Figure 144.

The sand bank and tidal channel area of Chesapeake Bay entrance. See Fig. 143 for location.

at different places in the entrance. At the surface, flood currents predominate over ebb currents in the northern part of the entrance, in North Channel, and in Chesapeake Channel. In the seaward opening channel between Inner Middle Ground and Nine Foot Shoal, on the average, the surface flood at strength is 93 cm sec^{-1}; whereas the ebb at strength is 57 cm sec^{-1}. Surface ebb currents predominate over flood currents in the approaches to Thimble Shoal Channel, in the southern part of the entrance.

At the bottom, the pattern of ebb or flood dominance is somewhat altered. Flood currents are of greater duration at the bottom than at the surface and reach their local peak velocities sooner. Thus some stations that exhibit minor ebb dominance at the surface are shifted to flood dominance at the bottom. This is especially the case in the deep main channel in the southern part of the entrance. A provisional pattern of ebb-flood dominance for the entire entrance area at the bottom has been adduced by Ludwick (1970b).

At two specific locations, one in North Channel and the other atop a sand bank, water motion effective in moving local sediment is seen to be substantially different (Fig. 145). In North Channel, effective flood-directed motion exceeds effective ebb-directed motion; whereas, atop the bank, effective ebb-directed motion dominates.

The gross circulation of the estuary averaged over time is that of a moderately stratified estuary (Pritchard, 1967). Tidal currents reverse with ebb and flood at all depths in the entrance area; however, the mean flow is such that there is a net outflow of water in the surface layers as a whole and a net inflow of water in the bottom layers. The depth of the zone of no net motion is presently estimated to be between 5 and 7 m.

The geographic location of a fixed survey line was chosen with regard to achieving perpendicularity to the trend direction of the crests of sand waves as determined during previous surveys (Ludwick, 1970a). It was also intended that the line should be located in such a position that, if direction and magnitude of sand wave migration were detected, some new light might be shed on shoal construction processes, on the role of tidal currents acting in North Channel, the master channel of the northern entrance area, and on the effect of tidal currents on bank tops. The locations of the three points, A, B, and C, that define the fixed survey line are shown in Figure 144. The A-B distance is 2630 m; the B-C distance is 3619 m. Total line length is 6.2 km.

Each day, before the fixed line was occupied for the purpose of resurveying, 8 buoys were set out along the line. Every effort was made to set a buoy exactly on point B. The other 7 buoys were set on or very near to the line, but not necessarily at predetermined places along the line. Average distance between buoys was 910 m, a distance that permitted the sighting of 1 to 4 buoys on the line ahead of the vessel. Average water depth along the line

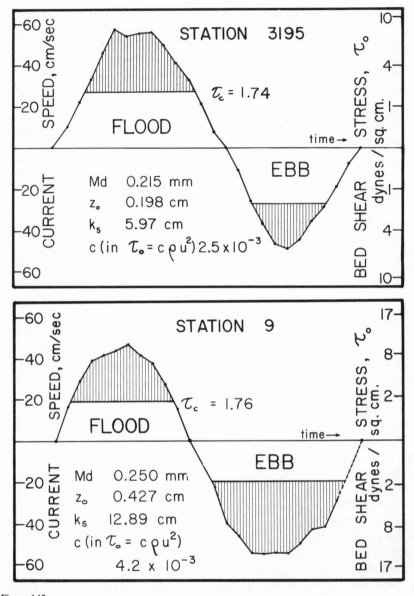

Figure 145.

Tidal current speeds in North Channel (Station 3195, October 16-19, 1963) and atop a sand bank (Station 9, September 15-22, 1952). See Fig. 144 for locations.

Total depth at Station 3195 is 56 ft; current data are for a point 18.5 ft above the bed. Total depth at Station 9 is 16 ft; current data are for a point 5 ft above the bed. Estimated bed shear stress, τ_0, is given. Observed speeds were corrected to mean tidal range and averaged over 6–12 tidal cycles. The Md is median diameter of the bed sediment; z_0 is the roughness length estimated from vertical velocity profiles, k_s is the height of bottom roughness elements deduced from z_0, and τ_c is the critical shear stress, calculated from Shield's entrainment diagram.

is approximately 7.6 m. A buoy tether line length of twice the water depth, or 15.2 m, was found satisfactory under most conditions of current. A set of four 3.6-kg rod-shaped iron weights linked together comprised the anchor for each buoy. When the currents were especially strong, a set of 6 weights was used.

After a buoy was provisionally set, the bow of the survey vessel was brought up to the buoy and an observer on the bow, using a sextant, measured horizontal angles between three known landmarks such as lighthouses, bridge spans, and radio navigation towers. Immediate plotting of the position using a metal, precision 3-arm protractor showed whether the buoy location was sufficiently close to the survey line or not. If the distance between the plotted buoy location and the survey line was 10 m or more, the buoy was dragged closer to the line, and the new position determined as before. In practical, realistic tests, it was repeatedly shown that a point in the water could be reoccupied to within 10 m or less using the sextant method, the same instrument, the same observer, and the same sighting points. Buoys were often checked to determine whether dragging had occurred. If so, the run was rejected. If it so happened that the tide turned after the buoys were set but before the line was run, a correction was made to the chart position of each buoy.

Water depth along the line of buoys was recorded using an Edo Model 578 precision depth recorder aboard the survey vessel. The paper record is made on rectangular coordinates at a depth scale of 1 cm (recording chart) per 1.28 m water depth. Chart speed through the instrument is 5.08 cm min^{-1} which corresponds to a horizontal scale of 1 cm (recording chart) per 6.09 m ground distance at a ship speed of 10 knots. Frequency of the sounding signal is 80 kHz. The acoustic cone angle between half power points is 8°. Draft adjustment, power supply frequency adjustment, and depth check were made prior to each run.

The line of buoys was followed by the survey vessel in making a run. Maximum use was made of multiple buoy sightings ahead of the ship. Compass heading was observed continuously and a run was not considered acceptable unless a constant angle of crab was maintained during a run. This last precaution insures that the ship does not drift systematically away from the fixed line between buoys due to cross currents. The strip chart was marked as each buoy was passed abeam. On each day of survey, the line was run twice.

SAND WAVES ON PROFILE A-B-C

Although comparisons among runs of the profile taken at different times reveal substantial differences, there are gross features of the bathymetry along the line that are more or less unchanging. It is these gross aspects that are to be described in this section. The changes with time will be treated in a following section. Among the gross aspects of the profile, the most obvious is the presence of larger sand waves on the landward half of the profile (Fig. 146). In the paragraphs below, the larger features are first described.

The trough to crest height of the largest seaward sand waves averages 1.9 m. The measurement refers to the vertical distance between the crest and the trough that is situated immediately seaward. Water depth to the shoalest sand-wave crest averages 5.0 m relative to MLW. Most of the sand waves on this part of the line are approximately the same height. There is no definite indication of an increase in height with distance along the seawards part of the transect.

Wavelength of the seaward sand waves was determined by means of spectral density analysis (Nordin and Algert, 1966; Nordin, 1968; Crickmore, 1970). The peak of the spectrum (Fig. 146) corresponds to a wavelength of 274 m. The mean of the spectrum, an alternative characterization, is 211 m. To obtain the spectral analysis, the original record of the run for December 29, 1969, was digitized at record intervals of 0.847 mm. Water depth was measured relative to MLW. In calculating ground distance adjustments were made for slight differences in speed of the survey vessel between each pair of buoys along the survey line.

Overall slope of the bottom along the whole line interferes with spectral analysis. A filtering procedure designed for slope removal was performed by taking a 101-point moving average of the water depths. In a final profile, elevations were calculated for each point relative to this mean surface. The resulting data set is characterized by a mean of zero. Elevations are standardized by setting the maximum deviation from the mean surface equal to unity, and changing all other values in the data set proportionally. The actual spectrum was obtained using the fast Fourier transform method embodied in a computer program written by my colleague, Dr. W. D. Stanley.

The asymmetry of the seaward sand waves is pronounced. Long gentle slopes on the backs of the features are inclined landwards; whereas shorter, steeper slopes forming the forefronts of the features are inclined seawards. Back-slopes average less than 1°; steeper forefront slopes average 1.5°, and occasionally reach 6°.

In contrast with the foregoing features, the sand waves on the landwards part of the profile are smaller and shorter. The average trough to crest height of the largest waves in this group is 1.4 m. Water depth to the crests of the shoaler sand wave averages 5.2 m below MLW. There is a rather strong indication that the height of the waves decreases with distance from B towards C, but the amount of decrease is less than 0.5 m. Sand waves essentially die out on the inner 0.5 km of the transect, except for a single feature near the inshore end of the line.

Wavelength of the inner sand waves was also determined by means of spectral analysis (Fig. 146). The peak of the spectrum indicates that maximum power

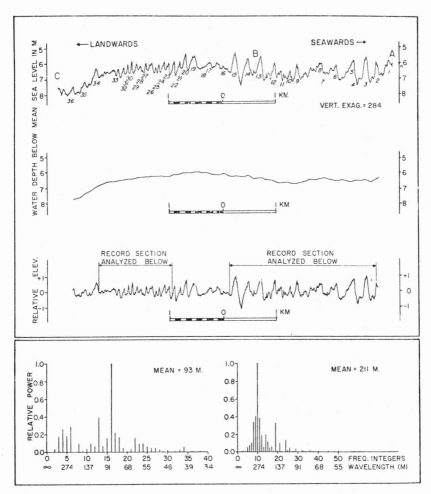

Figure 146.

Sand waves on profile A-B-C on December 29, 1969. Numbers designate sand wave crests and troughs for reference. The second profile is obtained by smoothing the first. The third profile is obtained by subtracting the second from the first. The lower diagrams are spectral density analyses of the indicated sections.

is contained in a sinusoidal component whose wavelength is 86 m. The mean of the spectrum is 93 m. The principal component dominates over other nearby frequencies to a greater extent than is the case for the peak frequency in the seawards part of the profile. The inner sand waves are more regular in spacing and height and better fitted by a single sine wave than are the larger sand waves on the outer part of the line.

The profile form of the inner waves is asymmetrical-trochoidal in contrast to the asymmetrical form of the outer sand waves. A few of the inner waves occasionally approach a perfectly symmetrical trochoidal form. When this is the case for a wave, both seawards and landwards slopes are concave upward, and the crest is quite peaked. This aspect of the features is emphasized by the vertical exaggeration in the field record, but nevertheless when allowance is made for this distortion, the slopes near pointed tops are seen to approach the angle of repose of the sediment. Most of the time, and over most of the profile, slopes of the inner sand waves are well below the angle of repose, as is also the case with the larger outer sand waves.

The contrast in sand wavelength, asymmetry, and height between the outer part of the whole profile and the inner part of the whole profile is marked. The line of division between the two sections is not sharp, but it appears to be located between 900 and 1000 m west of point B on the profile (Fig. 146). When the position of this zone of division is transferred to the transect profile obtained after smoothing, it is seen that the inner sand waves lie on one side of the high point, or divide, and that the outer larger sand waves lie on the other side of that point on the profile. The high point, or divide, separates North Channel from the unnamed channel to the south (Fig. 144). There is apparently some difference in near-bottom tidal current regime on either side of the divide.

SMALL SCALE FEATURES OF THE SAND WAVES

Certain small scale features of the sand waves are sufficiently common among the set of sequential profiles to warrant description. These features include the catback profile, the foreslope step and hole, and some very small order features discernable on the profile traces.

The term catback was first used by Van Veen (1935) to describe some unique profiles of sand waves surmounting the great tidal sand ridges of the North Sea. The term denotes a profile humped at one end and terminated at the other end by a small higher-pointed peak from which the bottom drops off abruptly into deeper water. Features of this form are common to both inner and outer sections of the present profiles (Figs. 147A1, D1, F1). The surmounting peak often rises 0.7 m or so above the top of a sand wave. Peaks of this size or thereabouts occur at the top of the foreslope of an asymmetrical sand wave, or in the center of a wave profile, or on the forefront slope itself. What is most commonly observed, by far, is a peak at the top of a foreslope. Presence of the features in identical positions on repeated profiles indicates that the peaks are not spurious records due to surface water-wave motion.

On the foreslopes of asymmetrical sand waves quite frequently there are steplike or terracelike features that interrupt the slope as it descends into the trough (Fig. 147B1, B2, E1, E2). The more or less horizontal surface of a step is approximately 150 m in width. The drop at the edge of this surface is approximately 0.6 m, beyond which the foreslope continues downwards into the trough. The edge of the step often bears a small peak. There are some instances in which the trough beyond a step appears to be abnormally deep, with steep sides so that the trough itself appears as a hole or notch in the profile (Fig. 147B1, B2, B3, E1, E2, F1). Troughs between waves of the outer profile section are often complex in form (Figs. 147E1, E2). They are the site of holes, compound smaller waves surmounted by small peaks, multiple steps, small sharply asymmetrical dunelike forms, undulations, and other features. Some waves of the outer profile, particularly those that are not strongly asymmetrical, display steps on both seaward and landward sides (Fig. 147C1, C2, C3, C4).

It is quite possible to separate waviform bottom features on the echo-sounding record from oscillations due to surface-water wave motion if the apparent periods of the two are substantially different. At an acoustic cone angle of 8° and a water depth of 6 m, a bottom area 0.8 m across is insonified. At a ship speed of 10 knots, this distance corresponds to a record length of 1.4 mm, or 1.65 sec. Thus features 0.8 m in wavelength on the bottom can be distinguished from the average surface water wave motion which has a period of 5 sec. When the ship moves in the direction of surface water wave propagation, as was usually the case, the effective period of the surface waves becomes 15 sec, and small features on the bottom are even less confounded.

It is evident that much of the profile commonly bears small waviform features on its surface (Figs. 147E1, E2, F1). These irregularities are approximately 10 m in wavelength and 0.3 m in height and may be small sand waves or dunes. Direct observation by divers will ultimately resolve this issue. The features are often strongly asymmetrical in profile. A common occurrence shows small features on one face of a large sand wave and an absence of similar small features on the other face which is usually the steeper side of the sand wave (Fig. 147B2, F1). Another occurrence shows small asymmetrical features on both sides of a larger sand wave (Fig. 147C1, E2). In this instance, the small features on both sides commonly face upslope in the direction of shoaling.

Bottom sediment along the profile is medium-grained sand. Coarsest quartz grains in each of 15 samples taken from the profile range from 3 to 7 mm in diameter with no systematic variation along the profile. The weight percentage of sediment finer grained than 0.062 mm ranges from 1.7 to 0.7 among the samples. Most of the samples contain 10% of coarse-grained broken worn shell fragments. Both the shell fragments and coarse quartz particles are iron-stained giving the samples a brownish coloration.

CHANGES IN SAND WAVES ON PROFILE A-B-C

The most obvious change with time in the sand waves on profile A-B-C is in the height of the features. Among the 21 successive profiles, 5 show

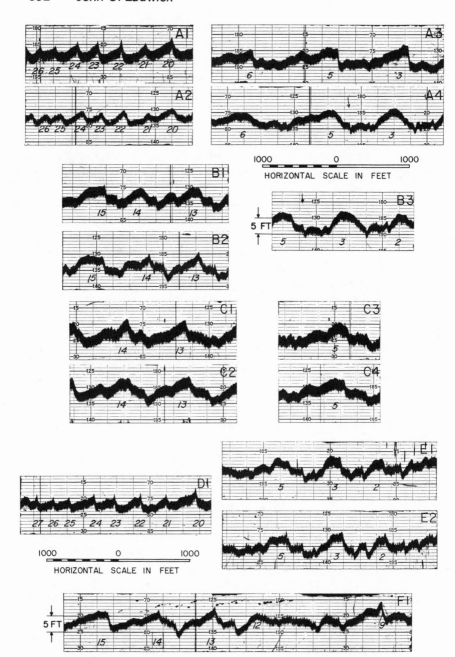

HORIZONTAL SCALE IN FEET

5 FT

1000   0   1000
HORIZONTAL SCALE IN FEET

5 FT

the bottom undulations to be low in relief, rounded in profile form particularly at sand wave crests, and free of irregularities along the surface. In contrast, 8 profiles show the bed waves to be high in relief, sharp and peaked particularly at sand wave crests, and frequently marked with sharp irregularities on foreslopes and in troughs. Steepest slopes found anywhere in the whole ensemble occur in this set. Some of the slopes among the forms in this group are as steep as 24°. These steep slopes are usually near the tops of sand waves and occur on the seaward side of the outer asymmetrical forms. On the inner asymmetrical-trochoidal features, the steep angle of repose slopes comprise the aforementioned sharp trochoidal peaks (Fig. 147A1).

Survey dates of records in the low rounded class, denoted by half-circles in Fig. 148, and dates of records in the high peaked class, denoted by triangles, are entered in the time-series plots of tidal height, tidal current strength, tidal current dominance, and wave or swell height (Fig. 148). It is readily seen that there is no correlation of the two classes with the tidal data. There is a strong correlation of the classes with the wave and swell data. Dates of the low rounded records correspond to that period of several months duration in which there are frequent episodes of high surface water waves or swells. A surface water wave height of 1.5 m (5 ft) appears to be the value that best separates high waves or swells from low waves or swells and, hence, low records from peaked records. The actual statistic plotted in the time-series diagram is the average, for each day, of eight 3-hourly observations of wave height and swell height. For each of the 3-hourly observations, used in computing the average, either wave height or swell height is used depending on which is larger.

An analysis of the dates of the low rounded group of profiles shows that, with one exception, they occur between October 6, 1969, and December 29, 1969. The one possible exception occurred on April 25, 1969, a date prior

---

Figure 147.

Depth recorder profiles from parts of survey line A-B-C (see Fig. 144). Numbers under sand waves correspond to those of Fig. 146. Vertical exaggeration is approximately 47X. Seaward is to the right, landward is to the left on each record: A1, asymmetrical-trochoidal highly peaked profiles of short wavelength sand waves, September 12, 1969; A2, same features on November 8, 1969, showing low rounded form; A3, asymmetrical profiles of long wave length, with 6° foreslopes, September 9, 1970; A4, same showing low rounded form on November 8, 1969; B1, step and hole on May 12, 1970; B2, same on June 27, 1969; B3, step and hole on November 25, 1969; C1, step on two sides of sand waves, on July 31, 1969; C2, same on August 28, 1969; C3, step on two sides of sand waves, on July 31, 1969; C4, same on August 28, 1969; D1, catbacks, on September 9, 1970; E1, complex trough topography on June 5, 1969; E2, same on June 5, 1969, second run of line; and F1, small scale sand waves and other minor features, on June 27, 1969.

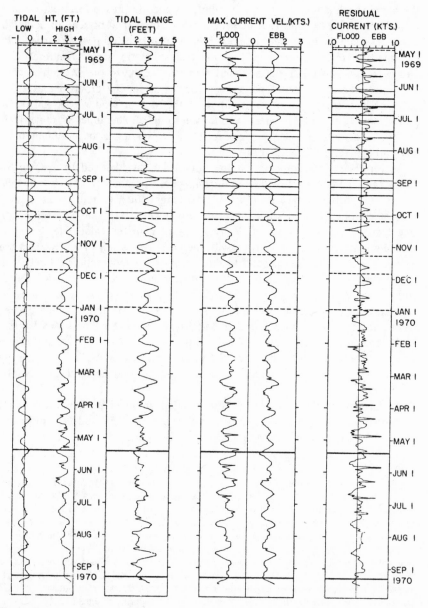

Figure 148.

The dynamic environment of Chesapeake Bay entrance during the survey period. Heavy horizontal lines mark survey dates on which sand waves on profile A-B-C were notably high in relief and sharply peaked. Triangles denote these lines; the lowest number indicating the most characteristic

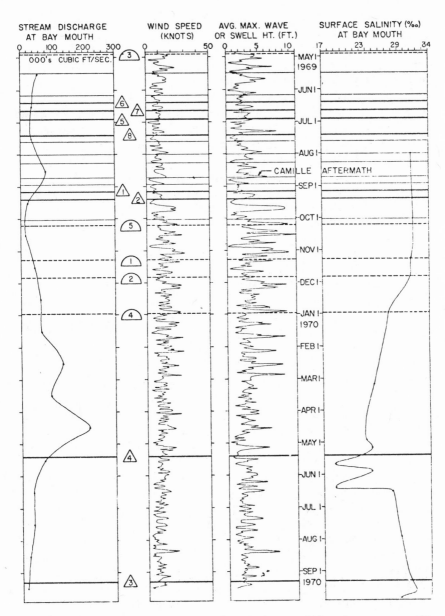

record of this class. Dashed horizontal lines mark survey dates on which sand waves were notably low in relief and rounded. Half circles denote these lines; the lowest number indicating the most characteristic record of this class. Light horizontal lines mark survey dates on which sand waves were neither notably high nor notably low in relief.

to which there is no continuous analyzed record of wave and swell heights for examination. The most characteristic member of the group is the record for November 8, 1969. Sand waves are lower and more rounded on this record than on any other. Examination of Fig. 148 shows that October 6, 1969, marks the first survey made following a 5-day period of unusually high waves and swells. No peaked, and, hence, contradictory, records were obtained during the period from the beginning of October to the end of December. It seems very likely that the period of low rounded profile types extends beyond December perhaps until late April, but no records were taken during this period owing to the unavailability of the survey ship.

Dates of records of the high peaked type range from June 12, 1969, to September 12, 1969. In 1970, records of this same type were obtained on May 12 and on September 9. Thus it would appear that this type of record occurs from May to September, a period of 5 mo. There is an indication that records obtained in May and June, although showing high peaked sand wave types, are not extreme; whereas records taken later on in September show the maximum development of high peaked forms. This latter month shows very few occurrences of surface water waves or swell height greater than 1.5 m.

The magnitude of the change in sand wave height during the year is considerable. First, with reference only to the seaward part of the profile, and with reference only to highest sand waves on each of the records, the height ranges from a minimum of 1.46 m to a maximum of 2.50 m, an increase of 71%. Second, with reference only to the landward part of the profile, and again with reference only to the highest waves on each of the records, the height ranges from a minimum of 0.92 m to a maximum of 2.13 m, an increase of 232%. The landward sand waves which are asymmetrical-trochoidal in profile undergo more than a seasonal doubling in height during the year and experience a greater change in height than do the seaward sand waves which are asymmetrical in profile.

A question that is closely related to change in sand wave height is whether an increase in height is due only to a buildup of the crest, or whether an increase in height is due only to a deepening of the trough, or whether the observed increase in height is due to both crest buildup and trough deepening. Examination of the data from the profiles indicates that both crest buildup and trough deepening occur when the height increases. Relative to MLW no sand wave was observed with less than 4 m of water atop its crest.

In addition to changes in sand-wave profile and height with time, there are also significant changes in sand-wave position during the 17-mo observation period. It was found possible to identify corresponding sand waves from record to record with essentially no uncertainty despite one time gap of four months. The average time gap between successive surveys was 24 days. Slowness of sand wave migration relative to elapsed time between surveys is what makes crest matching possible. The time-position history of 36 different sand-wave crests or troughs was followed during the total period of study.

Migration rates of sand waves on the profile were determined from plots of crest, or trough, position against time, so-called travel-time plots (Fig. 149). The slope of the best fit travel-time line is equal to the migration speed of a sand wave under study. Preliminary examination of the travel-time plots revealed that there was an irregular linear trend to the travel-time curves, and for this reason linear regression analysis was used to fit straight lines to the observed data. Linear regression analysis, as is well known (Sokal and Rohlf, 1969), assumes that one of the arguments, either time or position in this study, is error-free. Since the date of each successive survey was known, this argument was taken to be error-free, and hence is taken as the independent variable. Position of sand wave was taken as the dependent variable. Thus the line of fit that was obtained in each instance was the regression of sand wave position on time.

The relevant statistic obtained from the regression analysis is the regression coefficient, which is the slope of the line of regression of sand wave position on time. A regression coefficient, or slope, was obtained by the conventional least squares method for each of the 36 sand waves studied. The units of the regression coefficient are length per time. It was convenient to choose meters per year to express sand wave migration rate.

Examination of Tables XXVI and XXVII reveals that the maximum sand wave travel rate was 150 m yr^{-1}. The direction of travel was seawards. The sand wave involved in this motion was situated at the extreme seawards end of the profile. The rate may be somewhat in doubt because only four successive runs were available for this feature. It exited beyond point A on subsequent runs. The other extreme sand wave migration rate observed was 15 m yr^{-1} (Table XXVII). For this wave, the direction of travel was landwards. This wave is in the landwards section of the transect where the bottom features are asymmetrical-trochoidal in profile form. This wave is one of two on the entire transect that showed a landwards direction of travel. The average of all the 36 migration rates is 40.2 m yr^{-1}; and the direction is seawards.

It is seen in Tables XXVI and XXVII that sand wave travel rates differ between the landwards and seawards sections of the transect. Landwards of a point approximately 760 m west of point B on the transect, most sand wave migration rates are less than 34.8 m yr^{-1}. If valid, average migration rate of sand waves on the landward segment would be 22.2 m yr^{-1}; whereas average migration rate of sand waves on the seawards segment of the transect is 63.1 m yr^{-1}, or nearly three times as great. Direction of sand wave migration is seawards in both segments.

The question naturally arises as to whether there were a sufficient number of successive observations of the sand waves to warrant acceptance of the calcu-

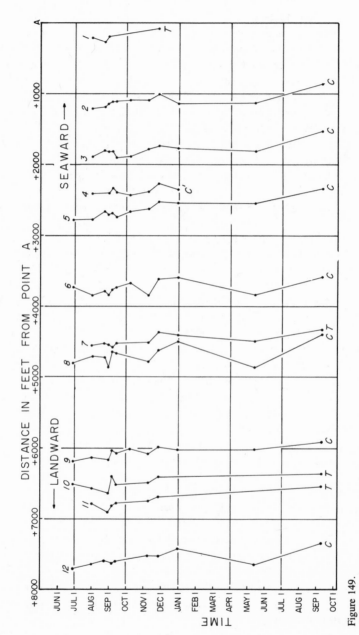

Figure 149.

Typical travel-time curves for sand waves on the outer end of profile line A-B-C.

Table XXVI
Migration Rates of Sand Waves on the Seaward Segment of Transect A-B-C in the Entrance
Area to Chesapeake Bay, Virginia[a]

Sand wave designation	Regression coefficient (m yr^{-1})	Significance $P =$	95% Confidence limits (m yr^{-1})	Position on Nov. 25, 1969 (m)
1 T	−150	<0.20	−102 - 402	$A +$ 21
2 C	− 61	<0.01	− 19 - 118	$A +$ 305
3 C	− 72	<0.01	− 31 - 112	$A +$ 530
4 C'	− 63	<0.20	− 37 - 163	$A +$ 690
5 C	− 89	<0.001	− 48 - 130	$A +$ 770
6 C	− 37	<0.20	− 18 - 93	$A +$ 1100
7 T	− 52	<0.02	− 14 - 90	$A +$ 1332
8 C	− 60	<0.02	− 19 - 140	$A +$ 1408
9 C	− 54	<0.01	− 24 - 85	$A +$ 1826
10 T	− 49	<0.10	− 10 - 108	$A +$ 1951
11 T	− 82	<0.01	− 37 - 127	$A +$ 2036
12 C	− 62	<0.05	− 11 - 113	$A +$ 2291
13 C	− 43	<0.20	− 12 - 99	$A +$ 2539
14 C	− 35	<0.40 NS	−	$B +$ 134
15 C	− 44	<0.20	− 11 - 99	$B +$ 378
16 T	− 53	<0.50 NS	−	$B +$ 570

[a] T = prominent low between sand waves; C = sand wave crest; C' = small crest between
larger sand waves; NS = not significant. Note: negative regression coefficient indicates seaward migration; positive regression coefficient indicates landward migration.

lated migration rates. A significance test for regression coefficients as given
by Sokal and Rohlf (p. 420) was used to answer this question. In the present
instance, the calculated regression coefficient was tested against a regression
coefficient of zero, i.e., zero migration rate. The significance test indicates
the probability that an observed difference in regression coefficient from zero
would occur by chance alone.

With reference only to the 16 features on the seaward segment of the transect
(Table XXVI), 8 of the regression coefficients obtained were significant at the
0.05 level or better; 14 were significant at the 0.20 level or better. That is
to say that there is only a 5% probability, or less, that a regression coefficient
as large as that obtained would occur by chance alone if the true regression
coefficient were actually zero. At least one-half the calculated migration rates,
and perhaps all but two, are deemed significant. It is concluded that the large
outer asymmetrical sand waves are actually migrating. Those closest to point
B yielded the least acceptable significance tests.

With reference only to 20 features on the landward segment of the transect
(Table XXVII), 16 of the regression coefficients obtained were considered not
to be significantly different than zero, the test yielding values of 0.40 or larger.
This figure indicates that 40% of the time one would expect to obtain regression

Table XXVII

Migration Rates of Sand Waves on the Landward Segment of Transect A-B-C in the Entrance Area to Chesapeake Bay, Virginia[a]

Sand wave designation	Regression coefficient (m yr^{-1})	Significance $P =$	95% Confidence limits (m yr^{-1})	Position on Nov. 25, 1969 (m)
17 C	−13	<0.90 NS	−	$B +$ 838
18 C'	+15	<0.90 NS	−	$B +$ 960
19 C	−15	<0.90 NS	−	$B +$ 1090
20 C	−25	<0.40 NS	−	$B +$ 1260
21 C	−21	<0.40 NS	−	$B +$ 1372
22 C	−27	<0.40 NS	−	$B +$ 1471
23 C	−27	<0.40 NS	−	$B +$ 1570
24 C	−25	<0.20	−17 - 66	$B +$ 1655
25 C	−33	<0.20	−14 - 81	$B +$ 1750
26 C	−27	<0.40 NS	−	$B +$ 1814
27 C	−27	<0.40 NS	−	$B +$ 1996
28 C	−34	<0.20	−23 - 91	$B +$ 2094
29 C	−39	<0.20	−24 - 103	$B +$ 2173
30 C	−31	<0.40 NS	−	$B +$ 2262
31 C	−20	<0.40 NS	−	$B +$ 2332
32 C	−14	<0.90 NS	−	$B +$ 2402
33 C	+ 2	<0.90 NS	−	$B +$ 2504
34 C	−34	<0.40 NS	−	$B +$ 2944
35 C	−26	<0.50 NS	−	$B +$ 3188
36 C	−24	<0.40 NS	−	$B +$ 3377

[a]C = sand wave crest; C' = small crest between large sand waves; NS = not significant.
Note; negative regression coefficient indicates seaward migration; positive regression coefficient indicates landward migration.

coefficients as large as those obtained if the true regression coefficient were actually zero. It is concluded that symmetrical-trochoidal sand waves found on the landwards part of the transect are not migrating, at least not within detectable limits of the experiment.

For those large asymmetrical sand waves that were found to migrate significantly, 95% confidence limits were calculated for the migration rate. It is seen in Tables XXVI and XXVII that the statistically justifiable conclusion is that these waves migrate at rates greater than, say, 15 m yr^{-1} but less than, say, 100 m yr^{-1}.

Detailed examination of the travel-time curves indicates the existence of some instances in which sand waves appear to move oppositely to their long-term seaward motion. Some of these instances are supported by several successive surveys, each of which shows a retrograde motion. The fact that not all the sand waves are affected at the same time lends credence to actual retrograde motion rather than an explanation based on positioning error. An example is shown in Figure 149, sand wave number 5, late August and September.

The fluid environment of the sand wave area is comprised of many factors all of which vary with time. In Figure 148 the variation during the 17-mo observation period has been shown for tidal height, tidal currents, fresh water runoff, wind speed, and wave or swell height. As regards tidal height, for each day the forecasted higher high-water and lower low-water elevations were obtained from tide tables for a point near the transect. As regards tidal range, for each day the larger range from high water to following low water was plotted. Synodic variations dominate the plot. As regards tidal currents, a plot is presented for the shoal top (Station 9, Fig. 144) of the variation in maximum forecasted ebb velocity at the water surface and maximum forecasted flood velocity at the water surface. There is a strong correlation between tidal range and tidal current speed. A plot is also presented of the difference, or residual, of ebb and flood tidal currents. Surface ebb currents are dominant at Station 9 on a time basis during the 17 mo and also in the magnitude of the residual velocity. Data are also plotted for the sum of the fresh water discharges of gaged rivers that enter Chesapeake Bay. Destructive flooding in the Appalachian Mountains following the rains from hurricane Camille are evidenced in August, 1969. Also shown is wind speed at a light tower station off the entrance to Chesapeake Bay. Individual data points are daily averages of eight 3-hr observations. The wave height or swell height data have been described in a previous paragraph. Finally, surface salinity is given for the entrance area.

DISCUSSION

The finding, in this study, of large migrating tidal sand waves whose slopes are everywhere substantially less than the angle of repose of the constituent sediment, requires an explanation at least partly different from that usually given for the mechanism of sand wave advance under unidirectional flow. With backslopes and foreslopes of 1°, or thereabouts, and at near-bottom flow rates less than 1 m sec^{-1}, significant boundary layer separation at a sand wave crest is unlikely, there is no lee eddy, there is no avalanche slope, and hence the usual mechanism of sand wave advance by the continued forward building of avalanche faces is infrequent or nonexistent.

Smith has presented a theory (1968, 1970) for unidirectional flow in which boundary layer separation, the lee eddy, and avalanche face are not necessary for sand-wave formation, and by implication, are not necessary for sand-wave migration. Observations from the present study appear to be consistent with this particular claim since the subject sand waves do not display avalanche slopes, and probably are too low in slope to produce boundary-layer separation and the associated lee eddy. And yet the subject sand waves do migrate.

By prior usage, these low-slope forms would be termed para-ripples (Bucher, 1919) if symmetrical or nearly symmetrical and lacking grain assortment. They would be termed accretion ripples (Imbrie and Buchanan, 1965) if asymmetrical in profile form, the term, accretion, being taken from Bagnold's (1941) term, accretion deposit, which denoted gently sloping, curved, tapering-aeolian cross-strata lacking conspicuous grain assortment. Although the last pair of authors did not observe the actual formation or movement of accretion ripples in their study of Bahamian deposits, they concluded from other evidence that accretion deposits with set thickness of 2.5 to 15.0 cm are formed by high velocity currents at least one-half meter thick moving down the lee side of an accretion ripple or other migrating embankment. They reasoned that accretion deposits, and hence accretion ripples, were formed at current speeds greater than those required to form avalanche faces but still in the lower flow regime. In the following paragraphs an alternative mechanism is presented for the migration of low-slope sand waves under reversing unequal tidal flows each of which are of low Froude number.

Near-bottom critical erosion velocity varies with bed slope (White, 1940; Vanoni et al., 1966), a lower downslope flow velocity being required to initiate sediment motion on a sloping bed than on a horizontal bed because the downslope component of gravity aids entrainment. The form of the relationship is

$$V_c = K_1 \ (\tan \alpha \cos \theta - \sin \theta)^{\frac{1}{2}}$$

where V_c is the critical flow velocity for initiation of sediment motion and is measured near the bed; K_1 is a dimensional constant that includes the effects of particle size, shape, and density as well as fluid density; α is the angle of repose of the sediment; and θ is the slope of the bed measured positively downward from horizontal in the direction of fluid flow.

As seen earlier in this paper, small bedform migration speed when related solely to fluid flow velocity, can be approximated by the form,

$$c_r = K_2 \ (V - V_c)^n, \qquad V \geqslant V_c$$

where c_r is bedform migration speed; K_2 is an empirical constant and depends on the choice of measurement units and/or the numerical value of n; V is the fluid flow velocity measured near the bottom; V_c is defined as in the preceding equation; and n is an empirically determined constant that appears to depend on size of the bedforms, particle size, and probably other factors.

If the previous expression for V_c is substituted in the equation above, small bedform migration speed, c_r, is then given in the resulting equation as a function, primarily, of bed slope and near-bed fluid flow velocity. Consistent comparison of bedform migration speed on beds of different slope can now be made when the various constants are assigned reasonable values. In the analysis that follows, α is taken as 30°, K_1 is taken as 0.4, which value gives V_c an acceptable magnitude in meters per second when $\theta = 0$ degrees, K_2 is taken as 200, which when coupled with $n = 2.5$ gives a relationship between bedform migration in meters per day and fluid velocity in meters per second not unlike those of some other workers (Fig. 142).

Now under tidal flow of known or assumed velocity near the bed, the equation can be used to estimate small bedform migration speed. To model sediment transport conditions on sand waves similar to those found on the subject survey transect, it is further assumed that near-bottom ebb velocity, V_e, is constant and greater than near-bottom flood velocity, V_f. Duration of ebb and duration of flood are each taken as 6 hr. The time plot of tidal current speed has the form of a square wave. Model sand waves for the area have a wavelength of 200 meters and foreslopes and backslopes are equally inclined at 1° from the horizontal. For purposes of analysis, foreslopes and backslopes are divided into serially numbered cells, each of which is 5 m in length.

The results of a sample computation using the equation described above are illustrated in Figure 150, where V_e is assumed to be 0.42 m sec^{-1} and V_f is assumed to be 0.41 m sec^{-1}. Residual migration is seaward on both slopes. This situation exists if ebb flow velocity exceeds flood flow velocity by more than 0.009 m sec^{-1}.

Under ebb flow, small bedforms that have been moving rapidly downslope are slowed when an upslope is met. This slowing-down, by itself, would be

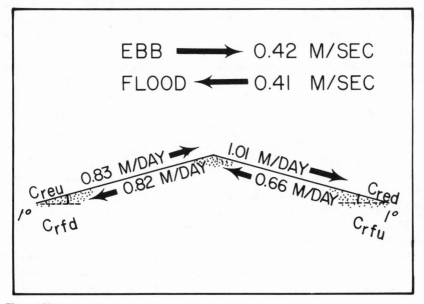

Figure 150.

Upslope and downslope estimated ripple migration rates for ebb and flood flows over an idealized sand wave.

expected to produce a decrease in crest-to-crest bedform spacing of the small bedforms on the upslope. This is tantamount to an increase in sediment thickness per unit length of upslope as compared to the thickness of moving sediment per unit length of downslope. Bumper-to-bumper automobile traffic occasioned by slow passage through a tunnel is an analogy. The ratio of the number of bedform crests per unit length of upslope to the number of bedform crests per unit length of downslope is given by c_{red}/c_{reul}, where the numerator is bedform migration rate on ebb flow on a downslope, and the denominator is bedform migration rate on ebb flow on an upslope. Under flood flow, the relative increase in sediment thickness on an upslope is correspondingly given by c_{rfd}/c_{rfu}.

It is inferred from the foregoing that the slowing-down of bedform migration rate on adverse slopes is responsible for the development of sediment steps seen on the sand-wave profiles that were described and figured earlier in this paper. Steps were found on both foreslopes and on backslopes of sand waves, giving the waves a "head-and-shoulders" appearance. The common complexity of the bottom seen in troughs between large sand waves is thought to be due to the change in bedform migration rate at that place, both on ebb and flood flows.

At the top of the large sand waves where the slope of the bottom is more or less level, the ebb-dominated tidal flow causes a net seawards sediment motion which, over time, fills the pocket on the seaward side of the feature between the crest and the seaward step. By these mechanisms, the large sand waves are believed to migrate seaward.

The explanation given above singles out small migrating bedforms as the main causal agent for movement of the large bedforms; however, observations by scuba divers have shown that there is, in addition to the small bedforms, a cloud 1 m thick, or thereabouts, of moving sediment and water in contact with the bed when tidal currents are flowing. This moving sediment probably also plays an important role in migration processes of the large sand waves. The equation presented above can be used to explore consequences of such movement by assigning a larger magnitude to the constant, K_2. This alteration allows for the much greater mobility of the sediment-water cloud as compared to small migrating bedforms and yet maintains the effects of upslopes and downslopes on speed of sediment motion.

Trajectories of particles for one ebb-flood cycle under different ebb and flood velocities were examined using the equation. One sediment particle was positioned initially in each cell of the model sand wave referred to above. The initial departure of a particle from a cell was denoted by the symbol, -1. The coming-to-rest of a particle in a cell at the end of one ebb-flood cycle was denoted by the symbol, $+1$. The algebraic sum of the losses and gains in each cell after one ebb-flood cycle indicates the net accumulation or net removal of sediment from that cell. Trajectories that begin on adjacent sand waves affect the outcome on a central sand wave to which interest was principally directed.

Results obtained indicate that for most reasonable combinations of ebb and flood velocities there is an accumulation of sediment on seaward-facing slopes and in troughs between sand waves. Minor erosion or zero net change occurs in cells positioned elsewhere along the sand wave profile.

The migration rates of the larger seaward sand waves, as well as the travel rates of the inner asymmetrical-trochoidal sand waves, are slow compared to the probable movement rates of sediment ripples and the even faster translation rates of the cloud of moving sediment and water near the bottom. This disparity in relative travel rate requires the existence in the bed of a traction layer at least 30 cm thick. It seems likely that there is net sediment transit through the sand wave field in this upper surface zone. In this sense the large sand waves may be likened to the kinematic waves described by cloud meteorologists, traffic engineers, and queuing theory mathematicians. The theory set forth in 1925 by Exner (Leliavksy, 1955, p. 24) for sand wave translation and deformation is a kinematic wave theory.

Kinematic waves, either stationary or moving, result from local bunching-up of particles, or automobiles, or unit parcels due to a change in forward velocity of the units. Nevertheless, the units pass through the waves. This characteristic of sediment passage through waves is a noteworthy distinction between low-slope sand waves and sand waves with avalanche slopes. On the latter, sediment is temporarily trapped; on the former features, sediment passes through.

The migration and profile form of the sand waves on profile A-B-C shed some new light on the configuration of ebb-dominated and flood-dominated tidal flows in the sand bank area of Chesapeake Bay entrance. Judged from findings presented above, the unnamed shoal on which profile A-B-C is situated (Fig. 144) is an embryonic flood parabola, in the usage of Van Veen (1936, 1950), i.e., a parabolic-shaped shoal open to flood tidal currents. The seaward-facing asymmetrical, faster-moving sand waves of limb A-B and part of B-C are apparently located on the flank of what Hayes (1969) has termed an ebb spit. This is a long trailing tail of a flood parabola (Fig. 151). This section is exposed to ebb currents which are quite strong because of the convergence of channel margins. Flood currents in this position outside the limbs of the parabola are much weaker than the ebb currents.

Limb B-C is chiefly through smaller asymmetrical-trochoidal sand waves that also face seawards. These waves apparently owe their more symmetric form and low migration rate to flood currents which somewhat offset the oppositely directed powerful ebb currents. The ebb currents are not as strong in this section as in section A-B because there is much less channel constriction.

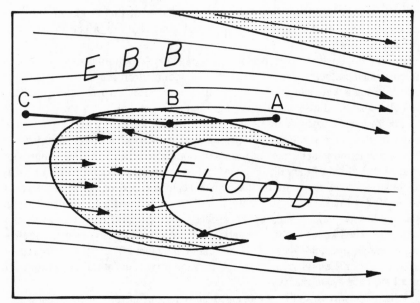

Figure 151.
The ebb-flood tidal hydraulic environment of profile A-B-C.

The flood currents are stronger over this section (B-C) as is consistently the case for flow between horns of flood parabolas. Nevertheless, as shown in Fig. 145, ebb-directed currents still exceed flood-directed currents.

The sand bank area in the entrance to Chesapeake Bay is contiguous with the sand ridge and swale bathymetry of the continental shelf off the eastern United States (Uchupi, 1968). The question of the origin of that bathymetry is not markedly clarified by the present study because the tidal currents insofar as they are known are so appreciably different in strength between the two areas. Findings of the present study relate chiefly to shoaling in tidal entrances or to areas like Nantucket Shoals or Georges Bank where bottom currents are strong enough to move substantial quantities of sand-sized sediment.

CONCLUSIONS

The following are the principal conclusions reached in this study:

1. In close proximity, on a shoal in the tidal entrance to Chesapeake Bay, there are large asymmetrical seaward-facing sand waves and smaller asymmetrical-trochoidal seaward-facing sand waves. The large features are approximately 270 m in wavelength and 1.9 m in average height. The smaller features are 85 m in wavelength and 1.4 m in neight. Both waves lack avalanche slopes; the average slope is 1° or less, but slopes up to 6° occur occasionally and in the extreme reach 20°.

2. The asymmetrical-trochoidal waves occur on one side of a bathymetric divide atop a shoal; the asymmetrical sand waves occur on the other side of the divide.

3. Profiles of the sand waves display catback shapes, sediment steps on one side or on both sides of central sand waves, and holes or other irregularities, especially in troughs between sand waves. Small sand waves, 10 m in length, 0.3 m in height, display avalanche slopes and commonly face upslope on both sides of large sand waves.

4. A survey line across the aforementioned features was surveyed 22 times over a 17-month period. The large asymmetrical sand waves migrate seawards at rates between 15 and 100 m yr^1 and average 63 m yr^1. These rates were shown to be significant statistically. The smaller asymmetrical-trochoidal sand waves could not be shown to be migrating within the detection limits of the experiment.

5. The sand waves studied were shown to be located on the margin of what is probably a flood parabola (Van Veen, 1936); the large asymmetrical sand waves were on the ebb-dominated flank of the ebb spit (Hayes, 1969); the small asymmetrical-trochoidal sand waves were on the ebb-dominated shoal top, but so situated as to be subject to strong flood flow emanating from a flood-dominated channel between the horns of the flood parabola.

6. The low-slope sand waves lack avalanche slopes and are believed to migrate under the existing subcritical flow conditions by the construction of sediment steps or secondary waves on sand wave flanks. These steps or secondary waves represent accumulations due to upslope slowing-down of small migrating bed-forms. Sediment transport on sand wave crests, by small migrating bedforms, fills the pocket between the crest and the step or secondary sand wave. Under ebb-dominated flow, this filling occurs more rapidly on the ebb side than on the flood side of the wave, thereby causing an apparent ebb migration of the entire large wave.

7. Sand waves were seen to undergo a pronounced seasonal change in profile form. From May to September, profiles showed relatively steepened, higher, peaked forms; whereas from October to April, profiles showed lower rounded forms for the same sand waves. The apparent cause is the near absence of surface water waves or swell higher than 1.5 m from May to September; whereas surface water waves and swell higher than 1.5 m occur frequently from October through April.

8. In shallow water under reversing tidal flow, the height and profile form of sand waves are determined by the action of surface water waves as well as by the velocity inequality of the tidal currents.

ACKNOWLEDGMENT

Several faculty members and graduate students of the Institute of Oceanography, Old Dominion University, participated in some of the studies reported in this chapter. They include Drs. D. J. P.

Swift and R. E. Johnson, and graduate students, W. C. Smith, J. T. Wells, and J. R. Melchor. Captains of R/V Albatross and R/V Linwood Holton were helpful in accomplishing the objectives.

REFERENCES

Allen, J. R. L. (1962). Asymmetrical ripple marks and the origin of cross-stratification. *Nature* **194**, 167-169.

Allen, J. R. L. (1963). Asymmetrical ripple marks and the origin of water-laid cosets of cross-strata. *Liverpool Manchester Geol. J.* **3**, 187-236.

Allen, J. R. L. (1965). Sedimentation to the lee of small underwater sand waves. *J. Geol.* **73**, 95-116.

Anonymous (1967). The northern limit of megaripples in the North Sea. *Hydrographic Newsletter* **1**, 339-340.

Ashida, K., and Tanaka, Y. (1967). A statistical study of sand waves. *Proc. XIIth Internat. Assoc. Hydraulic Res.* **2**, 103-110.

Bagnold., R. A. (1941). "The Physics of Blown Sands and Desert Dunes." 165 pp. Methuen, London.

Ballade, P. (1953). Etudes des fonds sableux en Loire Maritime, nature et evolution des ridens. *Comité de' Oceanographie Bull. d' Information* **5**, 163-176.

Belderson, R. H., and Stride, A. H. (1966). Tidal current fashioning of a basal bed. *Marine Geol.* **4**, 237-257.

Bucher, W. H. (1919). On ripples and related sedimentary surface forms and their paleographic interpretation. *Am. J. Sci.* **47**, 149-210, 241-269.

Carey, W. C., and Keller, M. D. (1957). Systematic changes in the beds of alluvial rivers. *Proc. Am. Soc. Civil Engrs.* **83(HY4)**, 1-24.

Cartwright, D. E. (1959). On submarine sand-waves and tidal lee-waves. *Proc. Roy. Soc. (London), Series A,* **253**, 218-241.

Chang, Y. L. (1939). Laboratory investigation of flume traction and transportation. *Trans. Am. Soc. Civil Engrs.* **104**, 1246-1284.

Cloet, R. L. (1954a). Sandwaves in the southern North Sea and in the Persian Gulf. *J. Inst. Navig.* **7**, 272-279.

Cloet, R. L. (1954b). Hydrographic analysis of the Goodwin Sands and the Brake Bank. *Geogr. J.* **120**, 203-215.

Cloet, R. L. (1954b). Hydrographic analysis of the Goodwin Sands and the Brake Bank. *Georgr. J.* **120**, 203-215.

Cornish, V. (1901). On sand waves in tidal currents. *Geogr. J. London* **18**, 170-202.

Crickmore, M. J. (1970). Effect of flume width on bed form characteristics. *Proc. Am. Soc. Civil Engrs., J. Hydraulics Div.* **96(HY2)**, 473-496.

Deacon, G. F. (1894). Discussion of "Estuaries" by H. L. Partiot. *Proc. Inst. of Civil Engrs.* **118**, 47-189.

Dingle, R. V. (1965). Sand waves in the North Sea mapped by continuous reflection profiling. *Marine Geol.* **3**, 391-400.

Forel, F. A. (1883). Les ridens de fond étudies dan le Lac Léman. *Archives des Sciences Physiques et Naturelles, 3e Sér.* **10**, 39-72.

Gilbert, G. K. (1914). The transportation of debris by running water. *U.S. Geol. Surv., Profess. Paper* **86**, 1-263.

Harms, J. C. (1969). Hydraulic significance of some sand ripples. *Geol. Soc. Am. Bull.* **80**, 363-396.

Harvey, J. G. (1966). Large sand waves in the Irish Sea. *Marine Geol.* **4**, 49-55.

Hayes, M. O. (1969). Forms of accumulation in estuaries. *In* "Coastal Environments of Northeastern Massachusetts and New Hampshire-Coastal Research Group," Contribution No. 1-CRG, Geology Dept., Univ. of Mass., Amherst, Massachusetts. 462 pp.

Houbolt, J. J. H. C. (1968). Recent sediments in the southern bight of the North Sea. *Geologie en Mijnbouw* **47**(4), 245-273.

Imbrie, J., and Buchanan, H. (1965). Sedimentary structures in modern carbonate sands of the Bahamas. *In* "Primary Sedimentary Structures and Their Hydrodynamic Significance" (G. V. Middleton, ed.), *Soc. Econ. Paleontologists and Mineralogists, Spec. Publ.* **12**, 149-172.

James, N. P., and Stanley, D. J. (1968). Sable Island Bank off Nova Scotia: sediment dispersal and recent history. *Am. Assoc. Pet. Geol. Bull.* **52**, 2208-2230.

Jones, N. S., Kain, J. M., and Stride, A. H. (1965). The movement of sand waves on Warts Bank, Isle of Man. *Marine Geol.* **3**, 329-336.

Jopling, A. V. (1965). Laboratory study of the distribution of grain sizes in cross-bedded deposits. *In* "Primary Sedimentary Structures and Their Hydrodynamic Significance" (G. V. Middleton, ed.), *Soc. Econ. Paleontologists and Mineralogists, Spec. Publ.* **12**, 34-52.

Jordan, G. F. (1962). Large submarine sand waves. *Science* **136**, 839-848.

Kennedy, J. F. (1963). The mechanics of dunes and antidunes in erodible-bed channels. *J. Fluid Mech.* **16**, 521-544.

Kennedy, J. F. (1969). The formation of sediment ripples, dunes, and antidunes. *Ann. Rev. Fluid Mechanics* (W. R. Sears, Ed.) **1**, 147-168.

Kenyon, N. H., and Stride, A. H. (1968). The crest length and sinuosity of some marine sand waves. *J. Sediment. Petrology* **38**, 255-259.

Klein, G. deV. (1970). Depositional and dispersal dynamics of intertidal sand bars. *J. Sediment. Petrology* **40**, 1095-1127.

Lane, E. W., and Eden, E. W. (1940). Sand waves in the lower Mississippi River. *J. Western Soc. Engr.* **45**, 281-291.

Langeraar, W. (1966). Sand waves in the North Sea. *Hydrographic Newsletter* **1**, 243-246.

Leliavsky, S. (1955). "An Introduction to Fluvial Hydraulics" 257 pp. Constable, London.

Ludwick, J. C. (1970a). Sand waves in the tidal entrance to Chesapeake Bay: preliminary observations. *Chesapeake Science* **11**, 98-110.

Ludwick, J. C. (1970b). Sand waves and tidal channels in the entrance to Chesapeake Bay. *Virginia J. Sci.* **21**, 178-184.

Nordin, C. F. (1968). "Statistical Properties of Dune Profiles." 137 pp. Ph.D. dissertation, Colorado State University, Ft. Collins, Colorado.

Nordin, C. F., and Algert, J. H. (1966). Spectral analysis of sand waves. *Proc. Am. Soc. Civil Engrs., J. Hydraulics Div.* **92**(HY5), 95-114.

Pritchard, D. W. (1967). Observations of circulation in coastal plain estuaries. *In* "Estuaries (G. H. Lauff, ed.), *Am. Assoc. Adv. Sci., Publ.* **83**, 37-51.

Salsman, G. G., Tolbert, W. H., and Villars, R. G. (1966). Sand-ridge migration in St. Andrew Bay, Florida. *Marine Geol.* **4**, 11-19.

Simons, D. B. and Richardson, E. V. (1960). Resistance to flow in alluvial channels. *Proc. Am. Soc. Civil Engrs., J. Hydraulics Div.* **86**(HY5), 73-99.

Simons, D. B. and Richardson, E. V. (1961). Forms of bed roughness in alluvial channels. *Proc. Am. Soc. Civil Engrs., J. Hydraulics Div.* **87**(HY3), 87-105.

Simons, D. B., Richardson, E. V., and Nordin, C. F., Jr. (1965). Sedimentary structures generated by flow in alluvial channels. *In* "Primary Sedimentary Structures and Their Hydrodynamic Significance (G. V. Middleton, ed.) *Soc. Econ. Paleontologists and Mineralogists, Spec. Publ.* **12**, 34-52.

Smith, J. D. (1968). "The Dynamics of Sand Waves and Sand Ridges." 78 pp. Ph.D. dissertation, The University of Chicago, Chicago, Illinois.

Smith, J. D. (1969). Geomorphology of a sand ridge. *J. Geol.* **77**, 39-55.

Smith, J. D. (1970). Stability of a sand bed subjected to a shear flow of low Froude number. *J. Geophys. Res.* **75**, 5928-5940.

Sokal, R. R., and Rohlf, F. J. (1969). "Biometry." 776 pp. Freeman, San Francisco.

Stride, A. H. (1963). Current-swept sea floors near the southern half of Great Britain. *Quart. J. Geol. Soc. London* **119**, 175-199.

Stride, A. H., and Cartwright, D. E. (1958). Sand transport at southern end of the North Sea. *Dock and Harbour Authority* **39**, 323-324.

Terwindt, J. H. J. (1971). Sand waves in the southern bight of the North Sea. *Marine Geol.* **10**, 51-67.

Uchupi, E. (1968). Atlantic continental shelf and slope of the United States - Physiography. *U.S. Geol. Survey Prof. Paper* **529-C**, 30 pp.

Vanoni, V. A. (1966). Sediment transportation mechanics: initiation of motion (Progress Report of Task Committee). *Proc. Am. Soc. Civil Engrs., J. Hydraulics Div.* **92(HY2)**, 291-314.

Van Straaten, L. M. J. U. (1950). Giant ripples in tidal channels. *Konikl. Ned. Aadr. Gen. Tijdschr.* **67**, 336-341.

Van Straaten, L. M. J. U. (1953). Megaripples in the Dutch Wadden Sea and in the Basin of Arachon (France). *Geologie en Mijnbouw* **15**, 1-11.

Van Veen, J. (1935). Sand waves in the North Sea. *Hydrographic Review* **12**, 21-29.

Van Veen, J. (1936). "Onderzoekingen in de Hoofden in verband met de gesteldheid der Nederlandsche kust." 252 pp. Ph.D. dissertation, Leiden University, Netherlands. Printed at The Hague.

Van Veen, J. (1950). Ebb- and flood-channel systems in the Dutch tidal waters. *Tidjdschr. Koninkl. Ned. Aardrijkskundig Genootschap* **67**, 303-325. A.T.S. Translation 132 DU.

Visher, G. S. (1965). Fluvial processes as interpreted from ancient and recent fluvial deposits. *In* "Primary Sedimentary Structures and Their Hydrodynamic Significance" (G. V. Middleton, ed.), *Soc. Econ. Paleontologists and Mineralogists, Spec. Publ.* **12**, 116-132.

White, C. M. (1940). The equilibrium of grains on the bed of a stream. *Proc. Roy. Soc. (London), Series A* **174**, 322-338.

CHAPTER 20

Water Circulation and Sedimentation at Estuary Entrances on the Georgia Coast

George F. Oertel, II, and James D. Howard

Skidaway Institute of Oceanography
Savannah, Georgia 31406

ABSTRACT

Estuary entrance-shoals extend seaward of all major entrances of the Georgia coast. Orientation and morphology of these shoals occur in response to (a) transient coastal-currents controlled by seasonal winds, (b) the relative magnitude of ebb-jets to coastal currents, and (c) hydraulic pressure gradients at entrances.

Northeasterly winds create high-energy swells which result in a southward diversion of longshore drift and a southeast aggradation of shoals on the north sides of inlets. These shoals are detached from the shoreline and are generally breached in several places by small tidal channels. Attached shoals on the south sides of inlets are sediment starved during these periods.

Southerly onshore-winds result in currents which cause a seaward prograding of shoals on the south sides of inlets. At the same time, sediment cells and gyres trap sediment on shoals on the north sides of inlets and accretion is multidirectional.

INTRODUCTION

Shoals which extend two to four miles seaward of estuary entrances characterize the major tidal inlets of the Georgia coast. Water and sediment circulation control the geometric configurations of these shoals, and hence, hydrodynamics and sedimentation associated with inlets and shoals should be considered when constructing theories of inlet and barrier island formation. An understanding of the relationship between sediment transport and hydrodynamics at barrier inlets is fundamental to the interpretation of barrier modification during transgressions and regressions. The effects of these relationships on barrier and inlet modification is pertinent for predicting strandline movements of geologic counterparts and to maintaining navigable inlets.

Literature specifically related to nearshore hydrodynamics and sedimentation at inlets includes Allen (1971), Bajorunas and Duane (1967), Hoyt and Henry, (1967), Ludwick (1970a, 1970b), Oertel (1971a), Robinson (1960), Todd (1968), and White (1966). Other literature with discussions of nearshore and offshore shoal-development includes Abbe (1895), de Beaumont (1845), Dietz (1963), Dolan and Ferm (1968), Johnson (1919), Otvos (1970), Pierce and Colquhoun (1970), Price (1964), Shepard (1960), Stewart and Jordan (1965), Swift (1969), Uchupi (1968), and Zenkovitch (1964). Bumpus (1955) and Kuroda (1969) have also discussed patterns of nearshore water-circulation which are relevant to this study. Many of these investigations are discussions of strictly stratigraphic, geomorphic, or hydrographic data. This study considers integrated characteristics of water circulation with patterns of sediment depletion, transportation, and accumulation.

GENERAL CHARACTERISTICS

All major estuarine inlets of the Georgia coast (Fig. 152) have been considered in this study and a detailed investigation of the hydrodynamics and sedimentation was conducted at the entrance of the Doboy Sound estuary. This estuary has not undergone any artificial modification by dredging or beach nourishments. Terminology used in this paper is illustrated in Fig. 153. Major tidal inlets along the Georgia coast are one to four miles wide. Maximum depths at the entrances vary from 15 to 35 m and are generally at the intersection of the inlet with the shoreline (see position of trough, Fig. 153). A ramp to the sea (Scruton, 1956) composed of texturally contrasting sedimentary sets extends three to four miles seaward of entrances to the shore face which varies from 5 to 6 m in depth. At each inlet, the ramp is flanked by shoals which separate it from the shore face. These shoals are partially exposed during low water and have slopes of 10°-30° toward the channel (Hoyt and Henry, 1967). Away from the ramp the shoals generally slope at angles less than 10°. Shoals along the north margins of the ramps have obvious similarities which distinguish them from shoals on the south sides of ramps. "North-margin shoals" are generally breached by tidal channels and runnels which are oblique or parallel to the

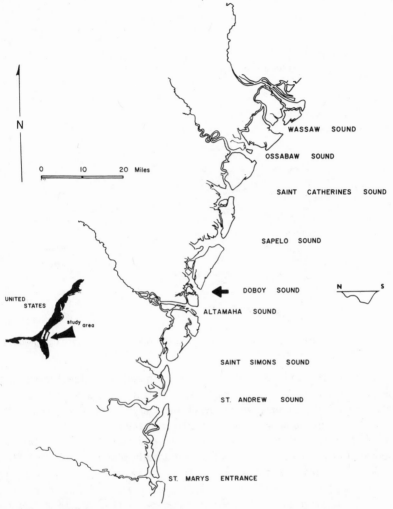

Figure 152.

Index map of Georgia coast. The Georgia coast is characterized by numerous barrier islands separated by major tidal inlets. General hydrography and sedimentation was studied at all entrances and a detailed study was conducted at the entrance of Doboy Sound.

beach. The deepest of these channels is closest to the shoreline and separates the shoals from the beach. Large sand-bars are commonly present along the seaward side of these channels. Small oblique-trending bars are also present along other smaller channels and runnels which breach the seaward shoal-trend. Hence, "north-margin shoals" are detached from the shoreline and have two

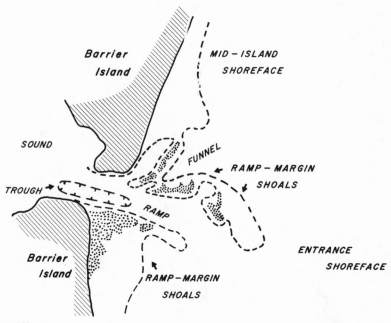

Figure 153.

Barrier-entrance terminology. Terminology of general characteristics of morphologic features of Georgia estuary-entrances are shown here.

major trends; one parallel or oblique to the beach and one parallel to the ramp to the sea. Shoals on the south side of the ramp to the sea are not detached from the shoreline. Runnels which breach these shoals are shallower and less common. "South-side shoals" are triangular in planview.

At the estuary entrances, there exists a variety of sediment textures. The shoreface is characterized by muddy fine sand locally containing abundant shells and shell fragments. Ramp-margin shoals are composed of clean, fine sand. Layers of coarse-to-fine sand are interbedded with layers of mud in the ramp-to-the-sea and the trough has a veneer of coarse sand to granules overlying Miocene limey sandstone.

The tidal range on the Georgia coast averages 2.3 m but exceeds 3 m at spring tides. Maximum storm swells are generally out of the northeast but are not prevalent. Currents over the offshore and inshore zones are transient and are related to seasonally variable wind conditions. Winds and swell statistics (U.S. Naval Oceanographic Office, 1963; U.S. Weather Bureau, 1969; Helle, 1958) indicate that wave-induced currents along the southeast coast of the United States are transient and related to seasonal weather patterns. Hydrographic data (Bumpus, 1955); Kuroda, 1969; Oertel, 1971a) show that coastal currents are also transient and are related to winds, runoff, and dynamic pressure-gradients.

Dynamic diversion (Todd, 1968) of tidal and transient currents generate sediment-circulation cells (Bajorunas and Duane, 1967; Ludwick, 1970a, 1970b; Robinson, 1960) and sediment gyres (Oertel, 1971b). These gyres and cells are mechanisms of sediment accumulation which are fundamental to the formation and maintenance of shoals extending seaward and marginal to estuary entrances.

DISCUSSION

Winds and Currents

Currents on the shallow shelf of the southeastern United States may be considered as two major types; tidal currents and transient coastal currents. Semidiurnal tidal-flow is approximately onshore and offshore. When transient coastal currents are added to this system, ebb or flood tidal-flow is either enhanced, inhibited, or diverted.

Winds are the most influential force producing transient currents. Longshore winds produce longshore wave-induced currents and longshore coastal currents. Onshore winds produce (a) currents in response to wave refraction, and (b) coastal currents which pile-up water on the shore and produce hydraulic currents. Based on United States Naval Oceanographic Office Report (1963) and from data collected at the Sapelo Island weather station in 1969, the frequency percent of winds which influence currents and sediment transport along the Georgia coast are clustered in three maximum-frequency groups (Fig. 155). The predominant frequency percent of winds during the winter months of December, January, and February is offshore. During the spring and summer months (March, April, May, June, July, and August), the predominant wind frequency is longshore from the southeast. During the autumn months (September, October, November) the highest frequency percent is longshore from the northeast and north-northeast. Throughout any given year there is a relatively high frequency-percent of onshore winds.

Onshore Northeast Winds

Prevailing autumn winds produce high-energy waves which result in south-flowing coastal and longshore currents. These waves produce a zone of longshore drift which at times extends several miles seaward to the zone of breaking waves. Hence, high wave-energy may be initially dissipated over the shallow shoreface before it impinges directly on the island beaches. South-flowing currents are diverted seaward when they come in contact with the influence of ramp-margin shoals and inlet ebb-jets (Todd, 1968). Initially, the extent and direction of seaward diversion depends on the relative intensities of the inlet ebb-jets and longshore currents. Once inlet-margin shoals have been establishehed, they become an important factor affecting wave refraction and longshore current

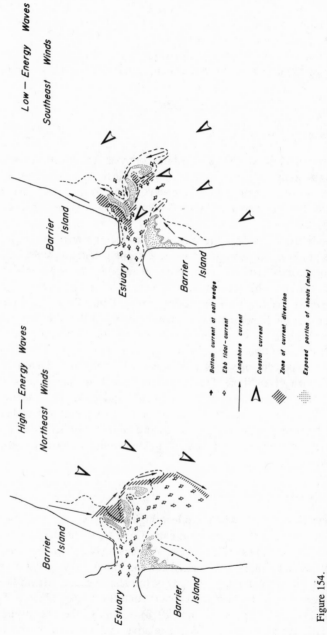

Figure 154.

Seasonal current patterns at Georgia estuary-entrance: (a) Patterns of ebb tidal-currents, longshore currents and coastal currents during northeast winds and resultant high-energy waves; (b) Patterns of ebb tidal-currents, longshore currents, and coastal currents during southeast winds and low-energy waves.

diversion. Onshore northeast swells wrap around seaward extending shoals and produce breaker zones with currents flowing toward the axis of the sounds. Northeast swells impinging on the beaches also produce a longshore drift flowing toward the sounds (Fig. 154a). Upon reaching the estuary entrance these wave-induced currents are diverted seaward by the ebb jets. The diverted flow is generally in a southeast direction along the axis of the ramp to the sea. During the early ebb-flow, dynamic diversion of longshore currents is initiated by ebb-jets. During the late ebb-flow, bathymetric shielding, by ramp-margin shoals, results in the confinement of flow to channels.

South-flowing longshore and coastal currents induce a southward sediment transport along the Georgia coast (Hoyt and Henry, 1967). Some of the sediment entrained in longshore currents is deposited on the south ends of barrier islands and results in a southward barrier progradation (Hoyt and Henry, 1967). However, most sediment is diverted seaward and is deposited along ramp-margin shoals. Sediment bypassing of the inlet takes place at a distal shoal on the seaward end of the ramp to the sea (between two and four miles offshore). Hence, sediment is diverted southward across inlets during notheast swells and is deposited a considerable distance seaward of the beaches on the north ends of barrier islands. Sediment entrainment in these patterns tends to form shoals and spits which are attached to the south ends of barrier islands. With this set of conditions, shoals prograde seaward and southward, and longshore drift may be of sufficient magnitude to cause encroachment of inlet channels (Johnson, 1919; Shepard, 1960; Hoyt and Henry, 1967) by margin shoals. Since sediment bypassing takes place several miles offshore, shoals on the south sides of ramps receive only small amounts of sediment and these shoals are apparently sediment starved and eroded by the high-energy waves.

Offshore Winds

During the winter months (December, January, February), prevailing winds are offshore. Offshore drift-currents which result from these winds enhance the ebb-jet flow out of estuary entrances. In the nearshore zone, longshore and onshore currents induced by waves are subordinate to these ebb currents. Current diversion is minimal and ebb-jet currents leaving constricted estuary entrances spill seaward in a radially distributed pattern. Sediment transport over the shallow shoreface is primarily accomplished by tidal currents. Shear between tidal currents and the sediment-water interface causes sediment migration in dune-shaped bedforms. The bedforms vary in size from ripples to sand waves and have orientations indicating tidal-flow patterns. Dune modification is common in shallow areas when onshore wind-shifts cause increased wave-activity and turbulence in the water column.

Onshore Southeast Winds

During spring and summer months, prevailing winds have predominant onshore and longshore components from the south. This long duration (6 mo) of southerly winds induces northward-flowing longshore-currents and coastal-currents as well as onshore drift-currents and onshore wave-currents. As indicated by Kuroda (1969), swells from this quadrant are of low energy (less than 5 ft). Longshore currents and onshore wave-currents produced by these swells are confined to a narrow zone bordering the shallow shoreface. Although these swells are low-energy waves, the frequency percent of winds out of the southeast quadrant is almost double the frequency percent of higher-energy swells out of the northeast (Fig. 155).

Onshore currents modify inlet circulation in two ways. Onshore drift currents pile-up water against the coastline and create hydraulic currents over the shallow shoreface. As swells wrap around ramp-margin shoals, the mass transport of waves refracting around ramp-margin shoals causes a convergence of water toward entrances. This onshore convergence of water masses toward the entrance is also enhanced by several other mechanisms. A weak salt wedge is present in the water column above the ramp and migrates from just inside entrances at high water to several miles seaward of entrances at low water. Landward-flowing bottom currents of the salt wedge inhibit the ebb flow during the falling tide and enhance the flood flow during the rising ride. The effect of salt-diffusion currents on the vertical differentiation of flow is readily apparent from current-velocity and current-direction data collected at the entrance to Doboy Sound (Fig. 156, 157). During periods of maximum ebb-velocities (90 cm sec^{-1}) at the surface of the water column at profile G, the bottom of the water column experienced "slack" water (Fig. 156). At profile D, ebb surface-velocities were also

	yearly	some longshore component	some onshore component	WINTER (Dec, Jan, Feb)	SPRING (Mar, Apr, May)	SUMMER (Jun, Jul, Aug)	AUTUMN (Sept, Oct, Nov)
Longshore (NNE, NE)	15.6	23.3		18.0	6.6	8.0	30.6
Longshore - onshore (ENE)	7.7		45.2	5.6	10.0	3.4	11.8
Onshore - longshore (E, ESE)	17.9	variable		13.5	20.9	20.5	16.5
Longshore - onshore (SE)	19.6	33.0		9.0	25.3	30.7	12.9
Longshore (SSE, S, SSW, SW)	13.4			18.0	18.7	22.7	5.9
Offshore (N, NNW, NW, WNW, W, WSW)	25.7			35.9	18.5	14.7	22.3

Figure 155.

Frequency percents of wind direction, Sapelo Island, Georgia. The left side of the figure illustrates the frequence percent of winds on a yearly basis and the frequence percent of the winds which have onshore or longshore components. The right side of the figure illustrates the frequency percent of winds on a seasonal basis.

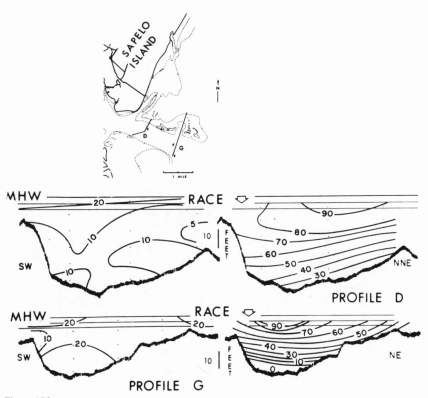

Figure 156.
Cross-sections of current velocity data collected at profiles D and G at the entrance to the Doboy Sound estuary. Profiles on the left illustrate the distribution of current velocities in the water column at mean high water. Profiles on the right illustrate the distribution of current velocities during maximum ebb surface-velocities. Contour intervals are 5, 10, 20, 30, 40, 50, 60, 70, 80, and 90 cm sec⁻¹. (Modified after Oertel, 1971a.)

90 cm sec⁻¹ while ebb flow at the bottom of the water column was inhibited by flooding bottom currents and reduced to a velocity of 30 cm sec⁻¹ (Fig. 156). At low water, water current reversals at the Doboy Sound entrance are also illustrated by the velocity data (Fig. 157). Ebb-jet currents leaving constricted estuary entrances are inhibited by the convergence of onshore currents toward the entrance. A pile-up of water at the entrances results in a dynamic-pressure gradient and in a dynamic diversion (Todd, 1968) of onshore, longshore, ebb, and hydraulic currents (Fig. 154b). The effects of the pile-up of water is exemplified by the current velocity data as well as the current direction data for the entrance of Doboy Sound. During the changing tide, the effect of the water pile is most obvious. At low water, surface currents (Fig. 158) flow seaward and bottom currents (Fig. 159) flow landward at profiles D and G. Blockage of the ebb

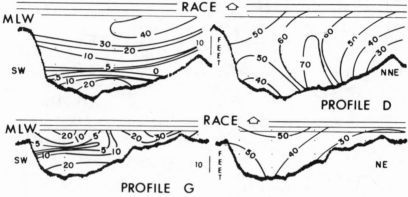

Figure 157.

Cross-sections of current velocity data collected at profiles D and G at the entrance to the Doboy Sound estuary. Profiles on the left illustrate the distribution of current velocities at mean low water. Profiles on the right illustrate the distribution of current velocities in the water column during maximum flood velocities. Contour intervals are 5, 10, 20, 30, 40, 50, 60, 70, 80, and 90 cm sec^{-1}. During mean low water at profile D, surface velocities are in a seaward direction and bottom velocities are in a landward direction. During mean low water at profile G, surface velocities in the southwest part of the channel are in a seaward direction, while currents in the remaining part of the channel are flowing landward. (Modified after Oertel, 1971a.)

flow, by bottom flood-currents causes a partial diversion of flow toward the northeast along the beaches. The resultant diversion is in the form of a mixed water-mass flowing in a clockwise-moving current at the bottom (Fig. 159). At high water, flood currents are present throughout the water column at profile D and G. Blockage of the ebb flow by onshore currents results in a northward deflection of flow along the beach and diversion of a mixed water is in a counter-clockwise flow at the entrance (Fig. 158). The resultant flow of ebb currents is toward the northeast along the barrier beach and over the ramp-margin shoals. As the tide continues to fall, flow is directed through spill-over channels which breach ramp-margin shoals.

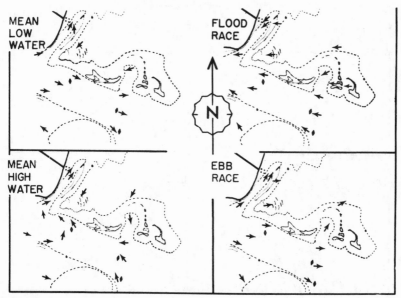

Figure 158.

Maps of surface velocity data. Data was collected at the entrance to Doboy Sound during four intervals of the tide (low water, maximum flood velocities or floor race, high water, maximum ebb velocities or ebb race). Arrows indicate the compass directions of data recorded at each interval of the tide. During low water, ebb currents are confined to deep channels and flood currents are present in shallow areas. During the flood race, all currents flow toward the inlet. During high water, ebb-jet currents at the entrance are diverted in a counterclockwise path to the north. During the ebb race, all currents have a seaward component of flow. (After Oertel, 1971a).

The general circulation pattern for all entrances indicates that during the late stages of the ebb (Fig. 160), bathymetric shielding by ramp-margin shoals restricts ebb flow to channels and at this time, the initial flood begins through the funnel-shaped channel between the major shoal-trends on the north side of the ramp. This flow-direction assymetry and flow-duration assymetry during the changing tide illustrate some of the mutually-evasive ebb channels and flood channels which result from patterns of dynamic diversion. Also, during the period of late flooding and early ebbing (Fig. 161), bottom and surface currents continue to flow landward over the ramp to the sea and the shoals.

Sedimentation accompanying dynamic diversion is related to ebb- and flood-channel differentiation. During the ebb, shoaling waves create sediment-laden onshore currents and longshore currents around ramp-margin shoals and beaches. Sediment accumulation takes place along the line of shear between these wave-induced currents and hyraulic and tidal currents spilling away from the entrance. Sediment entrained in tidal and hydraulic currents is deposited in dune-shaped bedforms. Sediment accumulation by wave-induced currents is deposited in swash bars (King, 1959) and swash platforms (Oertel, 1971b). However, a continuous

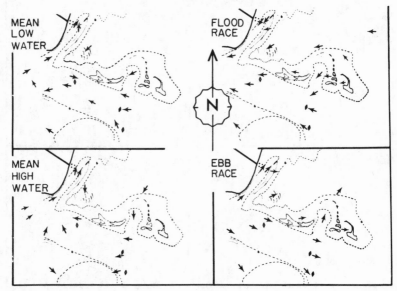

Figure 159.

Maps of bottom velocity data. Data was collected at the entrance to Doboy Sound during four intervals of the tide (low-water, maximum-flood velocities or flood-race, high-water, maximum-ebb velocities or ebb race). Arrows indicate compass directions of data recorded at each interval of the tide. During low water, ebb-jet currents at the entrance are diverted in a counter clockwise direction along the bottom to the north. During flood race, all bottom currents flow toward the inlet of Doboy Sound. During high water, ebb-jet currents at the entrance are diverted along the bottom to the north. During the ebb race, all current data indicate a seaward flow (after Oertel, 1971a).

series of wave-modified bedforms exists between bedforms of wave-current origin and tidal- or hydraulic-current origin. Wave processes of sediment accumulation take place on the seaward sides of spill-over channels which breach shoals. In that spill-over channels are more characteristic of north-margin shoals than south-margin shoals, secondary trends of shoal accretion are more obvious on north-margin shoals. On the south side of the ramp, northeast flowing longshore-currents are diverted seaward by a southeast-flowing branch of the ebb-jet. This relatively simple pattern of current diversion results in a reduction in the capacity of sediment-laden longshore currents. Sediment accumulation takes place along the line of shear between the two currents and results in seaward accretion of triangular-shaped shoals which remain attached to the shoreline.

Longshore trending shoals and offshore-trending shoals on the north side of ramps are separated by funnel-shaped flood channels. Sand waves in these channels have flood orientations throughout the entire tidal cycle. In that longshore-and offshore-trending shoals are interdigitate between ebb-dominated and flood-dominated channels, there is a net circulation of water and sediment in cells over these shoals (Ludwick, 1970a, b; Robinson, 1960). However,

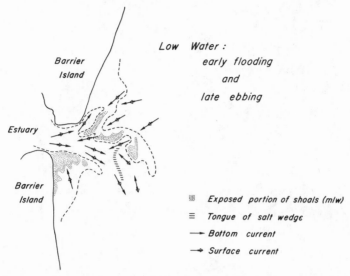

Figure 160.

Bottom and surface currents during low water (early flooding and late ebbing). During the late stages of the ebb, bathymetric shielding by ramp-margin shoals restricts ebb flow and, at the same time, the initial flood begins through the funnel-shaped channel between major shoal-trends on the north side of the ramp.

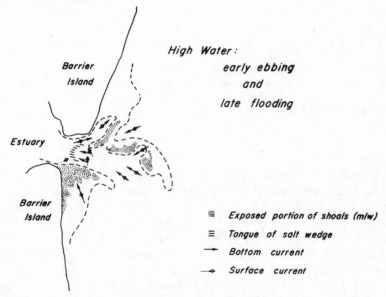

Figure 161.

Bottom and surface currents during high water (early ebbing and late flooding). During the period of late flooding and early ebbing, bottom and surface currents continue to flow landward over the ramp to the sea and the shoals (Fig. 155). At the same time, ebb currents are diverted northward in front of the beach.

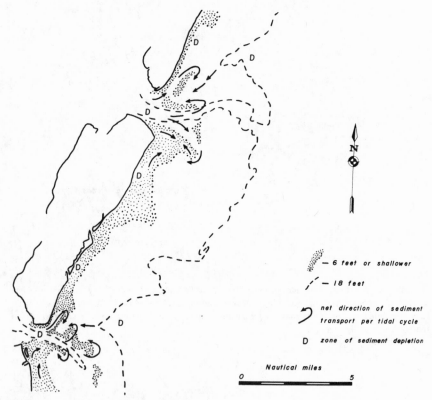

Figure 162.
Pattern of sediment movement and depletion adjacent estuary-entrance shoals. At estuary entrances, sediment becomes trapped in a closed system and shoal progradation takes place at the expense of the adjacent beach, shoreface, and tidal channels.

ebb and flood tidal currents are not the only currents affecting sediment transport over and around shoals. Results of fluorescent-tagged grain studies on shoals indicate that wave-induced currents entrain sediment in directions oblique or opposite to ebb flow (Oertel, 1971a). Hence, wave-induced currents produce sediment gyres (Oertel, 1971b) which are active during the ebbing stage of the tidal cycle. Net sediment-cells for a complete tidal cycle (Ludwick, 1970a, b; Robinson, 1960) and sediment gyres during the ebbing tide (Oertel, 1971b) serve to trap sediment in closed systems. The process of dynamic diversion (Todd, 1968) only diverts sediment transport and deposition results along lines of current diversion in response to current-shear and energy loss. At Georgia estuary entrances (Fig. 162), sediment is trapped in closed systems and shoal progradation takes place at the expense of the adjacent beach, shoreface and tidal channels. This is substantiated by sediment depletion at mid-island shorelines

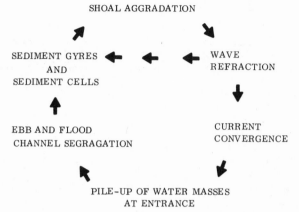

Figure 163.

Sequence of cumulative processes involved in ramp-margin shoal-development.

(Mikesh, Howard, Mayou, 1969), indentations of contour lines seaward of flood-dominated channels on the shallow shoreface, and erosion of indurated sandstone in troughs (Fig. 162) at estuary entrances. In that the sediment budget of this area permits the maintenance of offshore trending shoals in preference to longshore trending spits, it is suggested that there is essentially a balance between the effects that wave action and current diversion have on sedimentation.

CONCLUSIONS

Estuary inlets dominated by tidal and transient currents illustrate complex patterns of water circulation. Shoals on the north sides of ramps are breached by tidal channels in response to convergence of onshore currents which pile up water at the estuary entrance. This water pile-up results in hydraulic currents which are diverted to the northeast and spill across north ramp-margin shoals.

Ramp-margin shoal-aggradation along coastlines with a high frequency percent of onshore winds appears to be a self-perpetuating and cumulative process (Fig. 163). Annually predominant southerly onshore winds and waves result in a convergence of currents toward entrances. Diversion of these currents by tidal currents cause a pile up of water at entrances. Water-mass diversion results in mutually evasive paths of ebb flow and flood flow (Ludwick, 1970a,b; Robinson, 1960). Sediment cells (Ludwick, 1970a,b; Robinson, 1960) and sediment gyres (Oertel, 1971a) trap sediment in north ramp-margin shoals. Shoal aggradation of this type tends to be in several directions rather than in only one diversion direction. Thus, orientation and morphology of these shoals occur in response to (a) transient coastal-currents controlled by seasonal winds, (b) the relative magnitude of ebb-jets to coastal currents, and (c) hydraulic pressure gradients at entrances.

ACKNOWLEDGMENTS

We gratefully acknowledge the assistance of Sharon A. Greer, James Kirchhoffer, and George H. Remmer. This study was supported by U.S. Army Corps of Engineers Contract DACW 72-68-C-0030 and Oceanographic Section, National Science Foundation, NSF Grant GA-30565.

REFERENCES

Abbe, C. J. (1895). Remarks on the cuspate capes of the Carolina coast. *Bost. Soc., Natural Hist., Proc.* **26**, 489-497.

Allen, G. P. (1971). Relationship between grain size parameter distribution and current patterns in the Gironde estuary (France). *J. Sed. Pet.* **41**, 74-88.

Bajorunas, L., and Duane, D. B. (1967). Shifting offshore bars and harbor shoaling. *J. Geophys. Res.* **72**, 6195-6205.

Bumpus, D. F. (1955). The circulation over the continental shelf of Cape Hatteras. *Am. Geophys. Union, Trans.* **36**, 601-611.

de Beaumont, E. (1845). "Leconds de Geologie Pratique." p. 223-252. Paris.

Dietz, R. S. (1963). Wave base, marine profile of equilibrium, and wave built terraces: A critical appraisal. *Geol. Soc. Am. Bull.* **74**, 971-980.

Dolan, R., and Ferm, J. C. (1968). Cresentic landforms along the Atlantic coast of the United States. *Science* **159**, 627-629.

Helle, J. R. (1958). Surf statistics for the coasts of the United States. *Beach Erosion Board, Tech. Mem.* **10**.

Hoyt, J. H. and Henry, V. J. (1967). Influence of island migration on barrier island sedimentation. *Geol. Soc. Am. Bull.* **78**, 77-86.

Johnson, D. W. (1919). "Shore Processes and Shoreline Development." 584 p., Wiley, New York.

King, C. A. M. (1959). "Beaches and Coasts." 403 p. Edward Arnold, London.

Kuroda, Ryuyo (1969). Physical and chemical properties of the coastal waters off the middle of Georgia. Tech. Report Marine Institute, Sapelo Island, Ga.

Ludwick, J. C. (1970a). Sand waves in the tidal entrance to Chesapeake Bay: Preliminary Observations. *Chesapeake Science* **11** [2], 98-110.

Ludwick, J. C. (1970b). Sand waves and tidal channels in the entrance to Chesapeake Bay. *Virginia J. Science* **21**, 178-184.

Mikesh, D. L., Howard, J. D., and Mayou, T. V. (1968). Depositional characteristics of a washover fan; Sapelo Island, Georgia. p. 201. Program Geol. Soc. Am., Ann. Mtg., Mexico City.

Oertel, G. F. (1971a). "Sediment-Hydrodynamic Interrelationships at the entrance of the Doboy Sound Estuary, Sapelo Island, Georgia." Ph.D. dissertation, University of Iowa.

Oertel, G. F. (1971b). Bores of interfering waves: An important fantor in nearshore shoal development. *In* "Coastal and Shallow Water Research Conferences, Spons. by Geography programs, Office of Naval Research. (D. S. Gorsline, ed.) Vol. 2, p. 271.

Otvos, E. G., Jr. (1970). Development and migration of barrier islands, Northern Gulf of Mexico. *Geol. Soc. Am. Bull.* **81**, 241-246.

Pierce, J. W., and Colquhoun, D. J. (1970). Holocene evolution of a portion of the North Carolina coast. *Geol. Soc. Am. Bull.* **81**, 3697-3714.

Price, W. A. (1964). Cyclic cuspate sand spits and sediment transport efficiency. *J. Geol.* **71**, 876-880.

Robinson, A. H. W. (1960). Ebb-flood channel systems in sandy bays and estuaries. *Geography* **45**, 183-199.

Scruton, P. C. (1956). Oceanography of Mississippi delta sedimentary environments. *Bull. Am. Assoc. Pet. Geol.* **40**, 2864-2952.

Shepard, F. P. (1960). Gulf coast barriers. *In* "Recent Sediments, Northwest Gulf of Mexico" (F. P. Shepard, F. B. Pleger, and T. H. Van Andel, eds.), p. 197-220. Am. Assoc. Pet. Geol., Tulsa, Oklahoma.

Stewart, H. B., and Jordan, G. F. (1965). Underwater sand ridges on Georges shoal: 102-116. *In* "Paper in Marine Geology" (R. L. Miller, ed.), MacMillan, New York.

Swift, D. J. P. (1969). Shelf sedimentation lecture 4: *In* "D. J. Stanley, The New Concept of Continental Margin Sedimentation" Am. Geol. Inst., Short course lecture notes, November 7-9, 1969.

Todd, T. W. (1968). Dynamic diversion: influence of longshore current-tidal flow interaction on chenier and barrier island plains. *J. Sed. Pet.* **38**, 734-736.

Uchupi, E. (1968). The Atlantic continental shelf and slope of the United States: physiography. *U.S. Geol. Survey Prof. Paper* **529-C**, 1-20.

U.E. Naval Oceanographic Office (1963). Oceanographic Atlas of the North Atlantic Ocean, Pub. No. 700, Section IV: Sea and Swell.

U.S. Weather Bureau (1969). U.S. Weather Bureau daily wind and precipitation records, Sapelo Island weather station.

White, . A. (1966). Drainage assymetry and the Carolina capes. *Geol. Soc. Am. Bull.* **77**, 223-240.

Zenkovich, V. P. (1964). Discussion of: cyclic cuspate sand spits and sediment transport efficiency. *Jour. Geol.* **72**, 879-880.

CHAPTER 21

Onshore Transportation of Continental Shelf Sediment: Atlantic Southeastern United States

Orrin H. Pilkey* and Michael E. Field[†]

ABSTRACT

Several lines of evidence indicate beach and estuarine sands from the southeastern United States Atlantic coast are derived in part from the adjacent continental shelf. Abundance anomalies of phosphorite grains, total gold content, ooids, and various heavy minerals (in particular, epidote, staurolite, and garnet) on the shelf show a close correspondence to abundance anomalies in adjacent shoreline and near-shore environments. Carbonate content and textural parameters of beach and shelf deposits show a correlation between the two environments on a regional scale. Close correlation of shelf- and shore-sediment parameters may reflect ultimate derivation of sediment from similar sources or similar environments of deposition during Pleistocene sea-level fluctuations other than from onshore transportation. However, it can be shown that oolitic grains in central Florida beach sands and phosphorite grains on North Carolina beaches originate exclusively from continental shelf sources lying appreciable distances (up to 20 km) offshore, thus demonstrating onshore transportation in these cases. Furthermore, presence of easily abraded, shelf-derived oolite grains in quartz beach sands is evidence of onshore transportation occurring at present. Some of the inferred movement of sediment may have occurred actually by the simple process of landward migration of the surf wedge or shoreface during the last rise in sea level rather than by present-day processes. Possible mechanisms of present-day movement include asymmetry of shoaling waves, and tidal and storm-induced currents.

This close correlation between beach- and shelf-sediment characteristics may provide an excellent basis for low-cost mineral reconnaissance on the continental shelf by preliminary beach sampling.

*Department of Geology and Marine Laboratory, Duke University, Durham, North Carolina.

†U.S. Army Coastal Engineering Research Center, 5201 Little Falls Road, N.W., Washington, D.C. 20016.

Delineation of economic mineral deposits along the coast may facilitate location of similar deposits in adjacent offshore regions and reduce the necessity for costly regional shelf reconnaissance.

INTRODUCTION

During the past decade of research on continental shelf sedimentation processes, evidence has accumulated strongly indicating that beach and estuarine sands may be contributed in part by the onshore transportation of sand from the adjacent continental shelf. The evidence pointing to an offshore source has been largely indirect, that is, sediment budget calculations and comparison of shelf and coastal mineral assemblages. In general, it seems that the importance of onshore transport as a sediment source varies inversely with the magnitude of sediment contribution from the adjacent land mass. The exact extent of sediment contribution from the shelf in any given area has been an elusive figure, complicated by the fact that sands and muds must be considered separately because of the different modes and energies of transportation involved. It is important to emphasize that this report is concerned primarily with sand-size materials.

McMaster (1954), Shepard (1963), and Guilcher (1963) among others discuss the phenomenon of onshore transport for areas other than the Atlantic southeastern United States continental shelf. Postma (1967) briefly discusses possible mechanisms of shoreward movement of fine grained sediment on open shelves. Neiheisel (1959), Pevear and Pilkey (1966), Luternauer and Pilkey (1967), Meade (1969), and Windom et al. (1971) have presented evidence, primarily mineralogical, indicating onshore transport of offshore sources of beach or estuarine sand of the southeastern United States Atlantic shelf and also that rivers are not contributors of sand to the shelf at present. Through consideration of the sediment budget of a portion of the Carolina Outer Banks, Pierce (1969 estimated that 440,000 yd^3 of sand were supplied annually to the shoreline from sources on the continental shelf to meet the requirement for beach and barrier island maintenance.

To some extent the effect of shelf sand being contributed to beach areas may be deduced by elimination of other possible source areas. Unconsolidated sandy deposits along coastlines are derived from a limited number of sources, many of which can be shown to be ineffectual at this stage of geologic development of the study-area shoreline. Beach sands accumulate by direct littoral, fluvial, glacial, and aeolian transport; erosion of coastal formations and deposits; *in situ* organic and inorganic production of materials, and the onshore transport of shelf sediments. Along the low relief, humid, landward-retreating coastline of the Alantic southeastern United States many of these potential sources are inoperative. Littoral currents do not introduce new material to the system but rather redistribute the quantities already present. Glacial transport is nonexistent and aeolian transport is insignificant as is biogenic production, with the exception of the south Florida region. Most important, the numerous streams flowing into the region are no longer carrying appreciable quantities of sand (Meade, 1969). Thus, on a regional scale only coastal erosion and onshore transport of marine sands are potential sources for generation of present day southeastern United States coastal sands. Swift et al. (1971) have developed a model of "coastal erosion" for the Virginia-North Carolina shelf between Cape Henry and Cape Hatteras. Their concept is that erosion on a low retreating coast such as this occurs at the shore face, the base of which is typically 10-15 m deep and up to 10 km offshore.

To investigate the role that onshore transport may play in introducing shelf grains to the beach, we have reexamined and summarized data collected over the past 10 years, as well as evaluated new results from ongoing studies. These data include mineral analyses of many of the 2400 shelf grab samples collected aboard R. V. Eastward as well as several hundred beach and river samples also in the Duke University Department of Geology sample collection. Results from the central Florida region were compiled from analyses of over 150 vibratory cores (average length 10 ft) collected in conjunction with the Inner Continental Shelf (ICONS) Program of the Coastal Engineering Research Center. Laboratory treatment of samples for mineral analysis including techniques and procedures for separation, identification, and local distributions, are contained in the original references and are not discussed in detail here.

These beach and shelf sediments from along the 900-mi coastline of the southeastern United States have been compared from a standpoint of abundance anomalies of common mineral grains and presence or absence of unique grain types. In some areas there is no correlation between beach and shelf assemblages, and in other areas the correlation is subtle. Along much of the expanse of coastline, however, the similarities between beach and shelf mineral types is too striking to be disregarded, and this correspondence is interpreted as evidence that beach and estuarine sediments are derived in part by onshore transport of shelf sediments.

The regional sedimentary framework of the Atlantic continental shelf has been well established by a number of studies (e.g., Stetson, 1938; Gorsline, 1963; Uchupi, 1963 and Emery, 1965). Carbonate sedimentation in the study area has been investigated by Milliman et al. (1969) and by Pilkey et al. (1969).

RATIONALE

Correlation of the mineralogy of beach sands and that of the adjacent continental shelf does not in itself indicate that onshore transport has occurred, since in a given area other sources (rivers, coastal erosion) must be considered. Sands with related characteristics may simply reflect derivation from similar source areas or under similar physical conditions. For example, if the same regional trends in grain size exist on both the shelf and in beach sediments, the reason may be that the same regional wave-energy gradients present on modern beaches were in effect during deposition of the surficial sands on this relict shelf. Similarly,

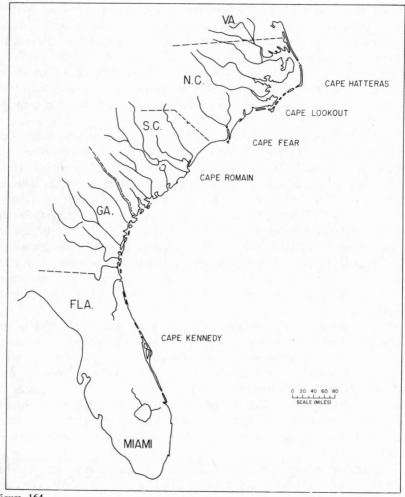

Figure 164.

Index map of the Atlantic southeastern United States indicating key physiographic features and names used in subsequent figures.

if the beach and shelf have closely related heavy or light mineralogy suites, the correlation could reflect similar provenance.

In order to uniquely demonstrate derivation of shoreline sands from the adjacent continental shelf by presently operating processes, it is necessary to (a) demonstrate a need based on sediment budget considerations, or (b) demonstrate the presence of unquestionable shelf-derived sediment components in beach sediments.

The evidence used in this paper for source of shoreline sediments can be categorized as follows. One category consists of the presence of sediment components in shoreline deposits that could have been derived only from the adjacent continental shelf. Included in this category are relict sediment components that are present in beach and shelf sands but are absent in both coastal formations exposed to active marine erosion, and modern fluvial sediments, that is, ooids and phosphorite grains. Following a similar line of reasoning, Postma (1967) and other North Sea workers have used the presence of open marine microfauna tests in estuarine sediments to demonstrate the role of an offshore source in estuarine sedimentation. Faunal constitutents of sediments used in the present study were not investigated, thus the application of their observations to interpretation of processes on the Atlantic southeast is not known.

A second category consists of the presence of sediment components in shoreline sediments that could have been derived from either the shelf or present day rivers. Since present day rivers are not contributing significant amounts of sand, these components are assumed to reflect an offshore source. This category includes light and heavy mineralogy of sands. This cannot be considered to be a conclusive line of evidence because of the problem of mechanisms of shoreline or shoreface erosion. This is discussed further in a later section.

SEDIMENT COMPONENTS DERIVED
EXCLUSIVELY FROM THE SHELF

Within the geographic scope of this paper two components that fall into the category of grains derived exclusively from shelf sources are phosphorite grains and calcareous ooids. In very localized areas both grain types may be derived locally from other than the shelf. For instance, phosphorite grains are probably contributed from eroding inlets of the Georgia coast, and some rivers are presently carrying very small amounts of phosphorite grains; however, volumes contributed by these sources are not considered significant. In continental shelf sediments these dark colored, shiny grains must be derived either from eroding outcrops, which can be shown to be the case in an area near Cape Fear, North Carolina (Pilkey and Luternauer, 1967), or from rivers carrying a load of different composition during times of lowered sea level.

Another example ot onshore transport of shelf derived grains can be demonstrated by a comparison of the central and south Florida regions. In the south Florida region (Palm Beach) the Miami Oolite formation is exposed along segments of the coast and provides a logical source through coastal erosion for the ooids present in beach and inner shelf sands. However, in central Florida (Cape Kennedy), coastal formations are devoid of ooids yet these grains are

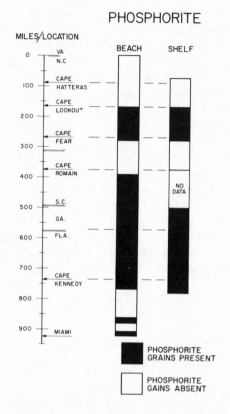

Figure 165.

Bar graph indicating the presence or absence of phosphorite grains in beach and shelf sands. Beach data are from Pevear and Pilkey (1966); shelf data north of Florida are from Luternauer and Pilkey (1967); and Florida shelf data are from unpublished studies at the Coastal Engineering Research Center.

ubiquitous in beach and shelf sands. In this instance, submarine erosion of Pleistocene ooid-bearing calcarenite and transport onto beaches is the dominant mechanism of supply.

Figure 165 is a bar graph showing the distribution of phosphorite grains (presence or absence) in shelf and adjacent beach sands. It is apparent that the two environments closely correspond to one another with regard to this component. The Georgia and North Carolina shelves and nearshore areas were studied in particular detail. Phosphorite is a ubiquitous component of Georgia shelf sands (always less than 5%; usually less than 1%) as it is also of Georgia beaches and estuaries. Phosphorite-free sand is found only in estuaries where relatively important rivers enter the marine environment. In such cases, the

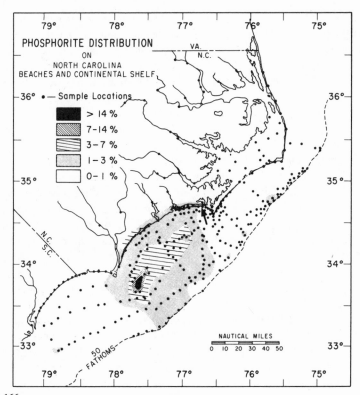

Figure 166.

Distribution of phosphorite in North Carolina river, beach, and shelf sands. The arrow points to Beaufort Inlet (see text for explanation). Data are taken from Luternauer and Pilkey (1967). Dots show sample locations on the shelf, beaches, and rivers.

fluvial sand is significantly diluted by marine sand before even reaching the inner continental shelf.

The North Carolina marine environment presents a different picture of the distribution of sand-size phosphorite grains. Figure 166 shows the phosphorite abundance in North Carolina shelf and beach sands. In this instance, phosphorite is not ubiquitous, distribution is restricted mainly to Onslow Bay, the portion of the shelf between Cape Lookout and Cape Fear. Distribution of phosphorite in beach sands parallels the distribution in shelf sands. The northern Onslow Bay inner shelf off Beaufort Inlet (see arrow in Figure 166) was examined in detail; in this locale the correlation between beach and shelf assemblages is very well defined. Shackleford Bank, the barrier island northeast of the inlet, contains nonphosphatic sands, whereas Bogue Bank, the barrier island southwest of the inlet, has slightly phosphatic sands. This same division is demonstrated on the adjacent continental shelf by a boundary between phosphatic and nonphosphatic sand which is marked by a line extending in a southerly direction from Beaufort Inlet. As in Georgia, the North Carolina river sands contain little sand-sized phosphorite so that in the Onslow Bay region a clear-cut distinction can be made of shelf versus river sand contributions in the estuarine environment.

In Onslow Bay, the phosphorite grains are derived in large part from dissolution of highly phosphatic Miocene limestones. Evidence for this is a close correlation between submarine outcrops of the parent limestone and phosphorite abundance anomalies in the unconsolidated sediment cover. Unpublished seismic studies conducted by the Coastal Engineering Research Center indicate Tertiary strata crop out or are only thinly veneered on much of the Cape Fear inner shelf.

Previous studies by Luternauer and Pilkey (1967) have delineated a particularly phosphorite-rich shelf area (up to 40% phosphorite by point count) lying 30 to 40 km ESE of Cape Fear. Data from adjacent beaches (Luternauer and Pilkey, 1967) indicates that the observed shelf concentrations are reflected in the adjacent beach sediments.

This close association of phosphorite-rich beach sands with phosphatic strata presently exposed on the adjacent shelf surface suggests a shoreward movement of sediment from the shelf. The alternative explanation, landward movement of the surf wedge by a continuously recycling beach, does not adequately explain the lateral anomalies in the beaches. The nearly-exposed Miocene strata both south and north of Cape Fear would seemingly have acted as a source for the entire region that would have resulted in an equal and ubiquitous distribution of the grains. The concentration anomalies of beach phosphorite evident in Figures 165 and 166 show this is not so.

In central Florida there is also evidence of onshore transport. Grains known to be derived only from shelf edge and central shelf areas (Macintyre and Milliman, 1970; Terlecky, 1967; Pilkey et al., 1969) also occur in nearshore and beach sands in minute quantities (Field and Duane, 1972). Ooids found in beach sediments resemble those collected from the inner shelf region both in grain characteristics and latitudinal distribution patterns. Individual grains are generally medium-sand size, poorly developed and partially abraded; nuclei are predominantly quartz grains, with mollusc fragments, microfauna, and on occasion a composite of grains. Distribution of oolitic grains in central Florida beach and shelf deposits is shown in Figure 167. Previous known landward extent of ooids as mapped by Terlecky (1967) is indicated in the diagram by a solid line. Outcrops shown in Figure 167 refer to areas mapped by Moe (1963) as areas of ledges and rocky bottom. Character and extent of the outcropping surface has been studied by Field and Duane (1972) and will be summarized later in this section.

North of False Cape, neither inner shelf nor beach sands contain ooids or the commonly associated rounded clasts of intergranular cement, both of which

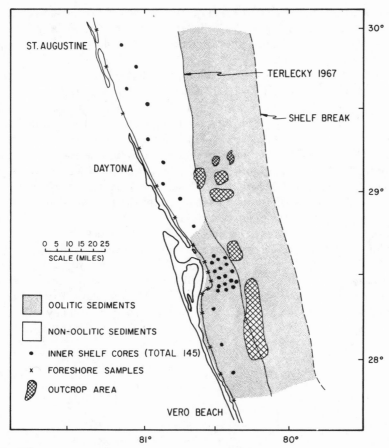

Figure 167.
Map of central Florida showing the presence or absence of ooids in shelf and beach sands. Each core location represents five vibratory cores averaging 10 ft in length. Outcrop areas are plotted from Moe (1963). Beach and inner shelf ooids are reported by Field and Duane (1972).

are present to the south in both environments. The base of the shoreface deepens north of False Cape to nearly 25 m; oolitic sands shown in this area on the map lie at greater depths (50 m) than those immediately south and are segregated from the shore by a belt of non-oolitic fine sand. Presumably oolitic sands are either not transported landward of their mapped boundary or they are intercepted and diluted by the belt of fine sand.

Ooids rarely exceed a concentration of 1% of total beach sediment or 3% of total nearshore sediment in the Cape Kennedy region, and in both environments the grains reflect the effects of abrasion and boring organisms. Oolitic grains on the beach especially show a series of stages of natural attrition from well-developed oolitic coatings to surficial coatings to simply quartz grains with layers of CaCo₃ in surface depressions. This evidence of loss through abrasion indicates that in order for even a minute quantity of ooids to be present in beach sands, onshore movement and continual replenishment of these grains must be an active process.

Although previous investigations have established a Holocene age for most central shelf oolites in surficial sediments lying at depths greater than 50 m (Terlecky, 1967; Macintyre and Milliman, 1970) we believe the beach deposits to be derived exclusively from Pleistocene nearshore sources. Milliman and Emery (1968) report that at least some of the Florida shelf ooids are late Wisconsin in age. Of 91 cores collected off Cape Kennedy, eight penetrated through the Holocene lagoonal and nearshore deposits to Pleistocene semilithified calcarenite (Field, Meisburger, and Duane, 1971). Contained within the subaerially formed blocky mosaic calcite cement are a small percent (5%) of ooids similar in all aspects to those described from the beaches. The upper surface of this oolite-bearing calcarenite has been defined by use of high-resolution seismic reflection profiling in addition to the core data. Based on stratigraphy and radio carbon dating (Field and Duane, 1972; Field et al., 1971), this surface is judged to mark the top of a late Pleistocene shallow water deposit that was inundated and eroded during the Holocene Transgression (Figure 168). The surface is at present exposed in some areas, and in other areas only thinly veneered, thus acting as a source of material for the adjacent littoral zone through bottom erosion and onshore transport.

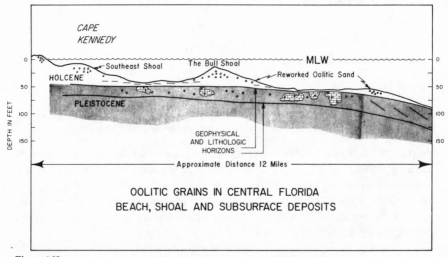

Figure 168.

Schematized cross-section of the Cape Kennedy inner continental shelf, based on high resolution continuous seismic reflection profiling and sediment coring (Field, Meisburger, and Duane, 1971; Field and Duane, 1972).

The highest concentrations of ooids observed (up to 16% of the total sample by grain count) in any portion of the southeastern United States Atlantic shelf are on the inner and central shelf in the northern portions of Onslow Bay, North Carolina (Terlecky, 1967). Unfortunately, adjacent beach sands have not been examined for this component, thus it is not known if onshore transport is further demonstrated by this parameter in North Carolina. It is apparent that the correlation between ooid presence or absence in beach versus shelf sands (Figure 167) is not as well defined as that observed with respect to phosphorite grains (Figure 166). This is due in part to dilution of north Florida beach sediments by material supplied by present day coastal erosion and southward littoral transport. It is also due to the relative lack of resistance to abrasion inherent in the aragonitic composition of ooids. Thus, these grains may not survive long distance transportation, and unless the source is close by they will be absent in quartz-rich beach sediments.

SEDIMENT COMPONENTS THAT OCCUR IN BOTH SHELF AND PRESENT-DAY FLUVIAL SANDS

Latitudinal distribution of minerals common to fluvial, beach, and shelf sands is displayed in Fig. 169. The percentages of heavy minerals shown in the bar graphs are based on counts of select size fractions of the nonopaque minerals. Shelf concentrations are approximations from the central shelf and are based on considerably fewer samples than the beach sand averages. Gold content of the samples used in Fig. 169 is based on analysis of the total heavy mineral fraction. Feldspar content is expressed as percentage of the light mineral fraction on a carbonate-free basis. Both feldspar and gold shelf data are based on all the shelf samples observed, rather than the central shelf only.

In most areas a general agreement between shelf and beach sand mineralogy is demonstrated by the figure. Both beach and shelf sands are enriched in staurolite between Cape Lookout and Cape Fear, North Carolina, and also in south Florida. Not all minerals show a direct relation between onshore and offshore sediments. For example, the data for epidote indicate that shelf and beach sands may be unrelated with respect to abundance anomalies of this mineral.

Except for two localized "highs," shelf garnet values closely approximate those on the beach, especially in regard to the garnet content north of Cape Lookout and in south Florida. Feldspar values show some correspondence between the beach and the shelf; higher concentrations are present both north of Cape Hatteras and adjacent to Georgia. Detectable gold extends from the northern limits of the study area south to below Cape Lookout on the continental shelf, whereas in beach deposits the last observed detectable gold is from beach sand obtained a short distance north of Cape Lookout.

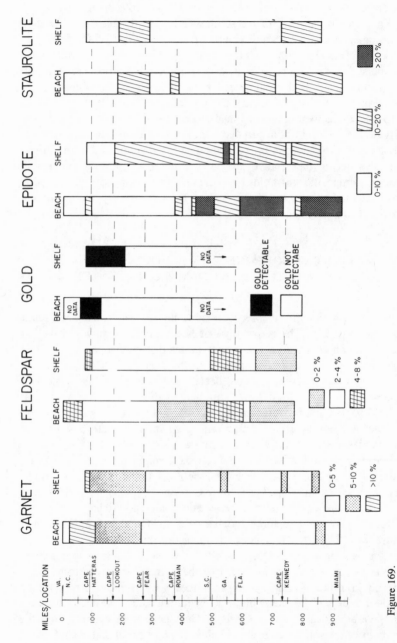

Figure 169.

Bar graphs showing the abundance of garnet, epidote, staurolite and feldspar, and presence/absence of gold in beach and shelf sediments. Sources of data for this figure are: heavy mineral beach values, Giles and Pilkey (1965); heavy mineral shelf values, Pilkey (1963); feldspar, Field and Pilkey (1969); and gold, Pilkey and Bornhold (1970). Detectable limit of gold varied from 0.02 to 0.10 PPB depending on the size of heavy mineral sample analyzed.

ECONOMIC IMPLICATIONS

Attention is now focusing on the continental shelves for potential exploitation of mineral reserves (Wang, 1970) and it appears the indicated onshore transportation through correlation of mineral types reported in this paper may provide an excellent basis for low cost preliminary mineral reconnaissance on the continental shelf through beach sampling. In a region such as the present study area it is suggested that investigation of beach sands would be the best preliminary approach to prospecting for continental shelf minerals expected to be present in the surficial unconsolidated sediment cover. Presence of economic minerals in beach deposits may reflect higher concentrations and greater volumes in adjacent shelf deposits. For example, the highly phosphatic area on the North Carolina shelf north of Cape Fear can be detected in adjacent beach sands (Fig. 166). Areas of the shelf that are essentially nonphosphatic and therefore a waste of time and effort to sample at sea are clearly indicated through study of adjacent beaches. The areal distribution of shelf gold (which is unlikely to be of economic significance in this area) apparently is reflected accurately in beach sediments as well. Perhaps in some instances, in other areas where beach deposits are of economic importance, the deposits are the result of onshore transport from mineable shelf deposits.

The simularity of total sediment properties, such as grain size and $CaCO_3$ content of beach and shelf sands, is also of economic importance. Due partly to increased land costs and diminution and depletion of previously used convenient sources, the inner Continental Shelf is being viewed with increasing interest as a source of sand for beach replenishment and industrial construction (Duane, 1968).

From a regional perspective, both beach and shelf sands decrease in size and carbonate content going toward the Georgia Bight from both South Florida and North Carolina. Grain size data for beach sands between Cape Hatteras and Miami have been reported by Giles and Pilkey (1965). Factors that control grain size include average wave energy and availability of local sources (river channels, outcrops, etc.).

Insufficient data are available on shelf sediment grain size to compare beach and shelf grain sizes over the entire study area. However, approximations of average grain sizes for continental shelf sediments off Georgia (Pilkey and Terlecky, 1966) and off North Carolina (Cleary, 1967; Pratt, 1970) indicate that the North Carolina shelf sediment averages slightly coarser than the Georgia shelf, as do the beach sediments. Detailed comparisons of beach and inner shelf textural and compositional parameters in Florida (Meisburger and Duane, 1971; Field and Duane, 1972) show a strong correlation between the two types of sand. These similarities were interpreted by the authors as an indication of exchange of sediment between the beach and offshore zone to depths of

Table XXVIII
Comparison of Percentages of Average Carbonate
Fraction Abundance of Beach and Shelf Sands[a]

	North Carolina	South Carolina	Georgia	Florida
Shelf	25	12	9	58
Beaches	4	9	2	23

[a]Shelf data are from Pilkey, et al. (1969) and beach data from Giles and Pilkey (1965).

at least 25 m. Texture alone is a poor criteria for judging the effects of onshore transport since current-induced transport is size selective. In addition, modification of beach granulometric properties by storms and localized erosion strongly masks any long-term effects.

Studies concerning the distribution and composition of the calcium carbonate fraction are more complete with regard to beach-shelf comparisons than are textural studies. The regional shelf variation of highest carbonate content off Florida, lowest off Georgia and intermediate values off the Carolinas was attributed by Pilkey et al., (1969) to differences in rate of sedimentation during the last sea-level transgression. Assuming constant productivity between Cape Hatteras and Cape Kennedy, the percentage of $CaCO_3$ becomes a function of the rate of dilution by noncarbonate materials, that is, controlled by the rate of terrigenous sedimentation which, during the Quarternary, was highest off Georgia and lowest off Florida.

Regional variations in beach carbonate content are accentuated by contributions from erosion of coastal formations. In Florida, Pleistocene coquinas are contributing significant quantities of shell material, whereas in Georgia and the Carolinas, eroding deposits are quartzose. The consistently higher carbonate content of shelf deposits relative to beach deposit shown in Table XXVIII is due somewhat to the fact that the shelf averages include the highly calcareous sediments of the outer shelves. Beach and shelf $CaCO_3$ abundance values might be much closer if only inner shelf sands were compared with adjacent beaches, as shown in the central Florida region (Field and Duane, 1972).

These regional trends in texture and carbonate content of southeastern United States beach and shelf sands are not judged to be a result of modern transport mechanisms. They exemplify regional onshore-offshore similarities that can be associated with other factors that are both contemporaneous (wave climate) and relict (extent of Holocene fluvial deposition). Factors such as these add difficulty to the task of assessing the relative importance of modern processes, such as landward transport, on a regional scale.

DISCUSSION

The evidence summarized in the previous sections shows that a reasonably close correlation exists between the nature of continental shelf and shoreline sands of the Atlantic southeastern United States. The similarity of offshore and nearshore sands is assumed to be due to the fact that the continental shelf, particularly the inner shelf, is acting as a source region for contribution of sand to coastal areas. Several lines of evidence contribute to this conclusion, the primary one being that the only possible source of some of the components (ooids and phosphorite) is the shelf. The heavy mineral suite observed in beach sands could be derived by fluvial transport rather than from the shelf, but since rivers are believed not to be contributing sand at present, the similarity of beach and shelf heavy mineral suites may reflect onshore movement of shelf sediments. At the very least, however, shelf-beach similarities are a requirement, if not proof, of onshore transport.

Following the model developed by Swift et al. (1971a) the continental shelf in some areas may be receiving fluvial sand by a continuous two-stage mechanism. Fluvial sand initially trapped in estuaries is ultimately released to the shelf as the shoreface retreats landward. Hence, on an eroding shoreline, such as much of the Atlantic southeastern United States, the sediment source on the shelf may be simply the eroding shoreface. The mapped sources for the two examples of shelf-derived grains (ooids and phosphorite) cited in this paper are well beyond the shoreface. The two-stage mechanism is still valid, especially with regard to phosphorite. Grains contributed to the barrier island complex by shoreface erosion at a time of lower sea level would conceivably serve a long residence time in the barrier as it migrates landward and upward. Long residence time in the barrier structure is dependent on the assumption that barriers retreat landward continuously and longshore transport is not significant in terms of quantities displaced. If barriers migrate discontinuously, then residence time of individual grains is greatly decreased. Similarly, appreciable longshore transport would result in dilution of the sediments from upcoast sources and significantly reduce the abundance of grains originally derived from the shelf source. There is little evidence presently available to evaluate the validity of either assumption. In any case, ooids, due to their fragile nature, very likely would have an insignificant residence time. Their presence in beach sediments, therefore, in Central Florida beach sands is considered very strong evidence of present-day onshore transportation.

Studies of mechanisms which transport sediments landward from the shelf are in the early stages of investigation, but progress is being made (Smith and Hopkins, 1972). With growing emphasis being placed on the processes of sediment transport, it seems likely that actual monitoring of shelf currents will become increasingly common during the next decade. Present-day net landward transport as inferred from our sediment studies may be generated by one of several known current and wave phenomena. Both the landward component of asymmetrical wave orbitals under shoaling conditions and the dominating tidal flood currents in shallow water are possible mechanisms for onshore transport. On portions of the Atlantic coast, the fair weather regime is not as influential as storm conditions in shaping inner-shelf morphology and texture (Swift et al., 1971b). It may be that this applies to the shelf as a whole and onshore transport is generated as a net response to storm induced bottom currents. Since onshore transport may occur over a long shoreline distance representing a variety of hydraulic regimes, it is likely that a variety of mechanisms are included in this active and continual process.

ACKNOWLEDGMENTS

A large portion of the data of this study was collected over a ten-year period of investigation of the southeastern United States Atlantic marine environment and many people and organizations have contributed ideas and information. We thank those students, researchers, and institutions who provided assistance in the field collection of data and discussion of results. Support for studies of continental shelf sands was given by the National Science Foundation and the United States Geological Survey. This study should be considered to be a contribution of four different organizations: The United States Army Coastal Engineering Research Center, the University of Georgia Marine Institute, the Duke University Department of Geology and Marine Laboratory, and the University of Puerto Rico Department of Marine Sciences (synthesis and analysis of the data and writing were carried out while the senior author was a visiting professor at the University of Puerto Rico). We wish to thank the Duke University Oceanographic Program for the use of R. V. Eastward for shelf sampling. Inner-shelf sediment cores and sparker records were obtained as part of the Inner Continental Shelf Program of the Coastal Engineering Research Center. Permission to publish this information was granted by the Chief of Engineers. We profited greatly from discussions with Donald Swift and Robert Carver on the contents of this paper.

REFERENCES

Cleary, W. J. (1967). Marine geology of Onslow Bay. Unpublished Master's thesis, 73 pp.

Duane, D. B. (1968). Sand deposits on the continental shelf: A presently exploitable resource. Natl. Symposium on Ocean Sciences and Engineering of the Atlantic Shelf. Marine Technology Society Program. pp. 289-297.

Emery, K. O. (1965). Geology of the continental margin off the eastern United States, *In* "Submarine geology and geophysics, Colston Papers" (W. F. Whittard and R. Bradshaw, Eds.), Vol. 17, pp. 1-20. Butterworths, London.

Field, M. E., and Pilkey, O. H. (1969). Feldspar in Atlantic Continental margin sands off the Southeastern United States. *Bull. Geol. Soc. Am.* **80**, 2097-2102.

Field, M. E., Meisburger, E. P., and Duane, D. B. (1971). "Late Pleistocene-Holocene Sedimentation History of Cape Kennedy Inner Continental Shelf." Abs. with program, p. 337-338. 45th Annual Mtg. of the Soc. of Econ. Paleontologists and Mineralogists.

Field, M. E., and Duane, D. B. (1972). Geomorphology and sediments of the Cape Kennedy inner continental shelf, U.S. Army Coastal Engrg. Res. Ctr. Technical Memorandum series.

Giles, R. T., and Pilkey, O. H. (1965). Atlantic beach and dune sediments of the southern U.S. *J. Sed. Petrology* **35**, 900-910.

Gorsline, O. S. (1963). Bottom sediments of the Atlantic shelf and slope off the southern U.S. *J. Geology,* **71**, 423-440.

Guilcher, A. (1963). Estuaries, deltas, shelf, slope. *in "The Sea," (M. N. Hill, ed.),* Vol. III. Interscience, New York.

Luternauer, J. L., and Pilkey, O. H. (1967). Phosphorite grains: Their application to the interpretation of North Carolona shelf sedimentation. *Marine Geology* **5**, 315-320.

Macintyre, I. G., and Milliman, J. D. (1970). Physiographic features on the outer shelf and upper slope, Atlantic continental margin, southeastern U.S. *Bull. Geol. Soc. Am.* **81**, 2577-2598.

McMaster, R. L. 1954). Petrography and genesis of the New Jersey beach sands, State of New Jersey Dept. of Conservation and Economic Development, Bull. 63, Geologic Series, 239 pp.

Meade, R. H. (1969). Landward transport of bottom sediments in estuaries of the Atlantic coastal plain. *J. Sed. Petrology* **39**, 222-234.

Meisburger, E. P., and Duane, D. B. (1971). Geomorphology and sediments of the inner continental shelf, Palm Beach to Cape Kennedy, Florida. U.S. Army Coastal Engrg Research Center Tech. Memorandum No. 34.

Milliman, J., and Emery, K. O. (1968). Sea levels during the past 35,000 years. *Science* **162**, 1121-1123.

Milliman, J., Pilkey, O. H., and Blackwelder, B. W. (1969). Carbonate sediments on the continental shelf, Cape Hatteras to Cape Romain. *Southeastern Geology* **9**, 245-267.

Moe, M. A., Jr. (1963). A survey of offshore fishing in Florida, Florida State Board of Conservation, Marine Laboratory, Prof. Papers Series No. 4.

Neiheisel, J. (1959). "Littoral drift in the vicinity of Charleston Harbor." *Am. Soc. Civil Engineers Proc. J., Waterways & Harbors Division* **85**, 99-113.

Pevear, D. R., and Pilkey, O. H. (1966). Phosphorite in Georgia continental shelf sediments. *Bull., Geol. Soc. Am.* **77**, 849-858.

Pierce, J. W. (1969). Sediment budget along a barrier island chain. *Sedimentary Geology* **3**, 5-16.

Pilkey, O. H. (1963). Heavy minerals of the U.S. south Atlantic continental shelf and slope. *Bull. Geol. Soc. Am.* **74**, 641-648.

Pilkey, O. H., and Terlecky, P. M. (1966). Distribution of surface sediments on the Georgia continental shelf. *in* "Pleistocene and Holocene sediments, Sapelo Island and Vicinity. (J. H. Hoyt, V. J. Henry, and J. D. Howard, eds.) p. 28-39. Geol. Soc. Amer., southeastern section, Field Trip No. 1, 1966.

Pilkey, O. H., and Luternauer, J. L. (1967). A North Carolina shelf phosphate deposit of possible commercial interest. *Southeastern Geology* **8**, 33-51.

Pilkey, O. H., Blackwelder, B. W., Doyle, L. J., Estes, E., and Terlecky, P. M. (1969). Aspects of carbonate sedimentation of the Atlantic continental shelf of the southern United States. *J. Sed. Petrology* **39**, 744-768.

Pilkey, O. H., and Bornhold, B. D. (1970). Gold distribution on the Carolina continental margin—A preliminary report. U.S. Geol. Survey Prof. Paper **700-c**, c30-c34.

Postma. H. (1967). Sediment transport and sedimentation in the estuarine environment. *In* "Estuaries" (G. H. Lauff, ed., Am. Assoc. Advance. Sci., Wasnington, D.C.

Pratt, F. P. (1970). "Marine geology of Long Bay," 88 pp. unpublished Master's thesis, Duke University.

Shepard, F. P. (1963). "Submarine Geology," 557 pp. Harper & Row, New York.

Stetson, H. C. (1938). The sediments of the continental shelf off the eastern coast of the United States. Mass. Inst. Technology and Woods Hole Oceanographic Inst. papers in Phys. Ocean. and Meteorol. No. 5, 48 p.

Swift, D., Sanford, R. B., Dill, C. and Avignone, N. F. (1971a). Textural differentiation on the shore face during erosional retreat of an unconsolidated coast, Cape Henry to Cape Hatteras, Western North Atlantic shelf. *Sedimentology* **16**, 221-250.

Swift, D., Holliday, B., Avignone, N. and Shideler, G. (1971b). Anatomy of a shore face ridge system, False Cape, Virginia. *Marine Geol.* **12**, 59-84.

Terlecky, M. (1967). The nature and distribution of oolites on the Atlantic continental shelf of the southeastern U.S. unpublished Master's Thesis, Duke University, Durham, North Carolina.

Uchupi, E. (1963). Sediments on the continental margin off eastern United States, *U.S. Geol. Survey Prof. Paper* 475-cp.

Wang, Frank (1970). Mineral resources of the sea: United Nations Publication E. 70 II B.4, 49 pp.

Windom, H. L., Neal, W. J., and Beck, K. C. (1971). Mineralogy of sediments in three Georgia estuaries. *J. Sed. Petrology* **41**, 497-504.

CHAPTER 22

Linear Shoals on the Atlantic Inner Continental Shelf, Florida to Long Island

David B. Duane,* Michael E. Field,*
Edward P. Meisburger,* Donald J. P. Swift,†‡
and S. Jeffress Williams*

ABSTRACT

The inner Atlantic continental shelf from Long Island to Florida is characterized by fields of linear, northeast trending shoals. The shoals exhibit up to 30 ft of relief, have side slopes of a few degrees and extend for tens of miles. Clusters of linear shoals merge with the shoreface in water as shallow as 10 ft. Most of the shoals out to depths of 120 ft have been examined by means of seismic profiling, precision depth profiling, grab sampling, coring; current monitoring has been conducted on a few shoals. These inner-shelf sand bodies or shoals can be grouped as arcuate (inlet and cape-associated) and linear. Linear shoals may radiate from estuary mouths, as a second order structure on arcuate, inlet-associated shoals, or may occur on the open coast. Linear shoals on the open coast may be shoreface-connected or isolated. All linear shoals of the open coast form a small angle (most < 35°) with the coast line; most open northward regardless of presumed direction of net littoral drift. Seismic reflection profiles show the linear and arcuate cape-associated shoals to be plano-convex features, some exhibiting internal inclined bedding strcutures, resting upon a featureless, nearly horizontal stratum.

*Geology Branch, Coastal Engineering Research Center, Washington, D.C. 20016.

†Institute of Oceanography, Old Dominion University, Norfolk, Virginia 23508.

‡Present Address: Atlantic Oceanographic and Meteorological Laboratories, National Oceanic and Atmospheric Administration, 15 Rickenbacker Causeway, Miami, Florida 33149.

Marked differences exist between sediments of shoals and those of underlying acoustically defined strata, which also show local and regional changes in lithology. The underlying strata, which are occasionally exposed in troughs between ridges, range in age from 7,000 to 25,000 years old. These ages, obtained from radiocarbon analyses of shells, peat, and organic mud, indicate that formation of linear and cape-associated arcuate shoals post-dated the last transgression; hence they are all younger than at least 11,000 BP. Changes in mineralogy of the shoals parallel changes in the mineralogy of adjacent beaches. Shoreface-connected shoals have granulometric characteristics similar to the beach.

Shoreface-connected shoals appear to be presently forming in response to the interaction of south-trending, shore-parallel, wind set-up currents with wave-generated bottom currents during winter storms. In some areas sequences of shoreface-connected shoals seem to comprise effective evolutionary series whereby the process of shoal detachment by deepening and headward erosion of the landward trough is illustrated. Morphological and hydraulic evidence suggests that detached shoals continue to respond to the modern hyraulic regime of the inner shelf, with the helical flow that they induce in coast-parallel storm currents serving to aggrade their crests, and fair-weather wave surge serving to degrade them. Equilibrium shoal crest depth seems to be about 30 ft on the inner shelf; a second mode of crest depth at 50 ft may reflect a recent still-stand at that level.

INTRODUCTION

In the ultimate sense of Johnson's (1919) classic theory of marine planation, continental shelves of the world would be featureless. That the shelves are not simply a broad and gently sloping platform but are comprised of reefs, shoals, and deep channels was known to the earliest seafarers. However, it was not until the advent of acoustical hydrographic surveying techniques in the 1920s that those interested in the ocean and its bottom became fully aware of its ubiquitous irregular topography. Excellent reviews of early exploration of the shelf are provided by Schopf (1968) and by Emery (1966) who point out in particular the significance of early studies by Veatch and Smith (1939). The works of Emery (1965, 1966, 1967, 1968, 1969), and colleagues (Uchupi, 1968, 1970, Schlee and Pratt, 1972, Ross, 1970) are no less significant, for they represent the first integrated examination and interpretation of topography, surficial deposits, underlying bedrock, stream drainage, and Quaternary history.

In their pioneer analysis of the topography of the central Atlantic shelf, Veatch and Smith (1939) noted that it was traversed by channels normal to the shoreline which they interpreted as subaerial river channels. Their maps show between these channels a distinctive alternation of ridges and swales which tend to converge gradually with the shoreline toward the south. This ridge and swale topography is clearly evident in the series of bathymetric maps of the Atlantic shelf prepared by Uchupi (1968) (Figs. 171, 172, 173). Locations of these and other figures used in this paper are shown in Fig. 170. Although Uchupi's maps illustrate the dominance of the ridge and swale topography on the entire shelf (Figs. 171, 172, 173), the emphasis of this paper will be placed on the inner shelf as the zone of generation of this topography. Veatch and Smith (1939, P40) concluded that this topography "is due to wave planation, with terraces, succes-

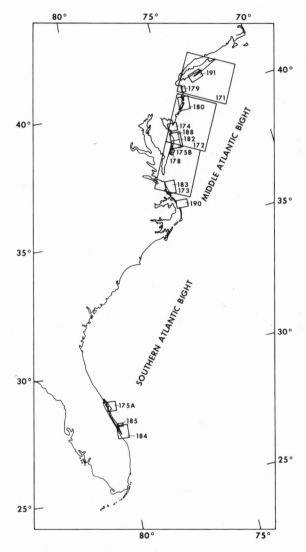

Figure 170.
Index of illustrations utilized in this paper.

sive barrier beaches and distinctive forms representing progressive steps of an advancing sea." This interpretation of the ridge and swale topography has been adhered to by many subsequent investigators, including Shepard (1963), Emery (1966), and Garrison and McMaster (1966).

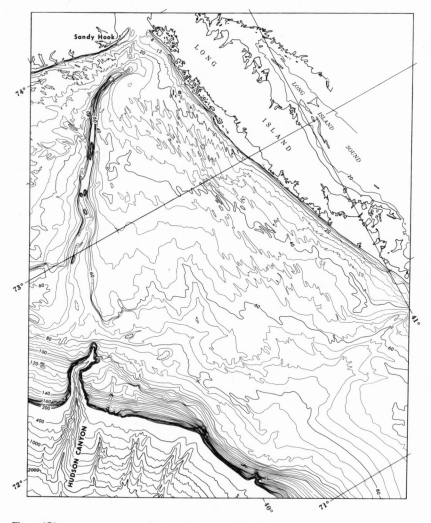

Figure 171.
Topography of the northern portion of the middle Atlantic Bight showing northeast trending bathymetric fabric off New Jersey and the southeast trending fabric south of Long Island. Contour interval is 4 m. From Uchupi (1968).

A variant hypothesis, concerned primarily with the inner-shelf ridges, correlates the ridges with subaerial beach ridges on the adjacent coast and interprets them as relict from a time of lower sea level earlier in the Pleistocene (Sanders, 1962; Kraft, 1971a, Fig. 6; Dietz, 1963, Fig. 13; Hyne and Goodell, 1967). Another school of thought has suggested that the ridge topography of the

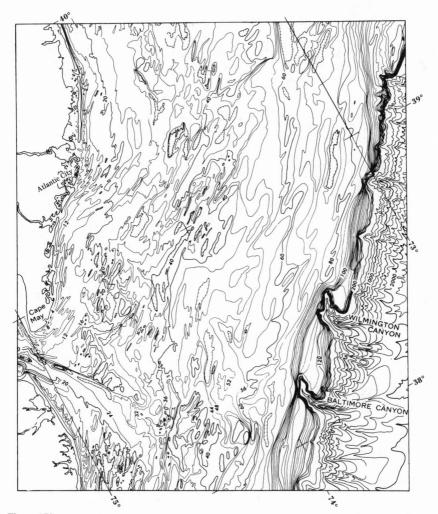

Figure 172.

Topography of the central portion of the Middle Atlantic Bight showing pervasive northeast bathymetric fabric. Contour interval is 4 m. From Uchupi (1968).

shelf is an assemblage of presently or recently active marine bedforms. Moody (1964) studied ridges on the shoreface of the Delaware coast and determined that they shifted position during the Ash Wednesday storm of 1962. He suggested that these ridges were of modern hydraulic origin, and that as the shoreface retreated, they were isolated and abandoned, to form the shelf ridge topography.

Uchupi (1968) commented in his study of both inner and outer shelf morphology that if the beach-ridge hypothesis was correct, then ''there has been a drastic

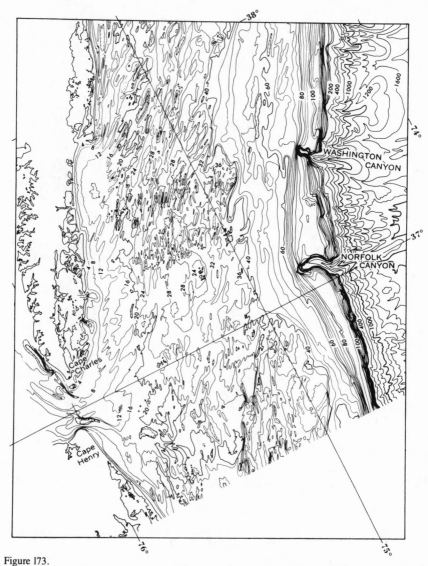

Figure 173.

Topography of the southern portion of the Middle Atlantic Bight showing pervasive northeast bathymetric fabric irrespective to changes in shoreline orientation. Contour interval is 4 m. From Uchupi (1968).

reorientation of the shoreline in recent times,'' since ridge trends intersect the modern shoreline at an angle. He also suggested that the topography was a response to the modern hydraulic regime. However, Uchupi (1970) has more recently returned to a tentative strand plain hypothesis. Schlee and Pratt (1972) refer to the "sets of sand ridges" (p. H33) and "systems of sand waves" (p. H36) and imply an hydraulic, post-transgression origin. McKinney and Friedman (1970), examining a portion of the Long Island shelf, concluded that since the majority of the ridges trended obliquely to the regional contours, they were instead stream interfluves.

It is clear from this brief review of pertinent studies that the physical characteristics, as well as relative age and mode of generation of these inner-shelf features are neither well defined nor agreed upon. Understanding the genesis and behavior of the ridge and swale topography is crucial to any hypothesis concerning the Holocene modification of the shelf surface and bedload transport. Effective planning for any use of the shelf surface also requires this understanding.

SCOPE AND TECHNIQUES

In 1964 the Inner Continental Shelf Sediment and Structure (ICONS) Program was initiated by the Coastal engineering Research Center (CERC) to survey subbottom structure and sediments of portions of the United States inner continental shelf. Objectives of the program are: delineation of sand bodies potentially available and suitable for beach nourishment; better understanding of sedimentation on the inner continental shelf as it pertains to supply of sand to beaches; changes in coastal and shelf morphology, longshore sediment transport, inlet migration-stabilization, and navigation; and better understanding of Quaternary shelf history. ICONS data pertinent to information contained in this paper are approximately 8000 mi of continuous seismic profiles and more than 800 10-30 ft cores from portions of the inner continental shelf of Long Island, New Jersey, Delaware, Maryland, Virginia, North Carolina, and eastern Florida. This coverage has permitted some examination of most linear sand bodies on the inner continental shelf of the eastern United States. Detailed study including measurement of water current velocity and direction along a portion of the Virginia Coast was carried out by one of the authors (Swift).

TYPES AND DEFINITION OF SHOALS

Our studies have been confined essentially to water depths of less than 130 ft (40 m). Within this depth range on the continental shelf there are numerous shoals, defined as "an elevation of the sea bottom comprised of any material except rock or coral . . ." (United States Army Corps of Engineers, 1966, p. A-31). We identify two broad classes: arcuate and linear. Two subgroups of each are also recognized: arcuate shoals we subdivide as inlet- and cape-associated (Fig. 174a, b), and linear shoals as shoreface-connected and isolated (Fig. 174c).

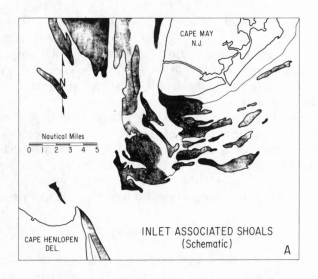

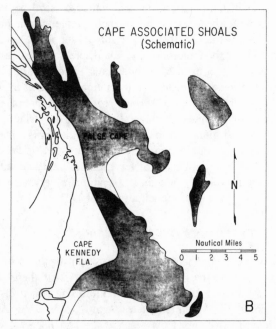

Figure 174.

Schematized plan view of major shoal types of the U.S. Atlantic coast. Types identified are :
(a) inlet-associated shoals; (b) cape-associated shoals; and (c) linear shoals. The linear type occur
as both shoreface-connected and isolated. (Shoal form enhanced by shading.)

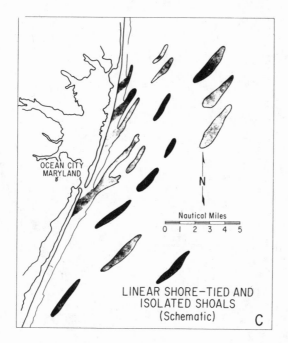

Arcuate inlet-associated shoals (Fig. 174a) are typified by those discussed by Ludwick (1972) and Oertel and Howard (1972). Arcuate cape-associated shoals are those such as Chester and southeast shoals at Cape Kennedy (Field and Duane, 1972) which have a hammerhead shape in plan view (Fig 174b). Linear shoals are defined as linear, positive features which exhibit at least 10 ft of relief between the crest and the surrounding surface. Shoreface tied shoals are those outlined by, but landward of, the isobath that defines the base of the shoreface (usually 30 ft). Isolated shoals are those lying on the inner continental shelf which display no topographic ties to the shoreface. Although linear shoals are known from other coasts, particularly the Gulf Coast, present discussion will be restricted to linear shoals on the United States east coast, south of Long Island. The distribution of linear and arcuate inlet and cape-associated shoals on the Atlantic Inner Continental Shelf is shown on Fig. 175.

CHARACTERISTICS OF LINEAR SHOALS: A SURVEY

General

At the present stage of our studies of linear shoals, we are concerned only with those shoals having a length of at least 3000 ft. Isolated shoals are those meeting the minimum length requirement and having a relief of at least 10 ft.

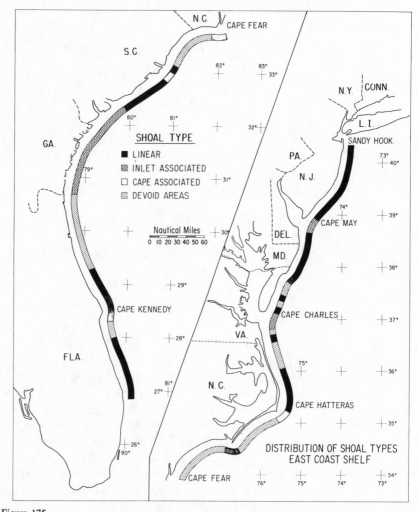

Figure 175.
Distribution of major shoal types on the United States Atlantic coast from Sandy Hook, New Jersey, to Palm Beach, Florida.

Shoals defined by a seaward excursion of the 30 ft depth contour of the shoreface are termed shoreface-connected.

All linear shoals in the area from northern New Jersey to Palm Beach, Florida form a small acute angle with the coastline. Nearly all shoals open northward regardless of presumed direction of net littoral drift. Seismic reflection profiles show that the linear cape-associated shoal complexes (such as those at Cape Kennedy, Florida and Cape Fear, North Carolina) to be plano-convex features resting upon a virtually featureless, nearly horizontal stratum. Marked differences exist between sediments of the shoal and those of the underlying sonic reflector, which itself undergoes local and regional changes in lithology. Sediment of the shoals is nearly totally silicate sand, except those at and south of Cape Kennedy, Florida, where carbonate skeletal sand becomes an important constituent.

In order to obtain to obtain a quantitative picture, data on six geometric properties were gathered using USC&GS 1200 Series Charts as data base. Characteristics, keyed to the highest point on the shoal, were: (a) water depth; (b) shoal length [of the 10 ft relief contour]; (c) shoal width (end points of a line passing through highest point and terminating on the 10 ft relief contour); (d) distance latitudinally from shore; (e) azimuth; and (f) the angle of the shoal with the adjacent shore-line trend.

Approximately 200 shoals were analyzed. In the course of the analysis, it became apparent that some of the operational definitions were redundant, and for some of the criteria, not precise enough to provide good repeatability among the five operators involved. Therefore, only clearly unambiguous criteria are discussed; that is, water depth, azimuth, and angle (preceding items a, e, f).

A histogram of water depth over shoal peaks is depicted in Fig. 176a. Two modes, 20 to 30 ft, and 40 to 55 ft are clearly evident. A third is suggested in water deeper than 80 ft, near the depth limitations imposed in this study. That the shoals continue into greater depth is clearly indicated in the following study by Swift et al. (1972), as well as by earlier workers such as Veatch and Smith (1939), Shepard (1963), and Uchupi (1968).

Major axis of the shoals has a mean azimuth value of 32° with a standard deviation of 25° as depicted by the historgram in Fig. 176b. The histogram suggests two modes, at apprximately 5° and 35°, with a third mode possibly located at approximately $-30°$. The correlation between water depth and shoal azimuth is nearly zero indicating no apparent linear dependence. Histograms showing the deviation of the shoal axis from shoreline axis and the azimuth of shorelines adjacent to observed shoals, shown in Figs. 176c and 176d respectively, will be referred to later.

Semi-diurnal tide range over the study area measured at the coast varies somewhat but is generally 3 to 7 ft. In the classification of Davies (1964) the tidal range discussed above lies within the microtidal range (upper limit of 6.5 ft). With a tide range in the micro scale or near the lower limit of the meso scale, Davies (1964) suggests that tide-induced currents will be of negligible significance in sedimentation processes. Some calculations based on open coast tide values (United States Dept. of Commerce, 1970) for Maryland, North Carolina, and central Florida, and the progression of the tidal wave, indicate open coast tidal currents are insignificant. This tends to corroborate the conclusion of Davies (1964) and would allow extrapolation of Davies' statements to other portions of the study area.

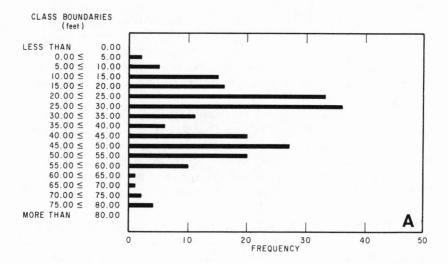

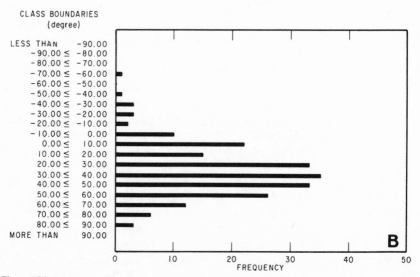

Figure 176.

Group characteristics of over 200 inner continental shelf linear shoals. Parameters plotted by frequency

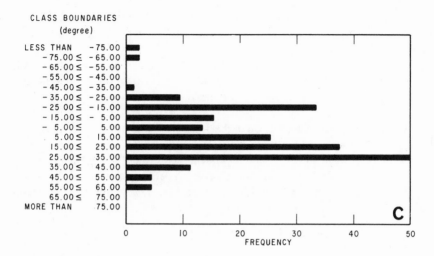

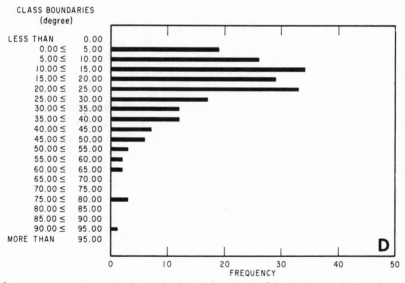

of occurrence are: (a) water depth over shoal crest; (b) azimuth of shoal axis; (c) azimuths of shore-lines adjacent to shoals; and (d) angular deviation of shoal axis trend with coastline trend.

Surface drift currents in the study area vary seasonally in intensity and direction (Bumpus and Lauzier, 1965). Maximum velocity reported is 0.86 ft sec^{-1} (26.2 cm sec^{-1}).

Wave climate on the inner shelf of the study area is discussed by Harris (1972). From his data it can be seen that for most of the area, wave heights in excess of 5 ft occur only 10% of the time, and waves higher than 8 ft occur only 1% of the time.

We will follow this introduction to the dimensions and hydraulic environment of the linear shoals with a more detailed description of selected shoal areas. Regional patterns of shoals in the best known sector, the Middle Atlantic Bight, will be surveyed. Following will be a discussion detailing those areas for which we have the most data: the shoal fields of central New Jersey, Maryland, southern Virginia, and the south and central east Florida coast.

Inner Shelf Shoals and Coastal Compartments
of the Middle Atlantic Bight

The portion of our study area for which the most bathymetric and geological data are available, from the literature and from our own efforts, is the Middle Atlantic Bight, between Cape Cod and Cape Hatteras. Within this sector it is useful to examine the related patterns of coastline and inner shelf morphology.

Shoreface-connected, isolated, cape-associated, and inlet-associated shoals occur in repeating patterns along four major coastal compartments of the Middle Atlantic Bight (Fig. 177). These compartments are delimited by the estuaries of Chesapeake Bay, Delaware Bay, New York Harbor, and Block Island Sound. Each compartment contains an eroding headland to the northeast. Because of the prevailing northeastern wave approach, recurved or cuspate barrier spits have grown on the north sides of the headlands. A seaward-convex barrier arc, consisting of a spit and successive barrier islands, has grown south from the headlands. Beyond the proximal barrier arc lies a less continuous distal barrier arc, generally seaward-concave, and separated from the proximal arc by a cuspate foreland, barrier-overlap inlet, or similar discontinuity.

Shoreface-connected shoals appear in this scheme as seaward extensions of the lower shoreface. The term shoreface is used herein as defined by Price (1954) as: "a relatively steep but short concave surface". Price's definition was made in reference to his idealized shelf equilibrium profile. Our usage of shoreface is geomorphic without necessarily implication as regards equilibrium. Despite local variation, a systematic scheme of distribution of shoreface ridge systems is discernable.

North-trending spits and cuspate forelands are associated with tide-dominated foreshores. An ebb tidal delta lies at the tip of Montauk Point (eastern end of Long Island Coastal Compartment). Sandy Hook and Cape Henlopen, the north-trending spits of the New Jersey and Delmarva Coastal Compartments, have linear, tide-built shoals extending southeast from their tips, between flood and ebb dominated channels. Hen and Chicken shoals, attached to Cape Henlopen

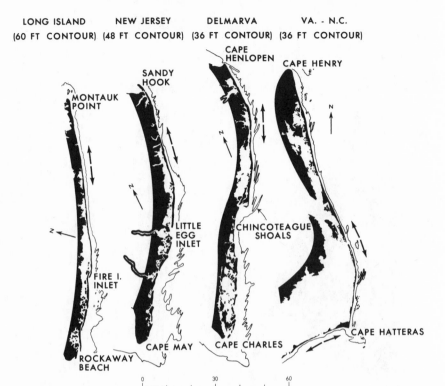

Figure 177.

Coastal compartments and shoreface-ridge systems of the Middle Atlantic Bight, as defined by the 60 ft contour off Long Island, the 48 ft contour off New Jersey, and 36 ft contour off the Delaware-Maryland-Virginia compartment and the Virginia-North Carolina compartment. Arrows indicate major littoral drift directions.

is large enough to appear in Fig. 177. A poorly defined ''submarine tidal delta'' extends east and south from Cape Henry (Payne, 1970).

The mainland beaches of the eroding headlands may be smooth (Long Island, North Carolina-Virginia coastal compartments), or may exhibit diffuse systems of wide angle (45° or greater) shore-connected shoals (New Jersey, Delmarva; Fig. 177). The proximal, seaward-convex barrier arcs tend to have shoreface-connected shoals grouped into more or less coherent fields. The fields tend to occur immediately south of gentle convexities in the shoreline (Fig. 177). Well-developed fields of shoreface-connected and isolated shoals generally occur adjacent to the barrier overlap, where inlet or cape-associated shoals separate the proximal from the distal barrier arcs (Figs. 177, 178). The shoal field corresponding to this category for the Virginia-North Carolina coastal compartment is Diamond Shoals, a classic cape-associated shoal system extending from Cape Lookout that will be discussed in the next chapter (Swift et al., 1972a).

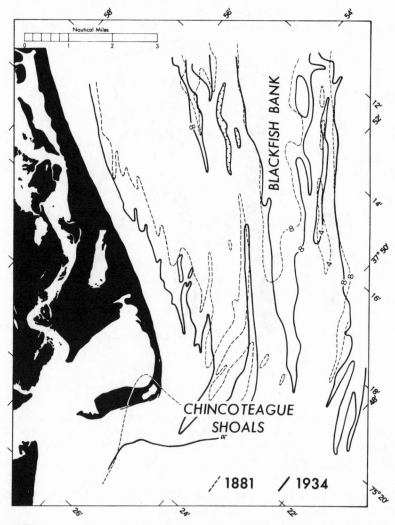

Figure 178.

Chincoteaque shoals off the southern end of Assateague, Md. Note the net southwestward movement as indicated by surveys in 1881 and 1934. Shoreline is from the 1934 survey. Isobaths are in fathoms.

The distal seaward-concave barrier arcs are characterized by a broad, gentle shoreface as indicated by the characteristic contour in Fig. 177. This is particularly true in the case of the look-alike New Jersey and Delaware-Maryland-Virginia (Delmarva) coastal compartments. The zig-zag nature of the contour indicates that linear shoals are present. However, they are so subdued (side slopes of less than 2° and heights of 10 ft or less) and the slope of the sea floor is so gentle, that it becomes difficult to characterize them as either shoreface-connected or isolated. The broad extent and gentle slopes of the shoreface of the distal barrier arcs suggest that these portions of the sea floor are comprised of markedly finer sand than are the steep narrow shorefaces of the coastal compartments northern sectors. Textural data supporting such a coast-wise grain-size segregation are available for the New Jersey coast (McMaster, 1954), and the inner Virginia Carolina Shelf (Swift et al., 1971).

The Long Island, New Jersey, and Delmarva coastal compartments terminate with inlet associated shoals which radiate seaward from the north sides of estuary mouths. In Chesapeake Bay, this is the Bay-Mouth Shoal described by Ludwick (1972). In Delaware Bay, it is Overfall Shoal (Fig. 174a) and associated shoals (Jordan, 1962). Here the ridges are structured by tidal currents into intricate ebb and flood channel systems. A detailed map of lower New York Bay (Veatch and Smith, 1939; plate 10) shows that Long Island also has an inlet-associated shoal off its western end. The distal portions of these tide-built ridges swing northeast to merge with the inner shelf ridge and swale topography.

New Jersey Inner Shelf Shoals

The northern coastal region of New Jersey has a rather steep, strikingly regular, and narrow shoreface (Fig. 179) with an average width of 2100 ft. Linear shoals are isolated from the coast by approximately 2.5 n mi; Shrewsbury Rocks shoal is an exception. It can be traced from the shoreface out to about 6.5 n mi where it is truncated by the head of the Hudson Canyon (Fig. 171). The shoals vary in length from 1 to 3 n mi with the exception of Shrewsbury Rocks, and have a mean width of 1500 feet. Water depth above shoal crests approximates 30 to 60 ft with overall shoal relief of 10-35 ft. All of the shoals have their long axis oriented east northeast. They form an angle of 30° to the coastline in the Barnegat section and 30° to 85° in the Asbury Park-Long Branch sector (Figs. 179, 180). The shoals tend to be nearly symmetrical. Where asymmetry occurs, southern flanks are usually steeper, except where shoreface-connected shoals join the shoreface; here the northern, inshore flanks are steeper. Isolated shoals tend to exhibit a longitudinal asymmetry as well, with peaks occurring toward the southern ends of the crestlines.

High resolution seismic reflection records show that shoals off the Shark River Inlet are underlain by two shallow, gently southeast-dipping strata. Poor record quality prohibits resolution of acoustic reflections below 100 ft and shoal interiors appear opaque and featureless. However, coast morphology and bathymetric fabric of northern New Jersey suggest that coastal plain Cretaceous and Tertiary strata crop out on the sea floor and are covered by only a relatively thin overburden of sand and gravel (Williams and Duane, in prep.). In the Long Branch-Asbury Park sector, a reticulate pattern of ridges appears (Fig. 179). East-

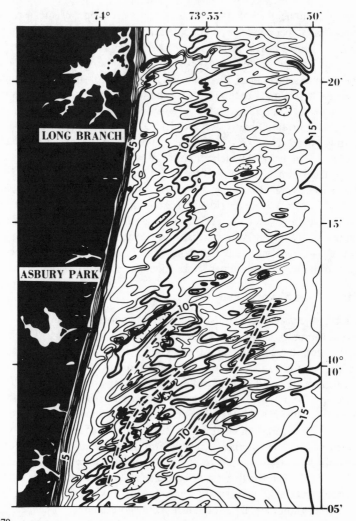

Figure 179.

Northern New Jersey shoreface region showing small linear ENE trending shoals superimposed on a NNE lineation (dashed lines) that reflect possible structural control. Contour interval is one fathom.

northeast trending ridges occur en echelon along at least two north-northeast axes. These axes are nearly parallel to the New Jersey coast and also to shelf-floor cuestas farther offshore (Swift, 1972). It seems probably that in this area, the ridges are at least in part attributable to an erosional origin. The erosion may have been marine, since the reticulate pattern bears no relationship to the adjacent sub-aerial drainage pattern (Fig. 171). Houbolt (1968) has described sand ridges of this sort on the floor of the North Sea that are maintained by marine processes and form a continuum from wholly constructional forms to wholly erosional, tide-sculpted forms.

Study of northern New Jersey surface and subsurface shoal sediments from cores indicates that the shoals are composed of medium-grained, polished, well-sorted, quartzose sand. These shoals are superimposed on a thin veneer of coarse, poorly sorted, iron-stained and pitted quartz and glauconite, overlying a substrate of fine-grained sands, silts, and clays. This substrate change occurs at approximately the depth of the uppermost sonic reflector. The unconsolidated sediment cover of these northern New Jersey ridges has been shown to be a combination of Pleistocene outwash from the north and material derived from land erosion to the west (McMasters, 1954; Williams and Field, 1971).

In marked contrast to northern New Jersey the linear shoals from Barnegat Inlet south to Cape May are longer and more abundant (Fig. 180). The shoreface is broader and more irregular than to the north and a number of the shoals are shore tied with the 20 or 30 ft contour. Mean depths to shoal crests are 25 to 30 ft and except for shoreface-connected shoals, are fairly uniform over the entire shoal length. Shoreface-connected shoals have a broad base anchored to the shore and gradually taper seaward for an average length of 3.5 n mi. All the shoals form an angle of $20°$ to $60°$ with the coast and maintain a northeast and east-northeast orientation independent of the change in shoreline orientation; they show about 10 to 20 ft of relief with respect to the adjacent relatively feature-less sea floor.

Geophysical coverage over the shoals offshore from Barnegat reveals the presence of an underlying, flat, featureless reflector dipping at a low angle to the southeast (Fig. 181). Selected cores on the shoal crests, flanks, and in the intervening troughs, show the shoals to be composed of clean, well-sorted, polished, medium-grained, quartz sand. Core penetration through the reflector shows it to be composed of a very coarse gravelly sand. The texture of the material and the presence of broken shell fragments suggest it is a lag deposit resulting from active marine reworking.

Field observations indicate that the Manasquan region is a littoral drift nodal area. South of Manasquan, longshore transport is southerly with an estimated volume of 50,000 yd^3 annually passing Barnegat Inlet (Fig. 180; Caldwell, 1966).

Maryland Inner Shelf Shoals

The coastal province of the Delaware-Maryland-Virginia (Delmarva) peninsula is a classic example of a headland-barrier island-spit complex that has developed through submergence, erosion, and extensive littoral transport (figs. 177, 182).

Natural borders to coastal Delmarva are the large Delaware and Chesapeake Bays to the north and south, respectively. Of the total 120 n mi of coastline, 20 mi belong to the low-lying headland-spit complex, which extends from

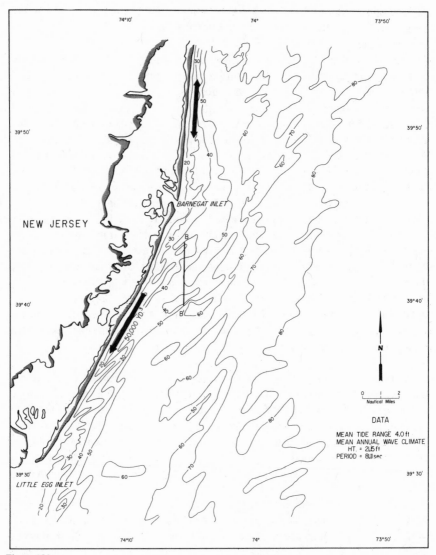

Figure 180.

Detailed bathymetry of the central New Jersey inner continental shelf at 10-ft contour interval. Direction and volume (yd³) of littoral transport, tidal range, and mean wave characteristics are indicated. Double-headed arrow indicates nodal zone. Line BB′ is the trackline location of the Barnegat geophysical profile in Fig. 181.

Bethany Beach, just north of the Maryland-Delaware border, to the tip of Cape Henlopen and includes the bay-mouth or washover barriers that enclose Indian River Bay and Rehoboth Bay and are separated by Indian River Inlet. The

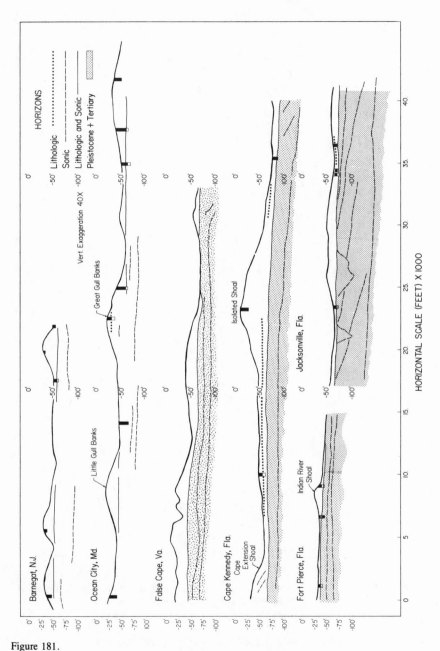

Figure 181.

Interpretative geophysical profiles across representative shoal areas of New Jersey, Maryland, Virginia, and Florida. Profile trackline locations are shown in Figs. 180, 182, 183, 184. Heavy vertical lines represent cores which define three indicated lithologic horizons.

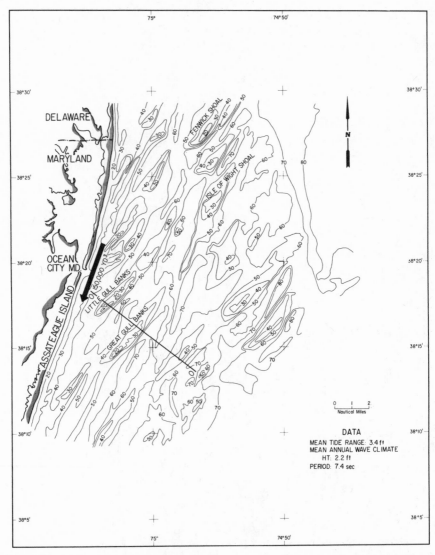

Figure 182.
Detailed bathymetry of the Maryland inner continental shelf at 10-ft contour interval. Direction and volume (yd³) of littoral transport, tidal range, and mean wave characteristics are indicated. Line 00′ is the trackline location of the Ocean City, Md. geophysical profile in Fig. 181.

proximal barrier arc (Fenwick Island, Ocean City and Asseteague Island) is approximately 40 mi long and stretches south from Bethany Beach. At present there exists only one major inlet (Ocean City) though many such inlets have existed within historical time. The 60 mi section of distal barrier arc south of Assateague to Chesapeake Bay entrance is dominated by numerous inlets and channels segregating small barrier islands and is similar in this regard to southern New Jersey and the sea island coast of Georgia (see Figs. 173, 177).

Littoral currents transport materials both southward and northward from the vicinity of Bethany Beach, Delaware. Going north toward Indian River Inlet, erosion has been occurring at a rate of 190,000 yd^3 annually; from the inlet north to Cape Henlopen the rate is five times as great. Littoral drift at Ocean City, Md. is estimated to be 150,000 yd^3 per annum in a southerly direction.

The central portion of the Delmarva Peninsula inner continental shelf is characterized by numerous long linear ridges or shoals trending northeast and opening at small acute angles with the shoreline (Fig. 172). From Ocean City to Indian River Inlet, the shoals become progressively smaller in scale, and become confined to the vicinity of the shoreface. Shore angles increase to 45° and greater in the Bethany Beach Shoals, while absolute orientation changes very little. A large ebb tidal shoal (Hen and Chicken Shoal) trends southeast from Cape Henlopen (Fig. 174a). South of the main Assateague Island barrier, inlet shoals are common; however the linear type under discussion in this paper are subdued and less abundant. Linear ridges between Fenwick Island and the southern end of Assateague Island display common characteristics in shape, depth, and orientation. Depths to crest range from slightly over 10 ft to nearly 50 ft; frequently shoals appear to be grouped en echelon (Fig. 181).

Transverse asymmetry (steeper southeastern sides) and longitudinal asymmetry (peaks toward the southwestern ends of crests) are better developed than on the New Jersey coast. The blunter southwestern ends of the ridges tend to curve toward the coast. The 46 shoreface-connected and isolated shoals in this region have a mean azimuth of 39° (standard deviation of 11.6°) and strike a mean angle of 18° (standard deviation of 8.9°) with the adjacent coast.

Typical subbottom structure and lithology of the shoals are demonstrated in the schematized cross section of the three well developed en echelon shoals (Little Gull Banks, Great Gull Banks and an unnamed shoal) shown in Fig. 181.

Continuous and discontinuous sonic reflectors are visible under the shoal region at shallow sediment depths (Fig. 181). At about −69 ft MLW near the shore face lies a nearly continuous reflector which dips very gently (5 ft mi^{-1} or 1/1000) seaward to about −90 ft MLW six miles offshore. Discontinuous sonic reflectors appear between the continuous horizon and the sediment-water interface. Frequently the upper horizons appear at or near the base of the shoals, suggesting a probable change in lithology. Cores penetrating to the depth of the upper horizon corroborate this lithologic change; sediments below the horizon are muds, while those above are comprised predominately of sand-sized material. Sediment distribution is related and attributable to local bottom morphology.

Inter-shoal flats and depressions contain silts and fine silty sands; shoals are composed of quartzose medium-grained sand in thicknesses up to 20 ft. Texture and composition of shoal sands are similar to adjacent beaches. Ocean City-Assateague beaches have a mean hydraulic (equivalent sphere) diameter of 0.440 mm or 1.18 phi (sieve diam. of 0.340 mm or 1.56 phi). Samples from the nearshore shoals are slightly finer, ranging in size from 0.203 mm to 0.400 mm (2.3 phi to 1.32 phi) in hydraulic diameter.

Age and environmental history of the surface and subsurface sediments are at present unknown; however, interpretation based on the literature and cursory examination of sediment data is possible. Based mainly on the sediment texture and to a lesser degree on indigenous fauna, the muds associated with the upper sonic reflector probably represent lagoonal-estuarine deposits. Lack of consolidation and desiccation in the fines suggest they represent deposition occurring landward of the barrier during the Holocene transgression. Supporting data are supplied by the radiocarbon dating of an organic mud deposit by Kraft (1971b). The sample was obtained offshore of Dewey Beach, Delaware, at a depth of −65 ft MLW and had a date of 7500 ± 135 yr BP. Depth of the sample correlates well with the depth range (−60 to −90 MLW) of the subsurface lithologic and sonic reflector from the Ocean City vicinity. We believe this strata may be a lagoonal deposit, and by correlation to Kraft's date, of Holocene age.

Detailed bathymetric mapping of the Ocean City area since 1850 has shown that the shoals have existed in their approximate location since that time. However, some migration to the south is apparent for most of the shoals. For instance, comparison of the 30 ft contour line of Little Gull Banks from surveys over the 80 yr period of 1850 to 1930 indicate the shoal center has shifted between 500 and 1500 ft to the southeast. At Chincoteaque (Fig. 178), north-south trending offshore shoals have extended themselves as much as 2 n mi to the south, while oblique, shoreface-connected shoals have migrated southeast.

False Cape, Virginia, Inner Shelf Shoals

The inner continental shelf off southeastern Virginia contains one of the best developed shoal fields between the south Florida field and Assateague Island, Maryland. Two large shore-connected shoals and one large isolated nearshore shoal off the False Cape, Virginia, area (Fig. 183) have been the subject of detailed study. North and east of these shoals the Virginia shelf as far north as Cape Henry contains many isolated shoals.

The False Cape shoals have a maximum relief of 20 ft with flank slopes of 2° or less. The main shoreface-connected shoals crest at around −20 and −30 ft MLW with the landward shoal being shallowest. The isolated shoal which lies farthest seaward crests at around −40 ft MLW. Shoal crests deepen and develop secondary crests toward the north. The long axes of the shoals are oriented on an azimuth of about 15° and form an angle with the adjacent

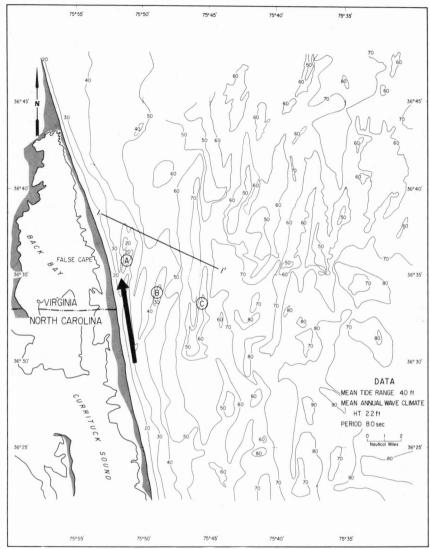

Figure 183.

Detailed bathymetry of the False Cape, Virginia, inner continental shelf at 10-ft contour interval. Direction of littoral transport, tidal range, and mean wave characteristics are indicated. Line 11′ is the trackline location of the False Cape, Virginia geophysical profile in Fig. 181. Shoals A, B & C are shown in greater detail in Fig. 186.

shore of about 25°. A profile across the False Cape shoals (Fig. 181) shows typical topography and subbottom horizons delineated by seismic reflection profiles and cores. The lower horizon is underlain by a greenish-gray muddy sand. A discontinuous layer of watery brown mud overlies the lower horizon and is in turn overlain by the surficial sand layer. Radiocarbon dates on carbonate material within the brown mud indicate that it is of mid-Wisconsin age. An articulated shell of *Mercenaria* sp recovered from a depth of 6 ft in the surficial sand unit was dated at 4220 ± 140 yr BP (Swift et al., 1972b).

Surficial sediment of the shoal crests is predominantly medium- to fine-grained quartz sand; shoal flanks, the adjacent shore face and margins of some intershoal troughs are characterized by fine- to very fine-grained sand. Along the central axis area of intershoal troughs a pebbly medium to coarse sand forms a thin and locally discontinuous layer covering compact greenish gray muddy sand and brown mud.

Florida Inner Shelf Shoals

General

The Florida inner continental shelf can be subdivided into three regions based on shoal morphology (Fig. 175). These regions are: South Florida which has a well developed shoal field similar to those of the Middle Atlantic Bight; Central Florida which is characterized by cape-associated and large linear shoals; and North Florida, which displays an anomalous morphology dominated by irregular banks of low-relief. Mean tide ranges and wave climate of the Ft. Pierce region are indicated in Fig. 184. Waves in the area come principally from the east and northeast giving rise to southerly littoral drift estimated to be 350,000 yd^3 yr^{-1} at Canaveral Harbor, 200,000 yd^3 yr^{-1} at Ft. Pierce Inlet, and 230,000 yd^3 yr^{-1} at St. Lucie Inlet (United States Army Corps of Engineers, 1967) (Fig. 184).

South Florida

The south Florida shoal field occupies the shelf off eastern Florida in the vicinity of Fort Pierce (Fig. 184). This field, the southernmost along the United States Atlantic Coast, contains well developed shoreface-connected, and isolated linear shoals with their long axes lying predominantly nearly north-south. Shoals in the south Florida field crest around −20 and −30 ft with some smaller shoals cresting at around −50 ft.

A definite and persistent acoustic reflector underlies the south Florida shoals at or near the elevation of shoal bases (Fig. 181). The sonic discontinuity is associated with marked changes in sediment characteristics.

Sediment comprising the shoals is characteristically well sorted biogenic medium- to coarse-grained sand with 15 to 30% quartz. Although the sand is occasionally fine or very coarse, poorly sorted, and varies in color from light brownish gray to medium gray, its constituents are similar. The biogenic fraction is composed mostly of barnacle plates, mollusk fragments and some

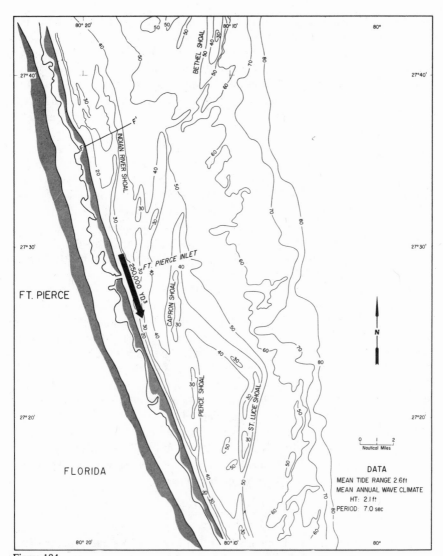

Figure 184.

Detailed bathymetry of the south Florida inner continental shelf at 10-ft contour interval. Direction and volume (yd³) of littoral transport, tidal range, and mean wave characteristics are indicated. Line FF' is the trackline location of the Ft. Pierce, Florida, geophysical profile in Fig. 181.

benthic foraminifera. The molluscan and foraminiferal assemblage is representative of one typical of a shallow water open marine environment similar to present conditions on the shoals. Meisburger and Duane (1971) present a more detailed description of sediment and geomorphology for this area.

Bethel Shoal situated near the seaward edge of the Fort Pierce shoal field is one of the largest shoals described in this study. It is also the only shoal of the Atlantic inner shelf area in which internal bedding can be delineated with the available seismic reflection profiles. Line profiles across Bethel Shoal showing internal acoustic reflectors are shown in Fig. 185. The internal bedding suggests a southeastward prograduation of the shoal coupled with upward growth. The present surface configuration of the shoal seems to be roughly paralleled by the major internal reflector surfaces.

Between the shoals the bottom is nearly flat and covered by a layer of biogenic sand similar in character to that comprising the shoals. In contrast, however, the material in the intershoal layer tends to be more poorly sorted, coarser, and more angular and is highly bored and incrusted by organisms.

Sediment at or below the acoustic reflector at the base of the shoals in the Fort Pierce area, like that described later at Cape Kennedy, is a silty shell gravel and partly lithified calcarenite. A radiocarbon date of 8640 yr BP from peat overlying the surface suggests that the reflector marks the Holocene-Pleistocene surface.

Hurricanes passing within 150 n mi of the coast have occurred with an average frequency of one every 3 yr for the past 70 yr. Their effects have often been devastating but of short duration; tides generated by the storms often exceed 5 ft. Northeast storms have often resulted in more severe effects particularly with regard to erosion of the shore. Historical data do not indicate significant changes in the axial location of the shoals; however, the survey control is not precise. The well rounded and polished condition of sediment recovered from the shoal tops indicates reworking and movement of grains must occur, though this may not entail any overall displacement of the shoals themselves.

Beach sands on the adjacent Florida coast are similar in size, sorting and biogenic constituents to the characteristic shoal material. However, the quartz content is significantly higher in the beach sands, comprising approximately 60% of the beach material (Meisburger and Duane, 1971).

Central Florida

The inner continental shelf topography adjacent to Cape Kennedy is dominated by large, cape-associated shoals (Fig. 174b) trending southeast from the peninsula and large isolated linear shoals set immediately seaward of the cape-associated shoal tips. On the north side of Chester Shoal, extending from False Cape, small linear shoals trend north from the base of the shoal (Fig. 174b). Historical surveys of Chester Shoal and Southeast Shoal, extending from Cape Kennedy, indicate they are active depositional sites for fine and medium-grained sands transported in the littoral zone. Morphology, textural characteristics and proces-

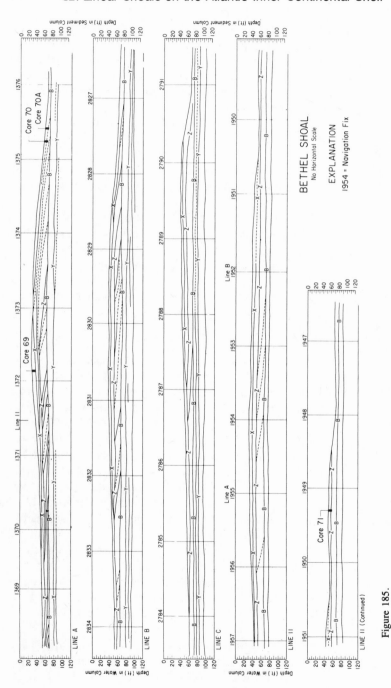

Figure 185.

Interpretative geophysical profiles of Bethel shoal, Florida, area showing internal structure (inclined bedding) and relationship to underlying stratigraphy. Lines A,B,C are east-west; line 11 is north-south. Area location shown in Fig. 184. From Meisburger and Duane, 1971.

ses of formation and maintenance of these arcuate features will be topics of future studies.

Two large, linear isolated shoals, Ohio-Hetzel shoal and The Bull, lie adjacent of Cape Kennedy. They exhibit structural and lithologic characteristics common to linear shoals in south Florida and the middle Atlantic Bight; however, their morphology and association with the cape shoals suggests a genesis that may not be common to these other regions. The northeast trending Bull shoal is 27,000 ft long, 7,800 ft wide and rises from a base of −60 ft MLW to within −15 ft of MLW. Ohio and Hetzel Shoals are separated by a shallow narrow trough; combined they represent a northwest trending shoal complex that rises to −20 ft MLW and is slightly asymmetrical in cross section, with steeper slopes on the landward flank.

Shallow structure of the isolated shoals based in interpretation of core borings and seismic reflection records is depicted in Fig. 181. A continuous and mappable subbottom acoustic reflector lies at depths of −60 to −90 ft MLW throughout much of the area and as far south as Fort Pierce. Beneath the shoals the reflector is essentially flat lying; on the flanks and in the adjacent depressions the reflector correlates with a major change in sediment lithology. Beneath the reflector, sediments are partially recrystalized and cemented; above the reflector sediments are completely unconsolidated, but their composition indicates derivation in part from the underlying layer (Field and Duane, 1972). A radiocarbon date of 7,320 ±140 yr BP was obtained from a peat layer (probable fresh water origin) overlying the acoustic reflector at a depth of −55 ft MLW several miles south of the shoal area (Field and Duane, 1972). One mile seaward of the shoal, shells characteristic of the intertidal zone (*Donax variabilis*) were obtained from the subsurface depth of the major reflector. A ^{14}C date of 23,500 yr BP suggests a mid-Pleistocene age for the materials below the sonic and lithologic horizon (Field, 1972). Both sediment characteristics and age dates thus indicate that the sonic reflector represents a Pleistocene surface, and that the overlying shoals are no older than 7000 years. Internal structure in the shoals is not evident from seismic or sediment studies. Cores penetrating to depths of 12 ft on the shoal contain well sorted medium to coarse mixtures of biogenic (chiefly molluscan) and terrigenous (chiefly quartz) sand.

The highly polished nature of the sand at 12 ft below surface suggests recent tractive movement. Both grain size and composition of the shoal sands are similar to deposits on adjacent beaches and both show major contribution from erosion of older deposits. This seems to be true also for beaches from Fort Pierce north to Cape Kennedy (Field and Duane, 1972; Meisburger and Duane, 1971).

Historical profile data collected between 1898 and 1965 show that the shoals have become several feet higher and shifted over a thousand feet to the southeast. Twenty-five miles to the north of the Cape Kennedy shoal field a small group of linear shoals occurs. These isolated linear bodies with the long axes oriented nearly parallel to shore are the most northerly shoals off the Florida coast which can be clearly related to the "accretional" linear and arcuate shoals occurring to the south.

North Florida

The typical shoal off north Florida is a broad nearly flat topped submerged bank which may be either roughly linear or totally irregular in outline. Predominant orientation of these shoal banks is northwest-southeast. Between many shoal banks there are deep northwest trending depressions which deepen progressively seaward.

Seismic reflection and sediment core data indicate that the nuclei of most large north Florida bank shoals may consist of pre-Holocene sediments (Meisburger and Field, 1972). The geometric pattern of the banks and intervening depressions suggests that the north Florida bank shoals are basically relict interfluves of a late Wisconsin subaereal erosion surface only slightly modified by the events of the Holocene. Thus, the north Florida "bank" shoals cannot be properly equated with the linear shelf shoals discussed elsewhere in this paper. Small linear shoals do exist, especially inshore, but these are generally ill defined on existing chart coverage and of insufficient relief to be included here.

LINEAR SHOALS AND COASTAL HYDRAULICS

Linear Shoals and the Wave Regime

The shoal fields of the inner Atlantic shelf lie at water depths which, during fair weather, are within the zone of wave surge. It seems reasonable, therefore, to consider experimental data pertaining to the generation of wavebuilt bars to determine the extent to which the shoals may be wavebuilt.

Kuelegan (1948), in a laboratory study of the formation of sand bars by waves impinging on a shoaling beach, developed dimensionless criteria relating the formation of bars to wave height and length and water depth to bar base. Concluding from his laboratory studies, Kuelegan (1948, p. 13) stated that: (a) "if water depth remains constant . . . the position of the bar, formed by a single system of waves, is a function of wave height and steepness; (b) if the water depth and wave steepness are held constant, an increase in the wave height will move the bar seaward; (c) if the water depth and wave height are held constant, an increase in wave steepness (shorter period) will move the bar shoreward; and (d) if the wave height and wave length are held constant, the depth of bar base below water will likewise remain constant, and any increase in the depth of water moves the bar shoreward."

Extrapolation of laboratory findings to field conditions is not always easy, nor direct. However, several aspects of Kuelegan's experiments have been confirmed in the field: Ratio of water depth over trough to depth over crest, approximately 1.9 in the lab, is 1.5 for Lake Michigan (Saylor and Hands, 1971) and 1.5 for Pacific Coast beaches (Shepard, 1950). Further, in the case of the Geat Lakes (where wave climate can be considered constant, or nearly so) increase in water depth (rise of lake levels) is associated with shoreward bar movement (Saylor and Hands, 1971) and increased erosion (Berg and Duane, 1968).

We have examined the bars in the four areas discussed in detail in the previous section, using water depth over bar base and wave characteristics. Ratios obtained from Atlantic inner shelf shoals show significant variation in the ratio of trough depth to crest depth. Range values for each of the major regions are as follows: Central New Jersey, 1.13 to 1.70; Maryland, 1.45 to 4.2; False Cape, Virginia, 1.3 to 1.7; and south Florida, 1.47 to 2.1. Although many of these values are within the limits established by Kuelegan (1948) and corroborated by field studies, the range of values indicates that it is unlikely that all the features described are wave-generated. The offshore bars from Lake Michigan and the Pacific Coast are nearly an order of magnitude smaller than the shoals under discussion here. Also, those Lake Michigan and Pacific Coast features display a remarkable linear continuity: they parallel the coast for distances in excess of tens of miles, unlike the relatively short, discontinuous, acute-angle shoals of the East Coast. Furthermore, using Kuelegan's data, we have calculated the theoretical parameters of waves required to form the shoals. Shoals lying at $-40'$ MLW and deeper would require waves having heights (H_0) ranging from $18'$ to $40'$ and periods (T) greater than 20 sec. The frequency of occurrence of waves having these heights is shown by Harris (1972) to be extremely low; also in shoaling water a wave will break when d/H is in the range of 0.78 to 1.0 (d is water depth). Hence, particular combinations of H_0, T required are unrealistic or theoretically impossible.

For shoaler features ($-25'$ MSL) certain combinations of H_0, T could conceivably occur and, therefore, result in bar construction. However, the vast majority of shoals require for their formation theoretical wave height and period combinations that are unrealistic for the United States Atlantic coast.

Linear Shoals and Storm-Generated Coastal Currents

The hydraulic regime of the inner shelf during storms offers more promise as a generative agent for the inner shelf shoal fields. Harrison et al. (1967) have summarized what is known concerning the effect of storms on the circulation of the Middle Atlantic Bight. During the winter, successive mid-latitude lows passing up the Alantic seaboard generate intense northeast winds. Repeated intense wind stress combined with falling temperatures serves to destroy the stratification of the shelf water during this period. Harrison et al. (1967, p. 72) note of individual storms that: "as the onshore component of the wind increases over the shelf and slope water zones of the northern part of the Middle Atlantic Bight, it may produce or augment a net southerly drift (permanent flow) owing to the effect of set-up on the southwesterly trending coast line, or to southward barotropic flow where the wind drift extends to the sea floor." They also note that such a southward flow would overcome the Ekman component of the drift. Wind-drift currents tend to develop speeds of one twentieth to one fiftieth that of surface and wind speeds (Weigel, 1964), and a northeaster blowing at 40 knots could generate a surface current of one or two knots (1.7 to 3.4 ft sec^{-1}) which, during several days of downward momentum transfer, might extend some portion of that velocity to the inner shelf floor.

In two of the most carefully studied systems of shoreface-connected linear shoals, a variety of textural and bathymetric data have been compared with direct hydraulic data. The combined data suggest that such storm currents and the wave trains associated with them are the dominant forces shaping the shoals. These systems occur at False Cape, Virginia (Swift et al., 1972b), and at Bethany Beach (Moody, 1964).

Process and Response in the False Cape System

Substrate Characteristics

Significant features of the linear shoals at False Cape which yield circumstantial evidence concerning the character of the hydraulic regime are the distribution of grain sizes and the geometry and orientation of the ridges themselves. Grain-size gradients at False Cape, Virginia, vary regularly with the topography (Fig. 186a, b). Crests consist of well-sorted medium- to fine-grained sand. Flanks are covered with fine- to very fine-grained sand. Stiff, pre-Holocene substrate is exposed intermittently in the troughs; locally it is covered with a few centimeters of coarse, pebbly sand which includes clay balls from the underlying substrate. A longitudinal gradient is superimposed on this transverse variation in grain size, in that toward the south the fine and very fine flank sands become more extensive and finally bridge over the coarse trough sands.

A comparison of graphic mean diameter with graphic inclusive standard deviation shows that trough and crest sands become better sorted with increasing grain size (Fig. 186b). This is interpreted as lag behavior in zones of higher energy; with time, more and more fines are removed, increasing the mean diameter and decreasing the size range. Flank sands, showing the reverse trend, are interpreted as the winnowed-out fractionates which, as they move intermittently away from the high-energy areas, lose successive coarsest fractions, thus decreasing the mean diameter while decreasing the size range.

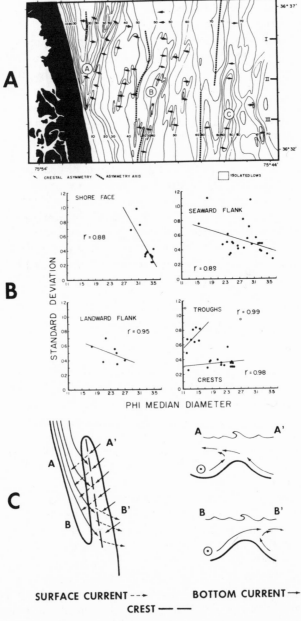

Figure 186.
Substrate response and hydraulic process for a narrow-angle, shore-connected shoal, False Cape, Virginia. (A) bathymetry, from precision navigation survey (Swift et al., 1972); (B) graphic mean diameter versus inclusive graphic standard deviation for False Cape sediment types (Swift et al., 1972); and (C) plan view (left) and cross-section (right) model for hydraulic regime over "A" ridge during a major storm.

The coarse trough sands are interpreted as a primary lag, generated by periodic storm currents which erode the troughs and sweep the sand up on the crests. Similar coarse lags have been noted in troughs of the tide-maintained ridge systems of the southern bight of the North Sea (Houbolt, 1968). In the North Sea the gain size continues to fine all the way to the ridge crests. The crestal grain-size reversal at False Cape suggests that in this area the operative currents are far more infrequent and that shoal crests are strongly modified by fair weather wave surge. Thus, a secondary lag is generated on the crests by swells whose surge winnows out fine to very fine sand and deposits it on the flanks.

The intermittent generating currents would appear to be south-trending. Crest and trough axes climb to the south. As each of the shoals joins the shoreface, its seaward flank becomes heavily mantled with fine sand—apparently swept out of the adjacent (landward) trough. There is an indication that the south-trending currents are structured into helical flow cells, as are the currents which form the shoals of the southern bight of the North Sea (Houbolt, 1968). The False Cape shoals bear asymmetrical, second-order ridges on their flanks, whose steeper sides face toward shoal crests and away from the "asymmetry axes" of major troughs (Fig. 187a). This geometry is compatible with the concept that, during the storms, the south-trending currents tend to descend in the troughs and sweep obliquely up the sides of the ridges to converge over the crests. To the extent that such helical flow may be a generating mechanism of water movement, analogies may be made to the helical flow patterns attributed by Dzulynski (1965) to experimental formation of bedforms; and to helical movements in the boundary layer of the air mass flowing over desert seif dunes (Folk, 1971). The most widely described helical flow phenomenon is Langmuir circulation observed above the thermocline at sea (Langmuir, 1938; Faller, 1969; Assaf et al., 1971). However differences exist between these examples and the linear shoals under discussion herein. For instance, seif dunes attain the angle of repose at least locally, and are generally of smaller wavelength. Langmuir cells tend to have a wavelength no greater than equal to the depth of the mixed layer (Assaf et al., 1971) and thus are nearly equant in cross section; the False Cape shoals are separated by distances of miles, but lie in 66 ft or less of water.

Hydraulic Data

Five direct observations of bottom currents have been made at False Cape with a Bendix Q-18 Orthogonal Current Meter (Holliday et al., in press). All but one were obtained during periods of calm weather. Weak north-trending currents generally less than 0.3 ft sec^{-1} (10 cm sec^{-1}) were measured for durations of 12 to 24 hr by 6 inch impellers; units were mounted on the sea floor. Currents observed may have been caused by the indraft of saline bottom water to Chesapeake Bay to the north. The currents were too weak to affect the troughs, but in combination with wave surge visibly affected shoal crests. Scuba dives to the station on the crest of "C" ridge (Fig. 186a) revealed sand ripples 4 to 6 inches (10 to 15 cm) high whose crests were activated by each wave surge and which slowly migrated to the northwest. The residual current for the 26-hr station was 0.12 ft sec^{-1} (3.6 cm sec^{-1}) to the north. However, when only the highest 10% of velocities were considered, the direction of the residual current swung to the northwest. This station apparently recorded the fair-weather situation that is interpreted to winnow and degrade the shoal and to create a secondary lag on their crests.

One of the current-meter stations in the most landward trough may have observed the sort of current that builds the shoals. Monitoring began under relatively calm conditions. Initially, currents of varying orientation and speeds less than 5 cm sec^{-1} were observed. After the passage of a cold front, a wind from the northeast built rapidly to 25 knots. A steady, south-trending bottom current with peak velocities in excess of 0.7 ft sec^{-1} (20 cm sec^{-1}) developed and persisted through the turn of the tide.

Wave-refraction studies at the False Cape area (Weinman, 1971) indicate that northeast and east-northeast waves (and to a lesser extent, east-southeast and southeast waves) concentrate energy on the False Cape area. During this particular station seas from the northeast at 4 to 6 ft began to break on the ridge; water pumped by surf over the crest of the ridge must have intensified the shore parallel current. The hydraulic head generated in the trough by the combination of this wave set-up and the wind drift current resulted in a large rip current which began to flow diagonally seaward across the head of the ridge, marked by flocks of terns hunting fish swept over the crest.

The pattern of bottom and surface currents at this time must have resembled that of Fig. 186c. Landward bottom currents associated with wave transport over the northern end and seaward flank of the ridge would become landward surface currents as the waves break on the crest and become bores. Indeed, the smooth, convex-up profile of the seaward flank of the shoal is nearly identical to the profile of the shoreface landward of the shoal. Bathymetric and grain-size profiles suggest that during a storm the shoal's seaward flank is an environment hydraulically analogous to zone of shoaling waves that shape the shore face (Swift et al., 1972b) and that its crest is an environment hydraulically analogous to the surf zone. With 9 to 11 second waves 4 to 6 ft high, a crest at 8 ft below sea level and a base at 15 ft below sea level, the shoal near its junction with the shore face would, by Keulegan's criteria, marginally have the response characteristics of a wavebuilt bar.

Southerly-trending wind set-up currents in the trough, amplified by water pumped over the crest of the shoal by mass transport effect of breaking waves, would become seaward surface currents as they pass over the base of the ridge

(Fig. 186c). Thus, during strong northeast winds, southerly bottom currents appear to converge over the ridge, maintaining it, and scouring out the trough.

As indicated in the cross-sections of Fig. 186c, the convergence generates helical flow half-cells on either side of the shoal. The two major shoals to the seaward may have had a similar genesis; if so, the most seaward shoal has broken contact with the shore face and is presumably maintained only by helical flow, which now is probably as much a consequence as a cause of the linear shoal's presence.

Process and Response in the Bethany Beach System

The distribution of grain sizes and the geometry of the ridges is rather different in the case of the Bethany Beach, Delaware, ridge system (Fig. 187a, b). Moody (1964) has shown that the Bethany Beach troughs are, like False Cape troughs, floored by pebbly, coarse-grained sand. The coarse sand becomes better sorted toward the next crest to the seaward and also finer, grading to medium-grained sand (Fig. 187c). The grain size decrease continues down the south slope, accompanied by a decrease in sorting, then increases abruptly to coarse sand when the next trough is reached. Thus, at Bethany Beach, the grain size distribution is more asymmetrical than at False Cape and more nearly resembles the distribution associated with a simple sand wave built by a unidirectional current, although slope angles are far below the angle of repose. The geometry of the shoals likewise more nearly approaches that of a transverse bedform than do the False Cape shoals; the shore angle is higher, ranging from 30° to 80° and averaging 45°, versus an average 25° for False Cape. Thus, the shoals are more nearly normal to coast-parallel currents. Relief, crestal continuity, and southward asymmetry increase to the south, which Moody, on the basis of his own and other surveys, shows to be the direction of migration (Table XXIX). Moody's migration figures show that, for a 42-yr period, rates were about 10 ft per year. However, the ridges moved up to 250 ft to the southeast between surveys of 1961 and 1963 which bracket the great Ash Wednesday Storm of 1962. Thus, as at False Cape, a dominant driving force appears to be the intense south-trending currents generated by major storms.

Also, as at False Cape, the wave regime appears to be a controlling force, although here in a rather different way. The mean trend of the ridges appears to be precisely that of the dominant northeast direction of wave approach. Moody (1964) prepared a series of wave-refraction diagrams which show that orthogonals tend to converge over crests. This by itself would tend to maintain the ridges, as near-bottom residual wave currents would converge obliquely shoreward toward the crests. Moody suggests that, in addition, they are generating mechan-

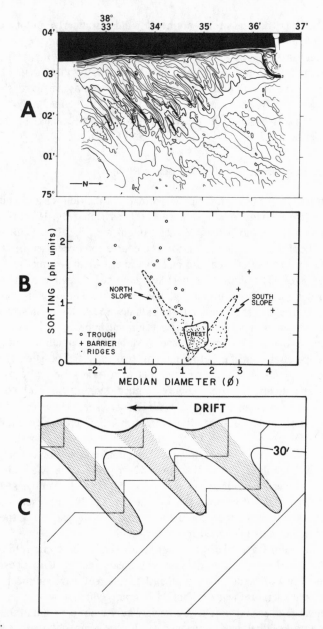

Figure 187.
Substrate response and hydraulic process for a wide-angle, shoreface-connected ridge, Bethany Beach, Delaware. (A) bathymetry, from Moody (1964); (B) median diameter versus inclusive graphic standard deviation for Bethany Beach sediment types [from Moody (1964)]; and (C) schematic model for the maintenance and generation of wide-angle ridges by wave refraction. Refracted waves converge toward ridge crests. Wave energy is concentrated at heads of troughs. Modified from Goldsmith and Colonell (1970).

Table XXIX

Rates of Crest Movement Offshore between Indian River Inlet and Bethany Beach[a]

Ridge	Rate of movement 1919-1961 (m yr^{-1})	Amount and direction of movement (m)			
		1919-1961		1961-1963	
D	0.0	0	–	55	SE
E	3.8	160	SE	30	NW
F	3.8	160	SE	10	SE
H	2.4 (NE end)	100	SE	25	SE
	2.6 (SW end)	110	NW	80	SE
J	3.9 (NE end)	165	SE	75	SE
	5.0	210	SE	40	SE
	2.9	120	SE	90	NW
	6.0 (SW end)	250	SE	50	NW
K	3.6	150	SE	55	SE
M	2.4 (NE end)	100	SE	54	SE
	–	0	–	30	SE
	3.7 (SW end)	155	SE	14	NW
N	1.4 (NE end)	60	SE	15	SE
	5.7	240	SE	20	SE
O	2.1	90	SE	20	SE

[a]From Moody, 1964, his Table 9.

ism for the shoal topography. The ridges, oriented parallel to the prevailing direction of wave approach, would serve (Fig. 187c) to focus wave energy rhythmically along the beach. Moody (1964, p. 113) notes that,

"although the pattern of crossed orthogonals and the amount of wave convergence differs according to wave direction, the wave-divergence zones on the shoreline remain about the same, slightly north of points where ridges meet the barrier. These areas should undergo greater amounts of erosion than adjacent sections of the shore."

Therefore, the ridges would tend to perpetuate themselves during coastal retreat by a sort of feedback mechanism. Goldsmith and Colonell (1970) documented this process on Monomoy Barrier Island, Cape Cod. While more work needs to be done before the situation at Bethany Beach is clarified, the parallelism of the ridges with the dominant direction of wave approach would require this or a similar pattern of process and response.

Thus, the False Cape and Bethany Beach systems of linear shore-connected shoals are interpreted to comprise two related, yet distinct, responses of the inner shelf substrate to a storm-dominated hydraulic regime. Storm-generated coast-parallel currents interact with waves in two different ways, and yet other modes of interaction may be represented in the spectrum of inner shelf shoals described in this paper. In both of the cases discussed above, the bottom topography appears to be a resultant response between the storm hydraulic regime,

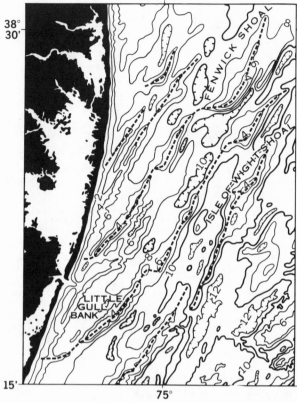

Figure 188.

Detail of Delaware-Maryland inner shelf bathymetry, showing shoreface-connected shoals in different stages of detachment. Stepped crestlines (dashed lines) of offshore shoal sequences record successive detachments of that series. Contoured at 2 fathom interval. See USC&GS 0807N-57 for one fathom resolution.

which tends to aggrade the ridges, and a fair-weather hydraulic regime of northwest-trending swells which tends to degrade the ridges.

SHOREFACE-CONNECTED TO ISOLATED SHOALS: SHOAL DETACHMENT AND EVOLUTION

Continuity between Shoreface-Connected and Shelf Shoal Fields

The apparently gradational transition between shoreface-connected and isolated shoals of the Atlantic shelf suggests that a genetic relationship exists between the two categories. The patterns of inner shelf shoals can be traced for up to 36 n mi across the continental shelves of Long Island, New Jersey, and Delaware (Figs. 171, 172, 173) to terminate in shoreface-connected shoals of apparently hydraulic origin. Brigantine Shoals off Atlantic City, New Jersey, and Chincoteague Shoals, south of Ocean City, Maryland, are located at the point of separation of the proximal and distal barrier arcs of their respective headlands. At these points, the arcuate trend of the innermost shoals is initiated by shoals as far seaward as the 120 ft isobath.

Significance of Shoal Orientation

Our shoal orientation statistics are useful in assessing the relation of shore face to inner shelf shoals. Generally speaking, the shoals maintain a constant angular relationship with the shoreline of about 20°, regardless of the shoreline orientation. Note that the histograms of azimuth of shoal axis (Fig. 176b) and azimuth of shoreline adjacent to shoals (Fig. 176c) are very similar, suggesting strongly that the orientation of the shoals follows that of the shoreline. However, it is significant that the shoals are not parallel to the shoreline (Fig. 167d). The mean angle (measured clockwise) of shoal with shoreline trend is 22°; standard deviation is 16°; and there are almost no negative values ($> 90°$).

This relationship breaks down for the wide angle-shoals of the northern New Jersey and northern Delaware coasts. In these two areas, the angle between the dominant northeast to east wave trend and the due north shoreline is apparently so large as to lead to development of shoals transverse to, rather than nearly parallel to, coastal storm currents. Additionally, morphological development in both areas may be influenced by subbottom structure and preexisting morphology (this paper, Fig. 179; Kraft, 1971a, Fig. 3). Both Delaware and northern New Jersey exhibit a northward transition in shoal orientation and symmetry. South of the transition zone shoals are symmetrical and open at approximately 20° with the coast, such as those at Ocean City, Md. and Barnegat, N.J. To the north of these areas shoals become progressively asymmetrical; shoal azimuths remain constant but the shoreline trend is more northerly.

From this it is possible to conclude that the characteristic 20° shore angle of the shoals is a consequence of the dynamics of shoal formation. As described in the previous section, shoreface-connected shoals appear to align themselves at this angle to the current which excavates their landward troughs and sweeps obliquely over their crests. Only when the angle between wave approach and shoreline is greater than 40° do the shoals apparently tend to orient themselves with the absolute direction of wave approach.

Shoreface-connected shoals seem to have been generated by nearshore processes. For isolated shoals though it means that they are now "relict" having been abandoned as the shore retreated in response to sea-level rise. The orientation of the shoals and correlation to the shoreline suggests also that during the retreat, the shoreline has retained essentially the same orientation it now has. The distribution of minerals on the shelf and beaches described by Pilkey and Field (1972) might be interpreted to record such a retreat.

It should be emphasized here that these shoreface-connected shoals are very large features and our data do not indicate these shoals ever move large distances. In this regard, the shoreface-connected shoals should not be construed to be the same as shallow water or swash bars which are usually described as 1-3 ft high, laterally continuous, and which through relatively short periods of time move landward, become welded to the beach, and in effect, prograde it (Bajorunas and Duane, 1967); Hayes and Boothroyd, 1969; Saylor and Hands, 1970). We judge, therefore, the formative processes for the "swash bars" and the linear shoals are also distinctly different.

Evolutionary Shoal Series

The relationship between shoreface-connected and isolated shoals is more explicitly inferred from an examination of four shoal fields whose constituent members appear to form evolutionary series. Moody (1964) had called attention to the apparent sequential relationship of the Bethany Beach ridges. As noted, his bathymetric time series indicates that these wide-angle shoals are migrating to the southeast, maintaining contact with the shore face and extending their crest lines as they do so. Height and asymmetry also increase to the southeast. Moody suggested that, as the shore face retreated and the water deepened over the ridges, faster ridges overtook slower ones, with the result that large shoals were isolated and left behind on the sea floor. He called attention to Fenwick shoal (Figs. 182, 188) as a possible example.

The Bethany Beach shoal system appears to be acceptable as an evolutionary series of wide-angle, shoreface-connected shoals, but shoreface-connected and isolated shoals to the south off Ocean City (Fig. 182) appear instead to be narrow-angle shoals of the False Cape type. Here the shoreface-connected shoals appear to form a typological or apparent evolutionary sequence of their own, from a wholly shoreface-connected, unnamed ridge in the north through ridges with saddles and various degrees of isolation to the wholly isolated Great Gull Bank to the south (Figs. 182, 188).

The westward hook of the base of Fenwick shoal and the shoal immediately to the south of it is reflected by the westward-skewed orientation of the saddles that connect some of the shoals to the shoreface and may relate to the manner in which the saddles are deepened and cut through during the isolation process. The en echelon nature of shoal peaks, and the stepwise trend of crestlines (Figs. 182, 188) suggest that a shoreface-connected ridge developed by headward erosion and continued deepening of its landward trough, with concomitant crestal aggradation, until a critical threshold is reached. The base of the ridge is then severed, and a new ridge begins to form inshore and down current from the initial segment. This sequence is schematized in Fig. 189.

The False Cape shoal system may also be envisioned as an evolutionary sequence (Fig. 183, 186a), with "A" shoal undergoing active accretion and "A" trough undergoing enlargement by headward and trough-floor erosion. Recent bathymetric observations of this sector indicate that a shallow trough separates "A" shoal from the shoreface, where during the northeaster a large-scale rip current was observed, and that the position of this saddle shifts from month to month. South of the saddle fine sand swept out of "A" trough has built up the shoreface 5 to 15 ft since 1922 (Swift et al., 1971). The shoreline south of the junction of the ridge with the shoreface has prograded up to 180 feet since 1922, while the shoreline north of the ridge junction has retreated 90 ft during this same period (Grafton, unpublished manuscript, Norfolk District, United States Army Corps of Engineers). Back barrier peat and clay in the floor of "A" trough is alternately covered with a few feet of coarse sand, then laid bare. In view of these systematic changes, it seems reasonable to surmise that, with time, first "B" shoal, then "A" shoal will sever their connections with

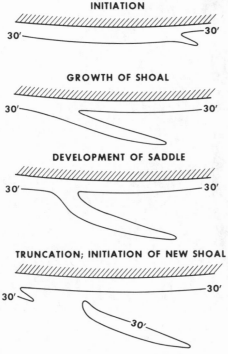

Figure 189.
Schematic of shoal detachment sequence.

the shore face, and that "C" shoal was once a shoreface-connected ridge like the other two.

A final morphological sequence with genetic implications occurs in the Platt shoals sector of the North Carolina coast adjacent to Oregon Inlet (see Fig. 190). A low, broad, linear shoal is attached to the shoreface north of Oregon Inlet. To the south lies the Platt shoals complex. Inner Platt shoals, cresting at 30 ft, is separated from the shore face by a trough over 60 ft deep. Unpublished vibracore data indicate that the trough is floored by a few feet of grav-

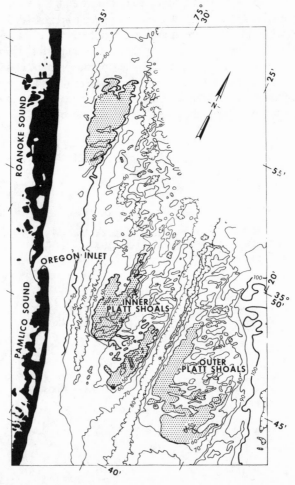

Figure 190.
Platt shoals sector of the North Carolina coast. Outer Platt shoals are interpreted as having undergone shoreface detachment. Contour interval is 10 ft.

el, with worn fragments of *Mercenaria* sp. as a prominent clast. The gravel overlies clean, well-sorted sand and is presumed to be a lag generated during the development of the trough.

Inner Platt shoals are still connected to the shore face by a low saddle; outer
Platt shoals to the south and seaward may once have been similarly connected,
but at present are entirely isolated. The linear, north-south-trending shoals
in this area have superimposed upon them small-scale, short-crested ridges
up to 10 ft high. They trend northeast, are nearly normal to the shoreline,
and are asymmetrical with steeper southeast flanks. Side slopes are much less
than the angle of repose. These features appear to be large transverse sand
waves that are activated during storms and, in this respect, are similar to the
wide-angle shoals of Bethany Beach.

Clearly, there is much to be learned from continued study and observation
of the sectors described. However, the shoal sequence of these sectors suggest
that the creation of isolated linear shoals from shoreface-connected shoals by
deepening and headward erosion of inner troughs is a presently occurring process
and one which in the past is judged to have occurred in concert with shoreline
retreat.

DISCUSSION

Past investigations of the morphology of the inner continental shelf have
resulted in a variety of genetic interpretations. Garrison and McMaster (1966)
attributed the ridge and swale topography of the shelf east of Long Island and
south of Rhode Island as primarily fluvial in origin. Knott and Hoskins (1962)
inferred that this shelf sector was shaped by fluvial and glacial processes. Mckinney and Friedman (1970) concluded that the topography of a corridor across the
continental shelf south of Long Island comprised a modified fluvial surface.

The nature of the central and outer shelf of the southern New England shelf
is beyond the scope of this paper and will be considered in future publications.
The Long Island shore face has not been examined in detail in this paper, because
our data on this sector are only partially processed. However, we note that the
linear, shoreface-connected shoals of the Fire Island sector (Fig. 191) are similar
in all important respects to the narrow-angle systems that we examined farther
south. In particular, the shoals exhibit the southward asymmetry which we
attribute to the action of coast-parallel, southeasterly storm-generated currents.
Consequently, we suspect that the shoreface-connected shoals of the Long Island
shelf are of modern hydraulic origin as well. However, the influence of Pleistocene
glacial outwash from Long Island in forming the existing surface or in controlling
subsurface morphology is not well known at this time. Certainly the proximity
and magnitude of the source are factors to be taken into consideration.

Sanders (1962) and Payne (1970) have pointed out that the False Cape, Virginia,

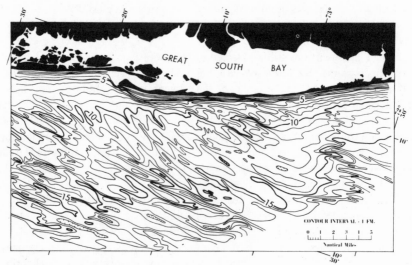

Figure 191.
Fire Island shoreface ridge system, south shore of Long Island, New York.

shoreface-connected shoals are on the same trend as the Late Sangamon raised beach ridges on the adjacent land and suggest that the shoals are relict Pleistocene features. Kraft (1971a, p. 21-35) notes that the Bethany Beach ridges are parallel to raised Pleistocene beach ridges on the adjacent shore. He suggests that "at least the loci of formation of the parallel offshore ridges must be related to pre-existing topography."

However, it is difficult to see how even this relationship between shoreface-connected shoals and raised Pleistocene beach ridges could exist. For one thing, these features occur at two different levels. The latter lie on a sub-horizontal, sub-aerial surface that is over 30 ft higher than the sub-horizontal inner shelf surface on which the shoals rest. These two "steps" are separated by the "riser" of the shoreface. If the shoals are remnants of seaward extensions of the subaerial beach ridges, then these seaward extensions must once have had a relief of 50 ft. No existing linear shoals and none of the classic barriers of the present United States coast have such relief. We are also doubtful of the stability of relict barriers in the high-energy zone of shoaling waves. The equilibrium profile of the shore face is typically that of a smooth, exponentially curved surface. This surface continually adjusts so as to absorb wave energy with maximum efficiency, and response time is extremely short, geologically speaking. Beaches may be stripped back 150 ft by a single storm, but most of the sand is returned during several weeks of fair weather as the beach adjusts to the reduced energy level. As noted by Fischer (1961), this profile must translate landward and upward during a marine transgression, and the result must be erosion of the shore face and destruction of its internal stratigraphy. Consequently, it is very difficult to accept the concept that large linear mounds of unconsolidated sand could exist on the shore face out of equilibrium with its hydraulic regime. We conclude that in the light of our studies and data, the alignment of raised subaerial Pleistocene beach ridges and nearby submarine shoreface-connected shoals is fortuitous and is not *a priori* evidence for either the age or origin of the shoals. Where hard data in the form of radiometric ages are available. the ridges prove to be younger than 11,000 yr.

A second hypothesis, first advanced by Veatch and Smith (1939) interprets the shelf surface as a relict strand plain, with ridges marking stillstands of the advancing Holocene sea. On close examination, however, it is difficult to fit the extensive inner shelf shoal field of the Atlantic shelf into this pattern. Subaerial barrier systems on the Atlantic coastal plain mark culminations of transgressions (Hails and Hoyt, 1969) or were formed during regressions (Colquhoun, 1969; Oaks and Coch, 1963). Only in the latter case does the spacing approach anything like the 1 to 2 mi of the inner shelf ridge fields. Furthermore, the shoal fields do not satisfy the stairstep criterion, first advanced by Gilbert (1899). Gilbert pointed out that, if a barrier is overstepped, the base of the next barrier to be built farther inland must lie at the same elevation as the head of the first. The shoal fields of Delmarva and New Jersey lie on a gently sloping surface which is parallel to a second gently sloping surface, defined by the nearly accordant summits of the closely spaced ridges. Instead of forming a sequence of steps and risers, the profile approximates a gently inclined sine wave. Each ridge has a shoreward trough nearly as deep as its seaward trough. Hence, any hypothesis interpreting these inner shelf ridges as overstepped barriers must acknowledge extreme modification of original morphology by modern hydraulic processes.

A third major hypothesis for the generation of linear, shelf-floor shoals has been proposed by Moody (1964). Basis of the hypothesis is that the ridges form in response to the modern hydraulic regime at the foot of the shore face and that they are isolated by the retreat of shoreline. Our survey of the central and southern inner Atlantic shelf is compatible with this hypothesis and adds supporting detail. Shoreface-connected ridges are seen as generated by the interaction of wind- and wave-generated currents with the substrate on the shore face during storms. Two major categories of response are envisaged; wide-angle (over 30°) and narrow-angle (less than 30°), depending on the angle made by the ridges to the shoreface. Narrow-angle ridges form normal to wave-generated bottom currents and nearly parallel to shore-parallel wind set-up currents, and during these initial stages have some of the response characteristics of wave-built bars. Wide-angle ridges with the reverse relationship appear to be more nearly large-scale, low-angle, intermittently active transverse bedforms with respect to coast parallel storm currents; with respect to wave generated currents they are viewed as longitudinal structures.

Our scattered hydraulic observations serve to provide only the most rudimentary and qualitative picture and have raised as many questions as they have answered. Clearly, the greater part of linear shoal dynamics is yet to be learned. We conclude that shoreface-connected shoals are generated by nearshore processes. For isolated shoals though it means that they are now "relict" having been abandoned as the shore retreated in response to sea level rise. Shoal sequences which appear to demonstrate various stages of the detachment process are present on all sections of the U.S. Atlantic coast. However, further detailed investigations of petrography, shallow stratigraphy and bathymetric time series of shoal system are required before this process will be completely understood. Nevertheless, it seems reasonable to state at this time that the linear shoal systems do record Holocene barrier retreat across the shelf surface, but in a manner more complex than originally envisaged by that hypothesis' proponents. The shoals represent neither the subaerial superstructure nor the submarine foundation of barriers. They are instead independent and distinct daughter forms, adjusted to the deeper water environment in which they are now located. The orientation of the shoals and correlation to the shoreline suggests furthermore that during the Holocene coastal retreat the shoreline has retained essentially the same orientation it now has.

CONCLUSIONS

The surface of the United States Atlantic continental shelf is characterized by numerous arcuate and northeast trending linear shoals and ridges which rise 10 to 40 ft above the surrounding sea floor and measure from hundreds to thousands of feet in length. The two broad shoal classes are further divisible as arcuate shoals: inlet-associated and cape-associated; and linear shoals: shoreface-connected and isolated. Regional patterns in shoal distribution parallel regional trends in coastline morphology. Thus inlet-associated shoals exist only at estuaries and other large inlets, and cape-associated shoals only at cuspate fore-lands and major barrier overlaps. Shoreface-connected linear shoals are present off headland areas; both shoreface-connected and isolated linear shoals are present adjacent to main barriers.

Marked similarities in external morphology of over 200 linear shoals on the Atlantic inner continental shelf between Long Island and Florida strongly suggest similar genesis of all linear shoals. All of the shoals open to the northeast forming small acute angles with the coastline (most less than 35°). Depth to shoal crests is strongly bimodal with modes existing at 20 to 30 ft and 40 to 55 ft; a third mode possibly exists in water deeper than 80 ft.

High resolution seismic reflection profiles and sediment cores from most shoals show them to be planoconvex bodies of sand resting upon a mappable, essentially featureless, acoustic horizon. The stratum associated with the acoustic reflector, which in some areas is exposed, varies from coarse unconsolidated sands to compacted muds to semiindurated calcarenites.

Samples of organic rich muds, peat, or shells lying at or just above the depth of the acoustic reflector upon which the shoals lie have radio-carbon ages of early to mid Holocene. Extrapolation of this information to other shoals which are morphologically and stratigraphically similar indicates that these other linear shoals are also of Holocene age. No evidence to support Pleistocene origin of these features was found.

Shoal sands are generally well-sorted, medium-grained sands that are similar in lithology to adjacent beaches and bear evidence of recent current and wave activity. Textural grading, historical records of shoal movement, and selected field hydraulic studies all show inner shelf shoals, particularly shoreface-connected ones, to be presently undergoing modification by storm currents and shoaling waves.

Available data indicate the mode of formation only in a general way. It seems clear, however, that the shoals are formed by nearshore processes. The consistency of shoal angle with the shoreline, internal structure (inclined bedding) and similarity in cross-sectional profile of shoals from Long Island to Florida, are judged proof that all shoals are genetically related by hydrodynamic processes that are operative on a regional scale.

In the area of study, shoreface-connected shoals exist which are believed to be in various stages of elongation, separation and isolation from the shore face as a result of coastal retreat. Therefore, isolated shoals on the shelf are judged to have been formerly shoreface-connected and subsequently detached during landward coastal retreat. If, as proposed, the linear shoals originate nearshore, then the similarity in orientation of both shoreface-connected and isolated shoals with respect to the shoreline indicates that shoreline orientation probably remained essentially constant as the Holocene sea transgressed the entire inner continental shelf. Presence of similar appearing shoal features on the outer shelf may indicate unchanging conditions for an even longer time period.

ACKNOWLEDGMENTS

The ideas discussed in this paper have benefited from discussions with so many of our colleagues that it is impossible to list them all. However, for particular help and provocative discussion, special thanks are extended to Dr. Wm. R. James of the Coastal Engineering Research Center, Dr. Ole S. Madsen of Massachusetts Institute of Technology, Dr. Robert Byrne of the Virginia Institute of Marine Science, and Dr. J. C. Ludwick of Old Dominion University.

Data obtained for this study, unless otherwise noted, resulted from the general research program of the Corps of Engineers' Coastal Engineering Research Center. Permission was granted by the Chief of Engineers to publish this information. Work on the Virginia coast was carried out under contract DACW 72-69-C-0016 between CERC and Old Dominion University. Additional funding of the Virginia coast work was provided by NSF grants GA-13831 and GA-27305 and by the United States Geological Survey. Portions of the field work were conducted aboard the Duke research vessel Eastward, supported by NSF grant GB-17545. The contribution by Swift to this paper is part of the COMSED project of NOAA-AOML, an investigation of sedimentation on the Atlantic continental margin of the United States.

REFERENCES

Assaf, G., Gerard, R., and Gordon, A. L. (1971). Some mechanisms of ocean mixing revealed in aerial photographs. *J. of Geophysical Res.* **76**, 6550-6572.

Bajorunas, L., and Duane, D. B. (1967). Shifting offshore bars and harbor shoaling. *J. of Geophysical Res.* **72**, 6195-6205.

Berg, D. W., and Duane, D. B. (1968). Effect of particle size and distribution on stability of artificially filled beach, Presque Isle Peninsula, Pa. *In* "Proc. 11th Conf. Great Lakes Res.," pp. 161-178. Internatl. Assoc. for Great Lakes Res., Ann Arbor, Michigan.

Bumpus, D. F., and Lauzier, L. M. (1965). Surface circulation on the continental shelf of eastern North America between Newfoundland and Florida. *In* "Serial Atlas of the Marine Environment." Folio 7, Am. Geographical Soc., New York.

Caldwell, Joseph M. (1966). Coastal processes and beach erosion. *J. Sc. Civil Eng.* **53**, 142-157.

Colquhoun, D. J. (1969). Coastal plain terraces in the Carolinas and Georgia, USA. *In* "Quaternary Geology and Climate." pp. 150-162. National Academy of Science, Washington, D.C.

Davies, J. L. (1964). A morphogenic approach to world shorelines. *Zeit. für Geomorphologie* **8**, 127-142.

Dietz, R. S. (1963). Wave-base, marine profile of equilibrium and wave built terraces: a critical appraisal. *Bull. Geol. Soc. Am.* **74**, 971-990.

Dzulynski, S. (1965). New data on experimental production of sedimentary structures. *J. Sed. Petrology* **35**, 196-212.

Emery, K. O. (1965). Characteristics of continental shelves and slopes. *Am. Assoc. Petrol. Geol.* **49**, 1379-1384.

Emery, K. O. (1966). Atlantic continental shelf and slope of the United States, geologic background. *U.S. Geol. Survey Prof. Paper* **529-A**, 1-23.

Emery, K. O. (1967). The Atlantic continental margin of the United States during the past 70 million years. *Geol. Assoc. Canada, Spec. Paper* **4**, 53-70.

Emery, K. O. (1968). Relict sediments on continental shelves of the world. *Am. Assoc. Petrol. Geol.* **52**, 445-464.

Emery, K. O. (1969). The contiental shelves. *Scientific Am.* **221**, 106-121.

Faller, A. J. 1969). The generation of Langmuir circulations by the eddy pressure of surface waves. *Limnology and Oceanography* **14**, 504-513.

Field, M. E. (1972). Buried strandline deposits on the central Florida inner continental shelf, 74 pp. Abstr. with Program, Geol. Soc. Amer. Southeastern Sect. Annual Mtg., **6**, 74.

Field, M. E., and Duane, D. B. (1972). Geomorphology and sediments of the inner continental shelf, Cape Kennedy, Florida. *U.S. Army Coastal Engr. Res. Ctr.*, Technical Memo.

Fischer, A. G. (1961). Stratigraphic record of transgressing seas in light of sedimentation on Atlantic coast of New Jersey. *Bull. Am. Assoc. of Pet. Geologists* **45**, 1656-1666.

Folk, R. L. (1971). Longitudinal dunes of the northwestern edge of the Simpson desert, Northern Territory, Australia. *Sedimentology* **16**, 5-54.

Garrison, L. E., and McMaster, R. C. (1966). Sediments and geomorphology of the continental shelf off southern New England. *Mar. Geol.* **4**, 273-289.

Gilbert, G. K. (1890). Lake Bonneville. *U.S. Geol. Surv. Monograph 1*, 438 pp.

Goldsmith, V., and Colonell, J. M. (1970). Effects of nonuniform wave energy in the littoral zone. *Proc. 12th Coastal Engineering Conf. Amer. Soc. of Civil Engineers, New York* **2**, 767-785.

Hails, J. R., and Hoyt, J. H. (1969). An appraisal of the evolution of the lower Atlantic coastal plain of Georgia, USA. *Tran. Papers Inst. Brit. Geog.* **46**, 53-68.

Harris, D. L. (1972). Wave estimates for coastal regions. *In* "Shelf Sediment Transport: Process and Pattern" (D. J. P. Swift, D. B. Duane, and O. H. Pilkey, eds.). Dowden, Hutchinson & Ross, Stroudsburg, Pennsylvania.

Harrison, W., Norcross, J. J., Pore, N. A., and Stanly, E. M. (1967). Shelf waters off the Chesapeake Bight. *Environmental Sciences Services Administration, Prof. Paper* **3**, 1-82.

Hayes, M. O., and Bothroyd, J. C. (1969). Storms as modifying agents in the coastal environment. *In* Coastal Environments, NE Mass. and N.H.," 245-265. Coastal Res. Group, Univ. Mass., Amherst.

Hjulstrom, F. (1939). Transportation of detritus by moving water. *In* "Recent Marine Sediments," (P. D. Trask, ed.), AAPG, Tulsa, Oklahoma.

Holliday, B. W., McHone, J., Shielder, G., and Swift, D. J. P. (in preparation). Evolution of an inner shelf ridge system, Virginia Beach, Virginia.

Houbolt, J. H. C. (1968). Recent sediments in the southern bight of the North Sea. *Geol. Mijnbouw* **47**, 245-273.

Hyne, N. J., and Goodell, H. G. (1967). Origin of sediments and submarine geology of the inner continental shelf off Choctawatchee Bay, Florida. *Marine Geol.* **5**, 1125-1136.

Johnson, D. W. (1919). "Shore Processes and Shoreline Development." 584 pp. Hafner, New York.

Jordan, G. F. (1962). Large submarine sand waves. *Science* **136**, 839-848.

Knott, S. T., and Hoskins, H. (1968). Evidence of Pleistocene events in the structure of the continental shelf off the northeastern U.S. *Marine Geol.* **6**, 5-43.

Kraft, J. C. (1971a). Sedimentary facies patterns and geologic history of a Holocene marine transgression. *Bull. Geol. Soc. Am.* **82**, 2131-2158.

Kraft, J. C. (1971b). A guide to the geology of Delaware's coastal environments. Publ. 2GL039, 220 pp. College of Marine Studies, Univ. of Delaware, Newark, Delaware.

Kuelegan, G. H. (1948). An experimental study of submarine sand bars. *Beach Erosion Board, U.S. Army Corps of Engineers Tech. Rept* **3**, 40 pp.

Langmuir, I. (1938). Surface motion of water induced by wind. *Science* **87**, 119-123.

Ludwick, J. C. (1972). Migration of tidal sand waves in Chesapeake Bay entrance. *In* "Shelf Sediment Transport: Process and Pattern" (D. J. P. Swift, D. B. Duane, and O. H. Pilkey, eds.). Dowden, Hutchinson & Ross, Stroudsburg, Pennsylvania.

McKinney, T. F., and Friedman, G. M. (1970). Continental shelf sediments of Long Island, New York. *Jour. Sed. Petrol.* **40**, 213-248.

McMaster, R. L. (1954). Petrography and genesis of the New Jersey beach sands. *State of N.J., Dept. of Conserv. and Econ. Devel., Bull.* **63**, 239 pp.

Meisburger, E. P., and Field, M. E. (1972). Neogene sediments of the North Florida Atlantic inner shelf. Abstr. with Program, Geol. Soc. Am. Annual Mtg., **85**, 593.

Meisburger, E. P., and Duane, D. B. (1971). Geomorphology and sediments of the inner continental shelf Palm Beach to Cape Kennedy, Florida. *U.S. Army Coastal Engineering Center, Technical Memo.* **34**, 111 pp.

Moody, D. W. (1964). Coastal morphology and processes in relation to the development of submarine sand ridges off Bethany Beach, Delaware. 160 pp. Ph.D. thesis, The Johns Hopkins University, Baltimore, unpublished.

Oaks, R. Q., Jr., and Coch, K. (1963). Pleistocene sea levels, southeastern Virginia. *Science*

Oertel, G. F., and Howard, J. D. (1972). Water circulation and sedimentation at estuary entrances on the Georgia coast. *In* "Shelf Sediment Transport: Process and Pattern" (D. J. P. Swift, D. B. Duane, and O. H. Pilkey, eds.). Dowden, Hutchinson & Ross, Stroudsburg, Pennsylvania.

Payne, L. H. (1970). Sediments and morphology of the continental shelf off southeastern Virginia. 70 pp. Thesis, Columbia University, New York, unpublished.

Pilkey, O. H., and Field, M. E. (1972). Onshore transportation of continental shelf sediment: Atlantic southeastern United States. *In* "Shelf Sediment Transport: Process and Pattern" (D. J. P.

Swift, D. B. Duane, and O. H. Pilkey, eds.). Dowden, Hutchinson and Ross, Stroudsburg, Pennsylvania.

Price, W. A. (1954). Dynamic environments: reconnaissance mapping, geologic and geomorphic, of continental shelf of Gulf of Mexico. *Transactions Gulf Coast Assoc. of Geologic Societies, Houston, Texas,* **IV**, 78-107.

Ross, D. A. (1970). Atlantic continental shelf and slope of the United States - heavy minerals of the continental margin from southern Nova Scotia to northern New Jersey. *U.S. Geol. Survey Prof. Paper* **529-G**, 40 pp.

Sanders, J. F. (1962). North-south trending submarine ridge composed of coarse sand off False Cape, Virginia (Abstr.). *Bull. Amer. Assoc. Petrol. Geologists* **46**, 278.

Saylor, J. H., and Hands, E. B. (1970). Properties of longshore bars in the Great Lakes. *Proc. 12th Conf. on Coastal Engineering, Amer. Soc. Civil Engineers, New York* **2**, 838-853.

Schlee, J., and Pratt, R. (1972). Atlantic continental shelf and slope of the United States. *U.S. Geol. Survey Prof. Paper* **529-H**.

Schopf, T. J. M. (1968). Atlantic continental shelf and slope of the United States - Nineteenth Century exploration. *U.S. Geol. Survey Prof. Paper* **529-F**.

Shepard, F. P. (1950). Longshore-bars and longshore-troughs. *Beach Erosion Board, Corps of Engineers,* Washington, D.C., *Tech. Memo.* **15**, 32 pp.

Shepard, F. P. (1963). "Submarine Geology." 2nd Ed., 557 pp., Harper and Row, New York.

Swift, D. J. P., Sanford, R. B., Dill, C. E., Jr., and Avignone, N. F. (1971). Textural differentiation on the shore face during erosional retreat of an unconsolidated coast, Cape Henry to Cape Hatteras, western North Atlantic shelf. *Sedimentology* **16**, 221-250.

Swift, D. J. P., Holliday, B., Avignone, N., and Schideler, G. (1972a). Anatomy of a shore face ridge system, False Cape, Virginia. *Marine Geol.* **12**, 59-84.

Swift, D. J. P., Kofoed, J. W., Saulsbury, F. P., Sears, P. (1972). Holocene evolution of the shelf surface, central and southern Atlantic coast of North America. *In* "Shelf Sediment Transport: Process and Pattern" (D. J. P. Swift, D. B. Duane, and O. H. Pilkey, eds.). Dowden, Hutchinson & Ross, Stroudsburg, Pennsylvania.

Uchupi, E. (1968). The Atlantic continental shelf and slope of the United States (Physiography). *U.S. Geol. Survey, Prof. Paper* **529-C**, 30 pp.

Uchupi, E. (1970). Atlantic continental shelf and slope of the United States—shallow structure. *U.S. Geol. Survey Prof. Paper* **529-I**, 44 pp.

Veatch, A. C., and Smith, P. A. (1939). Atlantic submarine valleys of the United States and the Congo submarine valleys. *Geol. Soc. Am., Spec. Papers* **7**, 101 pp.

U.S. Army Corps of Engineers (1966). Shore Protection, Planning and Design. *Tech. Rept* **4** (3rd Ed.), 400 pp.

U.S. Army Corps of Engineers, Jacksonville District (1967). "Beach Erosion Control Study on Brevard Co., Florida." Jacksonville, Florida, 42 pp.

U.S. Army Corps of Engineers, Baltimore District (1970). "Beach Erosion Control Study Atlantic Coast of Maryland and Assateague Island, Virginia." Baltimore, Maryland, unpublished.

U.S. Dept. of Commerce (1970). *Tide Tables for 1971, East Coast of North and South America, Including Greenland.* Environmental Sciences Service Administration.

Weigel, R. L. (1964). "Oceanographical Engineering." 532 pp. Prentice-Hall, Englewood Cliffs, New Jersey.

Weinman, Z. H. (1971). "Analysis of Littoral Transport by Wave Energy: Cape Henry, Virginia to the Virginia, North Carolina border." Ph D Thesis, Old Dominion Univ., Norfolk, Virginia, unpublished.

Williams, S. J., and Field, M. E. (1971). Sediments and shallow structures of the inner continental shelf off Sandy Hook, New Jersey. *Abstr. with Program, Geol. Soc. Amer., Northeastern Sect. Mtg.,* **3**, 62.

Williams, S. J., and D. B. Duane (in prep.). Geomorphology and sediments of the inner New York Bight continental shelf. U.S. Army Coastal Engineering Center, Technical Memo.

CHAPTER 23

Holocene Evolution of the Shelf Surface, Central and Southern Atlantic Shelf of North America

Donald J. P. Swift,* John W. Kofoed,*
Francis P. Saulsbury,* and Phillip Sears†

ABSTRACT

The floor of the central and southern Atlantic shelf is a palimpsest or multiple imprint surface. An initial pattern is an erosional one consisting of major transverse shelf valleys and plateaulike interfluves. The dominant pattern is that of constructional topography formed at the foot of the shoreface. This constructional pattern is undergoing modification toward a third pattern in response to the modern hydraulic regime; therefore, the term "relict" does not seem an adequate descriptor. A unifying concept for the interpretation of Holocene shelf history is that of Bruun coastal retreat. This variant of the equilibrium profile hypothesis states that a rise in sea level over an unconsolidated coast results in shore-face erosion, equivalent to parallel slope retreat, and a concomitant aggradation of the adjacent sea floor. The resulting discontinuous debris mantle, the Holocene transgressive sand sheet, is only partly autochthonous with respect to the Holocene sedimentary cycle, since it incorporates Holocene fluvial deposits.

The surface of this sand sheet has been molded into a variety of morphologic elements. Where the sheet has been generated directly from the retreating shoreface, a ridge-and-swale topography has been empressed upon it. Off cuspate forelands, the convergence of littoral drift has resulted

*Atlantic Oceanographic and Meteorological Laboratories, National Oceanic and Atmospheric Administration, 15 Rickenbacker Causeway, Miami, Florida, 33149.
†Institute of Oceanography, Old Dominion University, Norfolk, Virginia 23508.

in cape-associated shoals. Off estuary mouths the intersection of littoral drift with the reversing estuary tide has created inlet-associated shoals. Seaward of each of these shoal types, earlier generations of the same shoals commonly occur. The resulting shelf-transverse sand bodies, formed by the progressive landward displacement of shoreline depositional centers, are shoal-retreat massifs.

The ridge-and-swale topography is impressed on shoal-retreat massifs, as well as on other sectors of the shelf floor and appears to be a stable end-configuration toward which a variety of depositional and erosional topographies tend to converge. The asymmetry of large-scale morphological elements and also of small-scale bedforms suggests that southward sediment transport in the Middle Atlantic bight intensifies toward shore and toward the south. The southward asymmetry of ridges and shoal complexes is seen as far south as Florida.

INTRODUCTION

Scope of Paper

This paper has been written on the premise that the existing data on the morphology of the North American Atlantic shelf floor, and to a lesser extent data on petrography, shallow structure and hydraulic regime, has reached a sort of critical mass, whereby this information can be systematically appraised and conclusions concerning the Holocene evolution of the shelf surface can be drawn, which could not be inferred from examination of the facts separately.

The preceding chapter has concerned itself with a morphological element that constitutes a major key to the nature of shoreface retreat and the morphological development of the shelf surface—namely, the linear shoals that dominate the inner-shelf surface. The preceding paper has maintained that, while many aspects of the process are obscure, the shoreface-connected shoals appear to be presently forming at the foot of the shoreface, and that the shoal fields of the inner shelf comprise a record of Holocene shoreface retreat. This paper explores ramifications of this hypothesis. It considers the more general problem of Holocene shoreface retreat of the central and southern Atlantic shelf and its effect on the shelf surface. It also considers large-scale morphological elements that themselves are compounded of the linear shoals described in the previous chapters, and attempts to determine their relevance to the Holocene history of the shelf. Lastly, it attempts to draw preliminary conclusions concerning the nature of Holocene shelf sediment transport in the study area.

Nature of Data: Approach

The central and southern Atlantic coast of North America was the first to be explored and is presently the most densely populated; consequently, bathymetric data are abundant for this sector (Schopf, 1968). Much of it dates from the advent of echo sounding in the third decade of the century. Veatch and Smith (1939) were the first to utilize echo-sounding data for geologic interpretation and prepared a detailed bathymetric map of the New Jersey shelf at a scale of 1 : 120,000, contoured at a one-fathom interval. They gave names to numerous large- and small-scale morphological features in this region.

Stearns (1967) prepared 15 bathymetric maps of the shelf between Cape Cod and the Delaware coast at a scale of 1 : 25,000, contoured at a one-fathom interval. The data points are considerably denser than those available to Veatch and Smith. The maps reveal a ridged surface with a relief of 5 to 10 m, whose complexity and variety must be seen to be believed.

Other significant maps used in this study include: "Map showing relationship of land and submarine topography, Nova Scotia to Florida," at a scale of 1 : 100,000, with a 20 m contour interval by Uchupi (1965, 1968), and "Bathymetric map of the slope, shelf, and rise between Nantucket Shoals and Chesapeake Bay," by Uchupi (1970), at a scale of 1 : 75,000 with a 4 m contour interval. Portions of this map have been presented in the previous chapter as Figs. 171 to 173.

Since some maps used in this paper use the English system of units, their values will be described in the text in both the metric and English systems. Elsewhere, distances will be reported in the metric system.

The net effect of these maps is to produce a wealth of bathymetric detail which is embarrassingly in advance of our knowledge of the shallow stratigraphy, petrography, and hydraulic regime for the same area. A full understanding of the genesis of the shelf morphology will require further acquisition and interpretation of data in all these fields. Supplementary stratigraphic, petrographic, and hydraulic data from the literature and from the author's own observations are employed in this paper wherever possible. However, the bulk of the analysis is based on unsupported bathymetric information. This procedure is employed in the hope that such a preliminary regional assessment will greatly aid the authors and readers in defining specific problems relating to evolution of the shelf surface, and water-substrate interaction at this interface, and in designating specific areas in which the substrate should be sampled and the hydraulic regime monitored.

A Topographic Inventory

The morphological elements of the central and southern Atlantic shelf can be conveniently arranged in a hierarchical array (Table XXX), in which higher-order elements tend to incorporate lower-order elements. The ridge-and-swale topography, examined on the inner shelf in the preceding paper, is the dominant first-order morphological element over many shelf sectors (Figs. 171, 172, 173, preceding chapter).

An inner shelf province consists of long, regular, subparallel ridges and swales aligned obliquely across the regional trend of the contours and converging with

Table XXX
Morphological Elements of the Central and Southern Atlantic Shelf

Small-scale elements
  Ripples and sand waves

Large-scale elements
  First order: Shoreface-connected ridges and swales
                    Isolated ridges and swales

  Second order: Cape-associated shoals
                    Inlet-associated shoals
                    Ridge fields

  Third order: Shoal-retreat massifs
                    Shelf-transverse valleys
                    Cuestas
                    Deltas
                    Scarps

the shoreline at an angle averaging 20°. Ridge spacing averages 2.1 km
(1.2 n mi); ridge amplitude ranges from 2 to 10 m (6 to 33 ft) and ridge length
varies from 9 to 56 km (5 to 3 n mi). See Table XXXI. Side slopes are generally
several degrees or less. Locally, clusters of ridges trend into the shoreface,
generally merging with it at depths of 10 m (33 ft) or less.

A second topographic province occurs on the outer shelf. The boundary on
the Long Island shelf occurs at variable depths and distances from shore, being
generally coincident with the change in slope separating the relict Long Island
drainage basin of the outer shelf from the ridged surface of the inner shelf
(Figs. 192, 193, 194). Typical inner-shelf ridge-and-swale topography extends
seaward on the promontory east of the drainage basin to depths of 76 m (225 ft)
at a distance of 102 km (55 n mi) from shore. Further west, on the Long Island
shelf, the boundary between the inner and outer topographic provinces coincides
with the 40 m (131 ft) contour, and this isobath continues to serve as the boundary
on the New Jersey shelf (Fig. 171, previous paper). To the south, however,
the boundary shoals; at Cape Hatteras it lies at 30 m (100 ft). See Fig. 196
and Fig. 173, 174, previous paper. The outer shelf ridge-and-swale topography
appears less complex, partly, perhaps, because bathymetric data is less dense.
Ridges are subparallel to the shelf edge, but still form northward-opening angles
with the regional trend of the contours and with the shelf edge. They tend
to be further apart, with an average spacing of 6 km (2.8 n mi). The pattern
made by the troughs suggests an incomplete trellis pattern (Figs. 194, 195).
Scarps of apparent littoral origin are more abundant.

Careful examination of existing maps indicates that the ridge-and-swale topography is superimposed on a framework of higher-order morphological elements
of both constructional and erosional origin. Ridges occur in groups as inner
shelf ridge fields, inlet-associated shoals, and cape-associated shoals (see pre-

Table XXXI
Descriptive Statistics for Ridge-and-Swale Topography

Map	NE Corner, 15' Quadrangle	Province	Mean ridge length (km)	Mean ridge spacing (km)	Mean relief (m)	Trough line km per km²
Long Island 0808N-54	40° 45' - 73° 00'	Shoreface	17.8	2.2	5.5	0.10
Long Island 0808N-54	40° 45' - 72° 15'	Inner shelf	30.7	2.0	3.4	0.10
Long Island 0808N-54	40° 30' - 72° 00'	Inner shelf	54.9	2.2	1.7	0.07
Long Island 0808N-54	40° 15' - 69° 45'	Inner shelf	18.5	1.7	3.8	0.10
New Jersey 0807N-55	39° 15' - 74° 15'	Shoreface	12.2	1.4	4.7	0.14
New Jersey 0807N-55	39° 15' - 74° 00'	Inner shelf	16.6	2.6	6.2	0.09
New Jersey 0807N-55	30° 00' - 73° 45'	Inner shelf	12.8	2.4	5.7	0.09
New Jersey	38° 45' - 73° 30'	Outer shelf	32.9	6.1	6.0	0.08

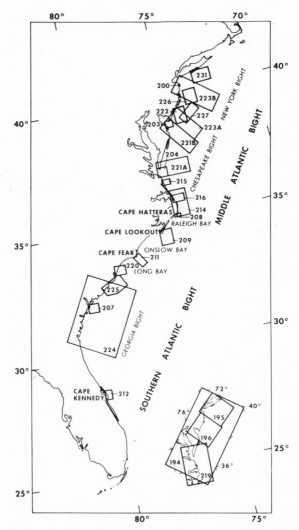

Figure 192.

Index map of topographic maps presented in this paper.

vious paper). Obvious erosional elements are the transverse shelf valleys (Fig. 193; Figs. 171, 172, 173, preceding paper), upon whose interfluves the ridge-and-swale topography has been developed. The Hudson shelf valley (Veatch and Smith, 1938), Block shelf valley (Veatch and Smith, 1938; Garrison and McMaster, 1966), and Long Island shelf valley (Garrison and McMaster, 1966; McKinney and Friedman, 1970) are known from the literature. Other shelf

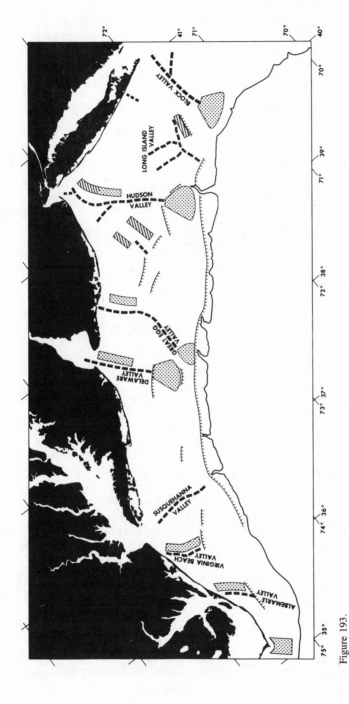

Figure 193.

Major morphological elements of the middle Atlantic bight. Dashed lines are shelf valleys. Hachured lines are scarps. Stippled areas are highs of probable constructional origin, including shoal-retreat massifs and stillstand deltas. Diagonally ruled areas are areas of probable erosional orgin, including cuestas.

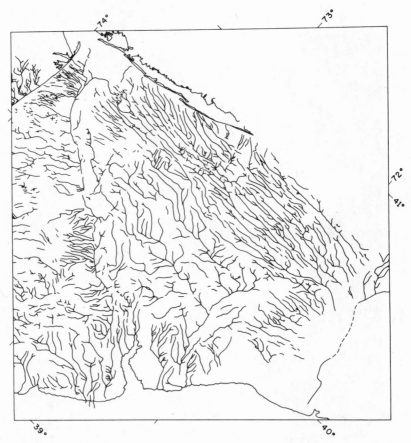

Figure194.
Pattern of linear lows on the Long Island shelf. Based on Stearns (1967).

valleys in Fig. 193 have been designated in this paper on the basis of existing and newly prepared bathymetric maps. Low, poorly defined large-scale, shelf-transverse ridges, often with ridge-and-swale topography superimposed on them, are believed to be in part shoal-retreat massifs or constructional features resulting from the retreat of nearshore depositional centers (cape- and inlet-associated shoals) with the transgressing shoreline (Fig. 193; Figs. 171, 172, 173, preceding paper). Other shelf ridges, particularly on the northern New Jersey and Long Island shelves, are interpreted as cuestas of erosional origin.

An additional higher-order morphological element consists of subcircular shelf-edge and mid-shelf plains, which may be either ridged or smooth. Their inner margins are depressed with respect to the surrounding shelf, and the contours here bend landward. Their outer margins tend to stand above the surrounding

Figure 195.
Pattern of linear lows on the New Jersey shelf. Based on Stearns (1967).

shelf, and the contours here bend seaward. Proximity to transverse shelf valleys indicates that these are shelf deltas. The Block and Hudson shelf deltas have been described in the literature (Veatch and Smith, 1938; Ewing and others, 1963; Garrison an McMaster, 1966; Knott and Hoskins, 1968). The Delaware and Great Egg shelf deltas have been placed in Fig. 193 on the basis of maps by Stearns (1967) and Uchupi (1970).

A final higher-order morphological element consists of the partly constructional and partly erosional shelf-parallel scarps that are interpreted as drowned shorelines. The Nichols and Franklin shorelines have been described by Veatch and Smith (1938) and Uchupi (1968, 1970). The boundary between the inner- and outer-shelf provinces is at many localities a well-defined scarp. It is designated in this paper as the mid-shelf shore. Merrill and others (1965) have noted a concentration of oyster shells at this depth, with dates suggesting a Holocene stillstand.

THE RETREATING SHORE FACE

Nature of the Problem

The genesis of the inner-shelf ridge systems and the evolution of the Holocene shelf surface as a whole are inextricably bound up in the problem of shore-face retreat, and it is necessary to consider this process in detail before attempting to interpret the morphological elements described in the preceding section. Three dominant hypotheses for the generation of the ridge-and-swale topography of the shelf have been presented in the previous paper. They imply three differing modes of shore-face retreat. The hypothesis that the ridges reflect stillstands of the advancing Holocene sea requires that the advancing sea deposit a blanket of littoral sand and that the terrestrial surface is preserved beneath it. The hypothesis that present submarine shelf surface is a drowned subaerial surface of relict fluvial or older marine origin requires that the advancing sea neither bury nor destroy the terrestrial surface. The hypothesis set forth by Moody (1964) requires that Holocene marine marginal deposits are consumed during retreat of the shore face and that the resulting debris is reshaped to form the ridge topography. In this case the advancing sea also deposits a blanket of sand, but it is the product of nearshore marine rather than littoral processes and the terrestrial surface has been truncated by the transgressive process itself.

Sloss (1962) and Allen (1964) have given some thought to the variables which determine the nature of shore-face retreat. They have listed them as the rate of relative sea-level rise (R), the character of the sediment (G), the rate of sediment input (S), and the energy input (E). It is convenient for expository purposes to group these variables in quasi-quantitative fashion:

$$\left(\frac{S}{E}\right)_G - R = K$$

to indicate that, for a given sediment character (for instance, grain size), the ratio between the rate of sediment input (S) and the energy available to disperse it (E) must be balanced by the relative rise in sea level (R); otherwise, the coastline will advance or retreat. Curray (1964) has described three cases of transgression involving successively larger negative values of K and corresponding to the three varieties of shore-face retreat listed above. In depositional transgression, subsidence is only slightly in excess of the rate of sediment input, but is more significant than the energy level. It corresponds to the first case, where a littoral sand blanket is deposited. In discontinuous depositional transgression, deposition is less able to keep up with the rate of sea-level rise and deposits

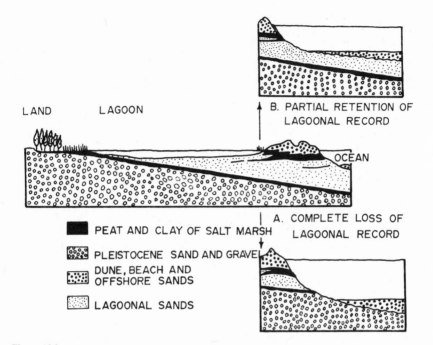

Figure 196.

Erosional (A) and depositional (B) transgressive stratigraphy, depending on the relative depth of excavation of the shore face. From Fischer, 1961.

are thin and discontinous or lacking. In erosional transgression, the energy level is presumably also significant and the deposits formed are the consequence of shore-face erosion. See Fig. 196.

The problem of shoreface retreat has also been attacked from a geometric point of view. The concept of the equilibrium profile was the focal point of the early, deductive school of shelf geomorphology exemplified by Gulliver (1899), Fenneman (1902), and Johnson (1919). The shelf was seen as a surface curved exponentially about an axis parallel to the shore and concave up, with the steep, nearshore sector comprising the shore face. The shore face was interpreted as a response to the regime of shoaling waves, whereby the slope at any point is the most stable one for purposes of absorbing wave energy. A change in the intensity or configuration of the long-term average energy field must be met by a compensating change in the equilibrium profile, by such erosion or deposition as is necessary.

The concept of the equilibrium profile has recently been restated by Bruun (1962) in a more restricted form that is helpful in interpreting coastal evolution. Bruun's rule, as amplified by Schwartz (1965, 1967, 1968), states that, on a low, unconsolidated coast, a rise in sea level results in shore-face erosion

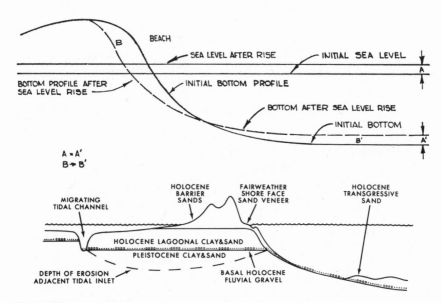

Figure 197.

(A) Schematic diagram of Bruun coastal retreat. During a rise in sea level, shore-face erosion and aggradation of the sea floor tend to balance. The resulting forward and upward translation of the shore profile results in a constant shore distance to water-depth relationship. From Schwartz, 1965. (B) Schematic stratigraphic model of North Carolina coast. Bruun coastal retreat has resulted in debris mantle (Holocene transgressive sand sheet) over resting disconformably on substrate. Size of beach greatly exaggerated. From Swift et al. (1971).

and an equal-volume aggradation on the adjacent sea floor. The result is a landward and upward translation of the equilibrium profile (Fig. 197), such that, in the immediate vicinity of the shore face, sea-level rise at a given point is balanced by deposition. The utility of the idea lies in the accretionary sediment blanket, which becomes the "relict" or transgressive sand blanket found on the present Atlantic Shelf floor, and the deposit on whose surface the ridge topography has formed. The concept emphasizes shore-normal sediment transport and, taken literally, requires that longshore transport be a steady-state process resulting no net change of sediment volume in the area under consideration. The concept as described is applicable primarily to a mainland beach or substructure of a coastal barrier. A model for barrier island retreat must be extended to include the behavior of the barrier superstructure, which tends rather to undergo retreat in cyclic, tank-tread fashion by a process of storm washover, burial, and re-emergence at the shore face (Dillon, 1970; Swift and others, 1971a, see Fig. 197, this paper). A mathematical model for shore-face erosion as a consequence of sea-level change has been presented by Scheidegger (1970, p. 324-328).

The concepts presented above will be used to examine, by means of case studies, the nature of coastal retreat along the central and southern Atlantic coast.

Erosional Transgression and Coastal Retreat
in the Middle Atlantic Bight

General

The middle Atlantic Bight, between Cape Cod and Cape Hatteras, would appear to be experiencing the necessary conditions for erosional transgression. The large intracoastal water bodies of Long Island Sound, New York Harbor, and Delaware and Chesapeake Bays suggest a strongly negative solution for the equation of transgressive variables. It has been generally assumed that these water bodies are effective bedload traps (Emery, 1965); hence, cannibalization of the coastal substrate would be the only significant sediment source. The repetitive pattern of coastal morphology, noted in the preceding paper, attests to the effectiveness of storm energy input. The wave climate of the area is moderate to high (Dolan and others, 1972). Various studies (Wicker, 1951; United States Army Corps of Engineers, 1955; Langfelder, 1968; Newman and Munsart, 1968) indicate that the open coast is receding at rates commonly in excess of a foot a year. There are at least three areas where bathymetric time series suggest that an erosional transgression is occurring, with the shore-face retreat tending to be of the type defined by Bruun.

Bathymetric time series, Virginia coast

The 47-yr time series available at False Cape, Virginia (Fig. 198), shows that the sea floor, deeper than about 18 m (60 ft) has aggraded while the shore face has eroded (Swift and others, 1972). The sediment budget has been distorted by the lateral transfer; the upper shore face to the south has aggraded somewhat in response to fine sand moving south through the ridge system (see previous paper). Here shore-face erosion has penetrated to lagoonal deposits of mid-Wisconsinan and early Wisconsinan age. The results of a heavy mineral study suggest that the shelf transgressive sand blanket and the shore-face veneer of fine sand was generated by fractionation of the Pleistocene substrate (Swift and others, 1971b).

Bathymetric time series on the Delaware Coast

Moody (1964, p. 142-154) presents a time series on the bathymetry of the Bethany Beach area which affords an insight into the complex short-term changes which result in shore-face erosion. "The barrier face steepens over a period

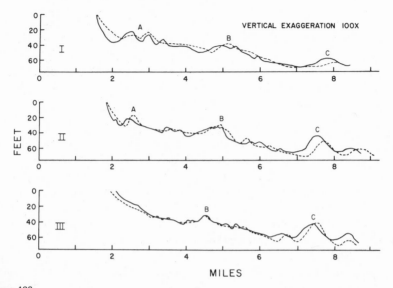

Figure 198.

Comparison of False Cape, Virginia, bathymetry from 1922 USCGS survey (dashed) and 1964 survey (Swift and others, 1972, solid lines). See Fig. 186a, previous paper, for location of lines. Bruun coastal retreat is occurring in I and II; upper shore face has aggraded in III due to coast-parallel transport.

of years to a critical slope, during which time the shoreline remains relatively stable." He notes that a groin system built in 1934-1938 "presumably trapped sand, causing the upper part of the barrier between mean low water and −3 m to build seaward. However, erosion continued offshore at depths between 6 or 7 m below mean low water." The steeping process is not continuous, but varies with the intensity and frequency of storms. "The slope of the [Bethany Beach] barrier steepened from 1:40 to 125 . . . between 1929 and 1954, but deposition on the barrier face between 1954 and 1961 regraded the slope to 1:40."

The steepening process is culminated by a major storm, during which time the gradient is reduced and a significant landward translation of the shoreline occurs. Moody (1964, p. 152) describes the Ash Wednesday storm of 1962 as having stalled for 72 hr off the central Atlantic coast. Its storm surge raised the surf 3 m above mean low water and brought the surf into the dunes of Bethany Beach for six successive high tides. The shoreline receded 18 to 75 m during the storm. While much of the sand was transported over the barrier building washover fans up to 4 m thick, much more was swept back onto the sea floor by large rip currents (revealed by aerial photographs taken during the storm; Moody, 1964, p. 114), and by the mid-depth return flow. Moody's time series shows that, over a thirty-two year period, shore-face erosion was

Table XXXII

Sediment Budget for the Barrier and Offshore areas Between Indian River Inlet and Bethany Beach, Delaware[*]

Area	Period	Average volumetric change $(m^3 \ yr^{-1})$
A. *Erosion*		
Shore face	1929-1961	− 148,000
Beach and dunes	1954-1961	− 100,000 (estimated)
Inner shelf floor (northwest sides of ridges)	1919-1961	− 100,000
Erosion from bay inside Indian River Inlet	−	− 69,000[†]
		− 417,000
B. *Deposition*		
Tidal delta	1939-1961	+ 120,000
Barrier south of Indian River Inlet	1939-1961	+ 5,700
Offshore accretion	1919-1961	+ 256,000
		+ 381,700
C. *Net Change*		
Total Erosion		− 417,000
Total Accretion		+ 318,000
Net (Erosion)		− 98,300 $m^3 \ yr^{-1}$

[*]Volumes of erosion (or accretion) are equal to the difference in area between 1929 and 1961 bottom contours multiplied by average change in depth. From Moody, 1964.
[†]Estimated by United States Corps of Engineers.

nearly compensated for by the growth of the ridges on the adjacent sea floor (Table XXXII), the difference here probably being accounted for partly by storm washovers and partly by the growth of the adjacent Cape Henlopen spit (Kraft, 1971b). The stratigraphic penetration of the shore face does not appear to be quite as deep in the Bethany Beach area as in the False Cape area. Coring and radiometric dating by Kraft (1971a) indicate that the Bethany Beach ridges are underlain by Holocene lagoonal deposits overridden by the retreating barrier. Moody (1964) cites heavy mineral data which leads him to conclude that the barrier and inner shelf sands are derived from the underlying substrates.

Bathymetric Time Series on the North New Jersey Coast

Harris (1954) examined the Long Branch, New Jersey, dredge dump site over a four-year period to determine whether or not the dumped sand had replenished the beach. Instead, his series of precision nagivation surveys (Fig. 199) showed that the shore face was undergoing erosional retreat and that the adjacent sea floor seaward of the 9 m (30 ft) contours had undergone little net change, thus inadvertently demonstrating Bruun coastal retreat. Local aggradation seaward of 9 m (30 ft) is nearly normal to shore, following the trend of the north New Jersey ridge system (see Fig. 179, previous chapter). Harris'

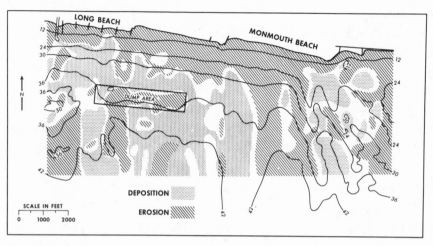

Figure 199.

Long Branch, N.J., coastal bathymetry, with erosion, deposition, or less than 0.4 ft of change indicated. Most changes were less than several feet. Dominance of erosion on shore face and near balance of sea-floor change relative to sea level is in accord with Bruun concept of coastal retreat. Modified from Harris (1954). Contour interval 6 ft.

data shows a net loss from his study area, partly because it does not extend sufficiently seaward to cover the entire system of sediment transfer, but mainly because the system is an open one with a net deficit as a consequence of littoral drift and the growth of nearby Sandy Hook spit (Caldwell, 1966). McMaster (1954) has concluded that in this area, the glauconitic heavy mineral suite has been derived from the underlying Tertiary substrate.

Evidence from the Pleistocene Record; Conclusions

The post-Pleistocene transgression of the middle Atlantic bight is one of a series of similar glacio-eustatic transgressions, and the subaerial coastal plain is veneered by a succession of deposits from previous cycles of Quaternary sedimentation. These comprise a time series of considerably greater scale and a significant test for hypotheses concerning the nature of the Holocene transgression. Coch (1963) has shown that major Pelistocene strand lines on the Virginia coastal plain are perched on the edge of steps in the underlying units cut by the respective marine transgressions, with the risers being old shore faces and the trends being marine erosional unconformities (Fig. 200). Colquhoun (1969) has mapped terrace deposits in South Carolina and notes tht only such deeply incised portions of the primary terrestrial erosional surface as drowned river channels survive the beveling process of the retreating shore face (Fig. 201).

Thus, Bruun coastal retreat appears to be widespread in time and space along the central Atlantic coast. The truncation of the subaerial surface and the continu-

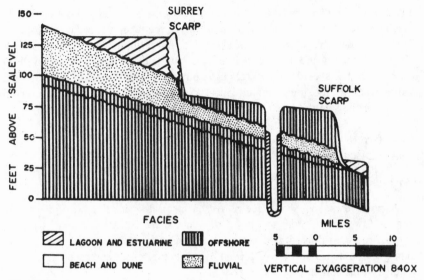

Figure 200.

Section through the Surrey and Suffolk scarps in southeastern Virginia, indicating erosional shore-face retreat. From Coch (1965).

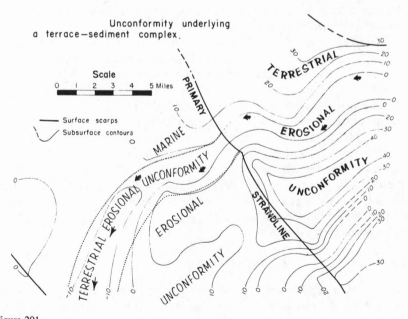

Figure 201.

Schematic diagram of the nature of marine erosional unconformities seen in raised Quaternary deposits of the South Carolina coast. From Colquhoun (1969).

ous destruction of marine marginal deposits by the advancing shoreline would preclude the survival of subaerial landforms, except under special conditions. In stratigraphic terms the result of shore-face retreat is the generation of a thin blanket of winnowed sand over a variable substrate of early Holocene and Pleistocene age. The sheet is not entirely autochthonous with respect to the Holocene sedimentary cycle, however, for it has locally incorporated Holocene fluvial deposits as they emerge at the foot of the shore face (Swift and others, 1971a).

Shore-face Retreat and Littoral Drift;
Coastal Discontinuities as Sediment Sinks

General

On previous pages, shore-face retreat has been described in two-dimensional terms. However, in order to fully understand shore-face retreat and its part in shaping the shelf floor, it is necessary to examine the shore face in three dimensions, and to consider the role played by projecting capes and reentrant estuary mouths in the retreat process.

As noted earlier, the Bruun concept of coastal retreat requires that littoral drift be treated as a steady-state process that does not effect the onshore-offshore mass balance. When areas of appreciable long-shore extent are considered, departures from this balance are to be expected as a consequence of loss of sediment to adjacent sinks for littoral drift.

The term littoral drift is commonly used to refer to the coast-parallel transport of sand by a wave-generated current in and landward of the breaker. This current is believed to be most intense beneath the breaker itself (Ingle, 1966). Very little is known about the behavior of the seaward margin of current. There is some reason to believe that with onshore or alongshore wind, the water column of the entire shore face and adjacent inner shelf is set into motion during storms. Murray (1970) has noted ''an unexpected seaward extension of the wave-driven longshore current,'' on the Louisiana coast during the onset of Hurricane Camille. The evidence presented in the preceding paper, and that presented by Harrison et al. (1965) suggests that during northeasters, wind and wave setup along the middle Atlantic bight generates in response a barotropic, southward-trending current strong enough to shape such large-scale bedforms as the inner shelf ridge fields.

Inner Shelf Ridge Fields and Littoral Drift

The relationship between littoral drift patterns and the distribution of shoreface-connected and isolated ridges is unclear. As noted in the previous paper, the ridges of the Atlantic coast open to the north regardless of the direction of littoral drift. Many shoreface-connected ridges merge with the shore face at depths as shoal as 4 m, and it seems probable that during intervals of long-period waves out of the southwest these features serve as natural groins, diverting littoral drift seaward along their crests. If so, then this circulation pattern, taken together with the storm pattern of Figure 186, previous paper, constitutes a closed loop in the sediment transport path. The shoreface-connected ridges would thus comprise sand circulation cells (Ludwick, 1970), and a mechanism would exist for nourishing the ridges from littoral drift prior to their detachment.

Inlet-Associated Shoals and Littoral Drift

The patterns formed by the intersection of coastal littoral drift cells with the reversing tidal regimes of barrier island inlets have long been known (Bruun and Gerritsen, 1960). The stable, large-scale, bed configuration is that of an arcuate, seaward-convex, sand shoal concentric about a scour trench. The curved crest of the shoal is generally high enough so that it breaks during heavy weather, or ordinary low water, and littoral drift is able to bypass the inlet on the shoal crest. If the littoral drift input is high enough, and the inlet wide enough, the shoal is transected by an intricate pattern of interdigitating ebb and flood channel systems stabilized by tidal currents (Ludwick, 1970). In each respective channel type, the ebb of flood current is longer or more intense, and for a brief portion of the tidal cycle currents of a given ebb-flood channel pair are going in opposite directions. Partitions between the channels are thus ebb-dominated on one side and flood-dominated on the other. They become circulating sand cells or closed loops in the sediment transport path (Ludwick, 1970), and grow upward until tidal aggradation is balanced by wave scour (Ludwick, 1972; Smith, 1969).

Many small shoals occur on the central and southern Atlantic coasts, associated with the small inlets between barrier islands. They play a role in shore-face retreat in that their associated scour trenches and back-barrier channels take part in beveling Holocene lagoonal and older Pleistocene deposits (Fig. 198). However, true barrier inlets are areally unimportant on the central and southern Atlantic coast. We are concerned here with the shoals associated with large estuaries, that have left a significant imprint as they have retreated across the Atlantic shelf floor.

Inlet-Associated Shoals of the Middle Atlantic Bight

As noted in the previous paper, the four estuary mouths which define the coastal compartments of the middle Atlantic bight, namely Block Island Sound, New York Harbor, Delaware Bay, and Chesapeake Bay, are sectors of tidal sedimentation. Delware and Chesapeake Bay (Figs. 202, 203) have large scale, flamboyantly patterned, inlet-associated shoals as a consequence of (a) the extremely large volumes of the tidal prisms of these enormous estuaries, resulting

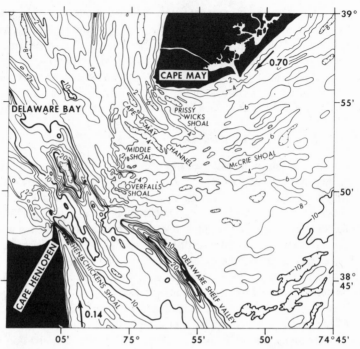

Figure 202.
Inlet associated shoal of Delaware Bay mouth. Flood-dominated channels deepen seaward, ebb-dominated channels deepen landward. Outer flood-dominated channel is continuous with Delaware shelf valley. Contour interval 2 fathoms. See USCGS Map 0807N-57 for 1 fathom resolution. Littoral drift in millions of cubic yards per year from Caldwell, 1964, and Moody, 1966.

in mid-tide surface velocities locally in excess of 200 cm sec^{-1}; (b) the broad nature of the estuary mouths (8 km or more); and (c) the high discharge of flanking littoral drift cells (volumes given in Figs. 202 and 203). The high lateral sediment input, however, is not enough to overcome the other two factors. The shoals are skeletal with respect to the typical barrier-inlet shoal and their axes are not appreciably seaward-convex. The north sides of the bay mouths receive the dominant littoral drift input, and probably also a significant input of fine sand from the intermittant, storm-driven transport system of the inner shelf. The repvious paper has noted that the shore face of the southern New Jersey and Delmarva coastal compartments consist of fine sand (Fig. 177; and text, previous paper). The material may have been winnowed from the massive ridge fields that characterize the central portions of these compartments. The subdued ridges and swales of the terminal sectors of the coastal compartments curve smoothly into large scale ridge fields of inlet-associated shoals themselves (Figs. 202 and 203), and may serve as sediment conduits.

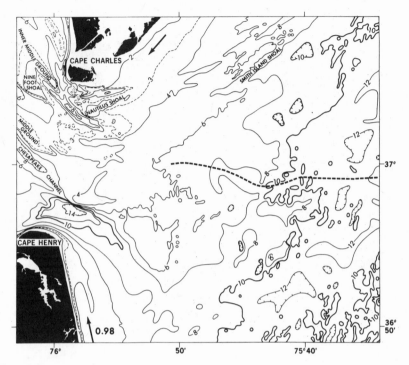

Figure 203.

Inlet-associated shoal of Chesapeake Bay mouth. Main scour channel is almost entirely contained within inlet-associated shoal. Older, relict channel is dashed. Well defined ridge and swale topography starts at 10 fathoms. From USCGS Chart 1222 contour interval 2 fathoms. Littoral drift at Virginia Beach in millions of yd³ yr¹ from Weinman (1971).

As a result of dominant sediment input from the north, the northern ends of the inlet-associated shoals are best developed, with intricate systems of mutually evasive ebb and flood channels, whose relief is on the order of 15 to 20 m (49 to 75 ft; see Figs. 203 and 204). The main scour channel in Chesapeake and Delaware Bays lies south of the drift-fed sand mass. In Chesapeake Bay, it sits athwart the crest of the inlet-associated shoal and its southward curving seaward terminus fades into a low smooth apron of sand referred to by Payne (1970) as a submarine "tidal delta." Typical ridge and swale topography begins seaward of the 15 m (45 ft) isobath. On the Delaware Bay Mouth's southern side, a couplet of mutually evasive ebb and flood channels comprises the main scour trench. The flood channel is the more seaward of the two, and is continuous with the Delaware shelf valley (Fig. 202).

The south sides of Delaware (Kraft, 1970b) and Chesapeake Bays (Fisher, 1967) developed cuspate spits during the Holocene. Cape Henlopen became a rapidly prograding recurved spit during historic times (Kraft, 1970b). It has

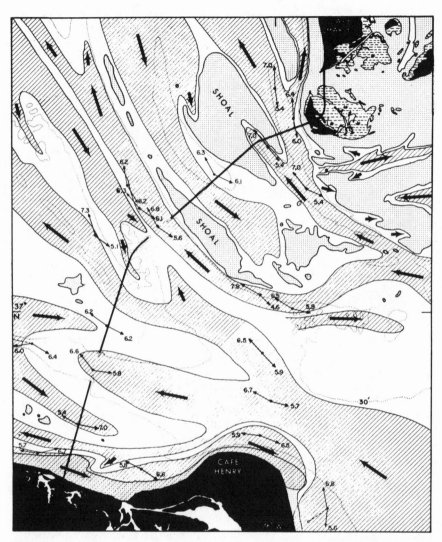

Figure 204.

An hydraulic and geomorphic interpretation of the net non-tidal (residual) flow pattern at the bottom in the entrance to Chesapeake Bay. Numbers are measured flood and ebb flow durations at the bottom in hours; small arrows show measured direction of near-bottom currents. Stippled areas are shoaler than 18 ft. Ruled areas are major areas of ebb or flood flow predominance. Large arrows indicate which is predominant. Nonpatterned areas are neutral, or are too finely structured for resolution. From Ludwick (1970).

advanced to the scour trench, and is presently feeding Hen and Chicken shoals, an ebb-dominated shoal. A small scale, subdued version of this shoal is seen off Cape Henry in Chesapeake Bay.

Shoreface Retreat on an Inlet-Dominated Coast,
Georgia Bight

The retreating estuary mouths of the middle Atlantic bight have had left significant traces on the Atlantic shelf floor as will be indicated on following pages. In the middle Atlantic bight, these estuary tracks are anomalous features of the shelf floor. In the southern Atlantic Bight, however, the closely spaced estuaries dominate the coast. Therefore, before considering the shelf floor, it is useful to briefly consider the dissimilar nature of shore-face retreat in the southern Atlantic bight. Comparison of the shorelines of the middle Atlantic and Georgia bights reveals very different patterns (Fig. 205). The valleys of the middle Atlantic bight were deeply eroded by glacial runoff, and the returning Holocene sea has drowned them for considerable distances inland (Meade, 1969). The result is a coastline whose solution for the equilibrium equation is strongly

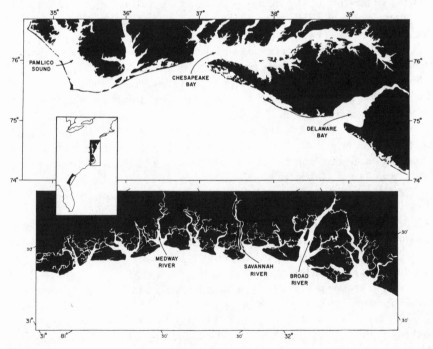

Figure 205.
Comparison of the central and southern Atlantic coasts.

negative. The large estuaries have dendritic outlines like those of recently dammed streams, indicating that the rate of sea level rise is great relative to the rate of sedimentation.

On the Georgia coast, runoff per unit area is less than it is further north due to greater evaporation, a larger growing season, and loss from aquifers directly to the sea floor (Manheim, 1967, Hidore, 1966; cited in Meade, 1969). Consequently, the southern estuaries are smaller and closer together, a fact partly masked in Figure 205 by the different scales. However, the small-scale stream net that drains the deeply weathered southern hinterland carries much more sediment than do the northern rivers, whose drainage basins are thinly veneered with compact, resistant till. The result is a nearly neutral (stillstand) coastline of estuaries whose simple, flaring mouths and meandering heads are primarily constructional rather than erosional forms (Leopold and others, 1964, p. 272). Their channels are presumably in equilibrium with their tidal prisms, a relationship of discharge to geometry which is indicated when the upstream intertidal volume of successive cross-sections forms a linear plot on log-log paper with the subtidal cross-sectional area (Weigel, 1964, p. 380). It seems probable that only a slight increase in the rate of sedimentation would be needed to reverse the sign of the solution for the equilibrium equation and cause it to prograde. Locally, as at the confluence of the Peedee nd Santee Rivers, the coast seems as nearly a deltaic as an estuarine coast.

Unlike the disequilibrium estuaries of the middle Atlantic bight the southern estuaries are potentially capable of bypassing sediment (Meade, 1969, p. 232). The presence of phosphorite in the Georgia estuaries (Pevear and Pilkey, 1966) indicates that the salinity underflow introduces littoral drift into the estuaries, but does not necessarily mean that fluvial sand does not escape when flood conditions push the salt wedge out (Meade, 1969). The phosphorite may be of quite local origin, since the scour trenches of the estuary mouths are up to 30 m (100 ft) deep and probably cut through to the surface of the underlying Miocene.

The Georgia coast differs in energy input as well as in sediment input. Mean annual wave heights are markedly lower than in the middle Atlantic bight (Tanner, 1961; Dolan and others, 1972), because of the orientation of the coast and its shoal nature. However, the shelf tide is concentrated by the bight and the spring tide is 2 to 3 m (6 to 10 ft). The result is efficient, coast-normal dispersion and fractionation of sediment by tidal currents.

The high sediment input and complex hydraulic structure of the transition zone has resulted in a lens of paralic sediment up to 40 km wide, which attains a maximum thickness of 15 to 30 m (49 to 100 ft) along the shoreline (Henry and Hoyt, 1968). The broad band of estuaries comprises an efficient sediment filter in which sand undergoes progressive sorting (Swift and Heron, 1968, p. 226; Swift, 1970, Fig. 16). Sand released to shore-face processes is fine-to very fine-grained. The shoreface is very broad and gentle, probably as a result of the low effective angle of repose of this material. The first 2 km offshore slope at 1.5 to 2 m per km, but this rate diminishes rapidly to 1 m per km, then 0.6 m per km in the next 4 km.

Shore-face retreat in the Georgia Bight is dominated by the scour trenches of the closely spaced estuary mouths (Fig. 206), which cut completely through the paralic lens. These commonly attain depths of 20 m (60 ft) and locally 30 m (100 ft) (Hoyt and Henry, 1965; Henry and Hoyt, 1969). A controlling element in their distribution is the Silver Bluff barrier chain, dating from a slightly higher than present stand of the sea at 25 to 30 thousand years ago (Hoyt and Hails, 1967). The modern barrier islands rest on the seaward flank of this older chain and the position of modern inlets are controlled by positions of Silver Bluff inlets. They tend to migrate southward (Henry and Hoyt, 1967), but apparently sea level has not been at its present height long enough for the inner Pleistocene barriers to be consumed and remade by migration of the modern inlets.

As the scour tenches advance landward, they erode leading-edge paralic deposits of basal fluvial sand and upper marsh sediments and probably also underlying Tertiary and Pleistocene materials. The debris of excavation together with littoral drift from the coastal compartment to the north is piled behind the scour trench as large, arcuate, inlet-associated shoals with transverse ridges and channels impressed upon them (Fig. 206). Crestal sectors of the arcuate shoals may be exposed at low water 7 to 9 km (3 to 4 n mi) offshore (Henry and Hoyt, 1969). The shallow crests of the shoals receive littoral drift from the north, and are scoured by south-trending storm currents. Hence, ramps and chutes are prominant on the north side; the central channel curves to the south, and the southeast sides are oversteepened. See Oertal and Howard, this volume, for a discussion of the hydraulic regime associated with the shoals.

If the coast is retreating more rapidly or as rapidly as tne inlets are migrating, or if this relationship was true in the past, then the Pleistocene "basement" of the shelf should be deeply furrowed with closely spaced tracks of inlet scour-trench retreat. These furrows would be filled with sand from the retreating inlet shoal; the narrow interfluves would be capped with a thinner veneer, also beveled by shoreface retreat, which might include Holocene back-barrier marsh deposits. The trough fill might stand higher on the sea floor than the adjacent interfluve, or, if a mutually evasive ebb-flood channel pattern (Ludwick, 1970, Fig. 1) has been impressed on the shoal, the seaward channel may have partially reexcavated the trough as the shoal retreated.

Hails and Hoyt (1969, p. 63) and Henry and Hoyt (1969, p. 201) have examined the Georgia coast in detail and have concluded that, like the central Atlantic coast, it is retreating according to the Bruun scheme. It seems to differ mainly in the greater role played by estuary mouths in generating coastal strati-

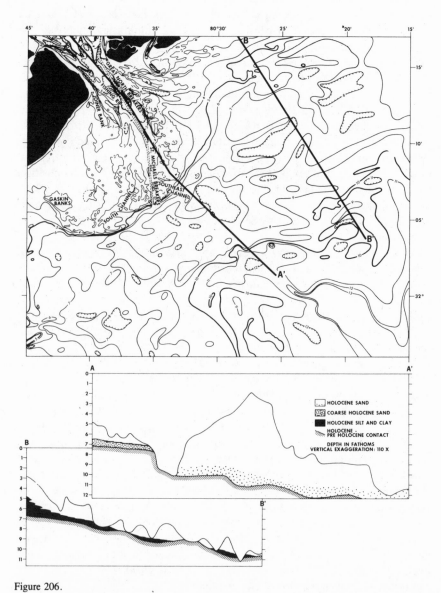

Figure 206.

Port Royal Sound, an inlet-associated shoal from the Georgia coast. Geological sections are hypothetical. From USCGS Chart 1240. Contour interval 1 fathom.

graphy, in the greater thickness of paralic deposits, and in the gentle slope of the shore face. These characteristics may result in a more nearly continuous record of paralic strata surviving shoreface retreat than is found further north; the continuous depositional transgression of Curray (1964). Ongoing studies by workers cited above will shortly test this hypothesis.

Cape-Associated Shoals and Littoral Drift

In addition to the estuary mouths, the cuspate forelands and associated shoals of the central and southern Atlantic coast are significant sinks for littoral drift, and their role in shore-face retreat must be considered, before examining the consequences on the shelf floor.

The discontinuities of Middle Atlantic Bight coastal compartments (see discussion, previous paper) tend to have prominent ridge fields associated with them; however, only the southernmost (Chicoteague Shoals) has a well developed cuspate foreland (Fig. 178, previous paper). The northern half of the southern Atlantic Bight has three classic, large-scale cuspate forelands (Fig. 192); Capes Hatteras, Lookout, and Fear. A fourth, Cape Romain, is nearly a delta. Cape Kennedy, midway down the Florida coast, is also a cuspate foreland.

Cuspate forelands appear to require for their formation (a) a high rate of sediment input in the littoral drift zone, (b) a high energy input, and (c) stabilizing or shielding points around which sand can accumulate. Given these factors, a feedback process begins. The waves refract around the shielding point and the shore face (if initially straight) rotates into two arcuate segments behind the shielding point, a configuration more stable in the face of wave attack. As the shore face rotates in this manner, the pattern of wave refraction adjusts to the new bottom configuration, and the bottom to wave refraction, until stability is attained.

Cuspate forelands as a class of coastal landforms are transitional with cuspate deltas, described by Curray (1969) as a characteristic morphological response of deltas to a coast with a high value of wave-energy input relative to sediment supply. The formation of cuspate deltas also seems to be favored by a high input ratio of bedload sediment to suspended sediment. Here the shielding point is the river mouth itself. For stability, littoral drift must trend from the apex toward the base on both sides.

While a deltaic origin has been proposed for the Carolina Capes (Hoyt and Henry, 1970) and since debated (Hopkins, 1971; Henry, 1971), close examination of the shoals seaward of the capes suggests that, if valid, it must have been so mainly at an early period in the capes' history. At present, the prevailing wave direction for the Carolina capes and Cape Kennedy, Florida, is from the northeast, oblique to the axes of the capes. The result is that littoral drift converges, rather than diverges, on the apices of the capes, since in the shadow zone to the south, waves from the southeast dominate. The large-scale, well-defined sand shoals which extend from the Carolina and Florida capes almost to the shelf edge (Figs. 207-211) must be a consequence of drift convergence. Either the shoals must have grown seaward from the tips of the forelands (Tanner, 1960; Shepard and Wanless, 1971), or the shoals must have grown landward with the retreat of the cape-apex depositional center. Consideration of available bathymetric information plus what little is presently known concerning the hydraulic regime in the vicinity of the capes permits an evaluation of these hypotheses.

Morphology of Cape-Associated Shoals

As noted in the previous chapter, the cape-associated shoals of the Carolina and Florida capes tend to be shelf-transverse, "hammer-headed," second-order features composed of first-order, seaward-convex ridges and swales. The latter are not quite symmetrical about the shoal axis; most troughs tend to become shallower to the south or southwest and may not extend completely through the shoal (Figs. 207-211). A minority of the arcuate swales become shallower in the reverse direction, however, and locally the two types alternate as apparent ebb-flood channel couplets of the sort seen in the entrances to tidal estuaries (Ludwick, this volume). The arcuate ridges tend to extend further from the shoal crest line on the shoal's northeast side than they do on the southeast; in other words, the large-scale shoals on which the ridges and swales are superimposed tend themselves to be asymmetrical, with steeper southwestern flanks. Thus, Lookout and Southeast shoals are comb-line shelf transverse structures with toothlike ridges projecting northeastward (Figs. 208, 21). Toward the base of the shoals the sequence of arcuate ridges may continue along the northeast mainland shore as shoreface-connected ridges.

A remarkable photomosaic of Lookout Shoals (El Ashry and Wanless, 1968; Shepard and Wanless, 1971; Fig. 209, this paper) shows in addition to a tendency toward longitudinal differentiation of the shoals. It indicates that the proximal third of Lookout Shoals consists of "overlapping" sand masses, which El Ashry and Wanless (1968) reasonably suggest represent major pulses of sand emplaced during storms. The central portion is flat, however, and the distal third is furrowed by small-scale features in the form of linear sand ridges, with amplitudes up to 4 m (12 ft) parallel to the main shoal axis. These features also appear on the more detailed bathymetry available for some of the capes (Fig. 210). Their axes tend to be gently convex toward the south, and on acoustic profiles a southward asymmetry is dominant, with steeper southward flanks locally attaining the angle of repose. A 6 m (19 ft) vibracore taken from a large feature immediately south of the Diamond Shoals Light consisted of coarse sand with southward-inclined cross-bedding at the base. The upper portion of the core was so loosely packed that it flowed within the core liner and no structures were retained.

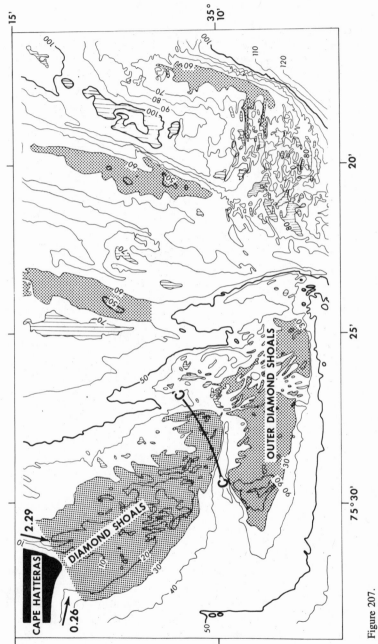

Figure 207.

Diamond Shoals, off Cape Hatteras, North Carolina. From USCGS surveys. Littoral drift in millions of cu yds/yr from Langfelder and others, 1968. Major isolated highs are stippled. Profile CC' in Fig. 231. Contour interval 10 ft.

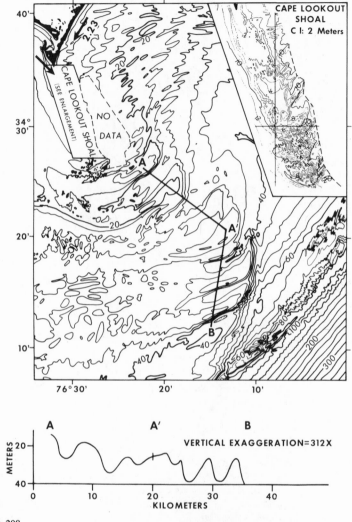

Figure 208.

Lookout Shoals, off Cape Lookout, North Carolina. From USCGS surveys. Littoral drift in millions of yd³ yr¹ from Langfelder et al. (1968). Contour interval 10 m out to 100 m, then 25 m.

These characteristics led us to conclude that the small-scale features are transverse sand waves. The southward asymmetry of the shoals and of the sand waves parallel to their crests indicates that the dominant sediment transport direction is across the shoals to the south. We therefore conclude that the shoals could not have grown seaward from their present bases; rather, the bases are the only growing portion and these are growing landward by the accretion of pulses of littoral drift received from the retreating foreland apex during storms.

Thus, forelands and shoals exist in a state of homeostatic equilibrium; as

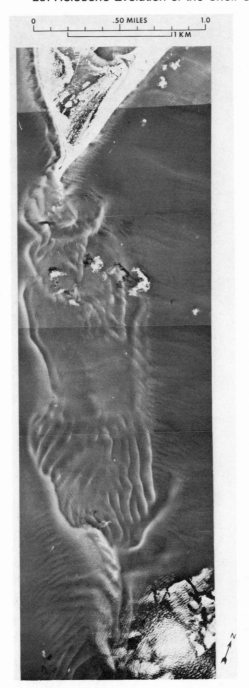

Figure 209.
Lookout Shoals, off Cape Lookout, North Carolina. From Shepard and Wanless (1971).

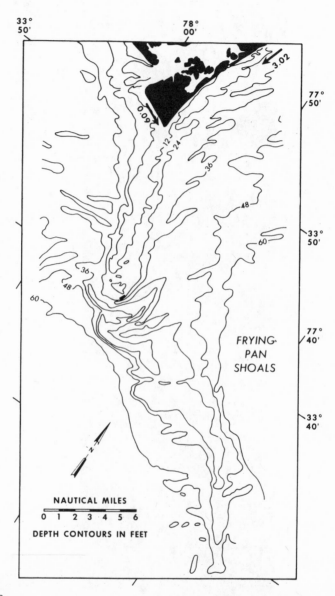

Figure 210.
Frying Pan Shoals, off Cape Fear, North Carolina. From Hopkins, 1971. Littoral drift in millions of yd³/yr⁻¹ from Langfelder and others, 1968. Contour interval 12 ft.

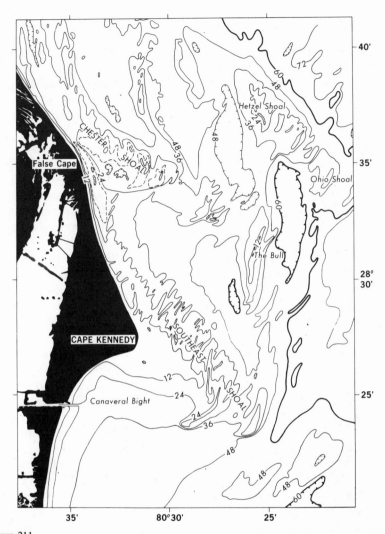

Figure 211.

Cape Kennedy and associated shoals, Florida. From USCGS bathymetric surveys. Countour interval 12 ft; dashed supplementary contours at 6 ft.

the shore face of the retreating foreland wastes away, the debris feeds the landward-prograding shoal base, so that its aggrading surface keeps pace with the rise in sea level in accord with the Bruun principle. The shoal, for its part, serves as the shield to which the foreland is anchored and a focal point for the wave energy which shapes the foreland. While the Carolina capes might have been initiated as deltaic (river-stabilized) forelands, they are at present shoal-stabilized forelands.

Hydraulic Regime of Cape-Associated Shoals

Scattered bits of hydraulic data, available in the literature, support this picture of seaward sediment transport at the shoal base and cross-shoal sediment transport over its central and distal portions. Transfer of littoral drift from foreland apices to shoal bases has been documented by the United States Army Corps of Engineers (El Ashry and Wanless, 1968, p. 360) for Cape Fear and Frying Pan shoals. At Frying Pan shoals, the eastern face of the apex has lost more sand than the western face has gained, the difference being gained by the shoal. The Corps of Engineers estimates an average shoal accretion of 24,500 yd³ a year for a 72-yr period.

Tides are weak over the shoals; surface tidal currents over Lookout Shoals average 0.07-0.18 knots (El Ashray and Wanless, 1968). However, there is evidence to indicate that south-trending, coast-parallel storm currents are a major factor in the shaping of the shoals. El Ashry and Wanless (1968) quote the United States Army Corps of Engineers to the effect that "surface-water velocities average 1.7 to 2.6 per cent of wind velocity." The Atlantic Coastal Pilot (1964) notes that, "during strong winds, currents set across the shoals with great violence." Stefansson et al. (1970) have noted surges of Virginia coastal water across inner Diamond shoals in periods of strong northeast winds during the winter. We infer that the south-shoaling arcuate troughs of the cape-associated shoals are spillways for these pulses and that the sand waves are also activated by them. The smooth surface of the inner shoals may indicate that, during storms, critical flow is attained in this shallow area. The regional asymmetry of the shoals may also be attributed to south-trending currents; a comparison of the 1953 and 1963 USCGS surveys of Diamond shoals indicates that north-facing slopes experienced up to 8 m (25 ft) of accretion, while south-facing slopes underwent an equivalent accretion (Fig. 212).

Wave refraction must be a second major force shaping the shoals and, in particular, the arcuate first-order ridges; the ridges are curved so as to be normal to wave orthogonals converging on the shoals. Tanner (1961) has noted a direct correlation between the grain size and topography on the San Blas and St. George cape-associated shoals, panhandle, Florida. Ridges are mantled with (or composed of) fine, well-sorted sands; troughs are floored with coarse, poorly sorted sands. The fine-grained sands of ridges are presumably analogs of the fine-grained sands of mainland shore faces and are the consequence of wave-winnowing, while the coarse trough sands are best understood as scour lags, generated in response to the coast-parallel, storm-generated currents. If Tanner's sampling had been more closely spaced, he might have found the secondary grain-size maxima on crests that occur as a response to wave winnowing in the False Cape ridge system (preceding chapter).

Thus the arcuate ridges and troughs of the cape-associated shoals appear to belong to the same basic class of features as the inner shelf ridge fields

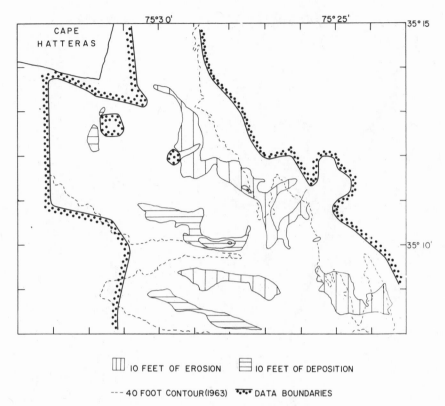

Figure 212.

Accretion and erosion at Diamond shoals between 1953 and 1963. From USCGS surveys.

with which they merge (Fig. 213); and like them are responses to south-trending storm currents and storm waves. Their arcuate nature is merely a result of the different boundary conditions under which they formed.

Shore-face Retreat and the Provenance of Coastal Sands

General

The preceeding analysis of shoreface retreat on the central and and southern Atlantic coast has been based primarily on stratigraphic and morphological considerations. However, petrographic studies of the provenance of Atlantic coast sands have yielded results which suggest a somewhat different model. In the southern Atlantic Bight (numerous studies summarized by Pilkey and Field, this volume) and again on the central and southern New Jersey coast (McMaster, 1954; Frank and Friedman, 1971) analysis indicates that coastal sands are petrographically dissimilar to modern river sands, and to the sands of the subaerial

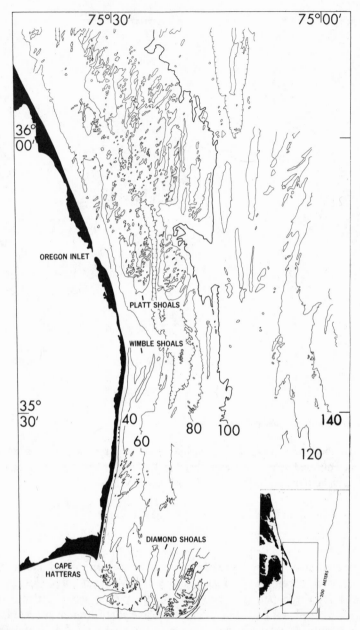

Figure 213.
Bathymetry of a portion of the North Carolina shelf. Contour interval 20 ft. From USCGS hydrographic surveys.

Pleistocene substrate, but are presumed, or are in fact similar to shelf sands. In these areas, the coastal sands have, therefore, been assumed to have moved "onshore" from a "continental shelf" source. These terms suggest a model diametrically opposed to Bruun coastal retreat, in which the response to a rising sea level is shelf floor erosion and coastal aggradation; a response difficult to visualize in dynamic terms. However, careful examination of the evidence suggests that despite the differing terminology, the Petrographic data comprises in some cases a converging line of evidence for shoreface retreat of the Bruun type, and in other cases, is at least compatible with it.

Provenance of New Jersey Coastal Sands

McMaster (1954) has investigated the provenance of New Jersey beach sands by means of detailed heavy-mineral analysis. His studies reveal a northern glauconitic province equivalent to the eroding headland sector described in the previous paper, an opaque heavy-mineral province equivalent to the proximal, seaward-convex barrier arc, and a hornblende province equivalent to the distal, seaward-concave barrier arc. He concludes that, within the glauconitic province, the beach is supplied by erosion of subjacent, glauconitic, Tertiary strata.

McMaster's study indicated that the genesis of the opaque heavy-mineral and hornblende provinces are less clear cut. The presumed source is the underlying Cape May Formation of Pleistocene age. A comparison of heavy mineral suites of the Cape May Formation versus those of the modern beach (McMaster, 1954, p. 178-185) is inconclusive. However, McMaster (1954, p. 187) notes that the Cape May mineralogy is quite variable. He surmises that the beach sands of the central and southern New Jersey shelf are derived from the offshore Cape May formation of the continental shelf rather than from the Cape May formation of the mainland. The conclusion appears to contradict the concept of beach nourishment by shore-face erosion. However, McMaster's (1954, p. 185) main argument is geographic rather than mineralogical:

Along the [New Jersey] seaboard, the possible submarine source contributors are restricted to the [Cape May Formation of the eroding headland sector] . . ., since a series of barrier bars guards the headlands . . . It is apparent that the major contribution of sediment for the beach opaque . . . and hornblende mineral zones cannot be located on the subaerial portion of the coastal plain. The only alternative must be the continental shelf.

Despite such terms as "continental shelf" and "offshore," McMaster seems nevertheless to be referring to the retreating shore face, since elsewhere (McMaster, 1954, p. 192) he comments more explicitly:

The present-day source for the beach sands in this area . . . is local. Therefore, all beach nourishment must be obtained in the vicinity of the beach itself. The beach-building material is derived from a reworking of bottom sediment in the immediate vicinity of [barriers] and [from] the sands moved by drift.

Frank and Friedman (1971) have recently compared landward lagoonal beach heavy-mineral suites with barrier beach heavy-mineral suites north of Atlantic City. They noted marked differences and concluded that the central New Jersey barrier formed from ''the adjacent continental shelf.'' They propose that the barrier retreated slowly landward by washover and concluded that ''this process would not result in massive shore-face erosion or aggradation of the adjacent sea floor, as has been proposed.'' Their argument is difficult to follow, since landward migration of a barrier requires ''massive shore-face erosion'' by definition. It is possible to envisage a variant of Bruun coastal retreat, whereby stratigraphic penetration of the shoreface is shallow as at Bethany Beach and the retrograding barrier retreats over a carpet of its own back-barrier muds (Fig. 196). Insulated from subjacent sands by the mud carpet, the barrier becomes a reservoir of relatively far-traveled and mineralogically distinctive sand, undergoing slow dilution by littoral drift during its cycle of washover, burial, and remergence. However, as noted by Pilkey (personal communication), the repeated subaerial exposure should render the heavy mineral suite more mature than that of its source. This variant model is compatible with McMaster's data, but not necessitated by it.

Provenance of Southern Coastal Sands

On the Georgia coast, the fine-grained shoreface sands extend to a feather edge 23 to 34 km (10 to 15 n mi) seaward in 12 to 15 m (40 to 49 ft) of water (Henry and Hoyt, 1969). Medium- and coarse-grained sands lie to the seaward; the contact between the two has been described as the contact between ''modern'' and ''relict'' sand by Pilkey and Frankenberg (1964). It has been suggested that the coarse sands are relict and residual sands and that the fine sand mantle of the broad, gentle shoreface is an out-winnowed fraction that has moved in with the advancing Holocene sea (Henry and Hoyt, 1969). The presence of trace amounts of phosphorite in the sand fraction of the Georgia shelf and beaches, and its absence in Georgia rivers, has been cited as evidence (Peavear and Pilkey, 1966). See also heavy mineral data of Pilkey and Field, 1972. However, it is possible to suggest an alternate hypothesis. The reversing jets at the estuary mouths must serve as efficient elutriators of sand, and the fine sand may be fractionated and spread across the shore face by such a mechanism. The leading edge of the paralic lens must consist of relatively coarse, fluvial sands at the fall line; these would be reworked by the advancing estuary mouth scours and incorporated into the bases of the arcuate shoals (hypothetical cross sections, Fig. 206). Retreat of the superstructure of the arcuate shoal with the shoreface would reexpose the coarse basal material, which thus would be allochthonous with respect to the Holocene sedimentry cycle rather than autochthonous, or ''relict.'' The anomalous trace minerals, as noted above, could be explained as mined from the underlying Tertiary by the estuary mouth scours.

However, it should be noted that these hypotheses are not necessarily mutually exclusive. Elsewhere in the southern Atlantic bight, notably in Onslow Bay (Pilkey and Field, 1972) long distance onshore movement of shelf sand seems to be the only viable hypothesis. Here the phosphorite component of the beach sands appears to be derived from offshore exposures of the Miocene Yorktown formation at a stratigraphically higher level than exists on land (Roberts and Pierce, 1964). The submarine outcrop area lies 20 km seaward of the modern in over 20 m of water. Here the phosphorite may have moved in with the recycling beach prism during the past millenia of sea level rise; or the barrier may indeed be exchanging sand with the adjacent sea floor.

THE PALIMPSEST SHELF FLOOR

General

In the previous chapter the evolution of shoreface-connected ridges into isolated inner shelf ridges was examined. The preceding section of this chapter has considered this transformation as part of a more general process of shore-face retreat. This section will attempt to explain the shelf-floor morphology in terms of shore-face retreat. To the extent that the shelf floor can be viewed as resulting from shore-face retreat, and modified by subsequent shelf processes, it is a palimpsest surface, or surface whose original (foot of shore face) features have been partially, but not wholly, obliterated by the subsequent (shelf) processes. However, detailed examination of the shelf suggests that the simple transition of shore-face ridge system to shelf ridge-field, developed in the preceding chapter, is only one of several modes of formation of the constructional, high-order morphological elements of the shelf floor.

Shelf-Transverse Elements I: Shoal-Retreat
Massifs from Inlet-Associated Shoals

The Problem

The hypothesis of detachment of shoreface-connected ridges to form isolated shelf-floor ridges, as set forth for the False Cape and Platt Shoals sector (previous chapter has internal consistency with respect to the areas discussed. It becomes less satisfactory when a relationship is sought between the ridges so ioslated and the transverse shelf valleys and shelf ridge systems that lie to the north of each of the detached ridge sets. The Delmarva inner-shelf ridge field which formed the main actualistic example in the previous paper is in fact an anomaly in that it is the only large-scale, inner-shelf ridge field of the middle Atlantic bight that is not associated with a transverse shelf valley. The other two examples, the Virginia Beach ridge field (Fig. 214) and the Platt shoals ridge field (Fig. 215), surround shelf valleys and attain their primary development on the northern sides of these shelf valleys. Any convincing model of ridge development must explain this association of shelf-floor ridge fields with transverse shelf valleys. Truncation of nearly coast-parallel inner-shelf ridges by shelf-transverse valleys seems at first glance to place contraints on the time relationships of the respective morphological elements. Payne (1970) has in fact suggested that the Virginia Beach shelf valley (Fig. 214) was cut by a pre-recent river, conceivably an ancestral Susquehanna, and that the flanking ridges which it appears to truncate may be relict Pleistocene barriers. However, close examination of flanking ridge systems raises major difficulties for this hypothesis.

Evolution of the Virginia-North Carolina Coast

The difficulties first appear when transverse profiles of the two shelf valleys are considered. These lack the bilateral symmetry to be expected if valley and ridges comprise a simple drainage basin (Fig. 216). Ridge crests and trough

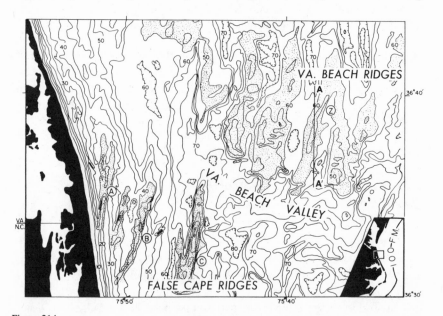

Figure 214.

Bathymetry of a portion of the Virginia shelf. Contour interval 10 ft. From Payne (1970). Profile AA' presented in Fig. 231. Dashed lines in profile are an acoustic reflector.

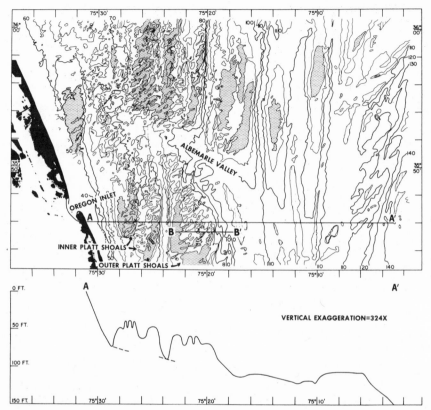

Figure 215.

Platt Shoals ridges and Albemarle shelf valley. Contour interval 10 ft. From USCGS hydrographic surveys. Profile BB′ presented in Fig. 231. Dashed lines in profile are an acoustic reflector.

axes climb to the south, both north and south of the valley (Figs. 214, 215, 216). Troughs north of the two valleys are separated from the valleys by low sills; thus the ridge set north of each valley comprise in each case a complex comblike transverse sand body with "teeth" pointing north, as do the cape-extension shoals previously discussed.

The pervasive southward asymmetry of the drowned valley systems appears to be a consequence of periodic, intense, south-trending currents generated by storms, as described in the preceeding paper (see also Swift et al., 1972; Shideler et al., 1972). These currents, probably structured into helicoidal flow cells, have scoured troughs down to the Pleistocene substrate, have generated small-scale transverse bedforms in them, and have swept finer sand south through the ridge systems, into the old river channels, and onto the flats south of the southern ridge sets. Thus, modern marine processes have resulted in extensive

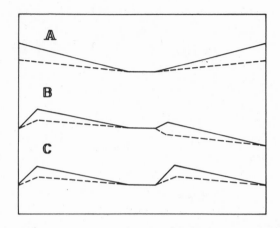

Figure 216.

Schematic coast-parallel profiles through inner shelf ridge channel complexes: (A) Bilateral symmetry of hypothetical subaerial drainage net; (B) Southward asymmetry of Great Egg shelf valley; (C) Southward asymmetry of Virginia Beach and Albemarle shelf valleys.

modification of an older, river-dominated topography. However, evolution of the ridges flanking the channels is still obscure. If the southern ridge sets are detached shore-face ridges, what are the northern sets and what is their relationship to the channels? These questions are best answered by recourse to our knowledge of late Holocene sea-level rise and its effects on continental shelf sedimentation.

Holocene sea-level rise is known to have undergone a reduction of rate between 4000 and 7000 BP, from a rapid to a gradual rise (Milliman and Emery, 1968). Presumably, the Middle Atlantic coast would have been undergoing erosional transgression (Curray, 1964, p. 200) prior to that time. As the rate of sea-level rise decreased relative to the rate of sedimentation, the coast stabilized or began to undergo depositional regression. Curray (1969) has suggested that many modern barrier systems with bases at 10 m (33 ft) began at this time, and the suggestion seems valid for the middle Atlantic coast (Newman and Munsart, 1968; Kraft, 1971b). The effective increase in the rate of sedimentation may have caused the barriers of the Carolina coast to prograde (Dolan, 1970). This coastal sector has since resumed its retreat (Langfelder and others, 1968). The resumption may have been a consequence of an increase in the rate of sea-level rise. However, even without such an increase the continued deepening of the water and upward growth of the barrier would have increased the submarine surface area of the barrier that needed nourishment (Dillon, 1970 and the concomitant widening of the lagoon would have separated it from fluvial sources of sand (Curray, 1964).

During the earlier period of rapid erosional retreat, the Virginia-northern North Carolina coast would have resembled the modern Georgia-South Carolina coast

Figure 217.
Schematic model for evolution of inlet-associated shoal into an inner shelf ridge system. Above: Estuary mouth with arcuate sand body. Based on the numerous similar estuarine shoals of the Georgia-South Carolina coast; Fig. 206, this paper. Below: Inner Shelf ridge system, modified from inlet-associated shoals during coastal retreat. Based on the Oregon Inlet area, with Platt Shoals, the closed Albemarle inlet, Currituck spit, and Roanoke and Currituck sounds on the North Carolina coast. Compare with Fig. 218. Contours in ft.

as described on previous pages, in that the estuarine regime would have extended from the river mouths across narrow marsh-filled lagoons and through inlets to build large, arcuate, inlet-associated shoals in the adjacent shoreface and inner shelf (Fig. 217). We propose that the Platt shoals complex and Virginia Beach-False Cape ridge complex were built as such shoals before the Ancestral Albemarle and James (or possibly Susquehanna) Rivers during this period.

However, the ancestral Albemarle and James' drainage basins lacked the innately high fluvial sediment input of the Georgia coast and their lower valleys had become deeply incised during the pleistocene (Meade, 1969). As sea-level continued to rise behind the stabilized barrier, they were unable to retain their direct outlets to the sea. The lower valleys become flooded almost to the fall line. The zones of maximum stratification and associated turbidity maxima withdrew to the upper portions of the resulting estuaries, which were now effective sediment traps. Roanoke and Currituck Sounds grew respectively north and south of lower Albermalre Sound into the sediment-starved back-barrier marshes, thus detaching the future Currituck Spit from the mainland (Fig. 218). A closed inlet in the vicinity of Kitty Hawk may be the last vestige of the former river mouth. If so, the inlet migrated north before closing, for recent seismic work (Riggs and O'Connor, personal communication) shows that the main channel passes under Currituck Spit a little to the south. Our interpretation of North Carolina coastal evolution differs from that of Pierce and Colquhoun (1970) in that we consider Currituck spit as well as Hatteras Island to have formed as a primary barrier (detached mainland beach; Hoyt, 1967) rather than as a secondary barrier, or true, genetic (coastwise prograding) spit.

This remarkable change in coastal regime, from erosional transgression to near standstill, has resulted in a complex of large-scale, intracoastal water bodies that contains one of the largest estuaries (Chesapeake Bay) and one of the largest lagoons (Pamlico Sound) in the world. The two-step cycle, taken as a unit, is tantamount to Curray's (1964) discontinuous depositional transgression in which barriers grow up vertically, only to be overstepped. Presumably, were it not for the stability afforded by Diamond Shoals as a focus for wave energy, the Carolina-Virginia barriers would be overstepped or driven back.

The ridge fields before the North Carolina barriers would thus appear to have been initiated as inlet-associated shoals. They can hardly be called relict, however, for the great bulk of their relief most post-date their abandonment by their respective rivers; the arcuate, inlet-associated shoals of the Georgia coast have only incipient first order ridges impressed upon them (Fig. 206), and these features are mainly on the north side, facing southward littoral and shore-face storm currents. Thus, the ridge systems north of the Albermarle and Virginia Beach shelf valleys may have started as ramps and chutes, but, since abandonment, have deepened and elongated their troughs and have aggraded their crests. Preliminary examination of the vibracores from the Virginia Beach ridge system indicates that the ridges contain cores of sand finer than the medium-grained sands of the crests.

Shelf valley segments appear to be ubiquitous on the inner shelf. Another example may be seen in Long Bay on the Carolina coast between Cape Fear and Cape Romain (Fig. 219). It is apparently a former channel of the Waccamaw

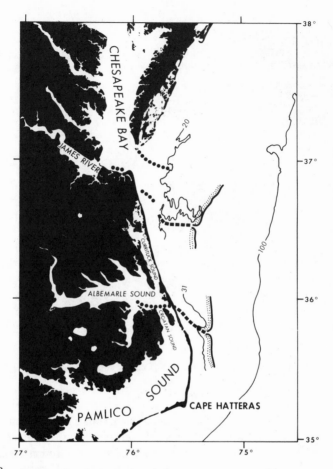

Figure 218.
North-Carolina-Virginia shelf, showing Albemarle and Virginia Beach shelf valleys (dashed). Dotted subsurface extension of the Albemarle Shelf Valley has been mapped by seismic profiler (Riggs and O'Connor, East Carolina University, personal communication). Dotted lines north of Virginia Beach Channel are subsurface channels traced by seismic profiler (Meisberger, 1972; Harrison et al., 1965). Depths in ft.

River. The Waccamaw presently flows toward the coast, then is deflected southward by a raised Pleistocene barrier. The shelf valley is aligned with a wind gap in the barrier near the area of the river's deflection. This shelf valley, like those to the north, has a thalweg at the same depth as the adjacent sea floor; it is defined not by an incised thalweg, but instead by two well-defined levees. The northern levee has transverse, coast-parallel ridges on its crest close to the shore and at the seaward end.

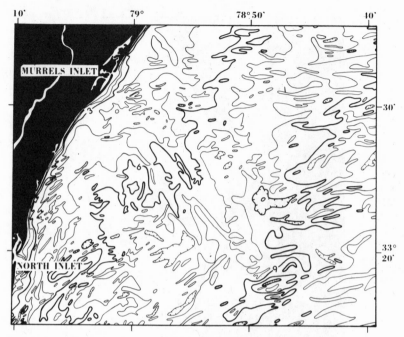

Figure 219.

Waccamaw shelf valley, Long Bay, Carolina shelf. Contour interval 1 fathom.

Shelf Valleys and Shoal Retreat Massifs

The transform of estuarine shoals into inner-shelf shoals may be most explicitly worked out for the short inner shelf channel segments. However, it also appears to explain a number of much larger-scale features. The Virginia Beach ridges extend from the area shown in Fig. 214 out to the mid-shelf shore (Fig. 218). The theory presented above would require that this entire shelf-transverse ridge, comprised of successive, nearly shore-parallel ridges, is a shoal-retreat massif left by the retreating James during a time when it occupied the Virginia Beach shelf valley.

The Susquehanna shelf valley to the north (Fig. 221a) has no such ridge in its northern flanks. Its inlet-associated shoal seems to have retreated directly up the valley, leaving it largely filled. Subdued transverse ridges interrupt the thalweg, suggesting continued post-transgression modification of the sea floor.

The Delaware shelf valley (Fig. 220b) consists of a well-defined channel paired with a well-defined shoal-retreat massif. The massif can be traced directly and without interruption into Delaware Bay-mouth shoal, as noted by Kraft (1971b). The channel can be traced directly into the seaward member of an ebb-flood channel couplet which breaches the bay-mouth shoal, and it is the

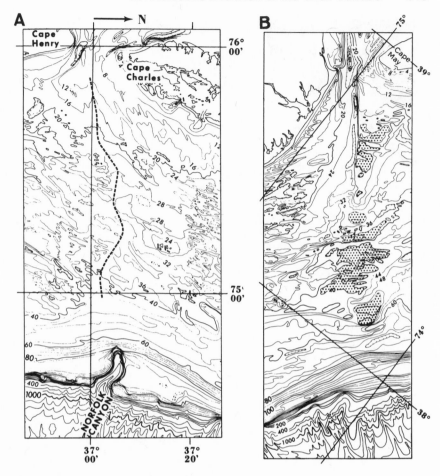

Figure 220.

Middle Atlantic bight shelf valleys. Contour interval 4 m. From Uchupi (1969). (A) Partially filled Susquehanna shelf valley. Thalweg (dashed line) locally interrupted by Post-transgression ridges or swales. (B) Delaware shelf valley. Paired estuarine scour channel and shoal retreat ridge. Isolated highs are stippled.

persistence through time of this tide-maintained channel on the retreating seaward face of the shoal that seems to have resulted in the sharply defined nature of the shelf valley (cf. Fig. 212).

The Great Egg shelf valley (Fig. 221, 222a) is the seaward extension of an insignificant modern river of that name. The headwaters of the Great Egg River are close to the right-angle bend in the Delaware, where it turns to parallel the fall line on its course toward Delaware Bay. It seems probable that at one time the Delaware River flowed without deviation across the Great Egg

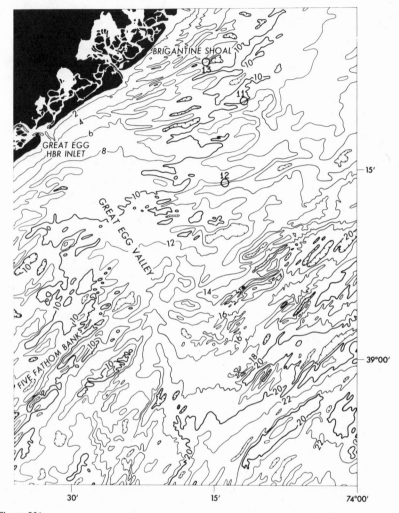

Figure 221.

Bathymetry of a portion of the New Jersey shelf. Contour interval 2 fathoms. From Stearns (1967). See source map, USCGS bathymetric map 0807N-55 for 1-fathom resolution. Numbered circles are submersible stations.

channel to the Baltimore canyon but since has been diverted by piracy (Widmer, 1964). This ancestral Delaware may have been joined by the Hudson along a fall-line valley since abandoned (Sanders, 1972, personal communication).

On the inner shelf, the Great Egg shelf valley (Figs. 221, 222a) exhibits an extreme form of southward asymmetry (Fig. 216); ridge crests to the north are little higher than the valley floor, while the northern trough axes are consider-

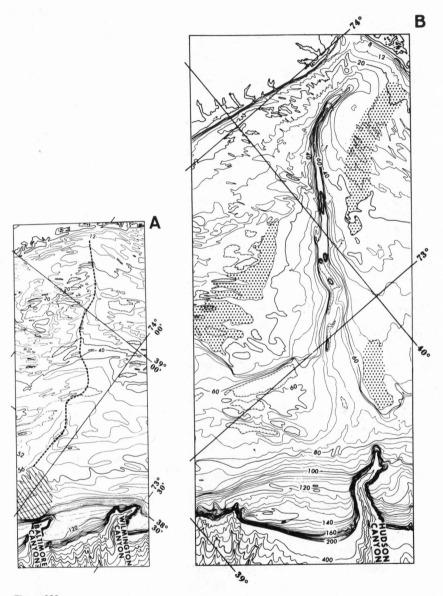

Figure 222.

Middle Atlantic bight shelf valleys. Contour interval 4 m. From Uchupi, 1969: (A) Great Egg shelf valley. Thalweg dashed; probable delta is diagonally ruled; (B) Hudson shelf valley. Highs are indicated by stippling.

ably below this level. To the south, trough axes continue at the same level as the valley floor, while the ridges rise above them. Thus, in profile, the inner valley is more nearly a complex sort of step than a true channel. This configuration is probably a result of valley fill and interfluve scour by south trending currents, after piracy of the Delaware and during subsequent development of the modern barrier shoreface under near stillstand conditions. A little further seaward, the northeastern side becomes a poorly developed shoal-retreat massif, out as far as the mid-shelf shoreline. Complex aspects of this massif will be described in a later section. Resolution deteriorates beyond this point, but the channel may be traced through a post-transgression ridge and swale topography to a deltalike bulge immediately above the Baltimore Canyon.

The highs on the southeast flank of the Hudson shelf valley (Fig. 222b) appear to be mainly of erosional origin and will be discussed in a later section. However, the more subdued northeast lip may be at least partly of shoal-retreat origin.

The shelf-transverse grain of the inner Gorgia coast (Fig. 223) is best interpreted as a primarily constructional shoal-retreat complex, formed by the retreat of closely spaced estuaries. Much of it is associated with the ancestral Altamaha drainage system (Pilkey and Giles, 1965).

Shelf-Transverse Elements II: Shoal-Retreat
Massifs from Cape-Associated Shoals

Cape-Associated Shoals and Shoal-Retreat Massifs

The frequency and dimensions of arcuate troughs on cape-associated shoals tend to increase toward their distal end, as though the water impounded on their north sides during northeasters can more easily incise spillways at points more distant from the point of sand input on the foreland apex, where such breaches are not readily repaired. Due east of outer Diamond shoals (Fig. 185) lies a zone of broken topography. Coast-parallel ridges up to 15 m high rest on a surface 15 m below the crest of Diamond shoals; the ridges terminate in a field of transverse sand waves similar to the field on top of outer Diamond shoals. This zone of broken topography appears to be an older, earlier, cape-associated shoal, which has been much dissected since formation, separated from the present Diamond shoals by a 10 m scarp.

The area between Lookout shoals proper and the shelf edge is occupied by a series of seaward-convex, arcuate ridges of much larger scale than those comprising the shoals proper (Fig. 186). Transverse asymmetry is reversed on this outer sector of Lookout shoals; the frontal scarps of the ridges curve to the east and north, causing this side of the outer shoal to be more sharply defined than the southwest. Possibly this is caused by the scour of bottom water flowing seaward around the shoal tip during storms.

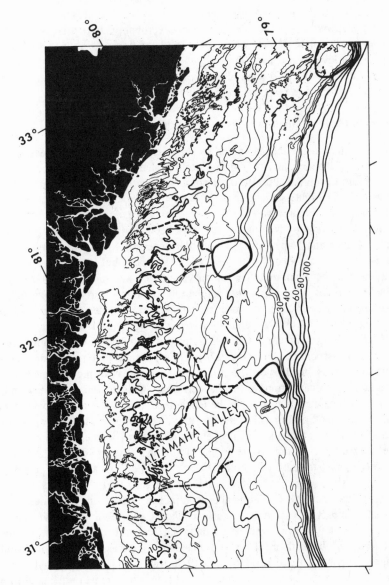

Figure 223.
Bathymetry of the Georgia shelf. From AAPG bathymetric maps of eastern continental margin.
Contours in fathoms. Circled areas are probable mid-shelf deltas.

The large-scale arcuate ridges appear to represent an older shoal sequence, dissected by cross-shoal erosion. The step from the outer to inner shoal may reflect a stepwise rise of sea level, or, more probably, two periods during a continuous rise in sea level, when shoal-top accretion was able to keep pace with the sea-level rise. Taken together, the sequence comprises a shoal-retreat massif with an actively growing inner terminus and a distant portion undergoing heavy modification in response to the outer-shelf hydraulic regime.

The South Carolina Shelf as a Shoal-Retreat Complex

Facts pertaining to the nature of cape-associated shoal-retreat massifs that have been gleaned from the North Carolina shelf can profitably be applied to the South Carolina shelf. Here the coast is transitional to the estuarine coast of the Georgia bight, which, in fact, begins at Cape Romain. The Carolina capes form a continuum in terms of river dominance, in which Cape Hatteras, during an early shelf-edge stage, may have been associated with the ancestral Pamlico River and Cape Lookout with the ancestral Neuse. Cape Fear (Fig. 210) is presently associated with the Cape Fear River, but is not a true cuspate delta, in that the river mouth is on the southern side and does not serve as a shielding point. Cape Romain, southernmost of the Carolina capes in the transitional zone, constitutes the near confluence of the Peedee and Santee Rivers and is as nearly a cuspate delta as a retrograding coastal landform can be. South of Cape Romain are a series of cuspate forelands of much smaller scale, each with its associated shoal (Fig. 224). The shoals are flatter and more poorly defined than those to the north, but the salient features are present; each shoal is marked by arcuate, transverse channels shallowing to the southwest and a scarp at the seaward margin. The better defined cape-associated shoals in the northwestern half of Fig. 224 are in fact shoal-retreat massifs composed of successive scarps and flats. The deeper, more seaward segments in some cases have seaward scarps that curve to the north and east, reversing the normal asymmetry, as is the case with outer Diamond and Lookout shoals. A prominent abandoned shoal segment on the northeast margin of Fig. 224 is apparently related to an ancestral Santee-Peedee confluence. However, like other cape-associated shoals, it is cut by arcuate channels and there is no evidence to indicate that it was ever a subaerial delta surface; true deltaic strata, if they exist, must lie beneath it. If so, the abandoned cape-associated shoal would comprise a special case of Scruton's (1960) destructional phase of delta building. Other, yet more poorly defined, shoal-retreat ridges may exist on the southwest end of Fig. 224, as suggested by the scalloped pattern of isobaths, with landward pointing cusps. However, no well-defined scarps are present to define individual segments; hence in this sector it may be more meaningful to talk in terms of a shoal-retreat sand sheet.

Figure 224.

Bathymetry of a portion of the South Carolina coast. Contour interval is 2 fathoms. Ruled areas are cape extension shoals, presently or formerly associated with cuspate forelands. Profiles were taken from original map contoured at 1 fathom. On letter map all separate ruled areas have, at least locally, seaward-facing scarps.

The mid-shelf delta of the ancestral Delaware (Fig. 220a) is traversed by arcuate, southward-shoaling troughs and, in its last phase, may have been a cape-associated shoal. After the establishment of the mid-shelf shoreline at approximately 40 m (132 ft) the coastal regime changed, such that the retreating bay mouth left not a cape shoal-retreat massif, but a scour trench paired with an inlet shoal-retreat massif.

Shelf-Transverse Elements III: Shelf Valleys and Cuestas

The Hudson shelf valley appears nearly gorgelike on maps (Fig. 222b). Although its slopes rarely exceed 10°, over 60 m (180 ft) of relief occurs along its upper portion. While the localized deeps within the Hudson shelf valley may also be relict estuarine scour trenches, something more is needed to account for the valley's remarkably deep incision. Veatch and Smith (1938) have suggested that during part of the late Wisconsinan ice retreat, the Hudson received the entire Great Lakes drainage and that the entrenchment of its shelf valley occurred then.

As a consequence of the Hudson's incisement, the flanking topography has become considerably dissected. Nearly coast-parallel highs of apparent erosional origin on the north New Jersey coast have been described in the previous paper. North-northeast trending highs also occur further seaward on the west flank of the Hudson shelf valley; the depression between them can be traced as far as the mid-shelf shore and will be referred to as the north New Jersey shelf valley (Fig. 225). The coastal and offshore highs are nearly parallel to Cholera Bank, a high on the east lip of the upper Hudson shelf valley which can be traced into the Long Island shore face. Locally there is direct evidence for an erosional origin for some of these highs. Divers have observed residual boulders at Shrewsbury Rocks on one of the north New Jersey coastal highs (personal communication, 1972, Charles Gibson, National Marine Fisheries Service, Sandy Hook Laboratory), and a submersible dive (senior author) to the Veatch and Smith scarp on the east flank of the north New Jersey shelf valley has revealed stiff, outcropping clay. It is more difficult to decide if these features are truly cuestas. The north to north-northeast trend of these highs is more northerly than the regional strike of outcropping strata on the subaerial coastal plain, but parallels isopach lines of the submarine Upper Cretaceous (Garrison, 1970, Fig. 7). Garrison's profiles suggest that at least the highs flanking the north New Jersey shelf valley are structurally controlled.

A northeast-trending high lies on the Long Island shelf between the Block and Long Island shelf valleys. A well-developed scarp on its seaward flank is a deeper northern member of the series of scarps collectively referred to in this report as the mid-shelf shore. The shielding action of this promontory during the drowning of the Long Island shelf valley was apparently responsible

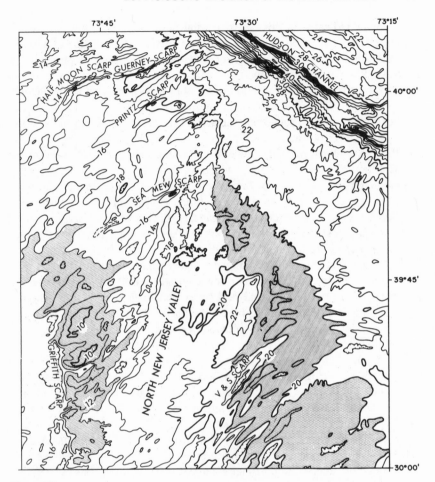

Figure 225.
South flank of the Hudson shelf valley. Contour interval is 2 fathoms. See source map, USCGS
bathymetric map 0807N-54, by Stearns (1967), for 1 fathom resolution. Cuestas bracketing north
New Jersey shelf valley are indicated by stipple.

for the amazing state of preservation of the valley's drainage net (Figs. 193,
194).

Shelf-Parallel Elements I: Ridge-
Capped Scarps as Overstepped Barriers

As noted in the previous paper, earlier theories of the origin of the Atlnatic
shelf ridge-and-swale topography saw it as a relict strand plain, with successive
ridges recording temporary stillstands of the Holocene transgression. However,

as previously indicated in this paper, coastal retreat of the Bruun type results in destruction of barrier deposits by shore-face erosion. Preservation of a barrier formed during a transgression requires that the barrier grow in place due to an abundant sand supply, so that its surface keeps pace with the rise of sea level. Eventually there is a reduction in the supply of river sand as the lagoons widen (Curray, 1964) or an increase in the rate of sea-level rise. These factors, or simply the increase in submarine surface area of the barrier requiring nourishment (Dillon, 1970), causes the barrier to starve and be overstepped by the sea. If the toe of the old shore face is below the effective wave base of the new sea floor, the overstep process will not completely destroy the barrier and some topographic expression of the feature should survive, although the result should be more nearly a scarp than a ridge. Gilbert (1890) has defined a morphological criterion for identifying barriers successively overstepped during a marine transgression; they should form a staircase sequence, with the crest of each barrier remnant lying at the altitude of the toe of the next barrier to the landward. Seaward flanks should be steeper than landward flanks, as is the case with active barriers. Intermittent coastal retreat by barrier overstep is a rather theoretical process and not a very actualistic one; no one has ever caught a barrier in the act of being overstepped, while coastal retreat of the Bruun type can be documented (previous section). Convincing examples of drowned barriers have been reported in the literature (Curray, 1961; McMaster and Garrison, 1966). However, the latter, a spit tied to a rock eminence on the southern New England shelf, is a special case. Such a rock-tied spit has literally nowhere to go as the shoreline retreats. Recently overstepped tombolos, rock-tied spits, and pocket beaches might indeed be sought on modern rugged coasts with a fully indurated (crystalline) substrate.

The successive scarps and terraces of the central and outer middle Atlantic shelf render this sector a reasonable place in which to seek overstepped barriers. One of the most interesting potential examples occurs on the New Jersey shelf in the area of complex, anastomosing inner-shelf ridges east of the Great Egg shelf valley (Figs. 221, 222a, 226), where it intersects scarps with toes at 40 and 46 m (131 and 150 ft). Two orders of ridges are apparent. Several large-scale ridges are 6 to 10 km (3 to 5 n mi) apart, with 10 to 20 m of relief. Their seaward flanks are generally steeper than landward flanks. Superimposed on these are ridges of smaller scale, with a spacing of 0.5 to 2.6 km (0.25 to 1.3 n mi) and a relief of 6 m (20 ft). Smaller-scale ridges are most abundant on the gentle northwest flanks of the large-scale features, where they climb toward the main crest line and tend to converge with it. Large-scale lows tend to be compartmentalized into en echelon small-scale lows, with up to 6 m (20 ft) of closure, that trend a few degrees more northerly than the main features.

It is possible to surmise that the large-scale features are barriers. Their widths of up to several miles are more comparable to the widths of raised Pleistocene

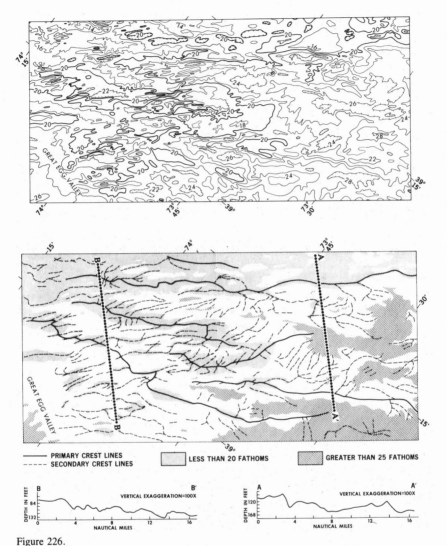

Figure 226.

Area of complex inner shelf topography off Atlantic City, New Jersey. Above: Map with 1 fathom interval. Below: Analysis. Possible overstepped barriers are outlined by 20- and 25-fathom contours. See USCGS bathymetric map 0807N-55, for 1 fathom resolution.

barriers on land (Colquhoun, 1969) and their steeper seaward faces, with 10 m (33 ft) or more of relief, are comparable to the modern shoreface. One such ridge sits squarely athwart the Little Egg shelf valley (Fig. 221); its resemblance to an estuary-mouth spit is diminished, however, by its reverse asymmetry; its landward face is steeper.

The two large-scale ridges crossed by profile AA' satisfy the stairstep criterion. However, the sequence of three large-scale ridges crossed by profile BB' does not, and simple intermittent transgression does not suffice to explain them. Perhaps sufficient sediment was available here in the vicinity of the Great Egg River mouth, so that depositional regression occurred during the 40 m stillstand of the mid-shelf shore and a northward-branching sequence of spits developed, similar to the Sandbridge spit sequence of the early Wisconsinan of Virginia (Oaks, 1964; Swift et al., 1971, Fig. 3); or possibly the barriers of the 40 m Holocene stillstand were localized by a preexisting Pleistocene shoreline, much as the modern Georgia barriers were localized by the Pleistocene Silver Bluff barriers when the advancing modern shoreline impinged on it (Hoyt and Hails, 1967). At any rate, the complex of barrierlike forms here on the flat central portion of the shelf between 30 and 60 m depth is compatible with Emery's (1967; 1968, Fig. 15) interpretation of shelf history. Emery suggests that, despite the low gradient, the advancing Holocene shoreline transversed this portion of the continental shelf slowly as a consequence of the low early Holocene rate of sea-level rise, and that, as a result, a great deal of river-transported sand was deposited as barriers on the central shelf. However, a more conservative explanation might call upon Emery's hypothesis simply as a reason for the presence of an unusually thick layer of loose sand over the older Pleistocene substrate and might interpret both large- and small-scale ridges to a response of the layer to the modern hydraulic regime.

The complex superimposition of small-scale ridges and swales at small angles to the large-scale form seems to have involved both the incision of the large-scale forms and the construction of attached tails and streamers of sand. Submersible dives in this sector reveal a distribution of grain sizes similar to that observed by the authors on the Virginia shelf (Swift et al, 1971). Troughs are floored by fine-grained sand; locally, a basal layer of coarse, pebbly sand over a stiff clay substrate is exposed. Fine-grained flank sands coarsen up the sides of ridges to a well-sorted, medium-grained sand on crests (Figs. 227, 228). As previously concluded by Donohue and others (1966), this topography has been heavily modified by the post-transgression hydraulic regime. It appears to be presently maintained by strong, intermittent currents which scour out troughs and fair-weather wave surge which winnows crests.

Frank (1971) has performed textural analyses on samples from this sector and concludes that relict littoral and nearshore deposits are present. His findings are compatible with our interpretation. However, the result of grain-size analyses must be used with caution, since it is not always possible to arrive at a unique solution. The hydraulic regime associated with the ridges very probably fractionates sand into types resembling those produced by the littoral hydraulic regime (Fig. 186b, preceding chapter).

Scarps of apparent littoral origin (Fig. 193) and large-scale ridges with spacings of barrier scale are common on the outer shelf, seaward of the 40 m (130 ft) isobath. The presence of orthogonal tributaries in the intervening lows, reminiscent of an incomplete dendritic drainage pattern (Figs. 194, 195), suggests a barrier topography with a superimposed subaerial drainage. However, the stairstep criterion is only locally satisfied and most profiles suggest that if these are barrier sequences, much post-transgression modification has occurred. Profiling of the North Carolina shelf (Shideler and Swift, 1972) indicates that the flanks of some large outer-shelf ridges truncate horizontal internal reflectors, indicating an erosional origin.

Within the middle Atlantic bight, the more northerly scarps tend to be found at successively deeper levels (Fig. 229). The reason for this is not clear. Veatch and Smith (1939) considered incomplete recovery of the northern sector from glacial loading, but believed that it was not entirely an adequate explanation. Fairbridge and Neuman (1967) suggest that geosynclinal subsidence may play a role. The mid-shelf shore appears to be a Holocene feature in that the shoal-retreat ridges and shelf valleys that extend to it are fresh and undissected, and oyster shells from its vicinity yield mid-Holocene dates (Merrill et al., 1965). If so, the tilting must have been fairly recent, or else the collection of scarps designated by this name is not a synchronous group.

Shelf-Parallel Elements II: Tributary Shelf Valleys and the Ridge-and-Swale Topography

Shelf-parallel notches are incised into the south flank of the Block shelf valley, into the south flank of the Hudson shelf valley, and into the cuestas on both sides of the north New Jersey shelf valley (Fig. 193, 225). These notches are separated by flat-topped divides and appear to be products of subaerial fluvial erosion. Notches tributary to the Hudson and northern New Jersey shelf valleys tend to have steeper northwest, seaward-facing sides. Many of these are capped by low ridges parallel to the scarps, with highest points immediately west of the steepest part of the scarp (Half Moon, Guerney, Printz, Sea Mew, and Veatch and Smith scarps, Fig. 225). It seems reasonable to interpret these seaward-facing scarps as wave-cut scarps and the ridges as remnants of barriers. However, the close correlation of ridge peak with steepest portion of scarp in Fig. 225 may indicate modification by modern currents sweeping sand west along the notches and diagonally up the north sides of the scarps. If so, the sand ridges are more nearly modern submarine levees than relict subaerial barriers. A dive by the senior author in the Perry submersible PC-8 to the Veatch and Smith scarp on the west flank of the north New Jersey shelf valley (Fig. 225) revealed bare gray clay substrate thinly veneered with coarse, rippled sand and

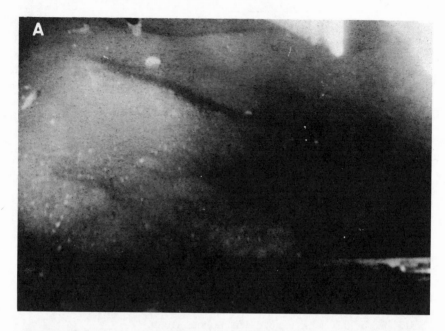

Figure 227.
Small-scale bedforms of the New Jersey shelf, photographed from the Perry submersible PC-8, on October 6 and 7, 1971. (A) Ripples, 15 cm high, in coarse sand of swale. Ripples are mantled with clay film. Station 11, Fig. 216. (B) Asymmetric ripples in medium sand on ridge crest, formed by a combination of wave surge and unidirectional flow, probably during Hurricane Ginger, 1971. Steep faces towards observer. Crests are armored with sand dollars. Station 11, Fig. 216. (C) Wave ripples in medium sand on ridge crest, 14 cm high. Small-scale current ripples at right angles are responses to tidal current.

clay pebbles from 46 to 38 m (150-125 ft), then a cap of sand, grading from fine sand at its base to medium sand at the crest at 30 m (100 ft). A dive several hours earlier to the nearby Sea Mew scarp had revealed active lingoid ripples of fine sand moving directly down the scarp in response to the rotary tidal current, then at its maximum value. But despite such intermittent movement of sand into the valleys, the sand caps have not been flattened. The valleys have not been filled, but merely veneered by coarse lag. Some sort of intermittent current has apparently been sweeping them out through the millenia since transgression.

On the Long Island shelf, valleys tributary to the Block shelf valley merge toward the west with the very regular ridge-and-swale topography of the Long Island inner shelf (Fig. 194, 230). Garrison and McMaster (1966) and McKinney

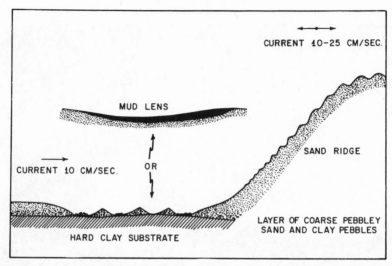

Figure 228.

Schematic model of an inner New Jersey shelf swale and adjacent ridge, based on submersible observations at stations 11-13, Fig. 216.

and Friedman (1970) have consequently deduced a fluvial origin for the topography, attributing the trend of the swales diagonally across the regional contours to subsequent tilting. However, the only subaerial "drainage" pattern that resembles this pattern of parallel ridges and troughs, with tuning-fork bifurcations (see Fig. 194) and numerous enclosed depressions (blowouts), is that of a seif dune desert (Folk, 1971a, b; Glennie, 1971). All other subaerial drainage patterns with dominantly parallel lows that were examined had at least occasional orthogonal (obsequent and resequent) first-order streams, forming an incipient trellis pattern. The inner shelf topography differs, however, from a seif dune topography in that ridge spacing is greater and side slopes are less.

Thus, the ridge-and-swale topography of the Long Island inner shelf comprises a sort of optical illusion, with a western margin of apparent relict fluvial channels, an ambiguous central sector, and an eastern margin where ridges, merging with the shoreface, are clearly of modern hydraulic origin. This surface may well have been initiated as a braided outwash plain draining eastward from the Long Island ice tongue toward the Block shelf valley. If so, its channels would have become incised during glacial rebound. We conclude that this sand deposit has been planed off by shore-face retreat and its drainage pattern partially erased, then subjected to an inner shelf hydraulic regime and partially rebuilt by it into a seiflike topography.

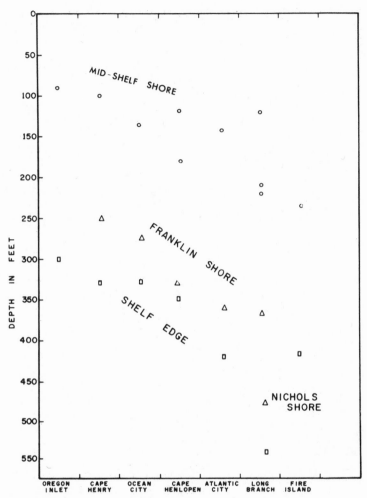

Figure 229.

Correlation of terraces, middle Atlantic bight.

SOUTHWARD SEDIMENT TRANSPORT
ON THE CENTRAL AND SOUTHERN ATLANTIC SHELF

Bedform Asymmetry and Southward Transport

Throughout the preceding discussion, much evidence has been presented which suggests that sediment is moving southward along the middle Atlantic bight.

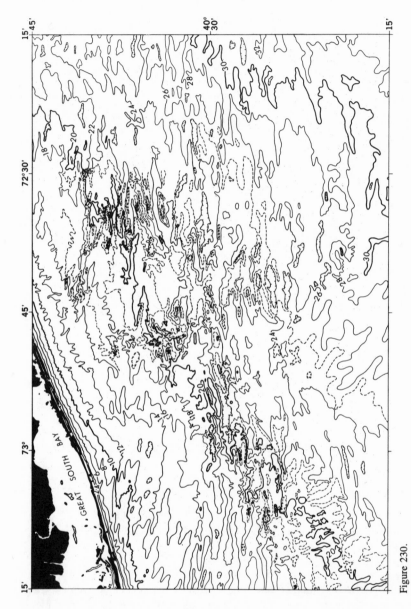

Figure 230.

Bathymetry of a portion of the Long Island shelf. Contour interval is 2 fathoms, with supplementary dashed contours at 1 fathom. From USCGS bathymetric map 0808N-54 by Stearns, 1967. See original for 1-fathom resolution.

This section will consider the evidence systematically and will introduce new information in support of the concept.

Evidence on the northern portion of the middle Atlantic bight is tenuous. To the extent that the inner shelf ridges, with their extreme parallelism and numerous enclosed lows, are seiflike bedforms, southward transport is occurring, for the ridges gain in relief and thalwegs climb and branch in this direction, as seif troughs do in the direction of transport. Evidence is stronger on the inner shelf. The ridges of the Fire Island shore face tend to be asymmetrical, with steeper faces to the southeast, as do the shoreface ridge systems of the New Jersey shelf (Figs. 179, 191, previous paper; Figs. 221, 227). In the North Sea, sand ridges of similar geometry are interpreted as large-scale current-parallel bedforms built by helical flow (Houbolt, 1968; see previous chapter). Transverse asymmetry of the sort noted above is taken as evidence that oblique, crest-converging currents on one side are stronger than on the other; and that the ridge, as well as extending longitudinally, is also shifting laterally. Thus, the asymmetry of inner Atlantic shelf ridges is compatible with the concept of southward bed-load transport. This asymmetry is locally reversed where large ridges join the shoreface in shallow water. Judging by the analogy of the inner False Cape ridge (previous chapter), intense scour in the heads of the troughs behind the ridges sweeps fine, shoreface sand out over the ridge base. The sand is fine enough to travel in suspension for considerable distances in the ridge-base rip current, with the result that the seaward flank of the ridge, where it joins the shore face, is aggraded to the point that asymmetry is reversed.

In the ridges of the inner shelf off Atlantic City (Fig. 221), New Jersey, longitudinal asymmetry becomes apparent; crestlines of isolated ridges climb gently to the southwest, then drop abruptly. Transverse and longitudinal asymmetry is well developed in the Fenwick ridge field of the Delaware coast (Fig. 188, previous chapter). A time series of 93 years on Chincoteague Shoals (Fig. 178, previous chapter) shows that the inner-shelf ridges are growing southward. On the Virginia-North Carolina shelf crest and trough lines of the ridge systems climb to the south. The regional distribution of grain sizes suggest that fine sand has been swept out of the troughs of the Virginia Beach, False Cape, and Platt shoals ridge systems and is being deposited in low areas south of these systems (Swift and others, 1972; Shideler and others, 1972). On the Virginia-North Carolina shelf a new element appears in the form of small current-transverse bedforms that are associated with the current-parallel sand ridges. South-facing sand waves with up to a meter of relief have been observed during summer scuba dives in the Virginia Beach ridge system. Clay lenses in troughs 10 cm (4 in) thick suggest that sand waves are relict from winter storms (Holliday et al., in press). Larger south-facing transverse forms with heights up to 3 m have been observed (Fig. 231), suggesting that the wavey appearance of the flanks of the Virginia Beach ridges in Fig. 214 is only partly a mapping artifact.

At Platt shoals (Fig. 215) and on outer Diamond shoals (Fig. 207) the crests tend to wrap around to the west at the ends of the ridges, swinging into parallelism with the small-scale, obliquely or transversely oriented, southward asymmetrical ridges that overlie the large-scale ridges. Tidal currents are too weak in these areas to generate these features. Presumably they are activated only by storm currents. Profiles (Fig. 231) indicate that their steeper southward slopes were not at the angle of respose during the summer of mapping.

Clear-cut evidence for sand transport in the form of sand waves and bathymetric changes occurs predominantly within the 25 m (78 ft) isobath. However, the flamboyant ridge-and-swale topography on the central New Jersey shelf east of the Great Egg shelf valley suggests that where a sufficient thickness of loose sand occurs, it will become current-molded out to depths of 40 m (125 ft). One of the few careful studies of outer shelf microtopography (Stanley and Kelling, 1968; Stanley and others, 1972) suggests that, at least in the vicinity of the Wilmington Submarine canyon, shelf-edge sediment transport is toward the southwest.

Thus, within the middle Atlantic bight, sediment transport becomes more intense with increasing proximity to shore and with increasing proximity to the southern end of the bight. It has been suggested in the previous paper that the dominant hydraulic element molding the middle Atlantic shelf floor are the southerly wind-drift currents generated by winter storms. It seems reasonable that their velocities and bottom shear stresses would become more intense

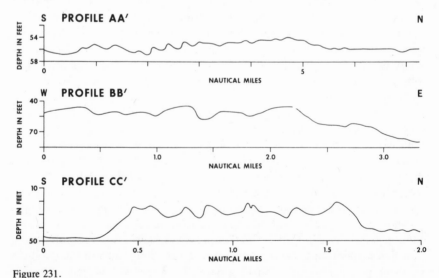

Figure 231.

Representative profiles of sand waves on the Virginia-North Carolina shelf. See Figs 204, 205, and 215 for locations. Top: Virginia Beach; middle: Platt shoals: bottom: Diamond Shoals.

as they moved toward shore and southward up the shoaling and narrowing shelf floor toward Cape Hatteras. The pattern of southward asymmetry of morphologic elements reflecting southward sediment transport continues through the series of shoal-retreat massifs which is associated with the Carolina capes. On the Georgia bight, arcuate, inlet-associated shoals are higher on the northern side and tend to have well developed ramps and chutes (Fig. 206). Southern and seaward margins are oversteepened. The channels within the shoals curve to the south, suggesting diversion by a south-trending coastal current. Southward asymmetry is clearly apparent at Cape Kennedy (Fig. 211).

Other papers in this volume indicate that this pattern of coastwise sediment transport is a common one, as indeed it must be, for reasons of hydraulic continuity. Coastwise advective transport of sediment is compatible with seaward diffusive transport; however, we are far from a point where our quantitative knowledge of the sediment budget for this coast will permit resolution of these components.

Bedform Hierarchies and the Ridge and Swale Topography

Allen (1968) has noted that several orders of bedforms tend to occur simultaneously and that such hierarchies of bedforms

arise because the quantities that determine flow are sufficiently numerous that several mutually unstable combinations can exist, each combination being expressed in terms of a bedform of a characteristic physical scale and orientation relative to flow.

Such a hierarchy appears to exist on the Atlantic shelf. Here coexisting forms include current, wave and combined flow ripples (small scale current-transverse bedforms whose scale is independant of water depth), sandwaves (intermediate scale, current-transverse bedforms with scale limited by flow depth), and sand ridges (large scale, current-parallel bedforms with scale limited by flow depth). The assemblage appears to result from Allen's *Condition 1*, in which stream power has a small or moderate value, such that the Froude number is substantially less than unity, and the sand bed is deep and continuous. Conditions on the Atlantic shelf, however, are more complex than this. For one thing, the major component of the flow field consists in most cases of storm-generated currents with durations of several days. Small-scale wave, current, and combined flow ripples (Fig. 227b) are obliterated between pulses by bioturbation or wave surge. Intermediate and large scale bedforms (sand waves and ridges) must represent time-integrated responses to these pulses, in which spacing, heights and side slopes are less than maximum, reflecting a compromise between storm aggradation and fair weather degradation. Second, the large-scale forms extend up into the wave-agitated zone. This, rather than the undirectional flow component limits their height, and to a certain extent controls their geometry.

Finally, the hydraulic regime itself is not stationary, but changes through the lives of the larger bedforms as a consequence of water deeping attendant on the Holocene transgression. Inner-shelf ridges are created entirely within the wave-agitated zone, as a consequence of the complex interaction of unidirectional storm currents and wave surge with each other and with the substrate (previous chapter). With detachment, they find themselves in a simpler inner shelf regime, for which 10 m appears to be the wave-limited crestal depth. Shelf ridge-fields with crests accordant at deeper depths may have retained this limit from shallower sea level stands, even though their troughs are being actively scoured.

The very different geometries of the ridge systems of cape-associated, inlet-associated, and inner shelf ridge fields tend to disguise their fundamental genetic relationship. All are large-scale, current-parallel bedforms whose spacing increases with flow depth (Fig. 232). Allen (1968) suggests that this geometry is a basic response of a mobile substrate to a flow regime with petubations transverse to the mean current direction. The parturbutions take the form of helical flow cells, in which bottom currents diverge over trough axes, and converge on crest axes. Thus the ridge and swale topography appears to be a stable end-configuration toward which a variety of erosional and depositional topographies converge.

Nature of Sediment Flux through the Ridge and Swale Topography

It is interesting to contrast the geomorphic evidence for southward sediment transport with textural evidence for sediment movement. In general, the patchy nature of sediments on the Atlantic continental shelf have led observers to assume

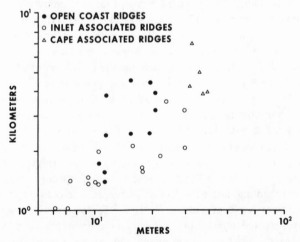

Figure 232.
Spacing against flow depth, ridge and swale topography of the central and southern Atlantic shelf.

that the surficial materials are "relict," that is, autochthonous with respect to the present sedimentary cycle. Milliman and others (1968, 1972) note that "the sharpness of assemblage boundaries and the patchy areal distribution of various carbonate parameters, both within and between (sectors of the Carolina shelf) indicate that transportation and redistribution by bottom currents has been local."

The apparent conflict of geomorphic and petrographic evidence can perhaps be resolved by the following considerations. The transgressive sand sheet of the Atlantic shelf has been molded into linear sand bodies. When careful attention is paid to the mechanics of sediment transport over their surfaces, most shallow marine sand bodies are resolved as the products of sand circulation cells, closed loops in sand transport paths (Houbolt, 1968; Ludwick, 1970, 1972; Oertel and Howard, 1972). This is certainly the case for the tide-maintained ridges of Chesapeake and Delaware Bay mouths (Ludwick, 1970). It may be true for some of the shore-face systems and may apply, or have applied at one time, to the shelf ridges. In such circulation cells the residence time of a sand grain in a loop is generally long with respect to the duration of the total excursion along the transport path. A grain trapped in such a cell stands a very high chance of being accepted for permanent deposition in the growing sand body, whose surface comprises the circulating sand cell. Thus, if the southerly transport paths of the middle Atlantic shelf have many such loops, then most bottom-sediment transport may consist merely of a sand grain migrating obliquely southward to the nearest ridge and being buried there, and long-distance transport may not be significant. However, it is worth noting that characteristic grain trajectories in sand circulation cells may vary greatly with grain size, and it is possible that long-distance transport may be occurring on the Atlantic shelf for fine- and very-fine sand classes. The observation has been made in the preceding paper that the shoreface before the distal, concave-seaward barrier arcs of the New Jersey and Delmarva coastal compartments are made of fine-grained sand. Much of this material may have been winnowed out of the ridge systems to the north. A similar coastwise segregation appears to be occurring between Cape Henry and Cape Hatteras on the Virginia-North Carolina coast (Swift and others, 1971a). This lateral fractionation by sediment size class has led to the statement that the Atlantic shelf is approaching textural equilibrium (Milliman and others, 1972) or that it is undergoing *in situ* or autochthonous grading (Swift, 1970).

EVOLUTION OF THE SHELF SURFACE

Analysis of existing bathymetric and geological information reveals a systematic distribution of morphologic elements on the central and southern Atlantic shelf and indicates the course of events which formed them. Bruun coastal retreat, whereby a rise in sea level results in shore-face erosion and concomitant aggradation of the adjacent sea floor, serves as a unifying concept by which a variety of constructional shelf-floor features may be traced back to an origin at the foot of the retreating shore face.

The simplest case appears to be the development of shelf-floor ridge fields by detachment during shoreface retreat as exemplified by the Delmarva ridge field, described in the previous paper. Two other major evolutionary sequences are those observed in inlet-associated, shoal-retreat massifs and cape-associated, shoal-retreat massifs. In these cases littoral drift plays a significant role, transporting sand to depositional centers which retreat through time with the shoreline.

The prominent ridge and swale topography of the central and southern Atlantic shelf appears to be a stable end-configuration toward which a variety of depositional topographies tend to converge. It may form directly at the leading edge of the residual Holocene sand sheet by the detachment of shoreface-connected ridges. Inlet-associated shoals may be relatively featureless initially or may bear an intricate ridge structure controlled by their tide-dominated hydraulic regimes. However, on abandonment by the shoreline, old ridges are rotated to conform to the coast-parallel storm currents and new ones are incised. Helical flow in these channels causes ridge crests to build to heights where aggradation is balanced by wave erosion, hence the process is constructional as well as erosional. Cape-associated shoals are progressively segmented into arcuate ridges separated by deep spillways. The fluvial topographies on the western margins of the Block and Hudson shelf valleys have been modified by southwest-flowing currents to fit the regional pattern; southwest-trending lows have been kept open by scour and locally deepened, while lows at other angles have been smoothed or filled. Southwest-trending highs have developed sand caps.

While shelf-floor constructional elements are relict in the sense of having survived from an earlier, different environment, the term is not an entirely satisfactory descriptor. They have responded to the changes in the hydraulic regime that built them by corresponding changes in their morphology. Hence they fit the basic definition of an equilibrium system; one that, when stressed, responds in such a way as to relieve the stress. If we are to gain further insight into the nature of shelf-floor sand bodies, we must undertake to monitor this sequence of hydraulic process and substrate response along lines indicated by earlier chapters in this volume.

Thus, the central and southern Atlantic shelf is a palimpsest or overprinted surface whose two major patterns are both submarine. The earlier of the two was formed at the foot of the shore face. Even the major shelf valleys may be viewed as tracks left by retreating estuary mouths. Bruun coastal retreat seems to have planed off the subaerial surface and destroyed all but the most resistant or deeply incised landforms. Where exceptions occur and subaerial features survive, the probable paleography includes promontories to shield the surviving features from wave attack, as in the case of the Long Island shelf valley.

Textural studies likewise stress the nearshore zone as the penultimate environment of modern shelf sands. Pilkey et al. (1969), after examining the carbonate fraction of Georgia bight sands, have concluded that "much of the present-day shelf sediment cover was deposited in the nearshore zone of the last transgressing sea, but the sediment was not derived from the beach environment itself, as indicated by particle roundness considerations." McKinney and Friedman (1970) note that, on the Long Island shelf, various sediment types can be recognized by grain-size criteria "mostly [relict] from the shallow . . . nearshore environments. Sediments relict [from] the beach swash zone environment, however, do not appear to be common. They apparently are destroyed by the advancing surf zone.

The stratigraphic consequence of this evolutionary course consists of a thin, discontinuous sand sheet of residual provenance, which might be said to result from "the destructional phase of shelf building." The sheet rests with marked disconformity on older Quaternary regressive deposits or, locally, remnants of Holocene lagoonal deposits (Kraft, 1971b; Swift and others, 1972; Shideler and Swift, 1972). The Holocene sand sheet is generally coarser, partly because its fines have been winnowed from it and partly because it has incorporated into its base Holocene fluvial sands which have passed through a resting stage in head-of-estuary sand bodies before emerging through the shoreface.

ACKNOWLEDGMENTS

This study was undertaken as a preliminary step in the NOAA-AOML Continental Margin Sedimentation project (COMSED). It incorporates data accumulated at Old Dominion University with the support of National Science Foundation Grants GA-13837 and GA-27305, and Coastal Engineering Research Center Contract DACW-72-69-C-0016. Some of the data was collected during Cruise E-21-71 of the Duke R.V. Eastward. The cooperative oceanographic program of Duke Marine Laboratory is supported by National Science Foundation Grant GB-17545. Some of the bathymetry presented in this paper was originally prepared by the late G. F. Jordan. The photographs of Fig. 227 and the data of Fig. 228 was collected from the Perry Submersible PC-8 under charter to the Manned Undersea Science and Technology Office of NOAA.

REFERENCES

Allen, J. R. L. (1968). The nature and origin of bedform hierarchies. *Sedimentology* **10**, 161-182.
Allen, P. (1964). Sedimentological models. *J. Sed. Petrol.* **34**, 289-293.
Anonymous (1964). United States Coastal Pilot. Atlantic Coast, Cape Henry to Key West, Vol. 4, 7th Ed., Natl. Ocean Surv., National Oceanic and Atmospheric Admin., Rockville, Maryland.

Bruun, P. (1962). Sea level rise as a cause of shore erosion. *J. Waterways and Harbors Div.*, *Am. Soc. Civ. Eng. Proc.* **88**, 117-130.

Bruun, P., and Gerritsen, F. (1960). "Stability of Coastal Inlets." 123 pp., North Holland Pub., Amsterdam.

Caldwell, J. M. (1966). Coastal processes and beach erosion. *J. Soc. Civ. Eng.* **53**, 142-157.

Coch, N. K. (1963). Post-Miocene Stratigraphy and Morphology, Inner Coastal Plain, Southeastern Virginia. *U.S. Office of Naval Research Tech. Rept.* **6**, 97 pp.

Colquhoun, D. J. (1969). Geomorphology of the Lower Coastal Plain of South Carolina. *Div. Geol. State Devel. Board*, 36 pp.

Curray, J. R. (1964). Transgressions and regressions. *In* "Papers in Marine Geology" (R. L. Miller, ed.) pp. 175-203. MacMillan, New York.

Curray, J. R. (1969). Shore zone sand bodies: barriers, cheniers, and beach ridges. *In* "The New Concepts of Continental Margin Sedimentation" (D. J. Stanley, ed.), pp. JC-2-1 to JC-2-19. Am. Geol. Inst., Washington, D.C.

Dillon, W. P. (1970). Submergence effects on a Rhode Island barrier and lagoon and inferences on migration of barriers. *J. Geol.* **78**, 94-106.

Dolan, R. (1970). Dune reddening along the outer banks of North Carolina. *J. Sed. Petrol.* **40**, 765.

Dolan, R., Hayden, B., Hornberger, G., Zeiman, G., and Vincent, M. (1972). "Classification of the Coastal Environments of the World, Part 1: The Americas." Washington, Office of Naval Research, 163 pp.

Donohue, J. G., Allen, R. C., and Heezen, B. C. (1966). Sediment size distribution profile on the continental shelf off New Jersey. *Sedimentology* **7**, 155-159.

El Ashry, M. T., and Wanless, H. R. (1968). Photo interpretation of shoreline changes between capes Hatteras and Fear (North Carolina). *Marine Geol.*, **6**, 347-379.

Emery, K. O. (1965). Geology of the continental margin off eastern United States. *In* "Submarine Geology and Geophycics" (W. F. Whittard and R. Bradshaw, eds.), p. 1-20. Butterworths, London.

Emery, K. O. (1967). Estuaries and lagoons in relationship to continental shelves. *In* "Estuaries" (G. H. Lauff, ed.), Washington Am. Assoc. Adv. Sci., 757 pp.

Emery, K. O. (1968). Relict sediments on continental shelves of world. *Am. Assoc. Petrol. Geol. Bull.* **52**, 445-464.

Ewing, J., Lepichon, X., and Ewing, M. (1963). Upper stratification of the Hudson Apron. *J. Geophys. Res.* **68**, 6303-6316.

Fairbridge, R. W., and Newman, W. S. (1967). Post-glacial crustal subsidence of the New York area. *Zeitschrift fur Geomorph.* **12**, 296-317.

Fenneman, N. M. (1902). Development of the profile of equilibrium of the subaqueous shore terrace. *J. Geol.* **10**, 31-38.

Fischer, A. G. (1961). Stratigraphic record of transgressing seas in the light of sedimentation on Atlantic coast of New Jersey. *Am. Assoc. Petrol. Geol. Bull.* **45**, 1656-1666.

Fisher, J. J. (1967). "Development Pattern of Relict Beach Ridges, Outer Banks Barrier Chain, North Carolina." 250 pp. Ph.D. thesis. Univ. North Carolina (unpublished).

Folk, R. L. (1971a). Genesis of longitudinal and Oghurd dunes elucidated by rolling upon grease. *Geol. Soc. Am. Bull.* **82**, 3461-3468.

Folk, R. L. (1971b). Longitudinal dunes of the northwestern edge of Simpson Desert, Northern Territory, Australia. Pt. 1: Geomorphology and grain size relationships. *Sedimentology* **16**, 5-54.

Frank, W. A., and Friedman, G. (1971). Barrier Island and migration: new evidence from New Jersey. *In* "Abstract Volume, Second National Coastal and Shallow Water Conference." 327 pp. University Press, Univ. Southern Calif., Los Angeles.

Frank. W. M. (1971). Continental-Shelf Sediments off New Jersey. Ph.D. thesis, 84 pp. Rensselaer Polytech. Inst. (unpublished).

Garrison, L. E. (1970). Development of the continental shelf south of New England. *Am. Assoc. Petroleum Geologists Bull.* **54**, 109-124.

Garrison, L. E., and McMaster, R. L. (1966). Sediments and geomorphology of the continental shelf off southern New England. *Marine Geol.* **4**, 273-289.

Gilbert, G. K. (1890). Lake Bonneville. *U.S. Geol. Surv. Mono.* **1**, 438 pp.

Glennie, K. W. (1971). "Desert Sedimentary Environments." 222 pp. Amsterdam, Elsevier.

Gulliver, F. (1899). Shoreline topography. *Am. Acad. Arts and Sciences Proc.* **34**, 151-258.

Hails, J. R., and Hoyt, J. H. (1969). An appraisal of the evolution of the lower Atlantic coastal plain of Georgia, U.S.A. *Trans. Papers Inst. Brit. Geog.* **46**, 53-68.

Harris, R. L. (1954). Restudy of Test-Shore Nourishment By Offshore Deposition of Sand, Long Branch, New Jersey. Beach Erosion Board Tech. Mem. **62**, 1-18.

Harrison, W., Malloy, R. J., Rusnak, G. A., and Terasmae, J. (1965). Possible late Pleistocene uplift, Chesapeake Bay entrance. *J. Geol.* **73**, 201-229.

Harrison, W., Norcross, J. J., Pore, N. A., and Stanley, E. M. (1967). Shelf waters off the Chesapeake Bight. *Environ. Sci. Services Administration Prof. Paper* **3**, 1-82.

Henry, V. J., Jr., and Hoyt, J. H. (1968). Quaternary paralic shelf sediments of Georgia. *Southeastern Geol.* **9**, 195-214.

Henry, V. J., Jr. (1971). Origin of capes and shoals along the southeastern coast of the United States: reply. *Geol. Soc. Am. Bull.* **82**, 3541-3542.

Hopkins, E. M. (1971). Origin of capes and shoals along the southeastern coast of the United States: discussion. *Geol. Soc. Am. Bull.* **82**, 3537-3540.

Houbolt, J. J. H. C. (1968). Recent sediments in the southern bight of the North Sea. *Geol. en Mijnbouw* **47**, 245-273.

Hoyt, J. H., and Hails, T. R. (1967). Deposition and modification of Pleistocene shoreline sediments in coastal Georgia. *Science,* 1541-1543.

Hoyt, J. H. (1967). Barrier Island Formation. *Geol. Soc. America Bull.* **78**, 1129-1136.

Hoyt, J. H., and Henry, V. J., Jr. (1965). Significance of inlet sedimentation in the recognition of ancient barrier islands. pp. 190-194. *In* Wyoming Geol. Assoc. 19th Field Conf.

Hoyt, J. H., and Henry, V. J., (1967). Influence of island migration on barrier-island sedimentation. *Geol.Soc. Am. Bull.* **76**, 77-86.

Hoyt, J. H., and Henry. V. J. (1971). Origin of the capes and shoals along the southeastern coast of the United States. *Geol. Soc. Am. Bull* **82**, 59-66.

Ingle, J. C. (1966). "The Movement of Beach Sand." 221 pp. Elsevier, Amsterdam.

Johnson, D. (1919) (1938, 2nd Ed.). "Shore Processes and Shoreline Development" Wiley, New York.

Knott, S. T., and Hopkins, H. (1968). Evidence of Pleistocene events in the structure of the continental shelf off the southeastern United States. *Marine Geol.* **6**, 5-26.

Kraft, J. C. (1971a). "A Guide to the Geology of Coastal Delaware." 220 pp. College of Marine Studies, Univ. Deleware, Newark.

Kraft, J. C. (1971b). Sedimentary facies patterns and geologic history of a Holocene marine transgression. *Geol. Soc. Am. Bull.* **82**, 2131-2158.

Langfelder, J., Stafford, D., and Amein, M. (1968). "A Reconnaissance of Coastal Erosion in North Carolina." 127 pp. Dept. Civ. Eng., North Carolina State Univ., Raleigh.

Leopold, L. B., Wolman, M. G., and Miller, J. P. (1964). "Fluvial Processes in Geomorphology." 522 pp. Freeman, San Francisco.

Ludwick, J. C. (1970). Sandwaves and tidal cannels in the entrance to Chesapeake Bay. *Virginia J. Sci.* **21**, 178-184.

Ludwick, J. C. (1972). Migration of tidal sand waves in Chesapeake Bay entrance. *In* "Shelf Sediment Transport: Process and Pattern" (D. J. P. Swift, D. B. Duane, and O. H. Pilkey, eds.). Dowden, Hutchinson & Ross, Stroudsburg, Pennsylvania.

McKinney, T. F., and Friedman, G. M. (1970). Continental shelf sediments of Long Island, New York. *J. Sed. Petrol.* **40**, 213-248.

McMaster, R. L. (1954). Petrography and genesis of New Jersey beach sands. *State of New Jersey Dept. Conservation and Econ. Devel., Geol. Surv. Bull.* **63**, 239 pp.

McMaster, R. L., and Garrison, L. E. (1967). A submerged Holocene shoreline near Block Island, Rhode Island. *Marine Geol.* **75**, 335-340.

Meade, R. H. (1969). Landward transport of bottom sediments in estuaries of Atlantic coastal plain. *J. Sed. Petrol.* **39**, 229-234.

Meisburger, E. P. (1972). Geomorphology and Sediments of the Chesapeake Bay Entance. *U.S. Army Corps of Engineers, Coastal Eng. Research Ctr., Tech. Memo* (in press).

Merrill, A. S., Emery, K. O., and Rubin, M. (1965). Oyster shells on the Atlantic continental shelf. *Science* **147**, 395-400.

Milliman, J. D., and Emery, K. O. (1968). Sea levels during the past 35,000 years. *Science* **162**, 1121-1123.

Milliman, J. D., Pilkey, O. H., and Blackwelder, B. W. (1968). Carbonate sediments on the continental shelf, Cape Hatteras to Cape Romain. *Southeastern Geol.* **9**, 245-268.

Milliman, J. D., Pilkey, O. H., and Ross, D. A. (1972). Sediments of the continental margin off the eastern United States. *Geol. Soc. Am. Bull.* **83**, 1315-1334.

Moody, D. W. (1964). "Coastal Morphology and Processes in Relation to the Development of Submarine Sand Ridges off Bethany Beach, Delaware." Ph.D. thesis, 167 pp. Johns Hopkins Univ., Baltimore (unpublished.)

Murray, S. P. (1970). Bottom currents near coast during Hurricane Camille. *J. Geophys. Res.* **75**, 4579-4582.

Newman, W. S., and Munsart, C. A. (1968). Holocene geology of the Wachapreague Lagoon, eastern shore peninsula, Virginia. *Marine Geol.* **6**, 81-105.

Oaks, R. Q., Jr. (1964). Post-Miocene stratigraphy and morphology, outer coastal plain, South-Eastern Virginia. *Office of Naval Research, Tech. Rept.* **5**, 240 pp.

Oertel, G. F., and Howard, J. D. (1972). Water circulation and sedimentation at estuary entrances on the Georgia coast. *In* "Shelf Sediment Transport: Process and Pattern" (D. J. P. Swift, D. B. Duane, and O. H. Pilkey, eds.). Dowden, Hutchinson & Ross, Stroudsburg, Pennsylvania.

Osmond, J. K., May, J. P., and Tanner, W. F. (1970). Age of the Cape Kennedy barrier-and-lagoon complex. *J. Geophys. Res.* **75**, 469-479.

Payne, L. H. (1970). "Sediments and Morphology of the Continental Shelf off Southeast Virginia." Ph.D. thesis, 70 pp. Columbia Univ., New York (unpublished).

Pevear, D. R., and Pilkey, O. H. (1966). Phosphorite in Georgia continental shelf sediments. *Geol. Soc. Am. Bull.* **73**, 365-374.

Pierce, J. W., and Colquhoun, D. J. (1970). Holocene Evolution of a portion of the North Carolina Coast. *Geol. Soc. America Bull.* **81**, 3697-3714.

Pilkey, O. H., and Field, M. E. (1972). Onshore transportation of continental shelf sediment: Atlantic southeastern United States. *In* "Shelf Sediment Transport: Process and Pattern" (D. J. P. Swift, D. B. Duane, and O. H. Pilkey, eds.). Dowden, Hutchinson & Ross, Stroudsburg, Pennsylvania.

Pilkey, O. H., and Frankenberg, J. (1940). The relict-recent sediment boundary on the Georgia continental shelf. *Georgia Acad. Sci. Bull.* **27**, 37-40.

Pilkey, O. H., and Giles, R. T. (1965). Bottom topography of the Georgia continental shelf. *Southeastern Geol.* **7**, 15-18.

Roberts, W. P., and Pierce, J. W. (1967). Outcrop of the Yorktown Formation (Upper Miocene) in Olslow Bay, North Carolina. *Southeastern Geology* **8**, 131-138.

Scheidegger, A. E. (1970). "Theoretical Geomorphology," 2nd Ed. 435 pp. Springer Verlag, New York.

Schopf, T. J. M. (1968). Atlantic continental slope and shelf of the United States—Nineteenth Century exploration. *U.S. Geol. Surv. Prof. Paper* **529-F**, 12 pp.

Schwartz, M. L. (1965). Laboratory study of sea level rise as a cause of shore erosion. *J. Geol.* **73**, 528-534.

Schwartz, M. L. (1967). The Bruun theory of sea level rise as a cause of shore erosion. *J. Geol.* **75**, 76-92.

Schwartz, M. L. (1968). The scale of shore erosion. *J. Geol.* **76**, 508-517.

Scruton, P. C. (1960). Delta building and the deltaic sequence. *In* "Recent Sediments, Northwest Gulf of Mexico" (F. P. Shepard and F. B. Phleger, eds.), pp. 82-102.

Shepard, F. P. (1963). "Submarine Geology." 557 pp. Harper and Row, New York.

Shepard, F. P., and Wanless, H. R. (1971). "Our Changing Coastlines." 579 pp. McGraw-Hill, New York.

Shideler, G. L., and Swift, D. J. P. (1972). Seismic reconnaissance of Quaternary deposits of the Middle Atlantic continental shelf—Cape Henry, Virginia, to Cape Hatteras, North Carolina. *Marine Geol.* **12**, 165-185.

Shideler, G. L., Swift, D. J. P., Johnson, G. H., and Holliday, B. W. (1972). Late Quaternary stratigraphy of the inner Virginia continental shelf: A proposed standard section. *Geol. Soc. Am. Bull.* (in press).

Sloss, L. L. (1962). Stratigraphic models in exploration. *J. Sed. Petrol.,* **32**, 415-422.

Smith, J. D. (1969). Geomorphology of a sand ridge. *J. Geol.* **77**, 39-55.

Stanley, D. J., Fenner, P., and Kelling, G. (1972). Currents and sediment transport at the Wilmington Canyon shelfbreak, as observed by underwater television. *In* "Shelf Sediment Transport: Process and Pattern" (D. J. P. Swift, D. B. Duane, and O. H. Pilkey, eds.). Dowden, Hutchinson & Ross, Stroudsburg, Pennsylvania.

Stanley, D. J., and Kelling, G. (1968). Sedimentation patterns in the Wilmington submarine canyon area. *In* "Transactions of the National Symposium on Oceanic and Sciences and Engineering of the Atlantic Shelf" (A. E. Margulies and R. C. Steere, eds.) 366 pp.

Stearns, F. (1967). "Bathymetric Maps of the New York Bight, Atlantic Continental Shelf of the United States, Scale 1:125,000." National Ocean Surv., National Oceanic and Amospheric Admin., Rockville, Md.

Stefansson, V., Atkinson, L. P., and Bumpus, D. F. (1971). Hydrographic properties and circulation of the North Carolina shelf and slope waters. *Deep-Sea Res.* **18**, 383-420.

Swift, D. J. P. (1968). Coastal erosion and transgressive stratigraphy. *J. Geol.* **76**, 444-456.

Swift, D. J. P. (1970). Quaternary shelves and the return to grade. *Marine Geol.* **8**, 5-30.

Swift, D. J. P., and Heron, S. D., Jr. (1969). Stratigraphy of the Carolina Cretaceous. *Southeastern Geol.* **10**, 201-245.

Swift, D. J. P., Dill, C. E., Jr., and McHone, J. (1971b). Hydraulic fractionation of heavy mineral suites on an onconsolidated retreating coast. *J. Sed. Petrol.* **41**, 683-690.

Swift, D. J. P., Holliday, B. W., Avignone, N. F., and Shideler, G. L. (1972). Anatomy of a shore-face ridge system, False Cape, Virginia. *Marine Geol.* **12**, 59-84.

Swift, D. J. P., Sanford, R. B., Dill, C. E., Jr., and Avignone, N. F. (1971a). Textural differentiation on the shore face during erosional retreat of an unconsolidated coast, Cape Henry to Cape Hatteras, western North Atlantic shelf. *Sedimentology* **16**, 221-250.

Tanner, W. F. (1960). Expanding shoals in areas of wave refraction. *Science* **132**, 1012-1013.

Tanner, W. F. (1961). Offshore shoals in an area of energy deficit. *J. Sed. Petrol.* **31**, 87-95.

U.S. Army Corps of Engineers (1955). New York District. Atlantic Coast of Long Island, New York, "Fire Island Inlet and Shore Westerly to Jones Inlet." Beach Erosion Control Rept. on Cooperative Study. 16 pp. (unpublished).

Uchupi, E. (1965). "Map Showing Relation of Land and Submarine Topography Nova Scotia to Florida." U.S. Geol. Surv., miscellaneous geological investigations map 1-451.

Uchupi, E. (1968). "Atlantic Continental Shelf and Slope of the United States—Physiography." U.S. Geol. Surv. Prof. Paper 529-C, 30 pp.

Uchupi, E. (1970). "Atlantic Continental Shelf and Slope of the United States—Shallow Structure."

U.S. Geol. Surv. Prof. Paper 529-1, 44 pp.

Veatch, A. C., and Smith, P. A. (1939). "Atlantic Submarine Valleys of the United States, and the Congo Submarine Valley." Geol. Soc. Am. Spec. Paper 7, 101 pp.

Weigel, R. L. (1964). "Oceanographical Engineering." 532 pp. Prentice Hall, Englewood Cliffs, New Jersey.

Weinman, Z. H. (1971). Analysis of littoral transport by wave energy: Cape Henry, Virginia to the Virginia, North Carolina Border. Masters thesis, Old Dominion University, Norfolk, Virginia (unpublished).

Wicker, C. F. (1951). History of New Jersey coastline. Berkeley, Calif., *Proc. First Conf. Coastal Eng.* pp. 299-319.

Widmer, K. (1964). "The Geology and Geography of New Jersey." 193 pp. Van Nostrand, New York.

CHAPTER 24

Sediment Textural Patterns on San Pedro Shelf, California (1951-1971): Reworking and Transport by Waves and Currents

D. S. Gorsline and D. J. Grant

Department of Geological Sciences
University of Southern California
Los Angeles, California 90007

ABSTRACT

The San Pedro shelf, off Los Angeles, California, is located in a region of intermediate wave energy, but is screened by offshore islands from most of the open ocean waves from the Pacific except for limited westerly and southerly approaches. Currents are tidal, wind-driven, and wave-generated long-shore drift from both south and west.

Trend surfaces of sediment-size parameters calculated as moment measures show patterns that can be related to long wave and local current regimes. An outer central zone of 20-40 m depth is probably affected by westerly and southerly long period storm wave convergence and prevailing wind-driven currents; a central area of outcropping Miocene rocks and relict sediments is being modified texturally by fine sediment drifted from the west; a southeastern coastal zone is dominated by north-flowing long-shore drift alongshore and by tidal exchange moving east from Los Angeles-Long Beach Harbor.

Comparison of recent (1969-1970) data with earlier work (1951) shows some differences that suggest that the surficial sediments of the entire shelf surface are subject to change in a period of a couple of decades. Changes include both intrusion of fine sediments into a stable coarse sediment and active drift of fine sands in depths of 20-40 m. Some modifications may extend to depths of as much as 100 m.

INTRODUCTION

The literature of the past 70 yr documents several cycles of hypothesizing, alternating with field research, concerning shelf sediments. It is interesting to see that the same basic ideas often reappear in early stages of each new cycle and are then modified or ignored as dogma is fixed until the next cycle begins. For example, a brief perusal of Barrell's writings early in the century (1906, 1912) shows that he cited records and observations of the strong effect of storm waves on shelves to depths of over 90 m (50 fathoms). He notes repeatedly (see 1912, p. 427-430) that the long storm waves would produce rippled and sorted sand lenses in water deeper than normal wave base. These would then be covered by graded finer sediments either as a result of settling of the winnowed particles after the storm, or as a result of the intervening lower energy sedimentary regime. Certainly these thoughts are a much generalized forerunner of the latest well-documented observations that have initiated the present cycle of renewed interest in shelf processes and products (see Swift, Stanley and Curray, 1971). As Swift and his associates have noted, the present philosophy has evolved from work by Shepard (1932) and later papers (see Shepard and Cohee, 1936; Emery, Butcher, Gould and Shepard, 1952; Emery, 1952) which laid to rest the prevailing ideas of the pre-1930s that assumed a model of graded sediments fining offshore away from the coastal sources.

These later ideas were then formalized in Emery's classification of shelf sediments (Emery, 1952) in which a relict sediment type was defined as sediments that are anomalous in terms of their present depth and lateral associations, typically are stained and coarse, contain worn fossil shells and shell fragments and often composed of mineral suites that are different from those of the contemporary beach and river contribution to the area. As Emery has noted (e.g., 1968) this is the important type areally on all but the narrowest continental shelves and indicates the relatively recent flooding of the shelves by the last rise in sealevel. Off California, these sediments are generally disposed in halos around rock outcrops standing above the general level of the central shelves (see Terry, Keesling and Uchupi, 1956), or in patches that form arclike trends subparallel with the present coast and at depths of 25-30 m corresponding to submerged terrace levels (see Uchupi and Gaal, 1963 for examples; also Emery, 1960; Gorsline et al., 1971).

A logical extrapolation from these observations and interpretations has been the generally held thought that these relict types are areas of nondeposition, or bypassing, but are inactive and not affected by present-day processes and agents. Hence, the interpretation of field studies may be strongly biased if viewed from the point of view stated above. The increasing emphasis on dynamic studies of processes *in situ* (see Stride, 1963; Cook, 1969; Loop, 1969; Palmer, 1969; Vernon, 1966; Swift, 1970; Clifton et al., 1971) has also produced data that opens the idea of static relict surficial deposits on the deeper parts of shelves to serious question. It also has required a change in outlook of the senior author.

Among the more thought-provoking work on the California shelf has been the observations by Cook (1969) and Loop (1969) of coarse sands and fine-medium sands lying side-by-side in depths of 20-30 m, apparently equally active and hydraulically equivalent. Cook offers the explanation that the different bedforms formed on the different textural groups (an observation also made by Loop) produce different bed roughnesses and, hence, different thresholds for grain movement under the same surge velocity conditions. Stated another way, fine and coarse patches have the same reaction to a given wave surge velocity at the same depth because of this differing roughness.

Certainly the concept of wave base as defined on the basis of the prevailing or "normal" wave (see Dietz, 1963) needs reevaluation (also see Komar et al., this volume). Where the normal limit of active wave movement of sands off the southern California coast is probably about 20-30 m, under long storm waves the entire shelf may come into a sufficiently strong wave surge regime as to produce marked effects in surficial sediment texture and bedform. It thus seems likely that the suggested term "palimpsest sediments" defined by Swift, Stanley and Curray (1971) may prove to be a more explicit description of the relict sediment type.

The following discussion is in large part speculative since we have found that dynamic measurements are surprisingly sparse even in this well known area. Thus, writing the paper has provided a basis for planning future research on the California shelf. The San Pedro shelf has been selected because it has been examined in some detail on at least four occasions during the past twenty years (see Moore, 1951; Hancock Foundation staff, 1965), and contains a variety of sedimentary types including relicts.

Figure 233 shows the location of the San Pedro shelf off Los Angeles, California. The southern California shelf is a narrow coastal terrace at the inner margin of the California Continental Borderland (Shepard and Emery, 1941). The narrow coastal terrace varies in width from less than 2 km off bold promontories to as much as 20 km off Santa Barbara, Santa Monica, and San Pedro. Each wide shelf unit is backed by coastal plains, bounded at each end by headlands and narrow shelves, and by steep fault-line (?) scarps seaward. As has been reviewed by Shepard and Dill (1966) these units are terminated at their southern ends by small submarine canyons which head close inshore and have been apparently cut by sliding and flowing sand dumped into their heads from longshore drift.

Emery (1960) has described the wide shelf segments as sediment cells which

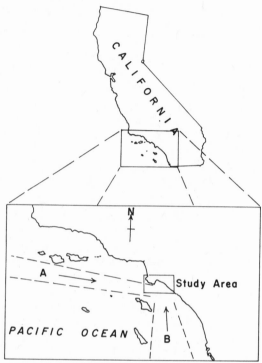

Figure 233.

Index map showing location of study area and open paths for open ocean wave approach to the San Pedro area. Arrows indicate direction of wave approach.

are partially insulated from adjoining cells by the canyons at their down-drift ends. Only minor leakage of sediment gets by these traps. Fine sediments tend to pass through the systems as suspension load and move to the basins as Drake et al., have described elsewhere in this volume. A major flood can sometimes overload the transport system and fines may then temporarily deposit on the shelf, but these are usually reworked into suspension and removed after a few months or a very few years.

In 1951, D. G. Moore sampled the San Pedro shelf in detail collecting about 150 samples to depths of about 40 m plus an additional small number of dredge samples to depths of up to 300 m. Station locations were based on visual bearings and are probably good to ±50 m. He used a clam shell grab that probably allowed some washing of silt and clay, but the general trends of sediment parameters and types are valid as proved by spot sampling in subsequent years. Some additional transects were sampled in 1958-1961 during a major program of shelf study done by the Hancock Foundation of the University of Southern California under State funding (Hancock Staff, 1965), but these are not sufficiently complete in their coverage of the entire shelf to be more than a control on sample validity.

In 1969, 65 Shipek grab samples were collected by the senior author to depths of about 100 m. These are also subject to minor washing effects as has been observed in the field and will be demonstrated later in this paper, but much less than the old clam shell samplers. Station location is of the same order of precision as in Moore's work.

In 1971, Dr. R. L. Kolpack, as part of the Sea Grant Program work at USC, collected an extensive net of box core samples from the southern California shelf of which about 100 are located in the San Pedro area (see Fig. 234) in depths to 100 m. Thus an opportunity presented itself to evaluate the effectiveness of several sampling methods as well as the basic aim of determining the presence and magnitude of changes in the surficial sediments. The Sea Grant samples are obtained by using a Reineck box core which recovers a sample without sediment loss and permits specific subsampling in terms of the stratigraphy of the upper 50-70 cm of sediment representing the *in situ* conditions.

Granulometric and mineralogic analyses were made using standard methods. Textural parameters are phi moment measures (note comments in Davis and Ehrlich, 1970). Laboratory work was done by the junior author as part of a thesis (Grant, in preparation). Trend surfaces were calculated using digital computer methods and the most effective surface (that is, one explaining the

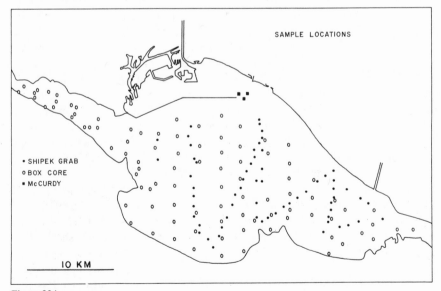

Figure 234.

Location of samples collected by Sea Grant personnel in 1971 (circles) and Shipek grabs collected by D. S. Gorsline in 1969 (black dots). Contour outlining shelf is 180 m.

largest percentage of variations) was drafted. In these instances the fifth order surfaces proved to be most efficient. A similar use of trends to determine the effects of wave and current action is reported by Booth (1971).

PHYSICAL SETTING

Bathymetry and Geologic Structure

Figure 235 shows the bottom topography of the San Pedro shelf and vicinity. The physiographic unit includes the narrow Palos Verdes shelf and is bounded to the west by Redondo Submarine Canyon at the southeastern end of the Santa Monica Bay cell. The southern boundary of the San Pedro unit is marked by Newport Submarine Canyon. Along the central shelf edge, the break in slope occurs in the depth zone from 75-150 m as a smooth roll-off rather than a sharply defined depth. The 180 m (100 fathom) contour is a good representation of the edge of the terrace and is used as the boundary in the figures of this paper.

Maximum width of the shelf is about 20 km south of Long Beach (see Fig. 235). The dip of the shelf is approximately 0°08'. Subterraces are cut into the general surface and are best defined off Palos Verdes at depths of 30, 60, and 80 m (Emery, 1958). These surfaces have been modified by mantling by later sediment (see examples in Moore, 1957 off Point Loma) but the main depth levels are correlative along the coast. Terrace levels at about 20 and 40 and possibly 60 m can be discerned in the San Pedro shelf.

Two reentrants in the edge of the shelf in symmetrical arrangement on each side of the main seaward projection of the shelf (Fig. 235) have been named San Pedro and San Gabriel Canyons by Shepard and Emery (1941) although seismic profiling indicates that they are more likely slump scars associated with fault trends cutting the dome structure that forms the bedrock structure for the shelf surface. The surface truncates the dome exposing Pliocene and Miocene sediments in outcrop at the stratigraphic crest of the structure about 5 km south of the Los Angeles breakwater. The structure is part of the Palos Verdes positive element and is structurally continuous with that high.

Other outcrops are present in the inner portions of the shelf and on the slopes of the seaward margin of the terrace. These do not form high relief features and are not easily seen on echograms. Wave action has effectively truncated the dome leaving only very low relief outcrop areas in the central shelf.

The sediment cover over the Tertiary sedimentary rocks is probably everywhere thinner than a few tens of meters. Seismic profiling indicates that the outer shelf has less cover than the width of the surface bubble pulse, or less than 5 m of sediment.

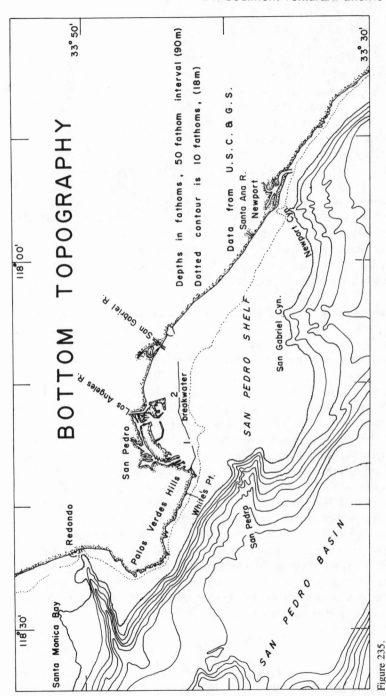

Figure 235.

Bottom topography of the San Pedro Shelf and vicinity based on contours from USC and G.S. Chart 5101. Depth interval is 90 m (50 fathoms) with the 18 meter (10 fathom) contour dotted. Latitude and longitude marked for 10' intervals.

Waves

Figure 233 illustrates the directions from which open ocean waves can approach the San Pedro area. Diffraction and refraction of incoming long waves by islands and banks effectively diminishes much of the wave attack from the west. A relatively open approach is available to southern swell although shallow submerged banks in that pathway may effect long southern wave trains. Data from Corps of Engineers publications (Horrer, 1950; O'Brien, 1950; Caldwell, 1956) show that long waves with periods of from 12-20 sec are common in the southern wave spectrum and these would be influenced by shelf topography to the depth of the shelf break (greater than 100 m). Refraction diagrams show that energy concentrations (convergences) occur along the eastern segment of the breakwater and refraction increases the energy over the two reentrants in the shelf edge (Figs. 235 and 236). A divergence of waves would theoretically occur over the central shelf. It is not likely that similar long waves would enter from the west, but if so, these would tend to diverge over the central shelf and converge along the southern edge of the San Pedro Canyon and over the outer shelf.

Calculations of the orbital velocity of water particles in waves with periods of 20 sec, open ocean lengths of about 620 m and heights of 2 m, all typical of hindcasting data for the area and from calculations based on general periodic wave equations (see Weigel, 1964, p. 64, and p. 207), indicate that the orbital velocity is about 25 cm sec^{-1} at a water depth of 100 m and about 40 cm sec^{-1} at a water depth of 50 m. The lower velocity approximates the velocity for erosion of fine sand particles (see the modified diagram after Inman, Hjulstrom and Sundborg in Allen, 1965). Thus, neglecting the effects of refraction, the magnitudes of the theoretical near-bottom velocities are appropriate for transport and resuspension of the fine sands and would keep finer particles in suspension at outer shelf depths. Note in the Sediment section that the sorting and mean diameters of sands in the 60-100 m range are appropriate.

Below about 100 m, the longer waves would have a negligible effect on the deposited sediment, but could still keep fine silt and clay in suspension.

Diving observations on the inner Palos Verdes shelf show that strong rippling of the sand surface often occurs at depths of 15-20 m, but that generally the sand at 20 m is not obviously rippled although biologic activity may rapidly destroy ripples formed by storm waves. Ripples are the typical bedform off the coast to depths of 20 m and are directly related in wavelength to the sand size (see Stone, Fishel and Summers, 1972).

Currents

Current patterns are not known in detail for the San Pedro Channel area although strong drift is typically noted at ship stations in the western and northern

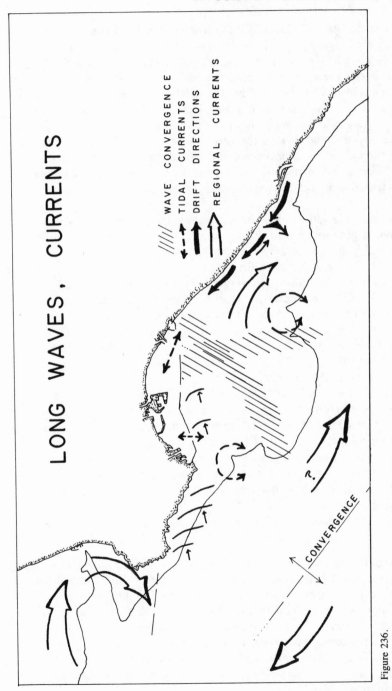

Figure 236.

Schematic presentation of known oceanographic factors. Wave convergences are areas of convergence of long wave (12–20 sec periods) orthogonals based on hindcasting methods. Contour outlining shelf is 90 m.

sections of the Channel. Bunnell (1969) has described hydrographic observations for Fall, 1967, and Spring, 1968. Hydrographic data available show a strong convergence to be present in the central San Pedro Channel that probably represents a boundary or shear zone between northerly (?) offshore flow and southerly (?) inshore flow. This inshore flow may affect the outer shelf and shelf edge sediments. It is also likely that seasonal changes occur in this system, but data are sparse and not yet sufficient to document small scale local patterns. The seasonal ebb and flow of the north-flowing Davidson current (California countercurrent) over the inner parts of the Borderland are well known and the strength and depth of its flow and the related local countercurrents over the adjacent shelf may also follow the large scale patterns of seasonal changes (Bunnell, 1969).

Bunnell's observations in a single "hydrographic year," 1967-1968, show by standard geopotential calculations, drogue tracks and drift card collections that in that year flow patterns were in harmony with the average conditions over the southern California Borderland. In fall and winter, 1967, the southerly countercurrent inshore of the convergence in the San Pedro Channel was well developed. Drift card return demonstrated that southerly flow was probably bounded near the mainland shore by a series of gyrals that may have flowed north within 1 or 2 km of shore. Major transport was southerly with the main north-flowing Davidson Current pushed offshore. Flow velocities in the upper 400 m ranged from 20-100 cm sec^{-1}. Drogues and drift cards showed surface velocities from 10-60 cm sec^{-1}.

In Spring, 1968, the north flow was strong and had compressed the coastal southerly countercurrent closer to the edge of the mainland shelf and velocities in the upper 400 m were 60-120 cm sec^{-1}. Drogues and drift cards showed surface (upper 5 m) velocities of from 5-30 cm sec^{-1}.

In summary, currents capable of moving fine sand are probably generated in the south-flowing countercurrent that probably is the prevailing flow over the outer shelf. This flow may break up into eddies as one approaches the inner shelf.

Tides

In general, tidal currents are a minor agent except in the harbor area. Spring tide ranges in the San Pedro area are of the order of 2 m. Limited data are available for the outer harbor and show an estuarine-like pattern with tidal waters moving into the eastern end of the outer harbor and also through the Los Angeles breakwater entrance to the west (see Fig. 235). The Long Beach opening in the central breakwater is apparently not a point of active exchange. A well defined salinity wedge shifts into and out of the inner channels with the changing tides. Exchange is probably most important to the east towards the Newport area (Fig. 235).

Some observations and theory suggest that the semimonthly cycle generates and then destroys large cellular mixed water units over irregular portions of the shelf. These cells build from neap to spring tide and then are broken up as the semimonthly cycle wanes. These gyral currents would be slow but might influence the transport of fine suspension load over the shelf, particularly in the vicinity of the canyons (see Leipper, 1955; Stevenson and Gorsline, 1956). Dr. D. E. Drake of USC, Department of Geological Sciences has collected some turbidity measurements that suggest cellular flows are a factor. In addition, current meter data of Dr. F. P. Shepard and Dr. Neil Marshall of Scripps Institution of Oceanography show currents of tidal period with typical maximum velocities of 8-10 cm sec^{-1} at depths of 135 m off La Jolla, California.

Figure 236 is a summary of the various oceanographic agents and their generalized areas or directions of action. Relationships of these patterns with the sediment characteristics will be discussed in the following section.

SEDIMENTS

Textural Groups

Figure 237 depicts the textural relationships of the sediments based on the 1969 and 1971 samples. Several subgroups form discrete trends on the diagram (also see Table XXXIII) and are plotted in Fig. 238, which shows a pattern that is similar to Moore's (1954) diagram although the method for separating the types is different in detail. Moore based his classes on percent sand-silt-clay as well as color. Also, the rock and rock fragment sediments on or near outcrops are not included in the present analysis. The relict sand of Emery (1960) and Moore's Pleistocene sand are represented by Group I type and its reworked mixtures by Group II, IV, and V types. Thus the coarsest sediments generally are orange-gray stained coarse sands that appear to have been deposited originally in beach or nearshore conditions. Plots based on sample color show these sands to be located in a rough arc around the rocky central shelf area and along the 20-30 m depth zone.

Group III sands and silty sands are the contemporary sediment type. The A and B subdivisions shown in the textural diagram are artificial and used to indicate roughly the trend of sorting in the group. Thus most of the shelf is covered by a laterally graded gray fine sand to silt with patches of coarser sediments surrounding the central tertiary outcrop zone and lying on a probable old shoreline of the 30 m submerged terrace, and at the inner edge of that zone. Coring around the periphery of these patches shows that the gray green sands and silty sands of Group III lap on the patches of orange-gray sands with some mixing of the two on the lapped edges. There is no obvious bathymetric expression of the coarse sand patches on echograms crossing the inner shelf

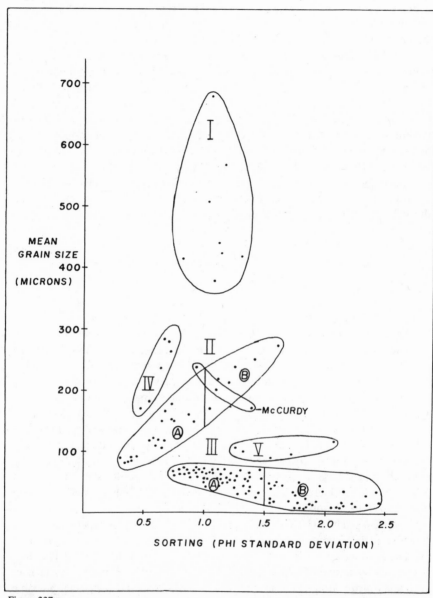

Figure 237.

Size relationships of shelf sediment samples from 1969 and 1971 collections. Three points outlined and labeled *McCurdy* are relict samples from a survey in 1963 and serve as indicators of variation due to differences in sampler.

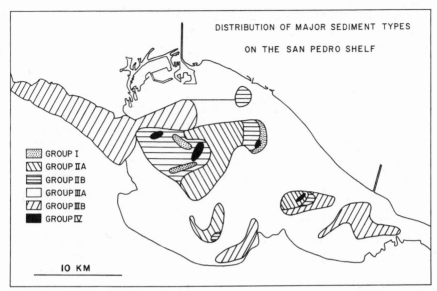

Figure 238.
Areal distribution of textural groups as defined by Fig. 237. Contour outlining shelf is 180 m.

Table XXXIII
Sediment Textural Groups

Group	Mean (μ)	Phi sorting	General description (Emery type)
I	400-700	0.8-1.4	Red-brown coarse sands: relict
II	60-300	0.4-1.7	Gray sandy silts to sands: mixed
III	20-60	0.7-2.5	Green-gray clay silt-sandy silt: detrital
IV	150-300	0.5-0.8	Gray fine to medium sands: mixed
V	100-110	1.3-2.2	Gray fine silty-sands and sands: mixed

and one must conclude that their original relief, if present, was planed off by wave action associated with a rising sea level.

Mineralogically, the various groups show important differences. Group I sands are distinct in that they have a high percentage of feldspars, garnets, opaques relative to the other groups in addition to coarser size and red stain. Mica and hornblende percentages are low.

Group II and the minor Groups IV and V (see Fig. 237) always are adjacent and peripheral to Group I. Typically they form a halo around the patches of

Group I sediments. They may be reworked and mixed from Group I. Mineralogically they contain less garnet, feldspar and opaques and higher percentages of mica and hornblende. They tend to be gray rather than red, orange or brown. Groups IV and V are similar and also appear to be reworked products of Group I plus Group III.

Group III is the dominant and probably contemporary sediment type. It typically ranges from fine poorly sorted greenish gray sand to greenish gray silty sand and sandy silt. These sediments are dominantly quartz in the light mineral fraction and dominantly micas in the heavy fraction. Hornblende is the next dominant and garnets and opaques occur rarely.

Examination of the textural parameters and mineralogical data (Gorsline, Booth and Grant, 1971) shows that the Los Angeles, San Gabriel and Santa Ana Rivers (Fig. 235) are probably contributors of Group III sediments together with a very minor contribution from the northern sources beyond the Palos Verdes Hills (Fig. 240), although most of the northern sediment is trapped in the Santa Monica cell.

Trend Surface

Trend surfaces for the major heavy mineral components were calculated. Of these, the epidote distribution was well defined by a fifth degree surface that was found to be the most efficient fit. Epidote percentages serve as a

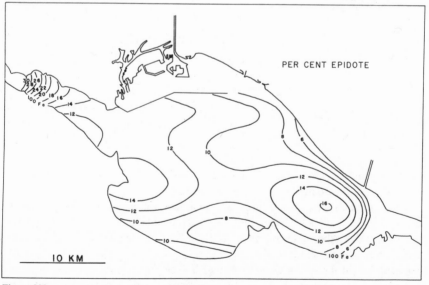

Figure 239.

Fifth degree trend surface fitted to epidote percentage of heavy mineral suites of sands. Contour used to outline shelf is 180 m (100 fathoms).

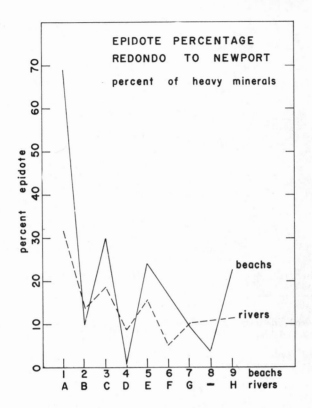

Figure 240.

Epidote percentages of heavy mineral fractions of sands from beaches and rivers in the coastal segment from Redondo Beach to Newport Beach (see Fig. 235). Beaches are: (1) Manhattan Beach, (2) Malaga Cove Beach, (3) Lunada Bay Beach, (4) Portuguese Bend Beach, (5) White's Point Beach, (6) Cabrillo Beach (breakwater approaches), (7) Long Beach, (8) Seal Beach, (9) Newport Beach. Rivers are: (A) Topanga Canyon, (B) Malaga Cove, (C) Agua Amarga Creek, (D) Palos Verdes Drive Creek, (E) Whites Point Creek, (F) Los Angeles River, (G) San Gabriel River, (H) Santa Ana River.

good tracer (Fig. 239). Note in the Figure that there is a well defined trend from the Palos Verdes shelf across the inner San Pedro Shelf to the eastern shore. Epidote is also contributed from Santa Ana River in large amounts (Fig. 240) and is being transported offshore by rip currents from a convergence in longshore drift near the Santa Ana River mouth (see Felix and Gorsline, 1971) and is the source of the concentration of epidote in the lobe of fine sand in the area north of Newport submarine canyon.

The offshore pattern of decreasing percentages is influenced by the much lower heavy mineral contents of silts and silty sand. Thus the percentage reflects the decreasing competence of the transport agents as depth increases.

Figures 241-244 illustrate trend surfaces for mean diameter, phi sorting (standard deviation), phi skewness and phi kurtosis. These surfaces often bring out

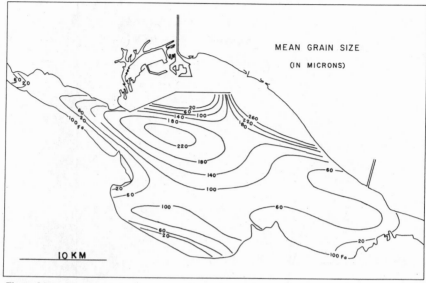

Figure 241.

Fifth degree trend surface fitted to mean diameter data from 1969 and 1971 sample suites. Outline contour is 180 m (100 fathoms).

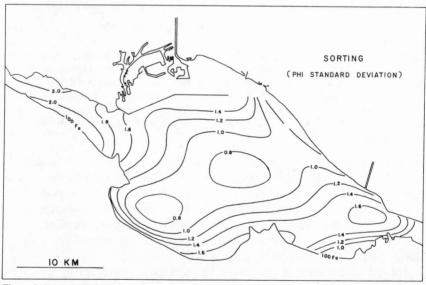

Figure 242.

Fifth degree trend surface fitted to phi standard deviations of size distributions from 1969 and 1971 sample suite. Contour outlining shelf is 180 m (100 fathoms).

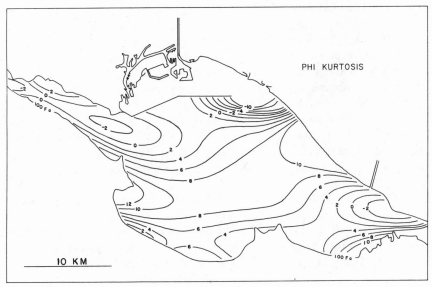

Figure 243.
Fifth degree trend surface fitted to phi skewness of size distributions of 1969 and 1971 sample suites. Outlining contour is 180 m.

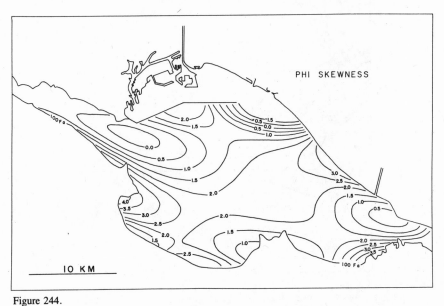

Figure 244.
Fifth degree trend surface fitted to phi kurtosis of size distributions of 1969 and 1971 sample suite. Contour outlining shelf is 180 m.

major trends in sediment parameters that are obscured by various sources of "noise" including mixing of types, sampling and analytical errors and artifacts of classification (see Booth, 1971). In all instances in this paper the trends shown are the fifth order approximations of the base data. That surface was found to be most efficient in explaining the variations of the individual parameters.

The mean diameter surface (Fig. 241) is obviously strongly influenced by the central rocky area and its halo of relict coarser sediments. Seaward of that area, the surface shows a band of fine sands with mean diameters in the range from 100 to 160 μ, and oriented diagonally across the shelf from west to east in average depths of 25 to 30 m. The sediment then rapidly becomes finer at the outer edge of the shelf and upper slope with mean diameters in the silt range.

Standard deviation (sorting) has a pattern (Fig. 242) that is generally similar to that shown in skewness (Fig. 243) and to kurtosis (Fig. 244) in that all three show a well defined saddle in the trend surfaces passing across the outer central shelf across the 20-40 m depth range in a roughly east-west alignment (Fig. 245). In this zone, sorting is better than in any other portion of the shelf

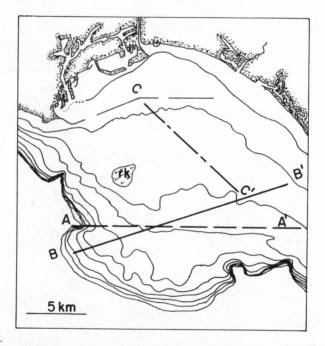

Figure 245.

Detailed bathymetry of San Pedro Shelf to depth of 50 fathoms (90 m). Contour interval is 9 m (5 fathoms). Area marked RK and stippled is crest of truncated dome forming basement of shelf surface. Trend A-A' shows axes of saddles in surface for mean diameter (Fig. 241). Trend C-C' shows weakly defined secondary saddle in surfaces of Figs. 241-244.

with most of the particles having diameters less than one Wentworth grade on either side of the mean. Skewness shows that this belt is strongly skewed towards the fines with the mean finer than the mode. This means that these sediments have a large input of finer particles.

Kurtosis also shows the same band pattern and is strongly leptokurtic, or very peaked. Hence a well sorted population of particles is tightly grouped around the mean diameter. This is also reflected in the excellent sorting (Fig. 242). It must also mean that the fines are present in amounts less than about thirty percent and decrease in percentage towards the middle of the band but are finer and in larger quantity at the western end. An increase in amount of fines is also shown near shore at the eastern shore. This either requires that a new source of fines must be introduced, or that a convergence in transport causes fine sedimentation. This may well be associated with suspension transport of fines from the Santa Ana River mouth and the Santa Ana sewer outfall to the southeast. Felix and Gorsline (1971) show that fine suspended sediment moves just over the bottom from the river mouth and outfall southeast to the Newport submarine canyon. Their data also show that the near bottom current is to the southeast and thus the area is not one where currents decrease. Therefore the decreased mean diameters of Fig. 241 must represent a new addition from the harbor and outfall. This fits the pattern of finer mean diameters with poorer sorting, platykurtic distributions and increasingly coarse-skewed distributions as the canyon is approached. Coarse skewness here indicates that the mean diameter is coarser than the median. Fines may also be transported out of the outer harbor by tidal currents. One can visualize a mean diameter increasingly influenced by the fine suspended fraction, finally coinciding with the mean diameter of that fraction and then becoming finer than the suspended load modal diameter as a result of trapping of flocculated clays and organics in Newport Canyon.This may indicate that the sand transport is mainly to the west with fine suspension transport dominantly to the east, along the eastern and southern shore.

The central shelf zone depicted by good sorting, fine skewed and leptokurtic fine sands (Figs. 242-243) would seem to represent wave-drifted sediment moving at depths well within the depth of effective action by long period swell. For example, a 12 sec wave, 200 m long and 2 m high would generate near bottom velocities of nearly 40 cm sec^{-1}.

There is a weakly defined trend in Figs. 242-243 that follows a line roughly from the eastern notch in the shelf to the central Long Beach breakwater (Fig. 245). This will be recalled later when we compare oceanographic factors with the shelf sedimentary patterns.

Skewness and kurtosis also include elements of the pattern of the mean diameter surface. Sorting, however, simply decreases towards the Palos Verdes shelf with no marked secondary anomalies. Skewness and kurtosis both show closures in the vicinity of the outcrop and relict area, but these maxima are offset to the west and north. Thus sorting improves towards the outcrop area, kurtosis is initially platykurtic then shifts progressively to leptokurtic (that is, changes from a broad flat distribution to a peaked one), while skewness shifts from an essentially normal pattern to a weakly fine skewed pattern (mean finer than mode). This may be a result of movement of fines across the area from the Palos Verdes shelf, passing over and depositing in increasingly coarser sediment, but with little movement of the coarser sediment component. The trend is apparently from west to east.

A patch of very fine sediment is located near the central breakwater in the 1969 and 1971 data. This may represent a surficial fine deposit laid down as a result of the February, 1969 floods that moved much fine suspended material from all southern California rivers and dumped it into the nearshore in quantities exceeding the capacity of the marine agents to remove it. The major difference discernible in comparison of Moore's 1951 data and the present data is in this area near the breakwater (Fig. 246). Moore's sediments are much coarser in that area.

Comparison of 1951 Data with 1971 Data

Moore's data also show coarser median diameter in the outcrop area. This may reflect drift of finer material into the central area since 1951 (perhaps as late as the 1969 flood which occurred prior to the field work of this paper). The outer band is less well defined in Moore's data, but this may be an artifact of the few samples in those depths of his sample net.

It must also be added that the generally coarser median diameters may reflect washing of the 1951 samples. The difference between Shipek and box core samplers for the 1969-1971 data is illustrated for a representative set of pairs from the same location in Fig. 247 and in Table XXXIV. Five of seven are indicative of better sorting and coarser mean diameters for Shipek samples. Only one Shipek sample is finer (No. 6), and has poorer sorting than the equivalent box core sample. Thus most show differences that are larger than analytical error for sorting but not necessarily for the means. Since the clam shell grab is probably more susceptible to washing than the Shipek, part of the coarser pattern of Moore's data may be an equipment artifact. In any event, the finer offshore band shown by Moore's data is probably real since it is contrary to the trend of sampler error. By the same token, the very appreciable difference in means near the breakwater is also probably larger than sampler effect which tends to be less on shallow casts because travel through the water is less.

Methods of analysis of the two sets of samples were similar. The main problem in comparison is that Moore used Trask parameters to describe the size distributions whereas we now use moment measures. For that reason, we calculated median diameter using the Trask parameter. Thus, the trends in Fig. 246 are probably not affected by analytical methods to any important degree.

Figure 246.
Comparison of fifth degree trend surfaces fitted to median diameters of samples from Moore, 1951 (top), and to median diameters of samples of 1969 and 1971 sample suite.

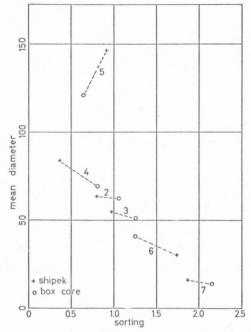

Figure 247.

Comparison of mean diameters and sorting for Reineck box core samples and Shipek grab samples at same locations. Pair number 1 is not plotted since the Shipek sample mean lies at 740 μ versus a Reineck core mean of 130 μ. See Table XXXIV.

Biological Effects

Undoubtedly infauna works over the upper few tens of centimeters of sediments over the entire shelf area. Studies of the benthic communities of this shelf have been conducted for many years by marine biologists of the Allan Hancock Foundation of the University of Southern California. The general pattern is probably quite stable and has been a more or less constant factor in the time span discussed in this paper. Undoubtedly the burrowers aid in working in the fine sediment deposited over the older coarse sands in the central shelf. This may well aid in holding the fines within the mixed mass and thus inhibit winnowing of the silts and clays by longer period waves.

Box cores and gravity cores in the areas peripheral to the "relict" sands show quite sharp contacts preserved between the contemporary greenish gray fine sands and the older orange gray coarse sands although some mixing has occurred laterally. Thus, the depth of borrowing and remixing by organisms is probably only a few tenths of a meter. The effect is probably local and small as compared with the quite large changes in median diameter in the area

Table XXXIV
Comparison of Sedimentary Parameters of Shipek Grab and Box Core Samples

Sample	Mean (μ)	Median (μ)	Sorting (ϕ units)	Skewness (ϕ units)	Kurtosis (ϕ units)	Coarse Fraction (%)
Group I						
13527	567	740	1.17	−0.04	0.20	99.0
15133	88	130	0.30	0.38	7.70	94.6
Group II						
13511	63	93	0.80	3.13	14.78	59.4
15140	62	107	1.07	3.13	10.70	68.4
Group III						
13494	55	87	0.99	2.59	9.09	49.5
15150	51	89	1.26	2.55	6.80	51.8
Group IV						
13522	84	142	0.39	0.78	3.11	88.2
15153	69	99	0.81	3.78	17.97	73.6
Group V						
13520	147	228	0.91	0.17	−0.25	89.1
15156	121	185	0.67	0.84	−0.47	89.6
Group VI						
13479	30	67	1.75	1.06	0.11	29.7
14765	41	67	1.25	1.43	2.55	30.2
Group VII						
13465	16	36	1.88	0.43	−1.08	11.7
14756	13	17	2.16	0.40	−0.88	15.5

near the central breakwater between 1951 and 1971 and in the area in the vicinity of Newport.

SUMMARY

Comparison of the schematic oceanography in Fig. 236 with the discussion of sedimentary parameters illustrated in Figs. 241-244 in the previous section shows some quite good relationships.

In the broad band across the outer central shelf, winddrift to the east, coastal countercurrents and westerly wave surge over the bottom toward the east beyond the lee of the Palos Verdes Hills may be the factors driving well-sorted fine sand and silts across the shelf. Note that the mean diameters are in the range of the most easily transported and eroded grain sizes (Inman, 1949; Allen, 1965).

The secondary and weaker trend may be the product of the convergence of southern long waves across the shelf (see Fig. 236). This intrusion of southern waves has occurred on occasions such as the 1930 and 1939 examples noted by Horrer (1950) and O'Brien (1950). A similar intrusion could eventually clean away the patch of fine sediment presently located off the breakwater. This fine silt deposit may be a product of the 1969 flood.

Southern swell dominates the southeastern shore's drift and contributes to the lobe of fine sand north of Newport Canyon that appears in the epidote patterns, mean diameter map and other size parameter plots. Just offshore of this area the currents to the east towards Newport Canyon apparently dominate the movement of fine sediment. Dr. D. E. Drake has noted strong movement of turbid layers into Newport Canyon from its adjacent shelves adding further evidence for this current flow.

Along the Palos Verdes inner shelf and along the inner San Pedro shelf, wave drift is apparently reworking fine sediment which then moves offshore to the central shelf outcrop area and mixes with the relict sediments. A lobe of fine sand is apparently moving east as indicated by the epidote distribution as well as by sorting and mean diameter trends.

The fine sediment patch next to the breakwater may have been moved via the tidal exchange from the harbor both through the Los Angeles Entrance and around the eastern end of the breakwater. Its source is tentatively assigned to suspension load from Los Angeles and San Gabriel River flood waters.

Since much critical data are still not available, the above relationships are tentative. However, the striking correlations of hydrography and sediment characteristics highlight some problems for further study. Bottom photography will be useful in checking the orientation of bedforms in the outer central shelf band. Current meter stations and sediment traps can help to verify or disprove the motion of fine sediments in the band. Felix and Gorsline's (1971) observations and the work in progress by D. E. Drake seem to fit the pattern and model suggested for the eastern portion of the shelf. Observations by J. Galloway of Scripps Institution of Oceanography on the heavy metal content of bottom sediments near the White's Point outfall, about 5 km west of the south end of the hills (see Fig. 235) indicate westward trends. Sand mineralogy gives evidence of transport to the east. Field observations also show black reduced sediment east of the outfall.

It is evident that the entire San Pedro shelf is an active sedimentary province in which surficial sediments are being modified to depths of at least 100 m by wave and current action within a period of twenty years and quite probably much less.

ACKNOWLEDGMENTS

Support for some of the work that contributed to this paper has come from contract 14-08-0001-10862 and 14-08-0001-12632 with the Office of Marine Geology and Hydrology of the United States Geological Survey. Research has also been supported from contract N00014-67-A-0269-0009C with the Ocean Science and Technology Division of the Office of Naval Research.

Shiptime has been paid for under grant GB-6319 of the National Science Foundation.

Samples were contributed by Dr. R. L. Kolpack from the Geological Sciences section of the Sea Grant Program at the University of Southern California. Dr. R. H. Osborne of the Department of Geological Sciences provided programs and advice on the computation of the trend surfaces and statistical operations on the data.

Field assistance of graduate students and marine technicians of the Department of Geological Sciences at USC is also gratefully acknowledged.

Drs. D. E. Drake and R. L. Kolpack read the manuscript and contributed valuable criticism. Drs. J. S. Creager and L. D. Kulm also critically read the manuscript and contributed numerous suggestions that have much improved the paper. The authors, of course, remain responsible for the interpretations. Mrs. Charlotte W. Newman typed the manuscript and proofed the references.

REFERENCES

Allen, J. R. L. (1965). A review of the origin and characteristics of recent alluvial sediments. *Sedimentology* **5**, 91-191.

Barrell, J. (1906). Relative geological importance of continental, littoral and marine sedimentation. *J. Geol.* **16**, 336-354, 430-457, 524-568.

Barrell, J. (1912). Criteria for the recognition of ancient delta deposits. *Geol. Soc. Am. Bull.* **23**, 377-447.

Booth, J. S. (1971). "Sediment dispersion in Northern Channel Island passages, California Continental Borderland." *Master's thesis, Geological Sciences, Univ. of Southern California*, Los Angeles (unpublished).

Bunnell, V. D. (1969). "The water structure of the San Pedro Basin, California Borderland." *Master's thesis, Geological Sciences, Univ. of Southern California*, Los Angeles (unpublished).

Caldwell, J. M. (1956). Wave action and sand movement near Anaheim Bay, California. *U.S. Army Corps of Engineers, Beach Erosion Board, Tech. Memo.* **68**, 21 p.

Clifton, H. E., Hunter, R. E., and Phillips, R. L. (1971). Depositional structures and processes in the non-barred high-energy nearshore. *J. Sedimentary Petrology* **41**, 651-670.

Davis, M. W., and Ehrlich, R. (1970). Relationship between measures of sediment-size-frequency distribution and the nature of sediments. *Geol. Soc. Am. Bull.* **81**, 3537-3548.

Dietz, R. S. (1963). Wave-base, marine profile of equilibrium, and wave-built terraces, a critical appraisal. *Geol. Soc. Am. Bull.* **74**, 971-990.

Drake, D. E., Kolpack, R. L., and Fischer, P. J. (1972). Sediment transport on the Santa Barbara-Oxnard shelf, Santa Barbara Channel, California. *In* "Shelf Sediment Transport: Process and Pattern" (D. J. P. Swift, D. B. Duane, and O. H. Pilkey, eds.). Dowden, Hutchinson & Ross, Stroudsburg, Pennsylvania.

Emery, K. O. (1952). Continental shelf sediments of southern California. *Geol. Soc. Am. Bull.* **63**, 39-60.

Emery, K. O. (1962). "The Sea off Southern California." Wiley, New York.

Emery, K. O. (1968). Relict sediments on continental shelves of the world. *Am. Assoc. Petroleum Geologists Bull.* **52**, 445-464.

Emery, K. O., Butcher, W. S., Gould, H. R., and Shepard, F. P. (1952). Submarine geology off San Diego, California. *J. Geol.* **60**, 511-548.

Felix, D. W., and Gorsline, D. S. (1971). Newport Submarine Canyon, California: An example of the effects of shifting loci of sand supply upon canyon position. *Marine Geol.* **10**, 177-198.

Gorsline, D. S., Brenninkmeyer, B. M., Meyer, W. C., Ploessel, M. R., Shiller, G. I., and Vonder Haar, S. P. (1971). General characteristics of nearshore sediments from El Capitan to Ventura, California, 1967-70. (R. L. Kolpack, ed.), *Biological and Oceanographical Survey of the Santa Barbara Channel Oil Spill, 1969-1970*, **2**, 296-317.

Gorsline, D. S., Grant, D. J., and Booth, J. S. (1971). Annual report on marine geologic research in the California Continental Borderland: Heavy mineral content of river, beach and nearshore sediments, Southern California. 76 p. Unpublished Tech. Rept USC-GEOL 71-5, *Depart. Geological Sciences, Univ. of Southern California,* Los Angeles.

Hancock Foundation Staff (1965). Oceanographic and biologic survey of the southern California Mainland Shelf. *Calif. State Water Quality Control Board, Publ.* **27**, 230 p.

Horrer, P. L. (1950). Southern hemisphere swell and waves from a tropical storm at Long Beach, California. *U.S. Army Corps of Engineers, Beach Erosion Board, Bull.* **4**[3], 1-18.

Inman, D. L. (1949). Sorting of sediments in the light of fluid mechanics. *J. Sedimentary Petrology* **19.** 51-70.

Komar, P. D., Neudeck, R. H., ana Kulm, L. D. (1972). Observations and significance of deep-water oscillatory ripple marks on the Oregon continental shelf. *In* "Shelf Sediment Transport: Process and Pattern" (D. J. P. Swift, D. B. Duane, and O. H. Pilkey, eds.). Dowden, Hutchinson & Ross, Stroudsburg, Pennsylvania.

Leipper, D. F. (1955). Sea temperature variations associated with tidal currents in stratified shallow water over an irregular bottom. *J. Marine Res.* **14**, 234-252.

Loop, T. H. (1969). Physical oceanography and sediment characteristics within Little and Shark Harbors, Santa Catalina Island, California. *Master's thesis, Geological Sciences, Univ. of Southern California.* 172 p. (unpublished).

Moore, D. G. (1951). Submarine geology of San Pedro Shelf, California. *Master's thesis, Geological Sciences, Univ. of Southern California* (unpublished).

Moore, D. G. (1954). Submarine geology of San Pedro Shelf. *J. Sedimentary Petrology* **24**, 162-181.

Moore, D. G. (1957). Acoustic sounding of Quaternary marine sediments off Point Loma, California. *U.S. Navy Electronics Lab. Rept.* **815**, 17 p.

Neumann, G., and Pierson, W. J. (1966). "Principles of Physical Oceanography." Prentice-Hall, Englewood Cliffs, New Jersey.

O'Brien, M. P. (1950). Wave refraction at Long Beach, and Santa Barbara, California. *U.S. Army Corps of Engineers, Beach Erosion Board, Bull.* **4**[1], 1-11.

Palmer. H. D. (1969). "Wave-Induced Scour around Natural and Artificial Objects." 172 p. *Ph.D. dissertation. Geological Sciences, Univ. of Southern California* (unpublished).

Shepard, F. P. (1932). Sediments on the continental shelves. *Geol. Soc. America Bull.* **43**, 1017-1039.

Shepard, F. P., and Cohee, G. V. (1936). Continental shelf sediments off the mid-Atlantic states. *Geol. Soc. America Bull.* **47**, 441-458.

Shepard, F. P., and Dill, R. F. (1966). "Submarine Canyons and Other Sea Valleys." Rand-McNally, New York.

Stevenson, R. E., and Gorsline, D. S. (1956). A shoreward movement of cool subsurface water. *Trans. Amer. Geophys. Union* **37**, 553-557.

Stone, R. O., and Summers, H. J. (1972). Study of subaqueous and subaerial sand ripples. *Rept. USC-GEOL 72-1, Geological Sciences, Univ. of Southern California.* 274 p. (unpublished).

Stride, A. H. (1963). Current swept sea floors near the southern half of Great Britain. *Quat. J. Geol. Soc. London* **III**, 175-199.

Swift, D. J. P. (1970). Quaternary shelves and the return to grade. *Marine Geol.* **79**, 322-346.

Terry, R. D., Keesling, S. A., and Uchupi, E. (1956). Submarine geology of Santa Monica Bay, California. *Hyperion Project, Geological Sciences, Univ. of Southern California,* 177 p. (unpublished report)

Uchupi, E., and Gaal, R. (1963). Sediments of the Palos Verdes Shelf. *In* "Essays in Marine Geology in honor of K. O. Emery" (T. Clements, ed.), 171-189.

Vernon, J. W. (1966). Shelf sediment transport system. *Ph.D. dissertation, Geological Sciences, Univ. of Southern California,* 135 p. (unpublished).

Wiegel, R. L. (1964). "Oceanographical Engineering." Prentice Hall, Englewood Cliffs, New Jersey.

CHAPTER 25

Observations and Significance of Deep-Water Oscillatory Ripple Marks on the Oregon Continental Shelf

Paul D. Komar,* Roger H. Neudeck,† and L. D. Kulm*

ABSTRACT

Bottom photographs of the Oregon continental shelf reveal symmetrical oscillatory ripple marks across the entire shelf out to water depths as great as 204 m. The ripples are principally of the trochoidal type, the crests trending north-south, parallel to the coast. Asymmetrical ripples are rare.

Using Airy wave theory, bottom orbital velocities are calculated for average and storm waves on the Oregon shelf. Velocities are sufficient to produce ripples out to depths of 100 m with occasional rippling to 200 m. Systematic variations in ripple length across the shelf are examined and it is concluded that the variations could be produced by a reasonable pattern of dispersed wave periods arriving from a distant-storm source.

Besides producing ripples, the surface waves appear to be important in stirring and reworking the shelf sediments. Sediment transport on the shelf is examined with the Bagnold model wherein the waves provide the power to place the bottom sediments in motion and a superimposed unidirectional

*School of Oceanography, Oregon State University, Corvallis, Oregon 97331
†3000 La Luz, Atascadero, California 93422

current produces a net sediment drift. It is concluded that the effects of surface waves, as well as unidirectional currents, must be included in estimates of sediment transport on the continental shelf.

INTRODUCTION

As early as the turn of the century it was recognized that the effects of surface waves on the oceans can extend to considerable depths. There were many stories such as gravel and large stones being washed into lobster pots by wave surge [see summary in Johnson (1919, p. 78-83)]. On this basis and on theoretical grounds, the belief was held that there is appreciable oscillatory wave motion down to depths of 600 ft (183 m) and perhaps deeper. More recently, studies such as Curray (1960) and Draper (1967) considered the wave-induced bottom surge and the frequency with which the bottom sediment would be stirred. Curray (1960) suggested that off the Texas coast the sediments would be moved by hurricane waves approximately once every five years on the shelf edge and more often in shallower water. Draper's (1967) data indicate that on the shelf edge west of Britain, wave action should be sufficiently intense to stir fine sand at a depth of 600 ft (183 m) for more than 20% of the year.

Due to the difficulty of observation, few reports have been made of ripple marks produced by surface waves in deep water. One of the earliest and most ingenious accounts of deep-water ripples is that of Siau (1841) [cited by Johnson (1919, p. 80-81)]. Using a weight coated with tallow, Siau obtained impressions of sea-bottom ripple shapes at a depth of 188 m near the Isle of Bourbon off the coast of Madagascar. Sediment adhered to the tallow, showing parallel bands of heavy particles concentrated in the troughs and the light particles collected on the crests. Although Siau attributed the ripples to wave action, it is possible that they were produced by unidirectional currents.

With the advent of underwater breathing apparatus, the direct study of ripple marks under wave action became more feasible. However, studies that have used this technique, such as Inman (1957), have been confined to fairly shallow water.

This study extends the observations of oscillatory ripple marks to the deeper waters of the continental shelf by utilizing a camera system to obtain stereo pairs of bottom photographs. The feasibility of such an approach to ripple investigations on the shelf was pointed out by the study of Owen et al. (1967). Our study is centered on the continental shelf off Oregon, an area noted for the severity of its wave climate. Symmetrical ripple marks, believed to be produced by surface waves, were observed in water depths as great as 204 m. We shall also consider the importance of surface waves in sediment transport on the shelf.

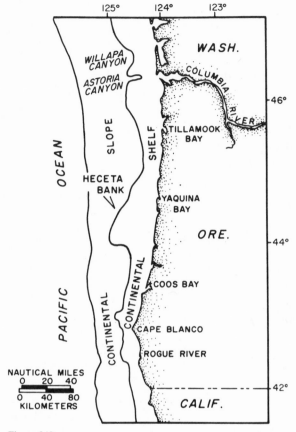

Figure 248.

Location map showing the Oregon continental shelf and related features.

SETTING OF STUDY

The continental shelf off Oregon (Fig. 248) is narrowest to the south, where it is some 16 km in width, and progressively widens to the north, reaching a maximum width of 75 km where it borders the shelf off Washington. The average shelf slope is a maximum of 12.5 m km^{-1} in the south, progressively decreasing to a minimum of 2.3 m km^{-1} in the north. The depth of the outer edge of the shelf ranges from 150 m in the north to 185 m in the south. Astoria Canyon, off the Columbia River, marks the northern boundary of the Oregon shelf. The Rogue Canyon, north of the Rogue River, is the only other major canyon system present. Submarine banks, such as Heceta and Coquille Banks, are the only other significant shelf irregularities.

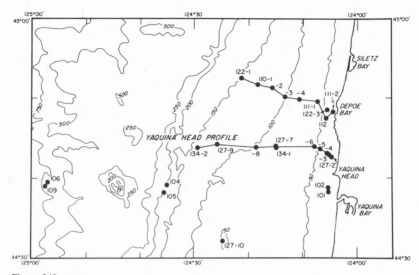

Figure 249.

Location map of camera stations off central Oregon. Depth contours are in meters.

METHODS

Bottom photographs were collected at a total of 72 stations on the Oregon continental shelf on seven cruises during 1967 and 1968 (Fig. 249). The photographs described in this study were obtained with an E.G.&G. stereo camera system. An underwater compass was suspended below the camera within the field of view to provide oriented photos. All directions and orientations were determined with respect to true north after making the correction for the magnetic declination and are usually accurate to within ±5° for individual frames.

Stereo viewing allowed the determination of the symmetry and wave height of the ripples as well as the ripple lengths. These were measured by standard photogrammetric methods.

Generally the best photographs were obtained in deeper water because of the clarity of the water. In the extremely shallow nearshore region, the water is so turbid that it is virtually impossible to determine the characteristics of the ripples.

Surface sediment samples were collected at some of the camera stations to supplement those available from the previous studies of Oregon shelf sediments. The particle-size parameters were computed according to Folk and Ward (1957).

RIPPLE CHARACTERISTICS AND PATTERNS

Ripples were observed in all sediment types on the Oregon continental shelf, but were best developed in the sandy facies. Ripples extended from the nearshore zone, across the entire shelf to depths of about 200 m near the shelf edge (Figs. 250, 251, 252). In all cases, the dominant set of ripple crests was oriented approximately north-south, nearly parallel to the coastline. Occasionally, in the nearshore region at depths less than 50 m, two sets of ripples with similar wave lengths and heights were superimposed to produce an ''interference pattern.'' The orientations of the two sets commonly differed by less than 30°.

The ripples in shallow water are generally fresher and better defined than those observed in deep water. In deeper water the ripples were observed to be in active formation, or very well preserved, only during the single winter cruise of February, 1968, a time of large surface waves. At other times of the year, the ripples show marked differences in the degree of preservation. Poor preservation or a complete absence of ripples was typical during the quiescent

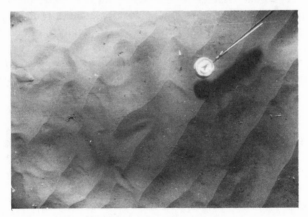

Figure 250.

Seasonal differences in ripple preservation at approximately the same geographic location. Top, station 134-1-11 (80 m) occupied May 1968. Bottom, station 127-7-24 (75 m) occupied February 1968. Last digits on station number refers to frame number on film.

Figure 251.
Rounded trochoidal ripples. Station 102.

periods of little surface wave action. Figure 250 shows some striking seasonal differences at the same station in 75 m of water off Newport.

Virtually all ripple profiles observed on the Oregon shelf are of the symmetrical type, typically produced by oscillatory currents. Asymmetrical ripples, as are produced by unidirectional flow or a combination of oscillatory currents with a superimposed net flow, were only rarely observed.

Inman (1957) classified the symmetrical oscillatory ripples into (a) solitary ripples, and (b) trochoidal ripples. Solitary ripples have flat troughs separating isolated crests. Inman found that this type is generally restricted to water shallower than 20 m. No solitary ripples were observed in our study, probably because it was confined to deeper water. Trochoidal ripples have rounded troughs and somewhat peaked crests that are either sharp or slightly rounded; this type is extremely common on the Oregon shelf. Figure 252 shows well-developed, sharp-crested trochoidal ripples that appear to have been in active formation. A third type of symmetrical ripple, the "rounded type," resembles a smooth sine curve (Fig. 251) and is also commonly observed off Oregon. Trochoidal ripples are generally the best preserved and have the most peaked and sharp crests. On the other hand, rounded ripples are generally more common among the less well-preserved ripples. Photographs such as Fig. 251 with many trails crossing the bottom suggest that the rounded ripples may result from the borrowing activities of benthic organisms partially destroying and smoothing primary ripples.

Inman (1957) classified the ripple pattern on the basis of the ratio of the average crest length to the average ripple wavelength. Under this classification, most of the ripples observed on the Oregon shelf were intermediate crested (crest length/ripple length = 3 to 8) or long crested (> 8). Short crested ripples

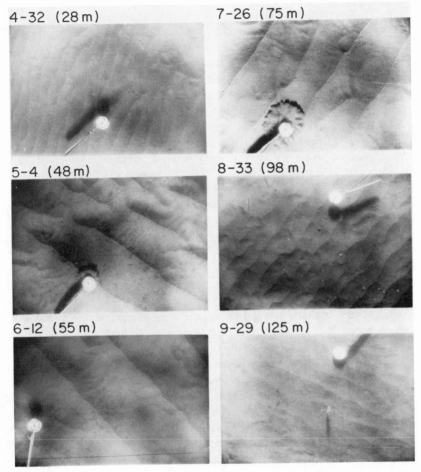

Figure 252.
Selected ripple patterns along the Yaquina Head camera profile (see Figure 249 for location), Stations 127-4, -5, -6, -7, -8, and -9. Ripple statistics are given in Table XXXV.

were occasionally observed within the troughs of larger ripples, but were always minor compared to the major ripple patterns.

YAQUINA HEAD TRAVERSE

During February 12 and 13, 1968, a camera traverse (Fig. 249, 127 series) was made across the continental shelf to a depth of 125 m. This profile is probably typical of the winter conditions on this high-energy shelf.

Well-preserved and well-defined sediment ripples, Fig. 252, were observed from the shallowest station, 127-4 (28 m) to the deepest station, 127-9 (125 m). Extremely turbid water at shallower depths precluded the observation of ripples, but they no doubt were present. The majority of the ripple crests were oriented in a north-northeast direction, approximately parallel to the coast.

The ripples along this traverse appeared to be in the process of formation out to depths of at least 100 m. At 125 m, the ripples were less well-preserved compared to those at 98 m. Very sharp-crested trochoidal ripples were present at all stations. The symmetrical ripple profile strongly suggests they were formed by relatively high frequency oscillatory currents that were oriented in an east-west direction. Such currents were most likely induced by surface waves and this suggestion will be discussed in a later section. At Station 127-9 (125 m), poorly developed but well-preserved asymmetrical ripples were present; their orientation suggests a current with a net westerly flow. The largest ripple wave lengths were observed at Stations 127-5 (48 m) and -6 (55 m) where they averaged 21 cm (Table XXXV). Shorter wave lengths were noted at both shallower and deeper stations.

Ripple wave lengths varied directly with ripple heights. The ripple index (length height^{-1}) varied between 3.0 and 7.3 (Table XXXV). However, it should be kept in mind that because the measured average ripple heights are only approximate, the ratios are also approximate. The mean diameter of the sediments ranges between 203 μ and 67 μ along the profile. At Stations 127-5, -6, -7, and -8 the sediments have essentially the same mean diameter, approximately 180 μ, although the wavelengths of the ripples vary considerably.

The following May, ripples were observed only out to the depth of Station 134-1 (80 m) occupied along the Yaquina Head profile. The ripples were oriented in the same approximate north-south direction as observed in February. In May, the sediment surface had a dirty, smoothed-over appearance and the ripples were very poorly preserved. The fine-grained sedimentary cover was marked with numerous small tracks, trails, and pits.

OREGON SHELF CURRENTS

The water circulation off the Oregon coast is dominated by the broad California current which flows south, parallel to the coast. During the winter months, the northward flowing Davidson current develops closer to shore, apparently driven by the strong gale force winds of that season (Burt and Wyatt, 1964). Both currents are shallow features, their velocities dropping appreciably with depth. Upwelling is common from late spring to early fall, producing onshore currents deeper than 50 to 75 m with offshore currents above (Smith et al., 1966). Currents associated with the upwelling are only of the order 0-2 cm sec^{-1}. Such currents may play a role in the transportation and distribution of fine sediments, but could not have produced the ripple marks observed in this study.

Table XXXV
Sediment Ripple Statistics for Yaquina Head Camera Profile

Station	Water depth (m)	Ave. wavelength (cm)	Ripple height (cm)	Ripple index
127-4	28	8	1.5	5.3
127-5	48	21	6.0	3.5
127-6	55	21	7.0	3.0
127-7	75	15	3.0	5.0
127-8	98	11	2.0	5.5
127-9	125	11	1.5	7.3

The Oregon coast has a spring tide range of about 3 m. Spectral analyses of current velocity records obtained off Oregon yield pronounced energy peaks at one and two cycles per day (Collins, 1968; Mooers, 1970; Harlett, 1972), indicating that tides are important in the generation of currents over the continental shelf off Oregon. The source may be internal tides (Mooers, 1970), as well as surface tides. Particularly pertinent to this study are Harlett's (1972) current measurements obtained with meters located 50 to 210 cm above the bottom. The strongest bottom currents measured were in the nearshore region at a depth of 36 m where velocity magnitudes over 40 cm sec^{-1} were attributed to surface waves by Harlett. At mid-shelf depths (90 m) and at the shelf-edge (165 m) the currents averaged about 10 cm sec^{-1}, ranging from 0 to 25 cm sec^{-1}. Spectral analyses indicated that they were in part tidally induced. The instrumentation for the deeper water measurements was such that surface wave effects could not be noted directly, even if present. Energy peaks noted in the spectra in the period range 2 to 10 min may reflect the effects of grouping of the surface waves, the well-known phenomenon of progressive variations in the observed wave height, where the waves become first larger with each successive wave and then smaller. Sediment transport and rippling could result from the higher velocities observed by Harlett so that tidal currents may be significant. The ripples so produced would be asymmetric rather than the symmetrical ripples that were almost universally observed in this study.

High frequency internal waves have not received much attention off the coast of Oregon. High frequency (7 to 144 cycles day^{-1}) water temperature spectra obtained by Collins (1967) showed significant peaks near 2 and 4 cycles hour^{-1}, suggesting the presence of internal waves. From the observed gradients of water density, Collins (1967) calculated a maximum value of 12 cycles hour^{-1} for the Väisälä frequency, thought to be the upper limit for the possible range of internal waves. Energy peaks between 2 and 7 cycles hour^{-1} are common in the near-bottom current spectra of Harlett (1972), also suggesting the possible presence of internal wave effects. Since internal waves are of such long periods (15 to 30 min), bottom currents they produce would be more akin to unidirectional than oscillatory currents and would generate asymmetrical current ripples. The production of asymmetrical ripples by internal waves has been noted by Southard (1971) in stratified wave tank studies.

Harlett (1972) obtained a series of measurements of the near bottom velocity profile using current meters 48, 75, 180, and 208 cm above the bottom. Half of the profiles indicate a velocity maximum of 15 to 20 cm sec^{-1} at 150 to 200 cm above the bottom. This tendency for a maximum velocity is more pronounced in the profiles obtained in deep water (73 m) than in shallower water (36.5 m). The flow is down slope, suggesting the possibility that this bottom seeking current was in part density driven, perhaps by high concentrations of suspended sediment stirred up from the bottom.

Strong currents, associated with intense winter storms, may occur even at mid-shelf depths. Smith and Hopkins (1971), for example, measure current speeds up to 70 cm sec^{-1} at 80 m depth off Washington. The exact origin of the currents is uncertain. They have fairly large offshore components and may be an important agent in sediment transport. Such currents probably also produce the asymmetrical ripple marks observed in this study.

OREGON SURFACE WAVES

The coast of Oregon is particularly noted for the severity of its wave conditions. Storms from the Gulf of Alaska routinely buffet the Oregon coast in winter. Offshore oil rigs equipped with automatic wave recording instruments, as well as having their vertical struts marked for visual observations, have reported the measurement of remarkable wave heights during intense winter storms. Watt and Faulkner (1968) reported waves up to 58 ft (17.7 m) with one 95 ft (29.0 m) wave generated by two separate storms. Rogers (1966) reports seas with waves of 50 ft (15.2 m) occurring under winds gusting up to 150 mph (Fig. 253). These observations do not represent average wave conditions during the severe storms, but are exceptional waves produced by the chance constructive interference of several large waves. Waves more routinely range from 10 to 20 ft (3.0 to 6.1 m).

Such sizeable storm waves should be capable of transporting and rippling sediments to considerable depths. Since our observations are limited to deep-water, we can apply the simple linear Airy wave theory to make some approximate calculations of the bottom orbital motion under waves found off the Oregon coast. Airy's (1845) wave theory yields an elliptical particle motion which becomes more circlelike as the surface is approached and more flattened with depth. At the bottom itself, the ellipse is reduced to a to-and-fro horizontal motion, the orbital diameter d_0 being given by

$$d_0 = \frac{H}{\sinh\left(2\pi h/L\right)} \tag{1}$$

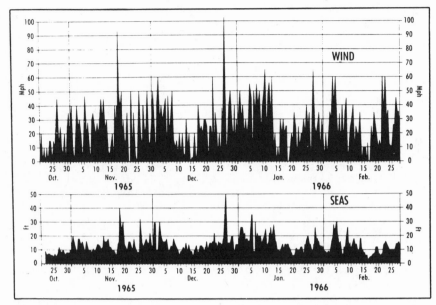

Figure 253.
Wind and wave conditions recorded from an oil rig off the coast of Oregon (Rogers, 1966).

where H is the wave height, h is the water depth, and L is the wave length. The maximum horizontal velocity associated with this to-and-fro motion at the bottom is then

$$u_m = \frac{\pi d_0}{T} = \frac{\pi H}{T \sinh (2\pi h/L)} \tag{2}$$

where T is the wave period. The wavelength L can be closely approximated with the relationship

$$L = L_\infty \left[\tanh \left(\frac{2\pi h}{L_\infty} \right) \right]^{\frac{1}{2}} \tag{3}$$

from Eckart (1952), where

$$L_\infty = \frac{g}{2\pi} T^2 \tag{4}$$

is the deep-water wavelength. The orbital diameter d_0 will, therefore, be strongly dependent on the wave period as well as on H and h.

Airy wave theory assumes no bottom friction. With friction, we will have the usual "no slip" condition in which the velocity actually equals zero at

Table XXXVI
Depth of Rippling for a Variety of Surface Wave Conditions[a]

1. Average Summer Waves
   $T = 12$ sec $h = 85$ m
   $H = 7$ ft $= 213$ cm $d_0 = 38.2$ dm
                                     $\lambda = 12.6$ cm

2. Average Winter Waves
   $T = 12$ sec $h = 99$ m
   $H = 10$ ft $= 305$ cm $d_0 = 38.2$ cm
                                     $\lambda = 12.6$ cm

3. Large Storm Conditions
   $T = 12$ sec $h = 138$ m
   $H = 30$ ft $= 914$ cm $d_0 = 38.2$ cm
                                     $\lambda = 12.6$ cm

4. Ultimate Maximum Conditions
   $T = 12$ sec $h = 149$ m
   $H = 40$ ft $= 1220$ cm $d_0 = 38.2$ cm
                                     $\lambda = 12.6$ cm

5. Lond Period Storm Waves
   $T = 15$ sec $h = 204$ m
   $H = 30$ ft $= 914$ cm $d_0 = 47.7$ cm
                                     $\lambda = 10.3$ cm

6. "Forerunners"
   $T = 20$ sec $h = 103$ m
   $H = 100$ cm $d_0 = 63.7$ cm
                                     $\lambda = 8.3$ cm

[a]Assuming a Threshold velocity of 10 cm sec^{-1}

the bottom. The velocity u_m given by Eq. (2) is commonly interpreted as the velocity at the top of the velocity gradient that develops near the bottom.

Measurements of the threshold velocity under orbital wave motion are almost entirely lacking. Threshold velocities can be expected to be somewhat less than those of unidirectional steady flow since accelerating currents produce more stress. Inman (1957), based on the laboratory data of Manohar (1955), used a value of $u_m = 10$ cm sec^{-1} for the threshold condition of fine sand and we shall use this value as well. Using this value in Eq. (2) and taking various combinations of wave height H and period T, we can calculate the water depths, h, to which rippling can be expected to occur under those conditions; the results are given in Table XXXVI. The first four calculations use $T = 12$ sec as an average period and demonstrate the effects of increasing wave height. In Calculation 4, the truly exceptional wave height of 40 ft (12.1 m) was used. In this case rippling would occur only to depths of about 149 m, nowhere near the maximum depth (200 m) to which ripples were observed. However, the orbital motion is more strongly dependent on the wave period than on the wave height, and in Calculation 5 we consider a storm wave with a 15 sec period. It is seen that with the higher period, rippling would occur down to a depth of

presented by Inman (1957) that fits the $\lambda = d_0$ line of Fig. 254 comes from such conditions, some of the data being from a wave tank study and some from an enclosed bay. If the continental shelves were covered with coarser sand, then maximum ripple lengths and reversals might be expected to occur more often.

Is there any reasonable explanation for the systematic changes in ripple lengths observed in the Yaquina Head profile? This question cannot be answered except to say that the observed changes can be produced by normal storm waves. It must be recognized that a storm generates waves with a wide spectrum of periods. Waves with the longest periods travel the fastest in deep water; as they travel from the area of generation they therefore become sorted out by period, a process known as wave dispersion. At a coast some distance from the generation area, the largest period waves would arrive first, followed by progressively shorter and shorter periods. The wave height similarly varies, generally first increasing as the period decreases, but finally decreasing as the last of the storm waves arrive. Sediment movement and rippling on the bottom would therefore be very complex, depending on the changes in T and H with time at a given depth. Since in general the wave period progressively decreases and the bottom orbital motion in deep water depends principally on the period, waves from a given storm will be able to ripple sands to progressively smaller and smaller depths as time passes. The level of competence would slowly migrate shoreward leaving ripples preserved in deeper water from an earlier, longer period phase of the storm waves.

The preserved ripples should approximately correspond to the threshold conditions of wave motion. If one uses Fig. 254 to obtain values of d_0 required to generate the observed ripple lengths of the Yaquina Profile, and uses the value $u_m = 10$ cm sec^{-1} for the threshold, then Eq. (2) can be used to calculate the period T which corresponds to the threshold condition. This was done and yielded the periods T of the third row of Table XXXVII. The corresponding wave heights H required to yield the orbital diameters d_0 are given in the fourth row. These approximate calculations demonstrate that a reasonable pattern of dispersal wave periods ranging from 7.5 to 14 sec could account for the ripples observed in water depths of 48 m and deeper. The results at 28 m only indicate

Table XXXVII
Wave Parameters Required for Ripples in the Yaquina Head Camera Profile

Depth (m)	28	48	55	75	98	125
λ (cm) =	8	21	21	15	11	11
d_0 (cm) =	68	24	24	32	45	45
T (sec) =	21.4	7.5	7.5	10.1	14.1	14.1
H (cm) =		374	602	313	172	289

that these ripples were probably active and that the period is anything less than 21.4 sec.

SURFACE WAVES AND SEDIMENT TRANSPORT ON THE SHELF

During times of winter waves it was noted that certain areas of the shelf that had previously been covered with a thin layer of silts and clays were winnowed to a clean sand by the wave induced oscillatory currents. Of course the surface waves alone could only place the sediments in suspension in a to-and-fro motion near the bottom, causing little net transport. However, if a linear current (e.g., tidal or inertial) is also present, superimposed on the wave orbital motion, a net drift could result. As the waves have already provided the power to place the sediment in motion, the magnitude of the linear current could even be below the threshold of sediment movement and still produce a net sediment drift. This suggests that surface waves may be an important agent in sediment transport on the continental shelf.

The model of Bagnold (1963, p. 518) applies to sediment transport produced by a coupling of wave action and superimposed linear currents. According to this model, i_θ, the immersed weight sediment transport rate per unit bed width, is given by the relationship

$$i_\theta = K \, \omega \, \frac{u_\theta}{u_0} \tag{5}$$

where ω is the power the waves expend in placing the sediment in motion near the bottom, and u_0 is the horizontal component of the orbital velocity at the bottom such that ω/u_0 becomes the stress exerted by the waves. u_θ is the superimposed unidirectional current flowing near the bed which produces the net drift of sediment and governs the direction of the net movement. The K is a dimensionless coefficient of proportionality.

Komar and Inman (1970) applied the model to sand transport on beaches, where u_θ became the longshore current, and found that it could successfully predict longshore sand drift. This indicates that the model is basically sound. As shown by the study of Inman and Bowen (1963), matters are somewhat more complicated in deeper water with a rippled sediment bottom. They superimposed currents up to 6 cm sec[-1] upon wave motion within a wave tank, the current flowing in the same direction as the wave travel. For very low superimposed currents the behavior was as expected, the sand transport along the wave tank increased with increasing velocity. However, at the higher velocities they obtained the peculiar result in which the sediment transport actually decreased with an increase in current, in one case the sand even moved up channel opposite to the direction of current and wave motion. This resulted from the complex periodic movement of the sand above the ripples. The effect of increasing the superimposed current was to reduce the symmetry of the system by increasing the effective orbital velocity in the direction of wave and current motion and to decrease it upwave. The vortex generated on the downwave face of the ripples became very strong and the material from the vortex was thrown up into suspension at a time when the wave orbital motion was again up tank (offshore), causing an offshore sediment drift. This occurred when the superimposed current, u_θ, was considerably less than u_0, the wave orbital motion. If u_θ were significantly increased the sediment transport would again be in the direction of the superimposed flow.

It is possible that the effects such as observed by Inman and Bowen (1963) in the wave tank are also important on the continental shelf. Fine sediments, such as clays and fine silts, that are placed entirely in suspension by the wave orbital motion would always be carried in the direction of flow of the superimposed current. Somewhat coarser material that participates in the vortex motion at the ripples may drift in the opposite direction to the superimposed current. It is conceivable that with an offshore current superimposed on wave motion, fine sediments would be carried offshore while at the same time coarser sediments are carried onshore. More studies, similar to those of Inman and Bowen (1963), are needed before we can achieve an understanding of sediment movement on continental shelves. It is apparent that the effects of surface waves on the drift must be included. Considering unidirectional currents alone may lead to quite erroneous conclusions.

CONCLUSION

Winter storm conditions off the Oregon coast generate waves that should commonly produce bottom orbital velocities capable of rippling the sediments to water depths of 150 to over 200 m. This conforms with bottom photograph observations of oscillatory ripple marks across the entire shelf out to depths as great as 204 m. Average summer waves produce ripples up to depths of 50 to 100 m.

It is concluded that surface waves produced the observed symmetrical ripple marks. Unidirectional currents probably produced the asymmetrical ripples observed in the study, but these were rare compared to the oscillatory ripples.

These observations indicate that surface waves are important in sediment transport on the shelf and in controlling the distribution of sediment facies. This is also indicated by the shoreward extent of mud on the shelf which is closely depth controlled, suggesting that it may be governed by a "wave effectiveness parameter" as defined by McCave (1971).

ACKNOWLEDGMENTS

This study was supported by the U.S. Geological Survey (Contracts 14-08-001-10766, -11941, and -12187) and by Sea Grant (Contract GH-97). We would like to thank B. T. Malfait and J. C. Harlett for their careful reviews of this manuscript.

REFERENCES

Airy, G. B. (1845). Tides and waves. *Ency. Metrop. Art.* **192**, 241-396.

Bagnold, R. A. (1963). Mechanics of Marine Sedimentation. *In* "The Sea" (M. N. Hill, ed.), Vol. 3, The Earth Beneath the Sea. Interscience, New York.

Burt, W. V., and Wyatt, B. (1964). Drift bottle observations of the Davidson Current off Oregon. *In* "Studies on Oceanography, a collection of papers dedicated to Kaji Hidaka, Seattle," Univ. Washington Press, Seattle.

Collins, C. A. (1968). "Description of Measurements of Current Velocity and Temperature over the Oregon Continental Shelf, July 1965-February 1966." 154 pp. Ph.D. thesis. Oregon State University, Corvallis.

Curray, J. R. (1960). Sediments and history of Holocene transgression, continental shelf, northwest Gulf of Mexico. *In* "Recent Sediments, Northwest Gulf of Mexico." (F. P. Shepard, F. B. Phleger, and Tj. H. van Andel, eds.), Amer. Assoc. Petrol. Geol. Tulsa, Oklahoma.

Draper, L. (1967). Wave activity at the sea bed around northwestern Europe. *Marine Geol.* **5**, 133-140.

Eckart, C. (1952). The propagation of waves from deep to shallow water. Gravity Waves. *Natl. Bureau of Standards* **521**, 165-173.

Folk, R. L., and Ward, W. C. (1957). Brazos river bar, a study in the significance of grain size parameters. *J. Sed. Pet.* **27**, 3-26.

Harlett, J. C. (1972). Sediment transport on the northern Oregon continental shelf. Ph.D. thesis. Oregon State University, Corvallis.

Inman, D. L. (1957). Wave-generated ripples in nearshore sands. *U.S. Army, Corps of Engineers, Tech. Memo.* **100**, 67 p.

Inman, D. L., and Bowen, A. J. (1963). Flume experiments on sand transport by waves and currents. *Proc. 8th Conf. Coastal Engr.* p. 137-150.

Johnson, D. W. (1919). Shore processes and shoreline development. Hafner, New York (1965 Facsimile, 584 p.).

Komar, P. D., and Inman, D. L. (1970). Longshore sand transport on beaches. *J. Geophys. Res.* **75** [30], 5914-5927.

Manohar, M. (1955). Mechanics of bottom sediment movement due to wave action. *Beach Erosion Board, Corps of Engineers, Tech. Memo.* **75**, 121 p.

McCave, I. N. (1971). Wave effectiveness at the sea bed and its relationship to bed-forms and deposition of mud. *J. Sed. Pet.* **41**[1], 89-96.

Mooers, C. N. (1970). The interaction of an internal tide with the frontal zone in a coastal upwelling region. 480 pp. Ph.D. thesis. Oregon State University, Corvallis.

Owen, D. M., Emery, K. O., and Hoadley, L. D. (1967). Effects of tidal currents on the sea floor shown by underwater time-lapse photography. *In* "Deep-Sea Photography" (John B. Hersey, ed.), p. 159-166. Johns Hopkins Press, Baltimore.

Rogers, L. C. (1966). Blue water 2 lives up to promise. *Oil and Gas J.* (August 15), 73-75.

Siau, A. (1841). Del'Action des Vagues a de Grandes Profondeurs. *Comptes Rendus de l'Académie des Sciences* **12**, 774-776.

Smith, J. D., and Hopkins, T. S. (1971). Sediment transport on the continental shelf off Washington and Oregon in light of recent current measurements. (Abstr.) *1971 Annual Mtg. of Geol. Soc. Am., Washington, D.C.* **3**, 710-711.

Smith, R. L., Pattulo, J. G., and Lane, R. K. (1966). An investigation of the early stage of upwelling along the Oregon coast. *J. Geophys. Res.* **71**, 1135-1140.

Southard, J. B. (1971). Sediment movement by breaking internal waves. (Abstr.) *1971 Annual Mtg. of Geol. Soc. Am., Washington, D.C.* **3**, 713-714.

Watts, J. S., and Faulkner, R. E. (1968). Designing a drilling rig for severe seas. *Ocean Industry* **3**, 28-37.

CHAPTER 26

Currents and Sediment Transport at the Wilmington Canyon Shelfbreak, as Observed by Underwater Television

**Daniel J. Stanley,* Peter Fenner,†
and Gilbert Kelling‡**

ABSTRACT

An underwater television survey was made of the outer shelf, shelfbreak, and upper slope in the vicinity of Wilmington canyon, about 175 km southeast of Delaware Bay and the Middle Atlantic States. It provides direct evidence of active currents and sediment transport at 26 stations, in water depths ranging from about 65 to 430 m. This technique for marine sedimentological studies can appropriately be used to examine conditions of the sea floor and overlying watermass of the outer margin. The 22 hr of videotape recorded for this study covered an area of about 400 km². In addition to the possibility of making continuous observation of the sea floor lithology and fauna along survey lines, this system allows documentation of current activity and sediment spill-over, that is, the movement of sand and silt from the outer shelf to deeper environments seaward of the shelfbreak.

Although the survey was made during a very calm period with low sea states, bottom currents competent to move sediment on the sea floor were observed throughout most of the survey area. Large amounts of suspended fine-grained material were observed in the near-bottom water and probably were scoured from the sea floor by these bottom currents and subsequently transported by movement of water masses in the vicinity of the shelfbreak. The regional current net in Wilmington

*Division of Sedimentology, Smithsonian Institution, Washington, D.C. 20560.
†Governors State University, Park Forest South, Illinois 60466.
‡Department of Geology, University of Wales at Swansea.

canyon shows a varied directional pattern of flow, presumably influenced by tides. Recorded velocities of up to 20 cm sec[-1] are of sufficient intensity to move fine sand and silt along the bottom and eventually off the shelf to the adjacent slope and canyon.

INTRODUCTION

The transport of sediment from continental shelves to deeper environments is a poorly understood phenomenon, particularly where the shelfbreak environment is in deep water, far from land. The wide Atlantic shelf off the northeastern coast of North America, one of the more intensively studied continental margins (Emery, 1968), is a case in point. It has been postulated that on this margin, as on most others, sediment movement from the continental shelf, across the shelfbreak, and onto the upper continental slope was most pronounced during Pleistocene low stands of sea level. The resulting net loss of coarse-grade material from the shelf has accelerated progradation of the outer continental margin (Uchupi and Emery, 1967). This off-shelf transport of the coarse fraction has been called spill-over (Stanley et al., 1972), and is inferred mostly from indirect observation of the outer continental margin.

This study demonstrates the use of underwater television in the direct examination of the sea floor on the Atlantic outer continental shelf margin about 175 km (95 n mi) southeast of Delaware Bay and the Middle Atlantic States. We show how this tool provides information on neocurrent (modern currents or those in the recent geological past) activity that results in sediment movement, including eventual spill-over, in the vicinity of the shelfbreak and the head of Wilmington canyon.

PREVIOUS STUDIES IN THE SURVEY AREA

Earlier studies by the authors, part of a long-term investigation of the continental margin between Norfolk and Wilmington canyons, presented some strong, but indirect, evidence that sediment has moved irreversibly from the shelf to the slope in the Wilmington canyon area during Recent time.

Observations made with an underwater camera in this region (Stanley and Kelling, 1968, their Fig. 8) have shown a predominant WSW current direction on the outer shelf east of the canyon head; this is based on ripple-mark orientation, aligned lenses of shell, and related current-produced features noted in photographs. Furthermore, this information suggests that sandy sediment carried toward the west would be intercepted by the NE-SW trending canyon head incised on this sector of the shelf. That the canyon does trap sand moving toward the west is substantiated by a study of the regional surficial sediment distribution:

bottom photographs and grab sample surveys show tongues of sand that drape from the shelf onto the east wall of the canyon (Stanley, 1970, his Fig. 15). Interpreted seismic profiles made across the canyon also indicate that tongues of surficial sediment are thicker on the east wall than on the west wall of the canyon (Kelling and Stanley, 1970, their Fig. 5).

Other evidence of current activity on the outer shelf and slope in this area is provided by a study of sediment in suspension above the bottom (Lyall et al., 1971). A survey of suspended sediment distribution, based on large water samples collected seaward of the shelfbreak, shows relatively high amounts (to >1 mg liter^{-1}) of silt- to clay-size matter. Data of this kind may help explain the very poor visibility of the bottom in the shelfbreak area as noted in the bottom photographs (Stanley and Kelling, 1968), and the presence of mud interfingering with sand within the upper reaches of the Wilmington canyon and adjacent slope. It has been postulated that material in suspension is eroded on the outer shelf by large-scale movement of water masses over the bottom in the shelfbreak region (Lyall et al., 1971).

Direct measurements of current activity in this region were obtained as a result of a current-meter survey (Fenner et al., 1971). The current meter data in the canyon proper and adjacent shelf are significant in that they document changes in current speed and direction over brief spans of time. Current direction and intensity data at each station plot as ellipses, attesting the influence of tides on water masses adjacent to and within the canyon. Current velocities to 19 cm sec^{-1} were measured.

The present paper is an outgrowth of a Markovian planning process, each cruise in the series having been planned in detail only after analysis of previous cruise results pointed to needs remaining. Thus, one aim of the cruise upon which this report is based was to add to the body of information about the region by directly observing the bottom and water mass by underwater television. This observational technique is still a relatively novel one for basic research, having been used mostly in applied problems such as search surveys and salvage operations (Stamp, 1953). An earlier research application involved the study of a small sector of the inner continental shelf off Georgia (Eddy et al., 1967) and Browns Bank (Drapeau, 1970).

In the present study, television was deemed appropriate for examining directly the bottom conditions at depths exceeding 100 m, and it allowed a broad areal coverage within the area of interest. The disadvantage of the relatively low image-resolution inherent in this system is more than offset by the advantages of obtaining a continuous time and space sequence of bottom and water mass conditions. Data from this survey, including descriptions of lithology and distribution maps of major sediment types and fauna are detailed elsewhere (Stanley and Fenner, in press).

AREA OF STUDY

The cruise described in the following account was made aboard the USCGC Rockaway (Wago 377, cruise RoS$_4$, July 11 to 18, 1968) during a period of very low wind and surface-wave activity and gentle swells. The locations of 26 television stations on the outermost shelf and upper slope bordering the head of Wilmington canyon are shown in Fig. 256. Loran-A was used for navigational control, with fixes taken at least every 15 min. Bathymetric records (PESR) were taken continuously during the course of the survey. Detailed discussions of the outer continental margin in this region, based on other cruises

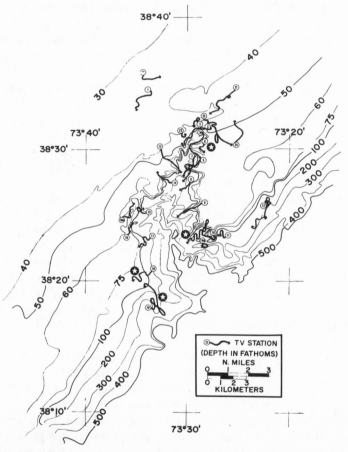

Figure 256.

Location of 26 television stations in the Wilmington canyon area east of the Middle Atlantic States occupied in July, 1968. Starred pattern at Stations 6, 10, 18, and 19 shows position of photographs in Figs. 259 to 262.

in the area, are found in Stanley and Kelling (1968) and Kelling and Stanley (1970).

The shelfbreak generally occurs at a depth ranging from 120 to 145 m (65 to 80 fms) but most frequently at about 130 m (70 fms). The head of Wilmington canyon, as arbitrarily defined by the 135 m (75 fms) isobath, has a north-south curvilinear trend and is incised about 18 km (10 n mi) into the shelf. The very head of the canyon pinches out near 38° 30' N latitude and 73° 30' W longitude; at its northernmost end the canyon is approximately one km in width and has a relief of more than 90 m (50 fms). From its head, the canyon bends south-southwest for nearly 13 km (7 mi) to an axial depth near 550 m (300 fms) and then veers southeasterly. This course is maintained to an axial depth of 1830 m (1000 fms). The canyon gradually increases in dimensions towards the regional shelfbreak, attaining both its maximum width (about 12 kms) and its greatest relief, approximately 915 m (500 fms), at the regional shelf edge. The walls of the canyon in this headward region are extensively modified by many depressions and secondary ridges, including several cirquelike tributary canyons as much as 1.5 km wide (Fig. 256).

METHODOLOGY

The basic underwater television rig used on this cruise comprised a Hydro Products Model TC 100 camera, 1000 w mercury vapor light, 460 m-long cable, power supply, video monitor, and Model EV 200 Sony videotape recorder. The camera and light source were mounted in a steel frame (120 cm × 120 × 120 cm) lowered and raised off the stern-mounted A-frame with 5/16 inch wire, using a steam winch. The cable relaying power to the camera and light was paid out by hand. A compass was suspended far enough from the the frame to be free from the effects of its magnetic field. In addition, a tapered stainless-steel rod 78 cm long was suspended from another arm. This penetrometer was hung sufficiently far below all other equipment to be in view when it hit the bottom.

Bathymetry and navigational fixes were continuously monitored during the entire television survey. In order to relate bottom depth and location to observations on videotape and records, time of day (GMT) was used as the basic reference for all systems. Notations on bathymetric logs and on the audiotape were synchronized by the time log. Close coordination between the scientists watching the television monitor and the winch operator was essential to keep the cage just above the sea floor in order to achieve the desired tolerance of ±20 cm. The depth limit of television stations was prescribed by the length of the camera- and light-power cable (approximately 460 m). Direct bottom recordings were made to depths of 233 fms (426 m). Continuous on-the-spot monitoring took place by teams of scientists at all stations (Fig. 256). A written log was kept to record observed features of the water mass (such as particulate matter and organisms) and the sedimentary and organic characteristics of the bottom. As prominent features appeared or as observed watermass or bottom characteristics changed, the videotape recorder was activated. While video recordings were being made, navigational, meteorological and bathymetric data, and running comments on observations were added to the audio record.

The camera and light source were oriented at an angle of about 30° from the vertical to obtain maximum field of view and relief and bottom detail. Of the 22 rolls of 1-inch, 63-min videotapes we obtained, two were studied on board ship and 20 were retained for subsequent study in the laboratory. The station net was designed to sample the outermost shelf and canyon around the entire head of the canyon. Locations were selected to make continuous transects from the shelf, across the shelfbreak and into the canyon (or vice versa). Few stations were also occupied on the outer shelf, primarily to observe large longitudinal sand ridges. The effect and direction of surface drift (0.1 to 0.25 knots) on the vessel were used to help plot, and then to traverse the television stations. Time on any one station ranged from 14 to 260 min, and the distance covered ranged from about 1.2 to 7.2 km.

The cage periodically was raised a short distance then dropped to the sea floor and allowed to rest on the bottom for as many as 30 sec or more by paying out cables as the ship continued to drift slowly. Thus the motion of particles brought into suspension by the currents produced by the dropping cage could be observed distinctly.

In addition to their functions as orientation and surface-consistency determinants, the suspended compass and the penetrometer proved a useful means of estimating the size of objects on the bottom. The dimensions of the shadow of the compass and its vane were also utilized for this purpose.

Photographs included in the present paper were made directly from a film kinescoped from our videotapes. It must be noted that these fifth-generation reproductions come nowhere near the quality of the original videotape recordings; this highlights one disadvantage of the television method: it works much better for the viewer than for someone who must examine printed results. Location of photographs made from videotape (Figs. 259 to 262) are indicated on Figure 256.

Of particular interest to the present study are the nature, orientation and dimensions of ripple marks and other features indicating current movement, including oriented organic structures; the direction and speed of suspended sediment moving along the bottom as a result of bottom currents; and the relative amount of suspended material. All data recorded from examination of videotapes have been tabularly entered on a summary form (Stanley and Fenner, in press, their Appendix).

SEA FLOOR LITHOLOGY AND FAUNA
MAPPED BY UNDERWATER TELEVISION

The distribution of major sediment types in the study area is based on an evaluation of firmness of the bottom, observations of the various lithological types making up the surficial sediments and a measure of the relative amount of shell.

Outcrops of bedrock are easily recognized where they occur on the west side of the canyon wall. There, large (extending for hundreds of meters) clifflike exposures of cavernous rock occur at depths ranging from 110 to 135 m (60 to 75 fms) and smaller outcrops down to 220 m (120 fms). Rock exposures on the eastern side of the canyon are somewhat different: they form less extensive patches whose surfaces are flatter and whose texture is burrowed but not cavernous. Texture of the bedrock appears to be medium to finer-grained. The rock appears to consist of the semiconsolidated strata of clayey sand and stiff clay characteristic of the adjacent subaerial coastal plain, and the rounded "cliffs" resemble the river- and ocean-cut bluffs found in that region. The cavernous hollows in these submarine exposures may be due in part to original subaerial erosion and in part to continued submarine erosion by physical and biological processes. In particular, lobsters and colonies of ling cod were observed inhabiting hollows and crannies in the outcrops. These, at the least, must maintain the hollows free of sediment and they may be capable of enlarging them.

A relative scale of bottom firmness was established on the basis of penetration of the scaled shaft: less than 6 inches, or 15 cm (sandy sediment); 6 to less than 10 inches, or 15 to 24 cm (mud); and more than 10 inches, or 24 cm (soft clayey or ooze bottom). Areas of low penetration (5 cm or less), indicative of a sandy or gravelly sand bottom, are restricted to the shelf and uppermost slope in water generally less than about 185 m (100 fms) deep. Primary sedimentary structures such as ripple marks, and organic tracks such as burrows and mounds (lebensspuren) provide additional evidence of the nature of the bottom sediment.

On the eastern sector of the canyon head, the firm, sandy appearing, surficial sediment extends to depths as great as 455 m (250 fms), indicating a draping of the sediment off the shelf. On the western margin of the canyon head, the firm surface generally terminates at lesser depths near the shelfbreak (that is, the mud line frequently occurs at less than 185 m, or 100 fms). The sea floor sediments are least firm (penetration greater than 12″ or 30 cm) near the axis of the canyon proper and in at least one distributory canyon. This surficial sediment distribution confirms earlier conclusions based on bottom photography (Stanley and Kelling, 1968).

The television survey shows that medium to fine sand and sandy silt are the predominant textural types comprising the upper canyon head fill. These grades are also common in areas of the shelf not covered by gravel and coarse sand. The muddy facies are located axially within the canyon and extend shelfward to depths of about 90 m (50 fms) at the very northward sector of the canyon head. Similarly, mud can be found in tributary canyons entering the canyon proper.

Shell, mostly whole or fragmented individual pelecypod valves, is a major component of the sandy surficial sediment on the continental shelf. Shell percent (0 to more than 30%) and depth generally were found to vary inversely with depth down to the shelfbreak approximately. Below this depth, shell generally forms a minor constituent of the canyon sediment. However, on the eastern canyon margin, the distribution pattern of shell-rich sediment coincides with that of the sand-rich sediment, draping from the shelf edge into deeper water.

Organisms and bioturbation were noted at each video station. The living forms observed were grouped into four generalized categories:

1. nonshelled swimming forms, mostly fish and, locally, squid;
2. shelly bottom dwellers, e.g., pelecypods, bryozoans, and echinoderms;
3. nonshelly bottom dwellers, e.g., sea pens, coelenterates, and worms;
4. crustaceans, including mainly crabs, shrimps, lobsters, and sea spiders.

The distribution and concentration of these four categories is discussed elsewhere (Stanley and Fenner, in press).

Worms are generally prevalent in the mud-rich sediment below the shelfbreak and within the canyon. The concentration of worm burrows was found to increase canyonward beyond the shelfbreak. These forms not only modify the strength of sediment (Stanley, 1971a), but also contribute to the addition of suspended sediment in the watermass. Worms were seen ejecting small clouds of mud as the cage moved slowly above the bottom. Fish are commonly observed stirring the bottom, thus modifying the sea floor and contributing in a small way to the sediment suspension (Stanley, 1971b).

SUSPENDED MATTER IN THE WATERMASS

During the course of the survey a considerable amount of suspended material was consistently observed in the watermass as the television cage was lowered to the sea floor. This matter was not evenly distributed through the water column. As the cage approached the sea floor a suspension-rich zone ten or more meters thick was observed, concentrated just above the bottom. At several stations near the shelfbreak, visibility was so restricted that technicians and scientists were concerned that the camera lens had clouded. Raising the rig to examine the camera lens established that the watermass was responsible, being almost opaque with suspended matter. This poor clarity of near-bottom water also helps explain the generally poor quality of bottom photographs noted in earlier photographic surveys in this region (Stanley and Kelling, 1968).

Only at two stations were low amounts of suspended matter seen above the bottom (Fig. 257). For the remainder of the stations, a very subjective classification of (a) medium and (b) heavy densities was established. This assessment is based on clarity of water and degree of sharpness of observed bottom features through a column of water 2 m thick (lens to sea-floor distance). We do not have sufficient data to attempt to find a relationship between areal extent of the suspended matter, depth of occurrence, and tidal cycles.

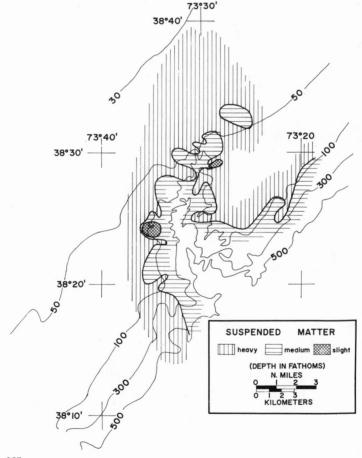

Figure 257.

Distribution of suspended sediment above the sea floor in the Wilmington canyon area as observed by television. Note relatively higher concentration of suspended matter in area landward of shelfbreak.

Despite the lack of objective criteria, a distinct pattern emerges: there is an increase in the amount of suspended matter essentially on the shelf side of the shelfbreak (Fig. 257). One area, upshelf but in line with the trend of the canyon head, also is characterized by only medium clarity. Furthermore, there is a suggestion of a slight westward offset of the medium clarity water from the canyon axis, particularly in the headward part. The amount of material in suspension locally exceeds 1 mg/l as measured on a subsequent cruise in this area by Lyall et al. (1971) and more than half of the suspensate is inorganic. The values measured are generally higher than on the mid-shelf and slope (Meade et al., 1970), strongly suggesting that at least some scour of the sea floor occurs in the vicinity of the shelfbreak in the study area.

BOTTOM-CURRENT ACTIVITY

The underwater television survey provided a unique opportunity for direct observation of the influence of current activity upon the sea floor. Ripple marks, aligned worm tubes, and aligned shells were used to infer neocurrent directions (Fig. 258). Observed fields of aligned polychaete worm tubes, *Hyalinoecia tubicola* (Müller), indicate that these organic forms are oriented by bottom currents: as the television cage moved above the tubes, sufficient turbidity was produced to roll them parallel to current flow (Fig. 259).

The operation of indigenous currents was noted on the basis of the heeling over of worm tubes, sea pen stalks, and anemone tentacles (Fig. 261). Direct current movement was observed by the following method: the cage was dropped onto the sea floor, stirring up bottom sediment (Fig. 260A); muddy sediment was noted moving away from the cage as a result of this artificially produced turbulence (Fig. 260B). After a short period (a few seconds), the effect of the true bottom current could be evaluated, that is, the direction and speed of the particles streaming above the bottom sometimes were seen to change after decay of the turbulence caused by the impact of the cage (Fig. 260C). Current directions are more or less consistent along a single station track (Table XXXVIII). Using this method, velocities to slightly over 20 cm sec^{-1} were directly measured by timing the movement of small particles past items of known length. Small turbidity currents were also initiated when the cage was dropped onto the soft sediment of the inclined walls of the canyon (Fig. 262).

Some evidence of current activity was noted at most of the stations in the study area but the pattern of observed sediment dispersal seems irregular (Fig. 258). Shells and worm tubes appear aligned roughly parallel to predominant current directions (Fig. 258), and ripple-mark crests are often aligned normal

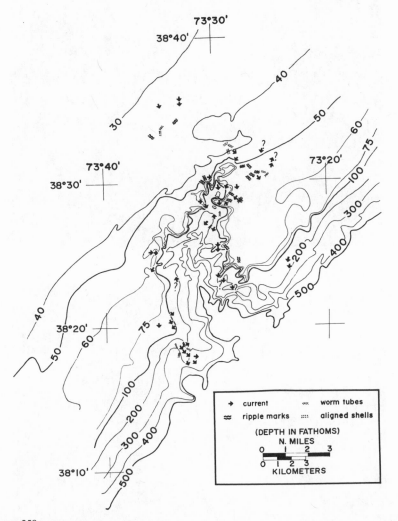

Figure 258.
Direct and indirect observations of bottom neocurrent activity near the shelf edge in the vicinity of Wilmington canyon. Arrow orientations are based on direct observation of sediment movement on or above the bottom, or orientation of sessile fauna. Alignment of ripple mark crests, worm tubes and aligned shells (providing indirect evidence of recent current activity) is also shown.

to the dominant current direction. Current flow at depths below the shelfbreak is almost invariably downslope; current patterns north of the canyon head tend to be oriented in a southerly direction.

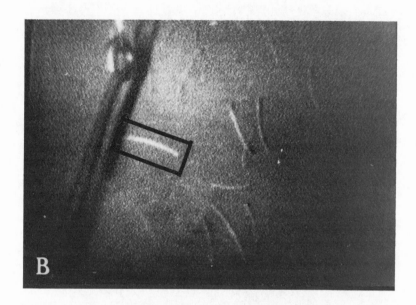

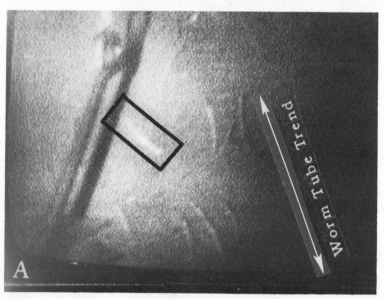

Figure 259.

Sequence of photographs showing aligned orientation of *Hyalinoecia tubicola* (Müller) polychaete tubes at Station 19 (346 m) on upper slope west of Wilmington canyon. Note realignment of tube in rectangle by turbulence caused by penetrometer and cage motion above sea floor.

Figure 260.

(A) Turbulence initiated by cage dropping on silty sea floor at Station 18 (133 m) on upper west wall of Wilmington Canyon; (B) Silt spreads over sea floor, covering compass; and (C) Several seconds later strong bottom current sweeps muddy cloud toward south.

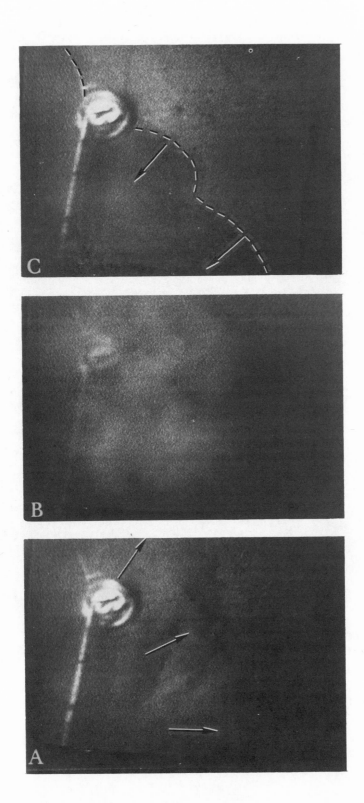

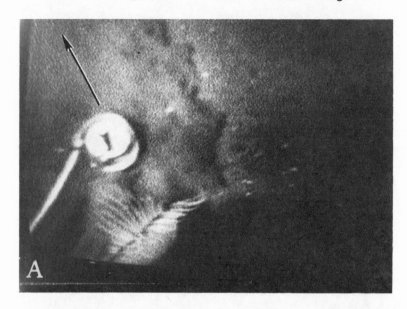

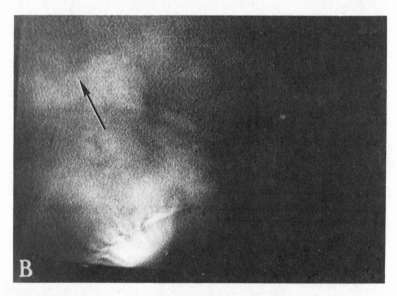

Figure 261.
Sequence of bottom photographs at Station 6 (137 m) on upper east wall of canyon, showing consistent current flow which tends to heel over sea-anemone toward SSE.

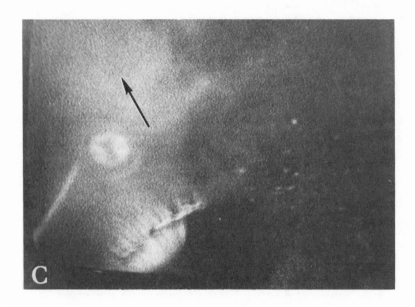

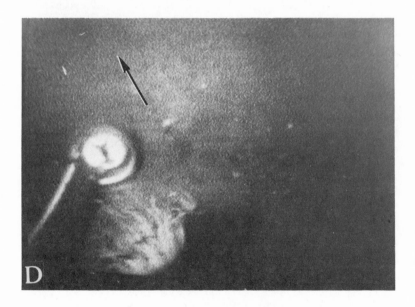

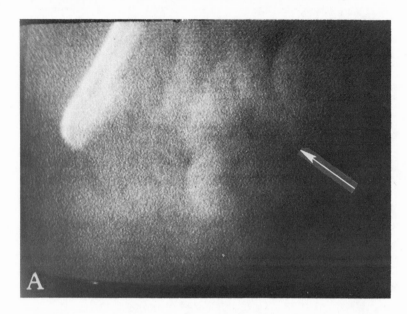

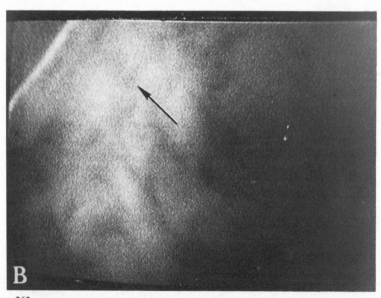

Figure 262.

Small turbidity current triggered by cage hitting soft bottom at Station 10 (362 m) on upper east wall of Wilmington canyon. Sequence shows muddy cloud advancing toward compass vane in upper left portion of photograph.

DISCUSSION

The presence on the outer shelf of strong bottom currents and oscillatory water motions has been amply demonstrated in a number of recent studies (see, for instance, Curray, 1960; Moore, 1963; Bumpus, 1965; Draper, 1967; Harrison et al., 1967; and Fisher, 1968; and Shepard and Marshall, 1969). Direct observation by television now brings into evidence the active dispersal of sediment at the head of Wilmington canyon and on the adjacent outermost shelf and upper slope. Of particular interest is that even under calm sea conditions, the shelfbreak is influenced by bottom currents capable of transporting dense suspensions of fine grained sediment. The large amounts of suspensates encountered above the sea floor and the direct evidence of currents along the bottom indicate that the relic surficial sediment cover is undergoing almost continuous reworking. Scouring of the sea floor is substantiated by the presence of bare rock outcrops and "cliffs" just below the shelfbreak, and the predominance of shelf-edge unconformities recorded in seismic surveys (Knott and Hoskins, 1968; Kelling and Stanley, 1970; and others). Our study area, like the shelfbreak in many other regions, is essentially one of nondeposition and bypassing of fines.

The more important processes governing sediment movement in the outer shelf and uppermost slope environments include those generated by wave-action of various types, others controlled by tidal motion and a final group of subpermanent currents created by oceanic circulation factors (summarized in Swift, 1969). Current patterns observed during the July, 1968, cruise probably resulted from several of these factors. Fleming and Revelle (1939) predicted that the strongest tidal currents are likely to occur near the edge of wide shelves.* Tidal currents are rotary but local bed-configurations or geostrophic efforts may induce a net shift of sediment in one or several predominant directions. Current-meter data collected in the study area on a later cruise show that currents in the study area are, at least to some extent, related to the effects of tides (Fenner et al., 1971). We believe that during the course of the television survey, the tidal effect was also one of the dominant factors resulting in the apparent irregularity of current paths plotted on the neocurrent diagram (Fig. 259) and in the transport of suspended sediment. It is doubtful, however, that tides alone are responsible for the transport of medium and coarse sand by shear exerted at the sediment-water interface.

At other times of the year, certain mechanisms are more likely to induce bedload movement as well as suspension of bottom sediment. Scour at the shelfbreak in the Wilmington canyon area, for instance, can be effected by large-scale movement of shelf and slope water masses, especially during periods

*As Fleming and Revelle (1939) have shown, $V_{max} \propto (x/d)$, where V_{max} is the maximum tidal velocity, x is the distance from shore, and d is the depth. Thus, the strongest tidal currents are likely to occur where x is large, near the shelf edge.

Table XXXVIII

Bottom-Current Activity at Selected Sites observed by Underwater Television[a]

Sta. (#)	Date July, 1968	Time (Z)	Depth (fms)	Direction	Intensity	Trend	Observation	Fig. No. this paper
1	12	0553	37			NNE-SSW		
						N-S	Aligned shells	
1	12	0605	37			NW-SE	Ripple crests	
2	12	0735	44			WNW-ESE		
						NNE-SSW	Aligned shells	
2	12	0748	44			WNW-ESE		
						NW-SE	Aligned shells	
3	12	1142	46	East[c]	>20 cm sec^{-1}			
3	12	1146	46	East				
3	12	1423	62			N-S	Ripple crests	
3	12	1423	62			130°-180°	Ripple crests	
4	12	1739	134				Ripple crests	
4	12	1739	138				Ripple crests	
5	12	1828	61				Sand waves	
5	12	1839	71				Ripple crests	
5	12	1903	104	Northwest	>10 cm sec^{-1}			
5	12	1928	153	South	>10 cm sec^{-1}			
6	13	0721	70	South				
				Southwest	<5 cm sec^{-1}			
6	13	0732	70	South and southwest				
6	13	0800	75	South southeast				261
6	13	0806	81	South				
6	13	0815	86	South				
6	13	0818	87	South				
6	13	0824	88	South southwest				
6	13	0921	128			NW-SE	Interference ripples	
6	13	0925	129				Interference ripples	
7	13	1109	105	South southwest				
8	13	1418	170	South southwest	5-10 cm sec^{-1}			
8	13	1421	175	Northwest				
8	13	1426	175	South				
8	13	1426	175	West southwest				
8	13	1429	190	Southwest				
8	13	1433	195	South southwest Southwest				
9	13	1750	67			NNW-SSE	Ripple crests	
9	13	1754	94	South				
9	13	1811	266	South				

Table XXXVIII *(continued)*

Sta (#)	Date July, 1968	Time (Z)	Depth (fms)	Direct current observations[b] Direction	Intensity	Indirect evidence Trend	Observation	Fig. No. this paper
9	13	1827	270	South				
10	13	2127	69			SSW-NNE	Ripple crests	
10	13	2135	69			SSW-NNE	Ripple crests	
10	13	2145	69				Aligned shells	
10	13	2248	168	South				
10	13	2300	198	South	>10 cm sec^{-1}			262
12	13	0851	170	South				
12	13	0900	171	South				
13	13	1445	163	East				
13	13	1452	166	South south-east				
13	13	1500	167	South south-east				
16	13	2214	72	Southwest	>10 cm sec^{-1}			
16A	15	0139	140	East north-east				
16A	15	0210	142	East				
16A	15	0220	140	East				
16A	15	0258	121	East				
18	15	0854	84	Southwest				
18	15	0904	84	East south-east				
18	15	0915	80	East south-east				
18	15	0932	78	East south-east				
18	15	0946	75	Southeast	>10 cm sec^{-1}			
18	15	1007	73	South				260
19	15	1221	189			WNW-ESE	Aligned worm tubes	
19	15	1226	189			NNW-SSE	Aligned worm tubes	259
19	15	1238	183			WSW-ENE	Aligned worm tubes	
19	15	1240	183			SW-NE	Aligned worm tubes	
19	15	1244	174			W-E	Aligned worm tubes	
19	15	1253	160			SW-NE	Aligned worm tubes	
19	15	1303	150			W-E	Aligned worm tubes	
19	15	1303	150	Northeast				
19	15	1313	148	Northeast				
19	15	1343	130	Southeast				
19	15	1352	124	Southeast	≥ 10 cm sec^{-1}			
19	15	1357	121	Southeast	>10 cm sec^{-1}			
19	15	1408	119	Southeast				
19	15	1410	117	Southeast				
19	15	1422	109	Southeast				
19	15	1426	109	Southeast				
19	15	1506	99	East south-east				

Table XXXVIII *(continued)*

Sta	Date	Time	Depth	Direct current observations[b]		Indirect evidence		Fig. No.
(#)	July, 1968	(Z)	(fms)	Direction	Intensity	Trend	Observation	this paper
20	16	0956	35	Southeast				
20	16	1021	36	South				
20	16	1033	36	South				
20	16	1046	37	South				
23	16	2016	44	Southeast				
23	16	2126	47			E-W	Ripple crests	
23	16	2145	47			E-W		
						SE-NW	Ripple crests	
23	16	2151	48			SE-NW	Ripple crests	
23	16	2154	48			NE-SW	Vague lineations	
23	16	2204	51	Southeast				
24	16	2230	51			NE-SW	Aligned worm tubes	
24	16	2345	49	North				

[a] Station location shown in Fig. 256. However, because of scale considerations Fig. 256 cannot show all observations included in this table.

[b] As determined by streaming of particles in water past object (compass or penetrometer) of known dimension, movement of sediment at bottom, or heeling over of sessile fauna.

[c] Observation with missing sense, or reported as probable, are omitted from this compilation.

of storms (Lyall, et al., 1971). In other regions, calculations indicate that wave-generated bottom flow velocities on the outer shelf can exceed 1 knot (5 cm sec^{-1}), although such a condition probably persists for a short period of perhaps only one day in each 1-5 yr (Moore, 1960; Hadley, 1964; Draper, 1967; Hayes, 1967). Surge flow of this type is almost invariably associated with hurricanes or exceptionally severe storms. However, even normal winddrift and wavedrift and wavedrift residuals are likely to interact with the outer shelf seabed, if sustained for sufficiently long periods (Swift, 1969). In this respect, attention should be called to large sand ridges (amplitudes of 2 to 7 m) observed during the course of several surveys on the outer shelf east and north of Wilmington canyon (Uchupi, 1968, his Fig. 14). The sand ridge crests are oriented NNE-SSW, and it has been suggested that these are longitudinal bedforms, which result from storm wave activity (Swift, 1969). The television observations show these features to be generally inactive, and it must again be noted that the survey was made during a particularly quiet period in summer. An event such as a hurricane, although rare in terms of human experience, may have great geological significance, especially on the outermost shelf (Moore and Curray, 1964; Gretener, 1967; Smith and Hopkins, 1972).

Theoretical and experimental studies also have suggested that water motion in the shelfbreak environment can be produced by internal waves shoaling over a sloping bottom and breaking against the uppermost slope (Cacchione, 1970; Southard and Cacchione, 1972). The importance of this factor during the period of the cruise is not known.

The seismic and visual observations furnish abundant evidence for the operation on the outer shelf of scouring mechanisms which may be attributed to one or several of the processes outlined above. The question which now arises concerns the path taken by these scoured materials in the vicinity of the shelfbreak. Bumpus (1965) has recorded offshore components of bottom flow on the outer Atlantic shelf, and his data indicate seasonal control of the bottom and surface current patterns, probably linked to onshore development of a saline wedge during periods of low runoff in the summer. Bottom drift of this type is presumably too weak to entrain tractional load of sand grade but it may significantly effect redistribution of silt and clay in suspension of the type observed in our investigation. Some of the fines moving seaward of the shelfbreak are transported at the surface and within the water masses as distinct nepheloid layers as described by Ewing and Thorndike (1965), Eittreim et al. (1969) and Lyall et al. (1971). This deposition of Holocene fines on the slope and rise (accumulating at a rate of about 10 cm per 1000 yr on the slope) is the deep-water equivalent of the Recent mud blanket of the inner and middle shelf. However, this by no means rules out the possible deposition on the shelf of some fines from slope water. Data shedding light on this problem in the study area is far too sparse for any evaluation at this time.

The recent entrapment of sands in the Wilmington canyon head and the draping of silt and sand on the slope below the shelfbreak in the adjacent region results in the "mud-line" being located well below the 200 m isobath and indicates that erosional activity at the shelfbreak is not limited to the clay and silt grades. Bottom-current activity, at least periodically, is capable of scouring the surficial sand cover and producing off-shelf movement of these coarser clastics. Once on the slope, gravity-controlled mass failure phenomena described by Dott (1963) and others promote the continued net downslope movement, or spill-over, of shelf-derived materials to deeper water. We can predict that increasing use of underwater television from surface vessels and research submersibles will assist the marine scientist in resolving problems of outer margin sedimentation.

CONCLUSIONS

From an underwater television survey of a 400 km² area about 170 km east of the Middle Atlantic States, we may conclude:

1. none of the 26 stations on the outermost continental shelf, shelfbreak and upper slope in the vicinity of Wilmington canyon was entirely free from current activity, in spite of the fact that the stations range in depth from 65 to 430 m and that the survey was made in summer under conditions of very calm weather and sea state;

2. current intensity ranged from imperceptible to about 20 cm sec⁻¹;

3. there is a suggestion that the irregularity of current direction in the study area reflects, at least in part, tidal influence on the water masses;

4. there is clearly a net southerly and downslope orientation to current flow observed within the Wilmington canyon;

5. the large amount of material concentrated in suspension above the bottom probably reflects scouring of the outer margin by this current activity;

6. the distribution pattern of suspended sediment shows a westward offset away from the canyon, which may reflect large scale motion of shelf and slope water in this region during the course of the survey;

7. long-term current activity at the shelfbreak has resulted in a net downslope transport of sand-size material further down the east wall of the canyon than the west wall;

8. the sum of the data shows that offshelf sedimentation and transfer of coarse sediment from the outer shelf onto the upper slope is taking place at present, and demonstrates that slow progradation of the outer continental margin is continuing as modern spillover and offshelf sedimentation processes continue to modify the shelfbreak environment.

ACKNOWLEDGMENTS

We would like to thank the United States Coast Guard and its Oceanographic Unit for providing us with ship time, and the Captain, officers, and men of the U.S.C.G.C. Rockaway for valuable help in the work at sea. Dr. D. J. P. Swift, Old Dominion University, and Mr. J. A. Good, Hydro Products, directly assisted us in the collection of video data as did Messrs. G. Bloomer, D. Eby, E. Gruenstein, and D. Kersey. The Department of Geology at the University of Illinois provided secretarial assistance. Thanks are due Mr. Harrison Sheng, Smithsonian Institution, for drafting. Funds to rent underwater television equipment and to complete this project were provided to one of us (DJS) by the Smithsonian Institution Research Foundation Grant Nos. 233770 FY 1969 and 234230 FY 1970. Gilbert Kelling acknowledges financial support provided by the United Kingdom Natural Environment Research Council Grant GR/3/899 for his continued participation in the canyons project. Peter Fenner acknowledges financial support provided by the Smithsonian Institution (Research Foundation Grant No. 233770), and Governors State University for continuing work on this project, and logistical support from the American Geological Institute during the field work phase.

REFERENCES

Bumpus, D. F. (1965). Residual drift along the bottom on the continental shelf in the Middle Atlantic Shelf area: Limnology and Oceanography, Suppl. to v. 3, p. 48-53.

Cacchione, D. A. (1970). Experimental study of internal gravity waves over a slope: *Mass. Inst. Tech. and Woods Hole Oceanogr. Inst. Rept* **70-6**, 226 p.

Curray, J. R. (1960). Sediments and history of Holocene transgression, continental shelf, northwest Gulf of Mexico. *In* "Recent sediments, Northwest Gulf of Mexico" (F. P. Shepard et al., eds.), p. 221-266, Am. Assoc. Petroleum Geol., Tulsa.

Dott, R. H. (1963). Dynamics of subaqueous gravity depositional processes. *Am. Assoc. Petroleum Geol., Bull.* **47**, p. 104-128.

Drapeau, G. (1970). Sand waves on Browns Bank observed from the Shelf Diver. *Maritime Sediments* **6**, 90-101.

Draper, L. (1967). Wave activity at the sea bed around northwestern Europe. *Marine Geology* **5**, 133-140.

Eddy, J. E., Henry, V. J., Hoyt, J., and Bradley, E. (1967). Description and use of an underwater television system on the Atlantic Contiental Shelf. *U.S. Geol. Surv. Prof. Paper* **575-c**, C72-C76.

Eittreim, S., Ewing, M., and Thorndike, M. (1969). Suspended matter along the continental margin of the North American Basin. *Deep-Sea Res.* **16**, 613-624.

Emery, K. O. (1968). Relict sediments on continental shelves of the world. *Am. Assoc. Petroleum Geologists Bull* **52**, 445-464.

Ewing, M., and Thorndike, E. M. (1965). Suspended matter in deep-ocean water. *Science* **147**, 1291-1294.

Fenner, P., Kelling, G., and Stanley, D. J. (1971). Bottom currents in Wilmington Submarine Canyon. *Nature, Physical Science* **229**, 52-54.

Fisher, A., Jr. (1968). ASWEPS shallow water investigation: Virginia capes area, February-March, 1967. *U.S. Naval Oceanog. Office ASWEPS Report* **14**, 19 p.

Fleming, R. H., and Revelle, R. (1939). Physical processes in the oceans, p. 48-141, *in* "Recent Marine Sediments" (P. D. Trask, ed.), 736 p., Amer. Assoc. Petroleum Geol.

Gretener, P. E. (1967). Significance of the rare event in geology. *Am. Assoc. Petroleum Geol., Bull* **51**, 2197-2206.

Hadley, L. M. (1964). Wave-induced bottom currents in the Celtic Sea: *Marine Geology* **2**, 192-206.

Harrison, W., Norcross, J. J., Pore, N. A., and Stanley, E. M. (1967). Circulation of shelf waters off the Chesapeake Bight. *Environ. Sci. Serv. Admin., Prof.* **3**, 82 p.

Hayes, M. O. (1967). Hurricanes as geological agents: case studies of hurricanes Carla, 1961, and Cindy, 1963. *Texas Bureau of Economic Geology, Rept. Investigation* **61**, 54 p.

Kelling, G., and Stanley, D. J. (1970). Morphology and structure of Wilmington and Baltimore submarine canyons, eastern United States. *J. Geology* **78**, 637-660.

Knott, S. T., and Hoskins, H. (1968). Evidence of Pleistocene events in the structure of the continental shelf off the Northeastern United States. *Marine Geology* **6**, 5-26.

Lyall, A., Stanley, D. J., Giles, H. N., and Fisher, A., Jr. (1971). Suspended sediment and transport at the shelf-break and on the slope:. *Marine Tech. Sc. J.* , 15-27.

Meade, R. H., Sachs, P. L., Manheim, F. T., and Spencer, D. W. (1970). Suspended matter between Cape Cod and Cape Hatteras. *In* Summary of Investigations Conducted in 1969. Woods Hole Oceanog. Inst. Ref. 70-11, p. 47-49.

Moore, D. G. (1960). Acoustic reflection studies of the continental shelf and slope off Southern California. *Geol. Soc. America, Bull.* **71**, 1121-1135.

Moore, D. G. (1963). Geological observations from the bathyscaph *Trieste* near the edge of the continental shelf off San Diego, California. *Geol. Soc. America, Bull.* **74**, 1057-1062.

Moore, D. G., and Curray, J. R. (1964). Wave-base, marine profile of equilibrium, and wave built terraces: discussion. *Geol. Soc. America, Bull.*, **74**, 1267-1274.

Shepard, F. P., and Marshall, N. F. (1969). Currents in La Jolla and Scripps submarine canyons. *Science* **165**, 177-178.

Smith, J. D., and Hopkins, T. S. (1972). Sediment transport on the continental shelf off Washington and Oregon in light of recent current measurements. *In* "Shelf Sediment Transport: Process and Pattern" (D. J. P. Swift, D. B. Duane, and O. H. Pilkey, eds.). Dowden, Hutchinson & Ross, Stroudsburg, Pennsylvania.

Southard, J. B., and Cacchione, D. A. (1972). Experiments on bottom sediment movement by breaking internal waves. *In* "Shelf Sediment Transport: Process and Pattern" (D. J. P. Swift, D. B. Duane, and O. H. Pilkey, eds.). Dowden, Hutchinson & Ross, Stroudsburg, Pennsylvania.

Stamp, W. R. (1953). Underwater television. *Scientific American* **188**, 32-37.

Stanley, D. J. (1970). Flyschoid sedimentation on the outer Atlantic margin off northeast North America. *Geol. Assoc. Canada Sp.* **7**, 179-210.

Stanley, D. J. (1971a). Bioturbation and sediment failure in some submarine canyons: Vie et Milieu [Suppl.] **22**, 541-555.

Stanley, D. J. (1971b). Fish-produced markings on the outer continental margin east of the Middle Atlantic States. *J. Sed. Petrology* **41**, 159-170.

Stanley, D. J., and Kelling, G. (1968). Photographic investigation of sediment texture, bottom current activity, and benthonic organisms in the Wilmington Submarine Canyon. *U.S. Coast Guard Oceanogr. Rept.* **22**, 95 p.

Stanley, D. J., Swift, D. J. P., Silverberg, N., James, N. P., and Sutton, R. G. Late Quaternary progradation and sand spillover on the outer continental margin off Nova Scotia, southeast Canada. *Contributions to the Earth Sciences*. Smithsonian Inst. Press, Washington. **8**, 1-88.

Stanely, D. J., and Fenner, P. (in press). An underwater television survey of the outer continental shelf and head of Wilmington Canyon. *Contributions to the Earth Sciences,* Smithsonian Inst. Press, Washington.

Swift, D. J. P. (1969). Outer shelf sedimentation: processes and products. *In* "The *New* Concepts of Continental Margin Sedimentation" (D. J. Stanley, ed.), Lecture 5, American Geol. Inst. Short Course, p. DS-5-1 — DS-5-26.

Uchupi, E., and Emery, K. O. (1967). Structure of the continental margin off the Atlantic coast of the United States. *Am. Assoc. Petrol. Geologists Bull.* **51**, 223-234.

Uchupi, E. (1968). Atlantic continental shelf and slope of the United States—Physiography. *U.S. Geol. Survey Prof. Paper* 529-C, 30 p.

CHAPTER 27

Nearshore Sedimentary Processes as Geologic Studies

James D. Howard

Skidaway Institute of Oceanography
Savannah, Georgia 31406
and
Department of Geology
University of Georgia

ABSTRACT

Studies of nearshore sedimentary environments and their processes can be greatly improved if those studying the modern and those investigating the ancient have a point of common interest. A definite possibility for this exists in the study of physical and biogenic sedimentary structures which represent the responses to processes.

In an offshore-to-beach sequence the spectrum of energy changes in a predictable way even though the energy maximums vary. As the energy varies, so do the processes, and responses to processes, which determine the preservable facies characteristics.

In such studies the Uniformitarian Principle is an obvious and necessary tool. More subtle and equally important is the opportunity to utilize the ancient as a key to understanding and predicting present-day processes.

INTRODUCTION

There is a gap in communication between studies of nearshore marine environments in the recent (emphasis generally on processes) and in the ancient (emphasis

645

generally on responses). It is hoped that by pointing out the concerns of the two principals there may result some attempt at establishing an interchange of information, knowledge, and experience. Obviously, this gap should not exist in the first place because papers and talks by both groups are presented in the same meetings and in the same journals. Yet, a void is present and the reason is that the two groups need a unifying interest which overlaps both areas of study.

PHYSICAL AND BIOGENIC SEDIMENTARY STRUCTURES

In theory, at least, students of modern and ancient marine sedimentation are both concerned with "happenings." In one case what *is* happening and in the other what *has* happened. Hopefully, both are interested in processes and responses to processes, but it is all too common that the student of the recent stops with the processes and his colleague studying the ancient is satisfied with a description of the response. If our combined work is to be exchangeable it is necessary to establish at least one point of mutual focus. This is conveniently represented by the physical and biogenic sedimentary structures of modern environments and ancient facies. These are chosen because they represent direct evidence of processes which occur or occurred in the environment when it was active. It is emphasized they must be considered as they relate to the geometry of the unit in which they occur and how they change laterally and vertically. It is recognized that physical and biogenic sedimentary structures are not the only factors worthy of consideration but it is believed that they represent the most obvious and most readily recognizable features of the environment or facies. And if they are present, one can be reasonably sure that they are indigenous to and characteristic of the maximum conditions contemporaneous or penicontemporaneous to the dynamic environment.

It is readily admitted that specific physical and biogenic sedimentary structures are not individually restricted to any particular facies or environment. Rather, they represent responses to processes which occurred and certainly processes are limited only by circumstances; this is probably more true in nearshore environments than elsewhere. On the other hand, sets of processes and responses and the lateral relationships of sets of processes and responses would seem to be limited by the general physical features of the overall environment. If the structures can be accepted as a source of common concern, it will require some concessions on the part of both parties. It necessitates that the "moderns" record, describe, and explain the nature of the substrate which results from the processes which they find and that they do this with some consideration of lateral relationships. It further requires that the "ancients" carefully record responses with consideration of their lateral and vertical relationships and, in addition, pay attention to what the "moderns" are saying about the processes.

OFFSHORE-TO-BEACH SEQUENCE

It is well known that present day coastlines differ greatly in the nature and the magnitudes of the energy which they receive due to variations such as wave-versus tide-dominated coasts, configuration of the coastline, bathymetric conditions of the area seaward of the shore, as well as the direct and indirect effects of geographic and seasonal meterological conditions. It should be anticipated that these factors were equally significant during the deposition of ancient sediments. Thus, in the detailed examination of modern and ancient depositional environments, the many problems which we encounter may be caused by spatial rather than temporal aspects. It is quite conceivable it would be more difficult to compare specific features of a low energy shoreline with a high energy shoreline in the Recent than to compare a Recent and Paleozoic shoreline which have very similar energy characteristics.

Yet in spite of the fact that shorelines vary greatly in their energy regimes, it is inevitable that the sediments will preserve a gradual change landward of increasing energy related to shallowing of the sea floor. Even though the maximum energy will vary significantly from one geographic setting to another, the geologic evidence or record of the physical energy which is imparted to the substrate is more obvious as one approaches shallow water. As the energy varies so also do the processes and responses to the processes which determine the preservable facies characteristics.

An overly simplified example of this is generally taught in beginning geology courses which show a reduction in grain size in a profile from source area to offshore area. It is easy to lose faith in such a simplified scheme due to such things as relict sediments, significant masking by biogenic reworking, and by our knowledge that storms probably leave a more profound imprint on the geologic record than do the day-to-day processes. Yet, if these factors are taken into consideration, it is possible to recognize the overall energy increase preserved in fossil sediments. In both modern and ancient sediments, we can see a change from offshore to onshore of a dominance of biogenic structures over physical structures passing landward into an alternation of the two responses and finally at the beach a dominance of physical over biogenic sedimentary structures. In addition, one can recognize variations in the nature of animal reworking related to the physical energy of the environment in this same sequence.

In the proximity of the shoreline, the rate at which facies changes occur will be significantly accelerated. Furthermore, these changes are seldom gradual or gradational in this sensitive and shallowing area. On the present day Georgia coast, marsh, tidal flat, beach dunes, backshore, foreshore, shoreface, shoals, bars, and tidal channels, each of which has very distinct physical and biogenic sedimentary structures, can all coexist laterally within one square mile area.

"REMEMBER HUTTON" AND OTHER CONCLUDING REMARKS

In the study of nearshore energy variations in modern and ancient sediments, the Uniformatarian Principle is an obvious and necessary aid. More subtle, but equally important, is the opportunity to utilize the ancient as a key to understanding and predicting present day processes. Thus, if in studying ancient sediments, a lateral or vertical change is seen in the physical and biogenic sedimentary structures (i.e., the response to processes), we should ask what sort of energy changes have occurred to permit this change in facies or subfacies. By the same token, when studying processes of present day environments, the investigator has an obligation as a geologist to determine what changes in response have occurred and how these changes are recorded.

Finally, in the study of the recent, information available from the sedimentary record should not be overlooked. Admittedly, the rock record may not at first seem too dramatic, yet it represents an opportunity to examine directly and in detail a multitude of sea floors. If in doing this the results and interpretations are not as expected based on readings recorded on salinometers, current meters, transmissometers, and various other black boxes, perhaps it would be worthwhile to reexamine the data or the method of data collection from the recent. After all the rocks didn't get there by accident.

Subject Index